Handbook
of Sensory Physiology

Volume I

Principles of
Receptor Physiology

By

R. A. Cone · G. M. Curry · M. E. Feinleib · Å. Flock
M. G. F. Fuortes · D. E. Goldman · H. Grundfest · M. Jacobson
A. Katchalsky · C. K. Knox · L. E. Lipetz · W. R. Loewenstein
B. L. Munger · D. Nachmansohn · A. Oplatka
D. Ottoson · W. L. Pak · G. M. Shepherd · S. S. Stevens · L. Tauc
T. Teorell · C. A. Terzuolo

Edited by

W. R. Loewenstein

With 262 Figures

Springer-Verlag Berlin · Heidelberg · New York 1971

ISBN-13: 978-3-642-65065-9   e-ISBN-13: 978-3-642-65063-5
DOI: 10.1007/978-3-642-65063-5

General Preface

Why should there be a handbook of sensory physiology, and if so, why now? The editors have asked this question, marshalled all of the arguments that seemed to speak against their project, and then discovered that most of these arguments really spoke in favor of it: there seemed to be no doubt that the attempt should be made and that it should be made now.

No complete overview of sensory physiology has been attempted since Bethe's "Handbuch der normalen und pathologischen Physiologie", nearly forty years ago. Since then, the field has evolved with unforeseen rapidity. Although electric probing of single peripheral nerve fibers was begun by ADRIAN and ZOTTERMAN as early as 1926, in the somatosensory system, and extended to single optic nerve fibers by HARTLINE in 1932, the real upsurge of such single-unit studies has only come during the last two decades. Single-cell electrophysiology has now been applied to all sensory modalities and on almost every conceivable phylogenetic level. It has begun to clarify peripheral receptor action and is adding to our understanding of the central processing of sensory information.

In parallel with these developments, there have been fundamental studies of the physics and chemistry of the receptors themselves: these studies are leading to insights into the mechanisms of energy transduction and nerve impulse initiation. One is also getting the first glimpses of those genetic and embryologic processes that attune receptors to their optimal stimuli and determine cellular recognition and interconnection. These studies touch on those concerned with synaptic mechanisms and their possible plasticity under natural or experimental variations of sensory input.

At the same time, and largely unnoticed by those less concerned with the behavioral aspects of sensory function, there have been major changes in our approach to psychophysics, the scaling of human sensation, sometimes combined with a reformulation of the classical problems of sensory thresholds in terms of modern detection theory. All this, together with refined methods for testing the sensory repertoires of animals, vertebrates and invertebrates alike, has defined an area of investigation so vast and so diverse in itself that a major handbook series entirely devoted to sensory physiology seemed to be needed in order to take stock of the present state of this field.

We had no illusions that the task would be simple. Even for any single sense modality the presently available information is so massive, and at the same time so widely dispersed that no one would attempt a handbook presentation, if he were concerned with all-inclusiveness or anxious to avoid the hazards of rapid obsolescence. The purpose of this handbook project, however, is not encyclopedic completeness, nor the sort of brief summaries provided by periodic annual reviews. Our aim is rather to help the investigator in coping with the bulk of information.

Thus, we hope to make a newcomer aware of the current state of his chosen domain in sensory physiology, by giving him access to a cross-section of those areas that seem to us representative and active, each area being described by one of its current practitioners. Similarly, we hope that this handbook will be helpful to those who have already colonized a particular corner of the field, by offering them a convenient introduction to other domains.

Current advances in sensory physiology are made possible by such a disparate set of specialized techniques, ranging from physics, chemistry and mathematics to molecular and macro-biology, and thence to psychophysics and behavioral science that no one could hope to achieve competence in all of them. The editors expect that the Handbook will reduce the gaps between these specialized approaches by providing complementary yet selective examples of their use in attacking common problems.

Though selective in detail, the plan of the Handbook includes a systematic coverage of all of the sensory systems and their functions in relation to structure. The series begins with a volume on general features of receptor processes in all modalities. It then progresses, in specialized volumes, to treatments of somesthesis, electroreception and chemical senses, audition and vestibular sensitivity, and vision in all of its aspects. A concluding volume is planned to deal with problems of perception, particularly in so far as they transcend the individual sensory modalities. Both in this concluding volume and in various places in the more specialized ones, questions of information theory and its application to the central issues of sensory physiology will be raised.

We are still very far from the goal of sensory physiology — an exact and complete description of the sensory process — but we believe that the field has entered a critical stage where the juxtaposition of information about all of the senses might lead to a forward leap in our understanding of each of them. Certain common principles that are now latent may well become visible by virtue of such a juxtaposition and irreducible differences among sensory systems might be disclosed.

Most of all, we are convinced that it is high time to attempt such a general overview. Unless we are able to pause and look at where we are, we cannot see where we are going. The enormous scope of this field may seem to be overwhelming but it is just because we do not want to be overwhelmed that we have embarked upon this task.

H. Autrum, R. Jung, W. R. Loewenstein,
D. M. Mackay, H.-L. Teuber

December 1970

Preface

Receptor biology, once the exclusive realm of the sensory physiologist and psychologist, is now a field onto which many disciplines have begun to converge. It has become a meeting ground for the physiologist, the psychologist, the biochemist, the biophysicist and the geneticist. The present volume, the first in this handbook series, reflects to some extent this state of things; this volume, unlike those to follow, is a collection of related topics rather than a continuous treatment of a subject matter. The articles are all directed to the central theme of sensory biology: how information contained in an environmental stimulus is converted into meaningful responses in an organism. The topics were selected from among those which seemed most likely to stimulate future research and to provide a guide for graduate courses and seminars. In a field as broad as receptor biology, such a selection is inevitably arbitrary. I can only hope that the gaps are not too many.

New York, August 1970   W. R. LOEWENSTEIN

Contents

List of Contributors

CONE, Richard A., The Thomas C. Jenkins Department of Biophysics, The Johns Hopkins University, Baltimore, Maryland 21218, USA

CURRY, George M., Biology Department, Acadia University, Wolfville, Nova Scotia, Canada

FEINLEIB, Mary Ella, Department of Biology, Tufts University, Medford, Massachusetts 02155, USA

FLOCK, Åke, King Gustaf V Research Institute, S-104 01 Stockholm 60, Sweden

FUORTES, M. G. F., Lab. of Neurophysiology, NINDS, National Institutes of Health, Bethesda, Maryland 20014, USA

GOLDMAN, David E., Department of Physiology and Biophysics, Medical College of Pennsylvania, Philadelphia, Pennsylvania 19129, USA

GRUNDFEST, Harry, Department of Neurology, Columbia University, College of Physicians and Surgeons, New York, New York 10032, USA

JACOBSON, Marcus, The Thomas C. Jenkins Department of Biophysics, The Johns Hopkins University, Baltimore, Maryland 21218, USA

KATCHALSKY, Aharon, The Weizmann Institute of Science, Rehovoth, Israel

KNOX, Charles K., Department of Physiology, University of Minnesota, Minneapolis, Minnesota 55455, USA

LIPETZ, Leo E., Academic Faculty of Biophysics, The Ohio State University, Columbus, Ohio 43210, USA

LOEWENSTEIN, Werner R., Department of Physiology, Columbia University, College of Physicians and Surgeons, New York, New York 10032, USA

MUNGER, Bryce L., Department of Anatomy, The Milton S. Hershey Medical Center of The Pennsylvania State University, Hershey, Pennsylvania 17033, USA

NACHMANSOHN, David, Department of Neurology, Columbia University, College of Physicians and Surgeons, New York, New York 10032, USA

OPLATKA, Avraham, The Weizmann Institute of Science, Rehovoth, Israel

OTTOSON, David G. R., Department of Physiology, Royal Veterinary College, S-104 05 Stockholm 50, Sweden

PAK, William L., Department of Biological Sciences, Purdue University, Lafayette, Indiana 47907, USA

SHEPHERD, Gordon M., Department of Physiology, Yale University School of Medicine, New Haven, Connecticut 06510, USA

STEVENS, S. S., Laboratory of Psychophysics, Harvard University, Cambridge, Massachusetts 02138, USA

TAUC, Ladislav, Laboratoire de Neurophysiologie cellulaire du Centre National de la Recherche Scientifique, F-75 Paris 16e, France

TEORELL, Torsten, Institute of Physiology and Medical Biophysics, Biomedical Center, S-751 23 Uppsala, Sweden

TERZUOLO, Carlo A., Department of Physiology, University of Minnesota, Minneapolis, Minnesota 55455, USA

Chapter 1

Mechano-Chemical Conversion

By

A. Katchalsky and A. Oplatka, Rehovoth (Israel)

With 11 Figures

Contents

A. Introduction

From the middle of the sixteenth century, when the concept of energy emerged and its practical utility began to be appreciated, energy conversion became a principal topic of applied science. The extensive use of heat engines was followed by the development of electromechanical and electrochemical devices. The search for new transformations of energy continues vigorously also in contemporary technology. Several modern methods of energy conversion differ *in principle* from the classical type represented by the reversible heat engine of Carnot. They are based on the *coupling* between irreversible flows, such as the coupling between heat flow and flow of electricity in thermoelectric generators. These energy conversions are of particular interest to the biologist, since all living systems derive their mechanical performance from coupling with irreversible metabolic processes. Thus we shall pay special attention to newly investigated transformations of energy.

The monumental work of J. W. Gibbs showed that chemical energy can be fitted into the general framework of energy conversion and that the chemical potential of any component represents its ability to produce mechanical work. It is, however, remarkable that almost a century after Gibbs' work, and half a century after the formulation of chemical thermodynamics by Lewis and Randall, no technological systems have been developed that are based on the direct conversion of chemical into mechanical energy. This is even more striking if we note that the only method of conversion practiced by all living creatures is an isothermal and *direct* conversion of metabolic energy into mechanical motility.

The physicochemical study of mechanochemical energy transformation was in a "dormant" state until it was suggested that the working constituents of the living organism are biopolymers, and until thermodynamic thinking was liberated from the confines of an equilibrium, reversible approach. Only the recent study of irreversible processes led the way to a theoretical consideration of coupled phenomena in living organisms and in corresponding man-made models.

The discussion here is devoted to the *physical principles* of mechanochemical conversions. It is not concerned with the *mechanism of biological motility*, but is intended to provide some of the general requirements which must be met by any mechanochemical converter, whether natural or synthetic. Our discussion of certain models, and, in particular, of that of regenerated collagen fibers, does not imply that their *performance* has a bearing on the operation of muscle, flagella, or cilia. Both the contractile fibers and the mechanochemical engine are used only for the demonstration of principles which have a validity wider than the models and which may be applied to the analysis of biological systems.

B. Contractile Models and Mechanochemical Engines

Before indulging in the sophisticated analysis of mechanochemical systems, it is advantageous to consider some simple models which will help in establishing the concepts applied below. The earliest models were devised some twenty years ago by KUHN (1949) and by KATCHALSKY (1949), during the study of dimensional changes accompanying the ionization of polyelectrolyte molecules. It was found that fibers prepared from cross-linked polycarboxylic acids expand appreciably after neutralization by strong alkali and contract violently when mineral acid is added. The overall process can be represented as follows:

Expansion: $R(COOH)_n + n\ NaOH \rightarrow R(COO^-Na^+)_n + n\ H_2O$ (1)

Contraction: $R(COO^-Na^+)_n + n\ HCl \rightarrow R(COOH)_n + n\ NaCl$ (2)

Total Chemical
Turnover: $n\ NaOH + n\ HCl \rightarrow n\ NaCl + n\ H_2O.$ (3)

There was no difficulty in interpreting the behavior of the polyelectrolyte fibers. During the neutralization (Eq. (1)), the polyacid chain is negatively charged. The electrostatic repulsion between the COO^- groups causes an extension of the molecular chains and a macroscopic elongation of the polyelectrolyte fiber. Moreover, since the ionization increases the osmotic pressure, the expansion process is fostered by a concomitant osmotic flow into the fiber. In contrast, the addition of HCl in Eq. (2) cancels the charges on the macromolecules and reduces the intramolecular repulsion. Brownian movement now comes into play and causes the polymolecules to curl up and contract. There is, in addition, the possibility of hydrogen bonding between the COOH groups, which helps contraction and fixes the contracted state.

If an extended ionized fiber is now subjected to a stretching force, f, it will assume a certain length, l. Upon addition of HCl, without changing the force, the fiber will contract and reach a length of $l - \Delta l$ (Fig. 1). The concentration performed useful mechanical work $(f\Delta l)$, on the account of the neutralization Eq. (2). This simple process is therefore an elementary mechanochemical conversion.

Despite the inviting simplicity of polyelectrolyte models, these systems suffer from many experimental disadvantages. No method has been found to prepare strong and reproducible fibers, which could provide consistent results in repeated work cycles. The rate of contraction was generally slow, but the major stumbling block was the pronounced irreversibility of the neutralization Eq. (3).

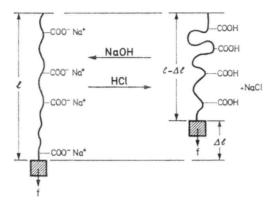

Fig. 1. Contraction and relaxation of a polymethacrylic acid fiber by HCl and NaOH, respectively

More promising results were obtained with ionized polyelectrolyte fibers subjected to an ion-exchange reaction with divalent ions (KATCHALSKY and ZWICK, 1955). It was found that when a sodium salt of a polyacid is converted into a calcium or barium salt, there is a critical point at which the fiber contracts violently. The contraction is reversible — upon exchange of some of the divalent by monovalent ions, the fiber relaxes to its original length. Fig. 2 represents the mechanism of this mechanochemical transformation based on ion exchange.

The reaction scheme for the contraction process is

$$R(COO^- Na^+)_n + \frac{n}{2} Ca^{++} \to R\left(COO^- \frac{Ca^{++}}{2}\right)_n + n\, Na^+ . \tag{4}$$

The reverse equation describes the expansion.

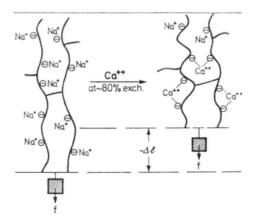

Fig. 2. Contraction and relaxation of a polyelectrolyte fiber by ion exchange

Although the reversibility of Eq. (4) is satisfactory, the sluggishness of the conversion process and the weakness of the fibers led us to seek for other and more effective models. There remains, however, the possibility that a better understanding of biological contractility based on interaction with Ca^{++}, such as found by Hoffmann-Berling (1958) in the stalks of *Vorticella*, will lead to more satisfactory man-made mechanochemical energy converters based on ion exchange.

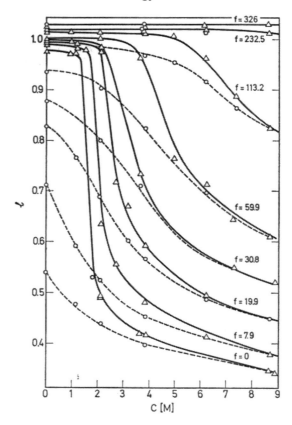

Fig. 3. Equilibrium length (l) vs. concentration (C) dependence of collagen fibers under the influence of various forces in aqueous solutions of KCNS. Force is given in kg/cm² of the original dry fiber under zero force. △ = first contraction, ○ = subsequent re-elongation

In 1956, Flory pointed out that polymeric fibers may develop appreciable mechanical force during phase transition, and if the transition is released by a chemical reaction, part of the chemical energy can be converted into mechanical work. An extensive screening carried out in this laboratory led to the conclusion that very favorable results may be obtained with reconstituted collagen fibers (product of Ethicon Co., Somerville, N.J.). Collagen fibers undergo a "chemical melting" when they interact with many electrolytes and non electrolytes. When attacked by strong contracting agents such as LiBr or KCNS, the collagen fibers may contract in a fraction of a second to about one-half their original length. During contraction, appreciable forces develop, about ten times greater than those developed by muscle when equal cross-sectional areas of contractile material are compared. Since surgical collagen may now be obtained in various forms — as fibers of different thickness, as woven material, or as films, — it is readily available for experimental study. Some results pertinent to our further discussion are reported below.

Fig. 3 represents the equilibrium length versus concentration of KCNS at different stretching forces for a highly oriented collagen fiber (YONATH, OPLATKA, and KATCHALSKY, 1965).

It will be pointed out that for each force (f) there exists a concentration range over which a sharp contraction sets in. The transition is sharper at lower forces and becomes less pronounced with increasing f. Moreover, the "melting concentration" of KCNS increases with f. The contraction represented in Fig. 3 is for the first melting process, and the fiber does not revert to its initial length if the concentration of the medium is reduced. To obtain reversible and reproducible contractions, cross-linked fibers were alternately treated with water and with a concentrated LiBr solution under zero force, in a cycle repeated several times.

There is good evidence from X-ray studies (SANTHANAM, 1959) that in the chemical melting process, the triple helical structure of native collagen undergoes a conformational change to a randomly coiled array of macromolecules.

Reich (REICH, KATCHALSKY, and OPLATKA, 1968) studied the dynamic elastic behavior of amorphous molten collagen in LiBr solutions and found that the stress-strain relations obey precisely the equation of an ideal rubber like material. According to the well-known theory of JAMES and GUTH (1943), a three-dimensional rubberlike network should obey the equation

$$\sigma = \frac{NkT}{3\beta}\{\mathscr{L}^{-1}(\alpha\beta) - \alpha^{-3/2}\mathscr{L}^{-1}(\alpha^{-1/2}\beta)\} \tag{5}$$

where σ is the stress, N the number of chain molecules per unit volume, α the extension of the fiber ($\alpha = l/l_0$, where l_0 is the initial length at zero force), and β denotes the ratio between the end-to-end distance of a single chain under zero force and that of the fully stretched one. $\mathscr{L}^{-1}(\alpha\beta)$ is the inverse Langevin function of the argument $\alpha\beta$. Values of β were determined for each LiBr concentration from characteristic frequency response curves, using the free vibration technique. At every salt concentration, σ was found to be a linear function of the expression in brackets in Eq. (5), thus giving substantial support to the notion of a randomly kinked molecular structure for the molten collagen network (Fig. 4).

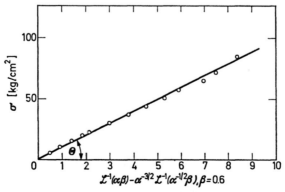

Fig. 4. Plot of the data according to Eq. (5) (collagen fiber treated once with 4.13 M LiBr solution)

The fraction of molten material increases with increasing salt concentration. It is therefore to be expected that the parameter β, representing the ratio of the contracted to the fully stretched chain, will decrease with increasing LiBr concentration; this is clearly shown in Fig. 5. The functional dependence of β on salt concentration is clearly related to the plot of length versus concentration (Fig. 3). (Fig. 3 describes the behavior in KCNS; the results in LiBr are very much the same).

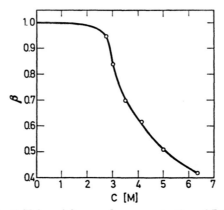

Fig. 5. Value of β vs. molar concentration of LiBr

The mechanism of the helix-coil transition of collagen molecules by interaction with salt is still under dispute by various research groups (BELLO and BELLO, 1962; ROBINSON and JENCKS, 1965). There is, however, some new evidence, provided by SHEREBRIN and OPLATKA (1968), which may help to elucidate the probelm of how LiBr or KCNS combines with the peptide chains and releases a strong contractile

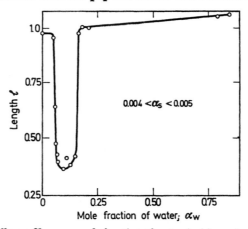

Fig. 6. Length (l) of collagen fibers vs. mole fraction of water (α_w) in acetone solutions containing mole fractions of LiBr (α_s) between 0.004 and 0.005

process. This study deals with the contractility of collagen fibers in acetone solution in the presence of LiBr — water as a denaturing agent. Surprisingly, it was found that the addition of LiBr to dry acetone had no effect on the length of the collagen fibers, whereas subsequent addition of small amounts of water caused a powerful contraction to about one-third of the original length. Even more surprising was the observation that the addition of more water brought about a relaxation of the fibers to their initial length. A typical experimental run is reproduced in Fig. 6.

More detailed analysis of various combinations of salt and water concentration in acetone showed that the range of contraction lies in a water-to-salt ratio range of 3:23. These findings seem to support ALEXANDER's hypothesis (1951) that partially hydrated Li+ ions draw into their coordination orbit the -NH or -OH groups from the peptide bond, causing distortion in the helical structure with an ultimate collapse of a random coil. In excess water, the hydration shell of the salt is complete and no interaction with the protein is expected. On the basis of these data, the need for high salt concentrations in aqueous media to contract collagen fibers is readily understood.

The data discussed above in this section were concerned only with equilibrium studies. As we pointed out in the introduction, the primary interest of the student of mechanochemical conversion lies, however, in the coupling of flows, so that an important contribution to the study of model systems is the kinetic measurements of YONATH and OPLATKA (1968) and OPLATKA and YONATH (1968). These investigators followed the contraction of collagen fibers under isometric and isotonic conditions on their transfer from water to salt solutions of various concentrations. A particularly instructive set of experiments was carried out on contraction at a constant rate. In this case, the fibers were prestretched to a chosen initial length and then equilibrated with a given salt solution. The fibers were then allowed to contract at a constant rate, and the force was recorded as a function of time. Since the length changes linearly with time, length represents time in Fig. 7. The reduced rates of contraction (expressed, as is the length, in units of the dry fiber length) ranged from 10^{-6} to 10^{-1} sec^{-1}.

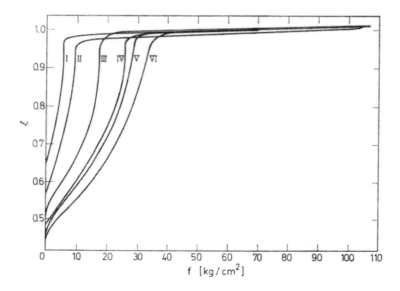

Fig. 7. Length (l) vs. force (f) curves for collagen fiber contractions at constant rates (V, sec^{-1}) in 3.18 M KCNS solution. V values: I 1.00×10^{-1}; II, 2×10^{-2}; III, 1.29×10^{-3}; IV, 2.52×10^{-4}; V, 1.30×10^{-4}; and VI, 2.12×10^{-5}

One can see that the curves exhibit three distinct regions of contraction. (1) an initial region exists in which the force drops rapidly from about 100 kg/cm² of the dry fiber under zero force, to 10 to 30 kg/cm², with a slight change in length of 3 to 5%. In this region, the fiber is oriented and crystalline, and its behavior corresponds to that of an almost rigid material. (2) There is a midregion, in which the length drops appreciably, at *constant force*. Since both the rate of contraction and the force in this region are constant, we encounter steady state behavior, in the sense of the thermodynamics of nonequilibrium processes. There is a striking similarity between the curves of Fig. 7 and the conventional run of regular phase transitions in equilibrium systems. By analogy, we would expect the steady state region to represent the realm of coexistence of crystalline and molten phases. Since the transition in this case occurs, however, in a nonequilibrium process, we do not expect a one-to-one relation between the melting concentration of the salt and the stationary force f. Indeed, the figure shows clearly that the stationary force decreases with the contraction velocity. As shown by Yonath and Oplatka (1968), the quantitative relation between the variables is → $f = A - B \ln V$ →, where A and B are constants depending on the salt concentration and temperature, and → V is the rate of contraction. (3) The third region is that of gradual contraction of an amorphous polymer, with a corresponding drop of the force to zero.

With these kinetic measurements, we would like to close the discussion of the factual data required for the utilization of model substances for the construction of mechanochemical converters and for the theoretical discussion which follows.

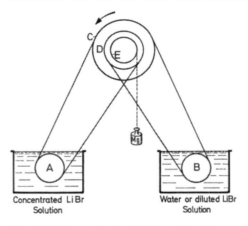

Fig. 8. Mechanochemical differential pulley

Interest in mechanochemical engines lies in the possibility of demonstrating that chemical energy can drive a mechanical device in a continuous fashion through numerous work cycles. These engines may also serve as models for facilitated transport of substances which move in biological systems with a velocity appreciably higher than expected for regular diffusional flows.

The engine demonstrated by Steinberg, Oplatka, and Katchalsky (1966) is a rotary device, based on the principle of a differential pulley (shown schematically in Fig. 8).

The problem solved by the differential pulley was the conversion of contractile forces into rotary movement. In the system represented in Fig. 8, the "prime mover" of the engine is a continuous collagen belt which, starting at wheel A, surrounds wheel D, continues to wheel B, proceeds from there to wheel C, and returns to A. The wheels C and D are rigidly coupled and constitute the pulley. Because wheel A dips into a concentrated salt solution, the fiber surrounding it will contract and exert *equal* forces (f) on wheels C and D. Because the radius of C (r_1) is larger than that of D (r_2), their rotary moments will be different and the resultant moment, which will exert a rotary torque on $C - D$, will be $f(r_1 - r_2)$. The rotation will, at the same time, bring collagen into bath A, create a new movement, and enable continuation of the rotation. Upon reaching wheel B, the salt is washed out, the fiber relaxes, and no opposing forces develop against the rotation due to the contraction in bath A. Evidently the driving process of the engine is the transport of salt from a concentrated to a dilute solution; the movement will stop when the concentration difference sinks below a certain level. The rotation may be utilized to perform mechanical work and thereby convert part of the chemical energy, dissipated in the transport, into useful work. In the construction of the engine one should take notice of the contractile parameters of collagen discussed in the previous paragraphs. Thus, to avoid an accumulation of collagen or an unsteady buildup of excessive tension, the ratio of the radii r_1 and r_2 should equal $r_1/r_2 = \bar{l}_{relax}/\bar{l}_{contract}$, where $\bar{l}_{relax}$ is the specific length of the relaxed fiber, and $\bar{l}_{contract}$ is the specific length of collagen contracted in the salt solution.

The model used for demonstration and educational purposes (Fig. 9) is a modified version of that shown in Fig. 8.

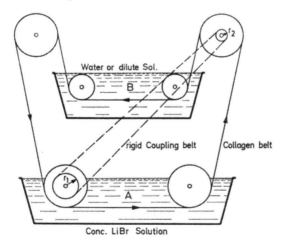

Fig. 9. Mechanochemical engine

C. Thermodynamic Consideration of Mechanochemical Conversion at Equilibrium

A straightforward treatment of the coupling between chemical and mechanical rate processes could be based on an explicit kinetic analysis of the change taking place in a contractile system. Such a treatment would, however, be limited to a specific model and could not lead to predictions of general validity. It is therefore advantageous to start our theoretical discussion with a more formal and less tangible approach, founded on thermodynamic considerations. Although thermodynamic conclusions are more difficult to visualize, we gain in generality what is lost in tangibility.

The appropriate discipline for the analysis of mechanochemical conversions is the thermodynamics of irreversible processes, for inspection of the examples treated above, as well as of the mechanochemical engines, shows that we are dealing with dissipative irreversible processes. There is, however, a logical advantage in introducing the equilibrium systems, which can be handled rigorously by the methods of classical thermodynamics (KATCHALSKY et al., 1960).

Let us begin with a differential *reversible* contraction (-dl) induced by a chemical reaction, e.g., by the application of acid to a charged polyacid or of salt to a collagen fiber. The work performed in the process is -fdl, and its relation to all other changes which take place in the fiber is expressed by the fundamental equation of GIBBS:

$$f dl = dU - TdS + pdV - \sum_k \mu_k \, dn_k. \tag{6}$$

In Eq. 6, dU denotes the change in the inner energy of the fiber; dS, the change in its entropic content; dV, the change in its volume; dn_k, the change in the number of moles of the kth permeating substance, (acid, salt, or water); and μ_k, the chemical potential of the permeant, which for ideal systems is given by $\mu_k = \mu_k^0 + RT \ln C_k$, where C_k is the concentration of the kth component.

It is evident that we cannot learn anything immediately from Eq. (6). The equation shows that the work obtained need not be the consequence of the chemical change ($\sum \mu_k d\mu_k$), but it may be due to the unknown changes in inner energy or entropy or to the volume expansion of the fiber. To make Eq. (6) useful, we take recourse to the old method of working in cycles. As is known, the cyclic integral of a total differential is zero ($\oint dX = 0$), and because dU, dS, and dV are total differentials for a reversible process, their integrals over a cycle must vanish. Now let us impose the condition that our mechanochemical process is *carried* out at $T = $ const and $p = $ const. Applying cyclic integration to Eq. (6), we obtain

$$\oint f dl = \oint dU - T \oint dS + p \oint dV - \oint \sum_k \mu_k dn_k \tag{7}$$

$$- \oint f dl = \oint \sum_k \mu_k dn_k. \tag{8}$$

In Eq. (8), the term on the left-hand side is the maximum mechanical work obtainable in a cycle, whereas the term on the right-hand side is the investment of chemical energy. The meaning of the equation is that in a reversible process all the free chemical energy is convertible into work. To make this conclusion more tangible, let us assume that we use only a single reactant. In order to obtain a cyclic performance, we dip the fiber first in a bath of reactant of chemical potential, μ_I and let it contract along the line 1—2 (Fig. 10). Contraction is then continued (2—3), outside the bath, as a result of constant composition rather than constant chemical potential. During this stage, the chemical potential of the reactant drops down to a lower value μ_{II}. Then the fiber is dipped into a second bath containing reactant at this potential. In the first two "strokes", the fiber contracts and performs work. Now it is stretched, and work is done on it in the two subsequent "strokes", 3—4 and 4—1, performed at constant potential and constant

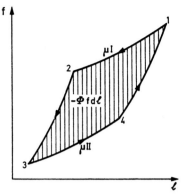

Fig. 10. Mechanical plane — force (f) vs. length (l)

composition, respectively. This completes the cycle and brings the fiber to its initial position. The work obtained is diagrammatically depicted in Fig. 10 on a mechanical plane, $f - l$, where the area of the Figs. 1—4 is equal to the mechanical performance of the fiber. During its contraction in bath I, the fiber combines with the reactant and takes up $\int\limits_{1}^{2} dn = \Delta n$ moles of substance. The same amount must be transferred to bath II to enable the cycle to be completed after it goes from 4 to 1; that is, $\int\limits_{3}^{4} dn = -\Delta n$. It is now easy to write the chemical input of the process

$$\oint \mu dn = \mu_I \int\limits_{1}^{2} dn + \mu_{II} \int\limits_{3}^{4} dn = (\mu_I - \mu_{II}) \Delta n . \qquad (9)$$

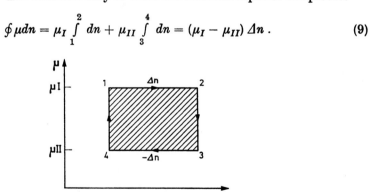

Fig. 11. Chemical plane — chemical potential (μ) vs. number of moles (n)

Eq. (9) may be diagrammatically depicted on a chemical plane of μ versus n (Fig. 11).

The application of Eq. (8) to the last two diagrams leads to the conclusion that the areas scanned by the process on the mechanical and chemical planes are equal.

It is worth noting in concluding this section that Eqs. (8) and (9) imply that no reversible cycle can produce work by interaction with a single bath of reactant.

The conversion requires that $\mu_I - \mu_{II} \neq 0$ or that a gradient of chemical potential exists for the realization of mechanical performance. This is a restatement in mechanochemical terms of KELVIN's famous formulation of the second law of thermodynamics. While the conversion of heat into mechanical work requires a difference in temperature, the conversion of chemical energy into work is based on the existence of a difference in chemical potentials.

Although the differential equation of GIBBS [Eq. (6)] proved of little immediate help, it is of inestimable value when it is integrated and transformed into numerous thermodynamic potentials, such as the well-known free energies. From the thermodynamic potentials, we may derive a wealth of cross relations, known as Maxwell relations, which are the major tools for the thermodynamic study of equilibrium systems.

The integral form of the equation of GIBBS is

$$U = TS - pV + fl + \sum \mu_k n_k \tag{10}$$

and a typical thermodynamic potential based on it is the Gibbsian free energy

$$G = U + pV - TS. \tag{11}$$

Let us demonstrate how Eq. (10) may be used for a quantitative analysis of chemical melting as shown by our collagen fibers in salt solution (KATCHALSKY and OPLATKA, 1965). It can be easily shown that upon differentiating Eq. (10) and inserting Eq. (6) the following result obtains:

$$S dT - V dp + l df + \sum n_k d\mu_k = 0. \tag{12}$$

Eq. (12) is the celebrated formula of GIBBS-DUHEM, which is extensively used in physical chemistry. Under isothermal and isotonic conditions, dT and dp are zero so that Eq. (12) reduces to

$$l df + \sum n_k d\mu_k = 0. \tag{13}$$

The collagen fiber comprises three components: water (w), salt (s), and protein (p) molecules; thus, Eq. (13) can be made more explicit:

$$l df + n_w d\mu_w + n_s d\mu_s + n_p d\mu_p = 0. \tag{14}$$

Or, if we consider reduced quantities, that is, per mole collagen, Eq. (14) is divided by n_p to give

$$\bar{l} df + \bar{n}_w d\mu_w + \bar{n}_s d\mu_s + d\mu_p = 0 \tag{15}$$

where the barred symbols represent the magnitudes per mole protein.

Now, a melting process means a state of equilibrium between the phases; in the cases considered here, this equilibrium is between the helical or crystalline phase (c) and the amorphous or random phase (a). A Gibbs-Duhem equation [Eq. (15)] may be assigned to each of the phases, but we must take note of the fact that at equilibrium all chemical potentials of the permeant components — as well as the mechanical force acting on both phases — are equal. Hence

$$\bar{l}_c df + \bar{n}_w^c d\mu_w + \bar{n}_s^c d\mu_s + d\mu_p = 0, \\ \bar{l}_a df + \bar{n}_w^a d\mu_w + \bar{n}_s^a d\mu_s + d\mu_p = 0 \tag{16}$$

which upon subtraction gives

$$(\bar{l}_a - \bar{l}_c) df + (\bar{n}_w^a - \bar{n}_w^c) d\mu_w + (\bar{n}_s^a - \bar{n}_s^c) d\mu_s = 0, \\ \Delta \bar{l} df + \Delta n_w d\mu_w + \Delta \mu_s d\mu_s = 0 \tag{17}$$

where Δ denotes the difference between the reduced quantities in the amorphous and helical regions. We have not yet taken into consideration that the melting fiber also maintains an equilibrium with the external salt solution. This solution too obeys the equation of GIBBS-DUHEM

$$n_w^0 d\mu_w + n_s^0 d\mu_s = 0 \tag{18}$$

and since salt and water permeate the fiber, $d\mu_w$ and $d\mu_s$ are the same in Eqs. (17) and (18).

From Eq. (18), we obtain $d\mu_w = -\dfrac{n_s^0}{n_w^0} d\mu_s$ which may be inserted into Eq. (17) to give

$$\Delta \bar{l}\, df + \left(\Delta \bar{n}_s - \frac{n_s^0}{n_w^0} \Delta \bar{n}_w\right) d\mu_s = 0 . \tag{19}$$

A simple calculation shows that $\Delta \bar{n}_s - \dfrac{n_s^0}{n_w^0} \Delta \bar{n}_w$ is the change in the number of bound salt molecules, per mole protein, accompanying the transition from the helical to the amorphous conformation. We denote this important quantity by $\Delta \bar{\varepsilon}$

$$\Delta \bar{\varepsilon} = \Delta \bar{n}_s - \frac{n_s^0}{n_w^0} \Delta \bar{n}_w . \tag{20}$$

Inserting Eq. (20) into Eq. (19) we obtain the working equation

$$\left(\frac{d\mu_s}{df}\right)_{\text{melting}} = -\frac{\Delta \bar{l}}{\Delta \bar{\varepsilon}} . \tag{21}$$

Recall that Fig. 3 showed that, with increasing force, a higher salt concentration was required for chemical melting. Now, a higher concentration means a higher chemical potential, so that μ_s increases with f and $(d\mu_s/df)_{\text{melting}} > 0$. On the other hand, the fiber contracts during the melting process so that $\Delta \bar{l} < 0$, and therefore, in order to obey Eq. (21), $\Delta \bar{\varepsilon}$ must be positive. In other words, the melting process is accompanied by increased binding of salt, or the amorphous protein carries more bound salt molecules than does the helical. Eq. (21) permits, however, not only a qualitative statement but also a quantitative evaluation of the salt binding from mechanical measurements. Because the activity of salt in the external solution is known, $(d\mu_s/df)_{\text{melting}}$ can be read off directly from Fig. 3. There is also no difficulty in estimating the extent of contraction $-\Delta \bar{l}$ and thereby to obtain $\Delta \bar{\varepsilon}$.

This example will suffice to demonstrate the power of thermodynamic treatment of an equilibrium mechanochemical system. We now consider the nonequilibrium thermodynamic treatment of the system in flow.

D. Nonequilibrium Treatment

The starting point for the analysis of irreversible phenomena is generally the *entropy production* or the *dissipation function* of the system. The second law of thermodynamics may be written

$$dS = d_e S + d_i S \qquad d_i S \geqq 0 \tag{22}$$

where the total change in entropy, dS, is expressed as a sum of the entropy exchanged with the surroundings, $d_e S$, and the inner entropy created in the system by irreversible processes, $d_i S$. Whereas $d_e S$ may be either positive or negative,

d_iS is either positive or zero. This is another expression for the fact that irreversible entropy is nonconservative and increases with the progress of nonequilibrium processes.

Entropy production is the increase in inner entropy per unit time, d_iS/dt, whereas the dissipation function is the degradation of free energy per unit time

$$\Phi = T\, d_iS/dt \,. \tag{23}$$

Since $d_iS/dt \geqq 0$, so is the dissipation function, or $\Phi \geqq 0$; it is larger than zero in irreversible processes and becomes zero where equilibrium is attained.

It is important to find the dependence of Φ on all the flows and forces operating in the system under consideration. To evaluate this dependence, it is postulated that for sufficiently slow processes, the equation of Gibbs holds locally, i.e., for every volume element. Local validity of Eq. (6) means that the temperature, pressure and chemical potential must be taken at their local values and not as the averages over the system as a whole. Moreover, although the parameters may change from point to point, a local equilibrium may be attributed to each volume element, so that well-defined values of p, T, and μ can be assigned to each volume element.

If the local Gibbs equation is processed by the methods of theoretical physics, it is found that the dissipation function equals the sum of the products of flows (J_i) and conjugated forces (X_i); i.e.,

$$\Phi = \sum J_i X_i \,. \tag{24}$$

Typical flows are, for instance, the flow of electricity (I), of diffusion (J_d), of entropy (J_s), or of a chemical reaction (J_r). The forces conjugated to these flows are, respectively, the electrical field (E), the negative gradient of chemical potential ($-d\mu/dx$), the negative gradient of temperature ($-dT/dx$), and the affinity of the chemical reaction (A). It might be recalled that the affinity of a chemical process is the sum of the chemical potentials of the reactants minus the corresponding sum of the products of reaction. Thus, for example, the affinity of the reaction $2A + 3B \rightarrow 4C$ is $A = 2\mu_A + 3\mu_B - 4\mu_C$.

With certain simplifying assumptions, the dissipation function for an irreversible mechanochemical transformation is given by

$$\Phi = f\, dl/dt + J_r A \tag{25}$$

where f is the stretching force; dl/dt is the change in length per unit time; J_r is the rate of chemical processes in the contractile fiber; and A is the affinity of the reaction.

Let us consider more closely the requirement $\Phi > 0$. On a priori grounds this seems to be a trivial condition, since we might expect every flow to follow the direction of the conjugated force. Thus, we expect to have for a positive affinity (A) a positive reaction (say from left to right) and for a negative affinity, a negative rate of reaction (i.e., from right to left). The product of either two positive or two negative quantities is positive, so that Φ might be expected to be greater than zero, without taking recourse to the second law of thermodynamics. When we consider, however, the first term on the right-hand side of Eq. (25), we find that this simple consideration is incorrect. It is true that in a simple mechanical system, in which no chemical reaction takes place, both dl/dt and f have the same sign, or a stretching force ($f > 0$) causes an elongation of the fiber ($dl/dt > 0$). However, in mechanochemical systems, despite the operation of a stretching force, the fiber contracts ($dl/dt < 0$) so that the product $f\,dl/dt$ is negative. This is a remarkable conclusion from a general point of view, since a negative term in the dissipation function means that *the process reduces the entropy of the system*, in apparent contradiction to the second law of thermodynamics. Indeed, Eq. (25) does not require both terms to be positive *separately*; it is only their *sum* which has to be greater than zero. Thus, the second law permits entropy-reducing processes, such as the mechanochemical contraction, if there exists a coupled process which provides the dissipation to make $\Phi > 0$. The entropy-producing process – in this case, the chemical reaction – is *driving* the entropy-reducing process, that of mechanical contraction, against the stretching force. It is readily recognized that the driven processes produce useful work, and hence the study of these cases is of primary interest for the understanding of practical devices based on irreversible phenomena. These are also the cases of mechanochemical importance to which we referred in the introduction.

The possibility of realizing a driven, entropy-reducing mechanochemical process depends on the existence of *coupling* between flows. To express coupling in a quantitative manner, we follow the procedure of ONSAGER, who assumed every flow to be linearly dependent on all the forces operating in the system. Thus, for a two-flow system, the Onsager equations are

$$J_1 = L_{11}X_1 + L_{12}X_2 \, ,$$
$$J_2 = L_{21}X_1 + L_{22}X_2 \, . \tag{26}$$

The meaning of the coefficients L_{ij} in Eq. (26) is the following: L_{11} is the *straight coefficient* which relates the flow J_1 to its conjugate force X_1, whereas L_{12} is the *coupling coefficient* which discloses the fact that the flow J_1 may be driven by the force X_2, even if $X_1 = 0$. Similarly, L_{22} is the straight coefficient for the dependence of J_2 on X_2, and L_{21} is the coupling coefficient relating J_2 to X_1. As shown by Onsager, the coupling coefficients L_{12} and L_{21} are equal; thus, the flow J_1 induced by force X_2 (at $X_1 = 0$) equals the flow J_2 induced by X_1 (at $X_2 = 0$), or,

$$(J_1/X_2)_{X_1 = 0} = L_{12} = L_{21} = (J_2/X_1)_{X_2 = 0} \, . \tag{27}$$

It is therefore evident that in order to obtain an irreversible mechanochemical conversion, $L_{12} \neq 0$. In a purely formal way, we may write, in accord with Eqs. (26) and (27):

$$\frac{dl}{dt} = L_{11}f + L_{12}A \, ,$$
$$J_r = L_{12}f + L_{22}A \, . \tag{28}$$

As shown by the nonequilibrium thermodynamics, L_{11} and L_{22} are always positive. Hence, if the affinity is zero ($A = 0$) and there is no driving force for the chemical process $dl/dt = L_{11}f$, a stretching force will cause an expansion of the fiber. On the other hand, if $f = 0$, $dl/dt = L_{12}A$, so that the chemical affinity will cause a change in the fiber dimensions. We know, however, that a positive affinity causes a contraction ($dl/dt < 0$), so that L_{12} must be negative to produce the expected mechanochemical effect.

A closer inspection of Eq. (28) leads to a series of interesting conclusions. Thus, we find that at $A = 0$ but at finite stretching force ($f \neq 0$), a chemical process

$(J_r)_{A\,=\,0} = L_{12}f$ will occur. However, since $L_{12} < 0$, the reaction will be negative; that is, stretching the fiber will cause a liberation of the reactant from the macromolecules, as is indeed found experimentally. An important case is that of isometric contraction in which force develops through chemical interaction at constant lengths. Under isometric conditions, $dl/dt = 0$, and hence from Eq. (28)

$$(f/A)_{dl/dt\,=\,0} = -\frac{L_{12}}{L_{11}} > 0. \tag{29}$$

That is, the force increases with the affinity of the reaction. Finally, it may be noted that Eq. (28) leads to several relations, the validity of which is independent of knowing the coefficients L_{ij}. Thus, in the case of $A = 0$, mentioned above,

$$\frac{dl}{dt} = L_{11}f \quad \text{and} \quad J_r = L_{12}f \quad \text{or} \quad \left(\frac{J_r}{dl/dt}\right)_{A\,=\,0} = \frac{L_{12}}{L_{11}}$$

whereas comparison with Eq. (29) shows that

$$\left(\frac{f}{A}\right)_{dl/dt\,=\,0} = \left(\frac{-J_r}{dl/dt}\right)_{A\,=\,0}. \tag{30}$$

Eq. (30) shows that the isometric force per unit affinity should equal the ratio of negative reaction to rate of extension at zero affinity, independent of our knowledge of the magnitude of the straight and coupling coefficients. Obviously, the above analysis of mechanochemical coupling carried out on page 15 is based on the assumption that $L_{12} \neq 0$.

The skeptical theoretician may, however, wonder whether this is really so. It has been known since the work of CURIE on piezoelectricity and its application by PRIGOGINE to flow phenomena that certain flows do not couple if the space is *isotropic*. Thus, vectorial flows (e.g., diffusional flows) do not couple with scalar flows (e.g., chemical reactions) in an isotropic space. To make such coupling possible, as in the case of the chemico-diffusional coupling of biological active transport, the membrane space must be anisotropic. Now, mechanochemical coupling seems to imply a vectorial flow of the fiber and a scalar chemical reaction both taking place in an isotropic fiber. Is this fact not in direct contradiction to the Curie-Prigogine principle? A more profound analysis shows, however, that the contradiction is only apparent since the mechanical flow is not vectorial. It can be easily grasped that the dissipative process of fiber contraction is due to frictional processes that take place in the fiber. The internal friction may be compared to the viscous dissipation in a flowing liquid and should be treated theoretically as tensors of the second order. The coupling which we assumed above (page 15) is, therefore, between a chemical and a frictional process; such coupling was shown to be "legal" from the point of view of nonequilibrium thermodynamics. Thus, it should be realized that the term $f\,dl/dt$ in Eq. (25) does not represent the external mechanical process but rather the inner frictional interaction which permits L_{12} to be nonzero.

In conclusion, a few words should be said about the coefficients L_{ij}. As is well known, thermodynamics cannot provide explicit values for characteristic material properties, so that the evaluation of L_{11}, L_{12}, and L_{22} requires an explicit model and kinetic treatment. Thus, it is possible to assign to the collagen fibers a viscoelastic model in which a chemical reaction takes place (*see* SPANGLER, OPLATKA, and KATCHALSKY, in preparation). Since the aim of this paper is, however, not to analyze collagen contractility but to discuss the principles of mechanochemical conversion, we shall not proceed with the kinetic calculations and will end our discussion with the general treatment by the methods of nonequilibrium thermodynamics.

Acknowledgements

The research reported in this manuscript has been sponsored by the Office of Scientific Research, OAR, under contract No. AF61(052)-919, through the European Office, Aerospace Research, U.S. Air Force.

References

ALEXANDER, P.: Changes in the physical properties in wool fibers produced by breaking hydrogen bonds with LiBr solutions. Ann. N. Y. Acad. Sci. **53**, 653—673 (1951).

BELLO, J., BELLO, H. R.: Evidence from model peptides relating to the denaturation of proteins by lithium salts. Nature (Lond.) **194**, 681—682 (1962).

FLORY, P. J.: The role of crystallization in polymers and proteins. Science **124**, 153—160 (1956).

HOFFMAN-BERLING, H.: Der Mechanismus eines neuen, von der Muskelkontraktion verschiedenen Kontraktionszyklus. Biochim. Biophys. Acta (Amst.) **27**, 247—255 (1958).

JAMES, H. M., GUTH, E.: Theory of elastic properties of rubber. J. Chem. Phys. **11**, 455—481 (1943).

KATCHALSKY, A.: Rapid swelling and deswelling of reversible gels of polymeric acids, by deionization. Experientia (Basel) **5**, 319—320 (1949).

— LIFSON, S., MICHAELI, I., ZWICK, M.: Elementary mechanochemical processes. In: Contractile polymers, pp. 1 — 40 London: Pergamon Press (1960).

— OPLATKA, A.: Mechanochemistry. p. 73. IV. Internat. Congress on Rheology (1965).

— ZWICK, M.: Mechanochemistry and ion exchange. J. Polymer Sci. **16**, 221—234 (1955).

KUHN, W.: Reversible Dehnung und Kontraktion bei Änderung der Ionisation eines Netzwerkes polyvalenter Faden molekülionen. Experientia (Basel) **5**, 318—319 (1949).

— HARGITAY, B., KATCHALSKY, A., EISENBERG, H.: Reversible dilation and contraction by changing the state of ionization of high polymer networks. Nature (Lond.) **765**, 514.

— RAND, A., WALTERS, D. H.: Contractile polymers, (H. Warner Ed.) p. 41. Oxford: Pergamon Press (1960).

OPLATKA, A., YONATH, J.: The mechanochemical melting of collagen fibers. II. Diffusion controlled contraction. Biopolymers **6**, 1147 (1968).

REICH, S., KATCHALSKY, A., OPLATKA, A.: A dynamic elastic investigation of the chemical denaturation of collagen fibers. Biopolymers **6**, 1159 (1968).

ROBINSON, D. R., JENCKS, W. P.: The effect of concentrated salt solution on the activity coefficients of acetyl-tetra-glycine-ethyl ester. J. Am. chem. Soc. **87**, 2470—2479 (1965).

SANTHANAM, M. S.: Proc. Indian Acad. Sci., A **49**, 215 (1959).

SHEREBRIN, M. H., OPLATKA, A.: Contraction-relaxation of collagen fibers in LiBr-acetone-water solutions. Biopolymers **6**, 1169 (1968).

STEINBERG, I. Z., OPLATKA, A., KATCHALSKY, A.: Mechanochemical engines. Nature (Lond.) **210**, 568—571 (1966).

YONATH, J., OPLATKA, A.: Mechanochemical melting of collagen fibers. I. Mechanical contraction. Biopolymers **6**, 1129 (1968).

— — KATCHALSKY, A.: Equilibrium mechanochemistry of collagen fibers. In: Structure and function of connective and skeletal tissue p. 381. London: Butterworth & Co. Ltd. 1965.

Chapter 2

Proteins in Bioelectricity.
Acetylcholine-Esterase and -Receptor[*]

By

DAVID NACHMANSOHN, New York (USA)

With 12 Figures

Contents

A. Excitable Membranes

The discovery, in the latter part of the 18th century, that the powerful shock of certain fish is an electric discharge immediately raised the question as to the mechanism by which living cells produce electricity. The importance of this

---

* This work has been supported, in part, by U.S. Public Health Service Grants NB-03304 and NB-07743, by National Science Foundation Grant GB-7149, and by the New York Heart Association, Inc.

problem for biology in general became apparent when it was firmly established during the 19th century that nerve impulses are propagated by electric currents. Thus, the understanding of one of the most vital functions of the organism became linked to the knowledge of the mechanism of bioelectricity. At the turn of this century, two notions were widely accepted. First, in a fluid system, such as the living cell, ions must be the carriers of electric currents. Since it was known that Na^+ ions are highly concentrated in the outer environment of cells whereas K^+ ion concentrations are high in the interior, OVERTON (1902) proposed, on the basis of simple experiments, that during activity Na^+ ions move into the cell interior and an equivalent number of K^+ ions flow to the outside. OVERTON's assumption was borne out when the availability of radioactive ions after World War II made it possible to measure the ion movements during rest and during activity. The second notion was concerned with the control of these ion movements; it was postulated that the cell membranes surrounding nerve and muscle fibers must be able to change their permeability to ions during activity.

Cell membranes at that time were not yet demonstrated; their existence was postulated on the basis of a variety of observations, e. g., that a barrier prevents the movements of small molecules and ions between cell interior and outer environment. During the last decade, however, the properties and function of membranes have been one of the most actively explored fields in biological sciences. Much information has accumulated owing to electron microscopy in combination with biochemical and biophysical methods. The notion of a "unit membrane" proposed by ROBERTSON (1960 a, b) and based essentially on the Danielli-Davson Model, assumed a structure 80A thick and formed by a bimolecular leaflet of phospholipids to which proteins are attached on the inside and outside by ionic forces. These views have been contested by powerful evidence (SJÖSTRAND, 1963; ELBERS, 1964; KORN, 1966; GREEN and PERDUE, 1966; GREEN and McLENNAN, 1969; SJÖSTRAND and BARAJAS, 1968). Membranes appear to be a mosaic of functional units (PALADE, 1963). The units are formed by lipoprotein complexes; the proteins apparently form the core of the complexes; phospholids are attached on the outside, probably by VAN DER WAALS' and coulombic forces and by hydrophobic bonds (BENSON, 1966). This idea has found much support, although it does not exclude the possibility of modifications, e. g., of lipid layers located between these complexes (KENNEDY, 1967; ROTHFIELD and FINKELSTEIN, 1968).

Although the precise molecular organization of cell membranes is far from elucidated, the most important result of recent developments has been a conceptual change. It is now well established that biomembranes are the site of many proteins and enzymes; they are highly organized and dynamic structures in which many chemical reactions take place, as illustrated by the well-explored mitochondrial membranes, but also by various others (GREEN and SILMAN, 1967; GREEN and McLENNAN, 1969; RACKER, 1965, 1967, 1969; KORN, 1969; LOEWENSTEIN, 1966; 'Membrane Proteins', 1969). The central role of proteins and enzymes in cell membranes accounts for their great diversity of function, their specificity and their remarkable efficiency more readily than did the notions prevailing for the last three decades which were essentially based on the physicochemical properties of phospholipids.

2*

The concept of the central role of proteins and enzymes in membrane function induced SJOESTRAND and BARAJAS (1968) to reevaluate the standard procedures used in electron microscopy for the preparation of the material. Using mitochondrial membranes, these authors found that the image of the unit membrane is created by an extensive denaturation. They applied methods aimed at protecting the native conformation of proteins. When the proteins are first stabilized by inter-molecular cross-linking with glutaraldehyde and then dehydrated by ethylene glycol, denaturation of the proteins is markedly reduced and their conformation appears to be relatively well preserved. The pictures of the mitochondrial membrane are quite different from those obtained with standard procedures; e. g., the thickness of the membrane is about 200—300 A and there is evidence for the presence of globular structures. The new approach is only a beginning, but obviously it may initiate important developments.

The special feature of excitable membranes is their ability to generate and propagate electrical currents generally attributed to a rapid and reversible change in permeability to ions. The mechanism of this process thus presents a key problem for the understanding of bioelectrogenesis and of the propagation of nerve impulses. In the concept of HODGKIN and HUXLEY referred to as the "ionic theory", a simple diffusion process following the concentration gradient is postulated (HODGKIN, 1951, 1964). The assumption of chemical reactions controlling this process is explicitly rejected; they are only accepted for the recovery process in which Na^+ ions are extruded to the exterior and K^+ ions flow into the interior *against* their concentration gradients. The notion of a simple diffusion process is difficult to reconcile with a variety of experimental facts. Drastic modifications of ion composition both in the interior of the axon and the outer environment have, for a considerable length of time, no effect on the electrical parameters contrary to the expectations of the theory (TASAKI, 1967). Another serious difficulty of their view is the strong heat produced and absorbed coinciding with electrical activity, as shown by Hill and his associates more than a decade ago (ABBOTT et al., 1958). In this paper and in a subsequent lecture (HILL, 1960), the authors emphatically stress the difficulty of seeing an alternative to the assumption that the early heat production and absorption roughly coinciding with electrical activity are due to chemical reactions associated with the permeability cycle.

In more recent measurements, the heat production coinciding with electrical activity was found to be diphasic: heat is produced during the rising phase of the action current and absorbed during its falling phase (HOWARTH, et al. 1968). The heat production and absorption are attributed by the authors to ion friction. This interpretation is open to question and will be discussed later in the light of the chemical theory proposed by the writer more than two decades ago and to be reviewed in this article.

The well-known theory of neurohumoral transmission associated acetylcholine (ACh) with nerve activity. It was assumed that, whereas electric currents propagate nerve impulses along nerve and muscle fibers, at the junctions (synapses) chemical compounds, ACh or others, are released from the nerve ending and, crossing the junctional gap, transmit the impulse to the second cell, nerve or muscle, thereby acting as chemical mediators. This idea was unacceptable to many prominent neurobiologists whose main objection was the similarity of many electrical properties of the membranes at junctions and in axons. This made it difficult, in their view, to interpret the observations reported in terms of neurohumoral transmission and to assume two basically different mechanisms of propagation, a purely physical one in the axonal membranes and a chemical one between the junctional membranes (*see*, for instance, ERLANGER, 1939).

The theory of neurohumoral transmission proposed in the 1930's was based on observations with classical methods of physiology and pharmacology. These methods are essential in studies aimed at higher levels of integration. They are inadequate for the analysis of the molecular mechanism responsible for the permeability changes controlling ion movements, events which take place in microseconds in a membrane of about 100 A thickness. However, the rapid advances of biochemistry and especially of protein and enzyme chemistry, and the development of highly refined techniques and methods such as the use of isotopes, electron microscopy, a great variety of optical methods, X-ray analysis, etc., have provided powerful tools for studying cellular mechanisms on cellular, subcellular and molecular levels. It appeared necessary to apply the new methods toward exploring the role of ACh and to find a more satisfactory answer to the mechanism of nerve impulse propagation. An analysis of the proteins and enzymes specifically involved in the function of ACh and their relation to the molecular processes which enable the excitable membranes to change their permeability appeared to be a promising line of approach and a prerequisite for the understanding of such a basic phenomenon of biology as bioelectricity.

However, membranes form an extremely small fraction of the cell mass. Even with all the refined methods and techniques, studies of proteins in membranes offer many difficulties. It is therefore, fortunate that a material is available which is particularly favorable for studying the proteins and enzymes associated with bioelectricity: the electric organs of electric fish. These organs are the most powerful bioelectric generators created by nature. The discharge of *Electrophorus*, the electric eel found in the Amazon River, amounts to 600 V. The electrical parameters of a single cell are similar to those in nerve and muscle fibers. It is only their arrangement in series, as in a voltaic pile, which is responsible for the power of the discharge. In the electric eel there are 5,000 to 6,000 cells, electroplax, arranged in series. But the most important feature of electric organs is their high specialization for their main function, bioelectrogenesis. Investigating the occurrence, distribution and concentration of ACh-esterase, the enzyme which hydrolyzes ACh, in a great variety of excitable tissues, nerve, muscle, brain, ganglia, etc., in many types of species, the writer came across, in 1937, the electric organs of *Torpedo marmorata*. Determinations of the concentrations of ACh-esterase in the electric tissue of the species showed that 1 kg of tissue (fresh weight) hydrolyzes 3 to 4 kg of ACh per hour, although the tissue is formed by 3% protein and 92% water. Similar values were obtained in 1938 with the electric tissue of *Electrophorus*. It was, therefore, immediately obvious that this material was uniquely favorable for studying the enzymes and proteins associated with ACh function and their relationship to bioelectrogenesis. In the 30 years since then, the material has proven to be of crucial importance for the advances accomplished and it is still one of the most useful, most efficient and indispensable materials for obtaining one of the specific proteins, ACh-esterase, in solution or for analyzing this and others in the intact membrane and exploring their function in bioelectricity. The earlier studies have been summarized in a monograph (Nachmansohn, 1959); the later developments have been described in several review articles (Nachmansohn, 1963a—c, 1964a—c; 1966a—c; 1967; 1968; 1969). In the present paper, several more recent advances in the study of the proteins are discussed with the main emphasis on the progress obtained with the ACh receptor.

B. Proteins in Bioelectricity

I. Role of ACh in Excitable Membranes

The biochemical studies based on the study of the proteins indicated very soon that the idea of neurohumoral transmission in its original form had to be modified. A large amount of data has accumulated during the last 30 years, and the evidence is continuously increasing, that ACh is not a mediator between two cells; rather it is the signal which, by its action within the excitable membranes, triggers off a series of molecular changes in the membrane which result in increased permeability. ACh is intrinsically associated with excitability and is an essential and integral part of the processes generating electric currents. Its role is fundamentally similar in the membranes of nerve and muscle fibers and in the membranes of the junctions, those of the nerve terminal and the postsynaptic membrane.

The main features on which this view is based may be briefly summarized: 1) ACh and the enzymes hydrolyzing and forming it, ACh esterase and choline-O-acetyltransferase (choline acetylase), have been shown to be present in all types of conducting fibers of nerve and muscle throughout the animal kingdom. They are found in motor and sensory, "cholinergic" and "adrenergic", peripheral and central fibers, in invertebrates and vertebrates, etc. 2) ACh-esterase, an enzyme relatively specific for ACh and distinctly different from other esterases, is localized in the excitable membranes of axons and muscle fibers as well as in those of junctions, i.e., in the membranes of nerve terminals and in the post-synaptic membranes. 3) ACh-esterase hydrolyzes ACh in a few microseconds, a prerequisite for attributing to the ester the proposed role as a trigger in the generation of bioelectric currents, since 1,000 or more impulses may be propagated per second, so that the trigger must be removed with sufficient rapidity for permitting recovery and passage of the next impulse. 4) The amazingly high concentration of ACh-esterase mentioned above in electric organs highly specialized for bioelectrogenesis suggests the close relationship of the enzyme with the primary function. Using electron microscopy combined with histochemical methods, Bloom and Barrnett (1966) found, in the electroplax of *Electrophorus*, strikingly similar distributions of the enzyme present in the membranes surrounding the small fibers innervating the electroplax, the pre- and postsynaptic membranes, and the conducting membranes between the synaptic junctions. 5) Specific and potent inhibitors of ACh-esterase have been shown to affect and eventually to block electrical activity in a great variety of nerve fibers, reversibly by physostigmine and other reversible inhibitors, irreversibly by so-called "irreversible" inhibitors, organophosphates such as diisopropylfluorophosphate (DFP) and Paraoxon. Table 1 shows the structure, chemical designation and the abbreviations of all organophosphates used in this article. Thus, electrical activity requires the activity of the enzyme present in the excitable membranes, supporting

Table 1. *Organophosphates mentioned in this article: abbreviations used, chemical names and structures*

Abbreviation	Chemical Name	Structure
DFP	Diisopropyl phosphofluoridate	$i\text{-}C_3H_7O$, $i\text{-}C_3H_7O$ — P $<^O_F$
Sarin	Isopropyl methylphosphonofluoridate	$i\text{-}C_3H_7O$, H_3C — P $<^O_F$
Paraoxon	Diethyl 4-nitrophenyl phosphate	C_2H_5O, C_2H_5O — P $<^O_{O-\langle\ \rangle-NO_2}$
TEPP	Tetraethylpyrophosphate	C_2H_5O, C_2H_5O — $\overset{O}{\overset{\|}{P}}$-$O$-$\overset{O}{\overset{\|}{P}}$ $<^{OC_2H_5}_{OC_2H_5}$
Malathion	O, O-Dimethyl S-(1, 2-dicarbethoxyethyl) phosphorodithioate	CH_3O, CH_3O — P $<^S_{S\cdot CHCOOC_2H_5}$ $\|$ $CH_2COOC_2H_5$
Isosystox	O, O-Diethyl S-(2-ethylthioethyl) phosphorothioate	C_2H_5O, C_2H_5O — P $<^O_{SCH_2CH_2SC_2H_5}$
Tetriso	O, O-Diisopropyl S-(2-diisopropylaminoethyl) phosphorothioate	$i\text{-}C_3H_7O$, $i\text{-}C_3H_7O$ — P $<^O_{SCH_2CH_2N<^{i\text{-}C_3H_7}_{i\text{-}C_3H_7}}$
Selenophos	O-ethyl Se-(2-diethylaminoethyl) ethylphosphonoselenoate	C_2H_5O, C_2H_5O — P $<^O_{SeCH_2CH_2N<^{C_2H_5}_{C_2H_5}}$
Pholpholine (tertiary)	O,O-Diethyl S-(2-dimethylaminoethyl) phosphorothioate	C_2H_5O, C_2H_5O — P $<^O_{SCH_2CH_2N(CH_3)_2}$
Methylfluoro-phosphorylcholine	Methyl-fluoro-phosphorylcholine iodide	$(CH_3)_3\overset{+}{N}CH_2CH_2O$, H_3C — P $<^O_F$

the assumption of its essential role in the permeability changes. 6) In axons exposed to organophosphates under conditions leading to an irreversible block of electrical activity, a powerful specific activator of the phosphorylated enzyme, pyridine-2-aldoxime methoidide, may, under appropriate conditions, restore electrical activity. 7) The presence of a protein distinctly different from ACh-esterase, the ACh-receptor, with which ACh reacts before being hydrolyzed by the enzyme, has been demonstrated in many axonal membranes. It has been shown that ACh and its congeners produce effects on the electrical parameters of axons similar to those previously observed on the junctional membranes, provided the structural barriers for lipid-insoluble quaternary nitrogen derivatives are inadequate to prevent the compounds from reaching the receptor or are reduced by chemical treatment. 8) Differences between the molecular groups of the active site of the ACh-receptor and those of the enzyme have been shown using the monocellular electroplax preparation of *Electrophorus*. With this preparation, evidence for the protein nature of the ACh-receptor has been obtained (*see* below). 9) Lipid-soluble inhibitors of the ACh-receptor that are analogs of ACh, such as certain local anesthetics, block electrical activity in all excitable membranes, thus demonstrating the essential role of the ACh-receptor protein in bioelectrogenesis and supplementing the evidence obtained for the essential role of the enzyme.

On the basis of accumulated data, the following picture of the role of ACh seems best to fit the available information. ACh is released within the membrane by excitation; it acts as a signal recognized within the membrane by a specific ACh-receptor protein. The reaction induces a conformational change, thereby possibly releasing Ca^{++} ions bound to carboxyl groups of the protein. Ca^{++} ions have been known for a long time to be involved in the excitability of nerve and muscle fibers; the Ca^{++} ions released may induce further conformational changes of phospholipids and other polyelectrolytes. The end result of this sequence of chemical reactions is the change of permeability to ions permitting the movement across the membrane of many thousands of ions, possibly as many as 20,000 to 40,000 in each direction, per molecule of ACh released. These reactions thus act as typical amplifiers of the signal given by ACh. ACh-Esterase, by hydrolyzing ACh, permits the return of the receptor to its original conformation; the barrier to the ion movement is thereby reestablished. ACh in its bound form and the two proteins reacting directly with the specific signal, ACh-receptor and ACh-esterase, are presumably linked together structurally as well as functionally and form a protein assembly in the excitable membrane in a way comparable to many other complex enzyme systems, for instance, the electron transfer system in the mitochondrial membrane. The existence of such protein assemblies is well established in biochemical thinking. The structural organization may account for the efficency and speed of the events following the release of ACh.

II. ACh-Esterase.
Pertinent Features of its Function

The chief aim of this article is to outline the properties and function of ACh-esterase and -receptor, the two proteins in excitable membranes linked directly with electrical activity. In the discussion of the enzyme, more recent advances will be emphasized because much of the information about this protein accumulated in three decades is readily available to the reader in many earlier reviews. Choline O-acetyl transferase (choline acetylase), the enzyme forming ACh, is active in the recovery process; summaries of this enzyme may be found elsewhere (NACHMANSOHN, 1959; 1963b; PRINCE, 1967).

a) **Localization.** For more than 20 years it was postulatet on the basis of a great variety of biochemical data that ACh-esterase is localized in the excitable membrane (for a summary, *see* NACHMANSOHN, 1959). This evidence was, of necessity, indirect. In the last 7 to 8 years, direct evidence for such a localization has been offered in many laboratories by electron microscopy combined with histochemical techniques, (*see*, e.g., BARRNETT, 1962; TORACK and BARRNETT, 1962; BRZIN, TENNYSON, and DUFFY, 1966; SCHLAEPFER and TORACK, 1966; LEWIS and SHUTE, 1966). In unmyelinated fibers evidence for the localization of the enzyme has been readily and repeatedly obtained. Fig. 1 shows, e. g., an electron micrograph obtained by SCHLAEPFER and TORACK (1966) with unmyelinated fibers of the sciatic nerve of rat tested for ACh-esterase.

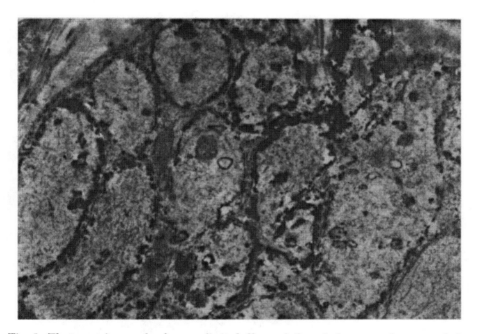

Fig. 1. Electron micrograph of unmyelinated fibers of the sciatic nerve of rat tested for ACh-esterase activity (from SCHLAEPFER and TORACK, 1966). The black deposits, which indicate the enzyme activity, are localized in the membranes surrounding the axoplasm and in the space which separates these membranes from those of the Schwann cell. Deposits are also found in some vesicles. The substrate used was acetylthiocholine. The thiocholine formed by the enzymatic hydrolysis reduces ferricyanide to ferrocyanide which precipitates with copper sulfate. × 19,000

In the membranes of myelinated fibers, the presence of enzyme could be occasionally seen, but it appeared to be irregular or even sometimes absent. This is illustrated in Fig. 2 taken from BRZIN's paper. Since even in a slice of 500 to 1,000 A thickness structural barriers may prevent or slow down the reactions between the enzyme within the membrane and the added compounds, BRZIN (1966) applied the detergent Triton 100 X to slices of an isolated sciatic nerve and found ACh-esterase in the plasma membrane,

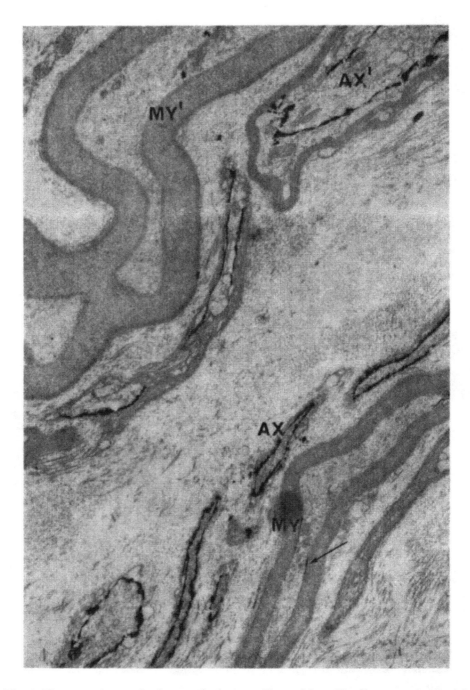

Fig. 2. Electron micrograph of sympathetic nerve fibers of frog. The fibers were incubated with a reaction mixture for demonstrating acetylcholinesterase activity. Acetylthiocholine was used as substrate. The hydrolysis product, thiocholine, reacts with Cu to form Cu-thiocholine which is converted into CuS. The dense end product is present at the axolemmal surface of unmyelinated axons *(AX)*. It is present in a myelinated axon *(AX')*, in which the membrane is detached from the myelin sheath. × 10,345. For details, see BRZIN (1966)

located between the myelin and the axoplasm (Fig. 3)[1]. A striking demonstration of the localization of the enzyme in the plasma membrane of the small diameter fibers of the stellar nerve of squid is shown in Fig. 4, an electron micrograph obtained by Dr. VIRGINIA TENNYSON.

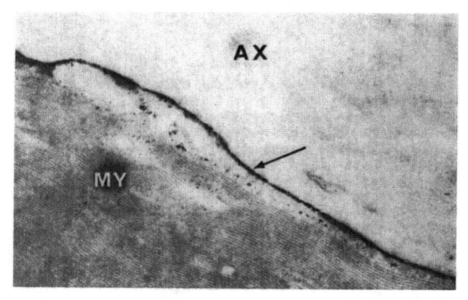

Fig. 3. Electron micrograph of a large myelinated *(MY)* ventral roots axon *(AX)*. Prior to the incubation for acetylcholinesterase activity the fiber was treated with Triton 100 X. Dense end product is present on the axolemmal membrane (arrow). × 43,345. (BRZIN, 1966)

Interesting observations were recently described by LUNDIN and HELLSTROEM (1968) applying a histochemical method combined with electron microscopy to the muscle cells of plaice: they found ACh-esterase to be localized in the sarcolemma (Fig. 5). When the preparation was exposed to DFP in low concentrations, 10^{-6} M, no enzyme activity was detectable. Since it had been previously found that ACh-esterase can be solubilized by a bacterial enzyme (LUNDIN, 1967), the authors incubated muscle pieces of this fish with the purified bacterial enzyme. They found that the enzyme activity had been removed from the membranes, although no morphological changes were visible. It seems noteworthy to the writer, that in this and other figures of the paper, the deposits indicating the enzyme activity in membranes seem to be not equally distributed along the membrane, but appear in patches. A distribution of the enzyme visibly in patches was

[1] These findings throws additional light on experiments reported during the last decade in a large number of publications from many laboratories. It has been claimed that ACh-esterase is absent in many excitable tissues, nerve and muscle fibers, and even electric organs, only on the basis of histochemical techniques combined with light microscopy. The failure to detect the enzyme with a method with so many known pitfalls, in additon to that just described, was considered as strong evidence against the postulated role of ACh and ACh-esterase in bioelectricity, in spite of the chemical data showing its precense in those tissues, in some of them even in high concentration.

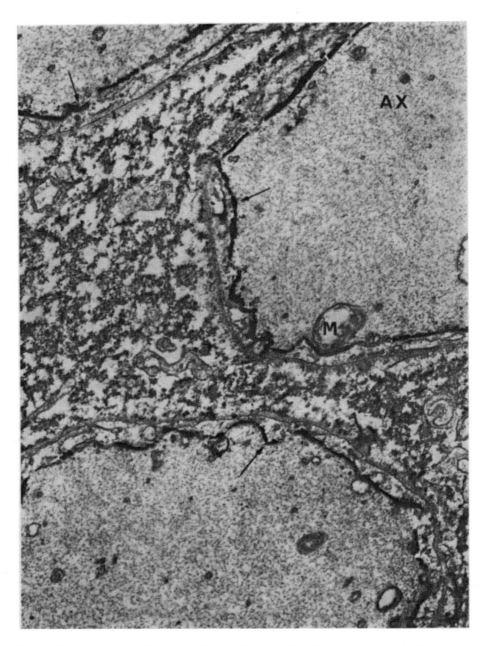

Fig. 4. Electron micrograph showing the localization of cholinesterase (arrows) at the axon-Schwann interface of small nerve fibers *(AX)* which accompany the giant fiber in the stellar nerve of the squid. A mitochondrion *(M)* is located close to the axolemmal surface. A basement membrane surrounds the Schwann cell, separating it from the fibrillar material of the inter-cellular matrix. Incubated in AThCh without inhibitors. × 20,500 (V. TENNYSON)

recently observed in the membranes of lobster nerves (DE LORENZO et al., 1968). The chemical data on ACh-esterase in the squid giant axon (BRZIN et al., 1965) support the assumption, which appears likely on the basis of many other facts, that the enzyme is not present all along the membrane but that there may be a considerable spatial separation between the molecules of this protein. Whether such a distribution is more readily detectable in some types of membranes and some animal species than in others or whether variations of experimental conditions and techniques are responsible remains to be seen.

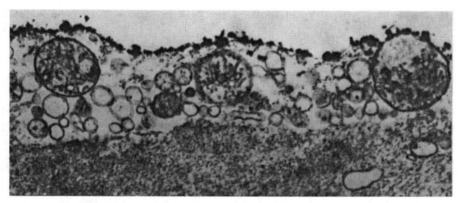

Fig. 5. Normal cholinesterase activity in a plaice muscle. Oblique section of a cell kept at 0° C in isotonic sucrose solution showing black precipitates along the sarcolemma as a result of cholinesterase activity. Substrate: acetylthiocholine for 45 min. at 0° C. Small amounts of precipitate can be discerned between the inner and outer membrane of one mitochondrion to the right. × 21,000. (From LUNDIN and HELLSTROEM, 1968)

b) Relationship between Electrical and Enzyme Activity Tested with Physostigmine. The localization of the enzyme in excitable membranes appears pertinent in relation to the postulated function, particularly in association with the high turnover number of the enzyme (ROTHENBERG and NACHMANSOHN, 1947; LAWLER, 1961; WILSON and HARRISON, 1961), and with the remarkably high enzyme concentration even in terms of activity per gram of tissue (fresh weight) in nerve fibers. The extremely high activity in the specialized electric tissue has been mentioned before. Since the actual activity should be referred to the membrane of 100 A thickness, it is evident that the actual values which were reported previously and were referred to gram fresh weight of tissue would be increased by several orders of magnitude.

These features of ACh-esterase — its localization, its high turnover number, and its extraordinarily high concentration — although prerequisites and suggestive for the postulated role of ACh, do not in themselves offer sufficient evidence in support of the theory. As has been stressed by the writer time and again, it is essential to demonstrate in addition a functional interdependence between electrical and chemical activity. If ACh is the signal initiating the permeability changes in axonal membranes and if the enzyme activity is essential for rapidly inactivating ACh by hydrolysis, the inhibition of the enzyme should affect and eventually block electrical activity.

The effect of potent and specific inhibitors of ACh-esterase has been tested on a great variety of different species and different types of nerve and muscle fibers. Several inhibitors have been shown to block electrical activity. Potent and competitive reversible inhibitors of the enzyme, e.g., physostigmine and the tertiary analog of neostigmine, block electrical activity reversibly. The concentrations of inhibitors required for affecting and blocking electrical activity vary greatly, and the rate at which the effect takes place may be slow, depending on the chemical properties of the inhibitor as well as on the structural features of the nerve (or muscle) fiber. In the early period of these studies, about 20 years ago, much discussion was provoked by the high concentrations of inhibitor required in many cases. At that time, there existed little information about ultrastructure and frequently little attention was paid to this factor. In order to block electrical activity of the frog sciatic nerve fibers, they must be exposed to 1 to 2 $\times$ 10^{-2} M (i.e., several mg/ml) physostigmine for a period of 30 to 60 min. This preparation is formed by several thousand heavily myelinated fibers and is surrounded by a sheath which forms a barrier for the entrance of many compounds. The effects on electrical activity of the preparation are readily reversible, and complete block may be even reproduced several times on the same fiber.

The explanation suggested for the reversible block of electrical activity was the reversible inhibition of ACh-esterase. The high concentration required might appear surprising in view of the potency of physostigmine as enzyme inhibitor in vitro, having a K_I of 10^{-7}. The discrepancy between the effect in solution and that on the intact structure is five orders of magnitude. The most plausible explanation for the necessity of high concentrations and low rates to produce the effect appeared to be the existence of many barriers through which physostigmine would necessarily have to pass before reaching the enzyme located in the plasma membrane. The explanation offered was vigorously rejected by some investigators because of the high concentration of inhibitor. However, the postulated role of ACh-esterase was borne out by observations made many years later when methods had been developed for testing electrical activity on single frog sciatic fibers. Dettbarn (1960a) exposed single fibers to physostigmine in low concentrations, 5×10^{-5} to 5×10^{-6} M, i.e., 1.5 to 15 μg/ml: effects on electrical activity measured at Ranvier nodes appeared rapidly, in seconds. Moreover, the electrical parameters in this single fiber preparation are modified in a way that has been postulated by electrophysiologists if the proposed theory were to be correct: first, an increased spike height and a prolonged descending phase were observed. On longer exposure, the amplitude increased but the prolongation remained or became even greater. At a concentration of 2×10^{-4} M (60 μg/ml), the spike height was first reduced and the duration was increased; after exposure of 15 to 20 min, block of conduction occurred. With 10^{-3} M (300 μg/ml), conduction was blocked in 25 sec. These modifications of the parameters could not possibly have been observed, for many reasons, on a preparation formed by many thousand fibers (Dettbarn, 1960a). Dettbarn also found that the effectiveness of physostigmine increases with increasing pH. A roughly linear relationship exists between the concentration of the conjugate base in solution and the percentage attenuation of the action potential, which is consistent with the assumption that the neutral molecule may readily penetrate; no such relationship was observed with the cationic form. It would be difficult to attribute the effects observed on the electrical activity of the single fiber preparation to some unknown or some unspecific effect in view of the very high affinity of physostigmine for ACh-esterase and the rapidity, effectiveness and the type of action. Moreover, when this single fiber preparation is exposed to curare, electrical activity is blocked, rapidly and reversibly, as previously described for the neuromuscular junction (DETTBARN, 1960b). These observations further support the assumption that the failure of certain compounds to act on axonal conduction is due to structural barriers, and is not due to two entirely different systems for controlling permeability changes of excitable membranes in axons and in junctions. This aspect will be discussed more fully later.

In the case of frog sciatic nerve fibers, the high concentrations of physostigmine and the long exposure time required for blocking electrical activity of the multifiber preparation have been explained in terms of gross barriers, but possibly also others, surrounding the excitable membranes. The single fiber, where the barriers are absent, shows the postulated effects clearly and unequivocally. Other nerve fiber preparations offer aspects which necessitate additional and different explanations. Complete and reversible block of electrical activity of the stellar and fin nerves of the squid (the former containing the giant axon and many small fibers) is obtained with 2 to 5×10^{-3} M physostigmine and with an exposure time of 15 to 30 min (BULLOCK et al., 1946). In contrast to physostigmine, neostigmine, also a structural analog of ACh and an equally potent competitive inhibitor of ACh-esterase in solution, has no effect on the electrical activity of the giant axon, even when the preparation is exposed to the compound in high concentrations, 10^{-2} M, for 5 hr. Other types of nerve fiber preparations tested were also found to be unaffected by neostigmine.

The giant axon offers the possibility of analyzing the penetration of compounds into the interior since the axoplasm may be extruded and tested for the presence of compounds applied on the outside. Thus, some information may be obtained about the existence and the nature of the permeability barriers. Physostigmine was found to have penetrated into the interior, whereas no neostigmine was found (BULLOCK et al., 1946). Physostigmine is a tertiary nitrogen derivative and lipid soluble in its conjugate, uncharged form; neostigmine is a quaternary nitrogen derivative, charged at every pH, and lipid insoluble. The failure of neostigmine to affect electrical activity, combined with its failure to penetrate, suggests that quaternary compounds, because of structural barriers, are unable to reach the excitable membrane, and that this may be the explanation why quaternary compounds, such as ACh and curare, are usually acting only at the junctions. Axonal membranes are usually surrounded by Schwann cells; they may form a structural barrier protecting the excitable membrane, although the nature of this "barrier" is still not clear. At the nerve terminal, the Schwann cell is frequently incomplete or absent. These findings thus offered the first reasonable explanation for the complete failure of compounds such as ACh and curare to act on the excitable membrane of axons in sharp contrast to their powerful action on nerve terminals. This explanation was directly tested and confirmed in experiments in which the squid giant axons were exposed to N^{15}-labeled ACh: no isotopic N was found in the extruded axoplasm after 60-min exposure. In contrast, when the preparation is exposed to the tertiary nitrogen derivative dimethylethanolamine, labeled in the same way, the compound penetrates into the interior and is in equilibrium with the outside concentration within 60 min (ROTHENBERG et al., 1947). Also pertinent in this connection is the observation that electrical activity of axons is blocked by the tertiary analog of neostigmine in concentrations similar to those of physostigmine, although this analog is an inhibitor of ACh-esterase which is a hundred times less potent than either neostigmine or physostigmine. Thus, the removal of one methyl group from neostigmine drastically changes its permeability properties.

The presence of structural barriers preventing lipid insoluble quaternary nitrogen derivatives from reaching and affecting excitable membranes of axons has been further confirmed and greatly extended in a series of investigations by ROSENBERG and his associates in which they used snake venoms for reducing these barriers (for a summary see ROSENBERG, 1966). They exposed squid giant axons to a few μgrams of venom for a limited time, about 15 to 30 min. After removal of the venom, the axons were exposed to curare and ACh. Electrical activity was rapidly and reversibly blocked in concentrations of 10^{-4} to 10^{-5} M. Testing the penetration of C^{14}-labeled ACh, curare and choline under the same experimental conditions, ROSENBERG and HOSKIN (1963) found that the compounds had penetrated into the axoplasm, in contrast to the controls where, without treatment, no radioactivity was observed in the interior. Neostigmine, however, even after snake venom treatment, neither affected electrical activity nor penetrated into the axon (BRZIN, et al., 1965). In view of the structurally similar features of these various compounds, it is difficult to understand why several of them are able to penetrate after snake venom treatment, whereas neostigmine still does not enter. The findings are another illustration how difficult it is to interpret activity of enzyme inhibitors, or lack of it, if the enzymes are localized in intact cell structures, and to extrapolate the reactions in solutions to the effects on cells.

Examination with electron microscopy of the squid giant axons exposed under the same experimental conditions to the venom of cottonmouth moccasin have revealed that the SCHWANN cell is markedly affected; no structural change of the membrane was noticeable (MARTIN and ROSENBERG, 1968). The active component of the snake venom proved to be phospholipase A (ROSENBERG and NG, 1963; CONDREA and ROSENBERG, 1968); the formation of lysolecithin by this enzyme has been shown to account fully for the effects observed (ROSENBERG and CONDREA, 1968).

While these studies have greatly contributed to the understanding of some of the problems of structural barriers, other questions still remain open. In the unmyelinated stellar nerve of squid, relatively high concentrations of physostigmine, 2 to 5 × 10^{-3} M, are required for blocking electrical activity, although one would assume that at least the gross barriers surrounding the excitable membranes are less protective. An important clue is the observation that at the time when electrical activity has disappeared, the concentration of physostigmine inside the giant axon is close to 10^{-3} M. Exposure of the axon to snake venom before the use of the inhibitor decreases the effective outisde concentration by two- to fourfold, but the inside concentration still remains about the same. Thus, the concentration of the inhibitor on the inside and the outside of the excitable membrane of this preparation is more than three orders of magnitude higher than one would have expected on the basis of the K_I found in soultion. This discrepancy raises the problem if some factors within the membrane structure, at least in this particular fiber, may be responsible for the inability to inhibit the enzyme activity to the degree which makes conduction impossible. Twenty years ago, when the early experiments on the effects on ACh-esterase inhibitors on axonal conduction were performed, virtually nothing was known about the structure of cell membranes. Today we know that cell membranes, although only 100 A thick, are structurally highly organized, even though details of the ultrastructure are far from being established. As mentioned before the lipoprotein complexes forming the membrane may be organized in such a way that the lipid is located on the outside of the complec and the protein on the inside BENSON, 1966). In any event, the complexity of the ultrastructure makes it extremely difficult to interpret the effects of compounds applied from the outside of the cell as to their reactions with the components of biological membranes. Within such tightly organized structures, it may have little meaning to speak of a concentration of a compound or of an enzyme. We do not know what the actual amounts of a given compound such as, for instance, physostigmine may be which are present within the 100 A thick membrane when electrical activity is affected, nor do we know, in what volume of fluid they are dissolved, no matter what the concentrations on the outside or the inside may be. We do not know the fraction of the enzyme which must be inhibited to obtain the first electrical signs of impairment, nor how much enzyme still remains active when electrical activity is abolished. On the other hand, there is increasing information indicating that there are very marked differences between reactions of a protein in solution and those in structures. For instance, with "insoluble enzymes", i.e., enzymes surrounded by charged polymers, KATCHALSKI and his associates (LEVIN et al., 1964; GOLDSTEIN et al., 1964) found striking differences in behavior; with the proper kind of charges, the K_m between substrate and enzyme may decrease by a factor of 100, the substrate being directed by the charges towards the active sites. This situation is relatively simple compared to the complexity of a structure such as a biological membrane (see also SILMAN, 1969; RACKER, 1967).

An excellent illustration for the fundamental differences between reactions in solutions and those within a structure offer the many data accumulated in the last decade on mitochondrial membranes; many reactions in the membrane do not take place in solution, others are markedly modified. The complexity of an intact structure must be kept in mind in the case of differences between K_I vallues of an enzyme inhibitor in solution and concentrations of the inhibitor applied on the outside to a structure. Differences of three to four orders of magnitude may have little significance. In the case of neostigmine, the difference between the tertiary and the quaternary analogs is infinite. The discrepancy between effective concentrations of the inhibitor in solution and those which have to be applied to intact cells, even if difficult to explain at present, is not surprising.

However, another possibility has to be considered. Physostigmine in 10^{-3} M, and even in 2 to 5 $\times$ 10^{-4} M, acts on the electroplax as a receptor inhibitor (HIGMAN and BARTELS, 1963; BARTELS, 1968). The reversible block of electric activity on the squid giant axon may, therefore, also have an effect on the receptor. Since both reactions are reversible, it is in this particular case impossible to decide whether the block of electrical activity is due to the inhibition of the ACh-receptor or of ACh-esterase. Thus, using this particular compound on this particular preparation does not permit one to demonstrate the essential role of ACh-esterase in the physiological removal of ACh during electrical activity. But neither can this case be used as an argument against this role, once that has been demonstrated with this compound on the single frog fiber preparation. It would be difficult to believe that ACh-esterase is essential for electrical activity in the isolated axons of frog sciatic fibers, because there physostigmine acts in the expected low concentrations, whereas in squid axons ACh-esterase, although present in the membrane, has a different function or none at all, solely because physostigmine acts only in higher concentrations and therefore possible blocks the receptor first.

The effect of physostigmine on electrical activity has been demonstrated on various types of axons, such as purely sensory fibers [the dorsal roots of ox (COHEN, 1956)], on the so-called "adrenergic" fibers [the splanchnic nerve of the bull frog (BULLOCK et al., 1947)], on the superior cervial sympathetic of the cat (GRUNDFEST et al., 1947), on different types of invertebrate fibers, on muscle fibers, and on others. The concentrations applied and the rates of action vary. In the lobster walking leg, for instance, physostigmine, 5×10^{-3} M, virtually blocks electrical activity in about 2 min (DETTBARN and DAVIS, 1963). In some preparations, periods of exposure ranging from a few up to 60 min and concentrations of 10^{-3} M or higher are required to produce this effect. In view of the great diversity of gross and, probably, fine structure of all these multifiber preparations, variations of inhibitor concentrations and of periods of exposure, which are effective, are to be expected.

c) Effects of Organophosphates on Conduction. Different proplems are raised by the effects of organophosphates on electrical activity of axons. This type of compounds was first assumed to be not only potent but also irreversible inhibitors of ACh-esterase. When the potent inhibitory action of these compounds became known some 25 years ago, many investigators thought, that exposure of axons to *low* concentrations of these compounds should affect and block electrical activity irreversibly if the postulated role of the enzyme is correct. Actually, the concentrations required are rather high and the periods of exposure for obtaining irreversible inhibition are frequently, although not always, quite long. After a short exposure, the electrical activity is usually rapidly affected and sometimes even blocked in a few minutes, but on return to saline solution the activity is restored, with the recovery periods usually longer than the periods of exposure. When, for instance, the squid stellar nerve is exposed to 5×10^{-3} M DFP, conduction is abolished in about 10 min, but it reappears on return of the fibers to sea water. It takes 30 to 60 min of exposure to block electrical activity irreversibly (BULLOCK et al., 1946). On the other hand, when the axons of the walking leg of lobster are exposed to Paraoxon in 5×10^{-3} M, electrical activity is very much decreased in 2 min, and in a few more min it is irreversibly abolished (DETTBARN and DAVIS, 1963).

When the squid stellar nerve was exposed to 5×10^{-3} M DFP and the penetration into the axoplasm of the giant axon was tested at the time when the action potential was abolished (about 30 min), only 0.1 μg of DFP was found inside per gram axoplasm. This was a surprising finding. DFP is water- and lipid-soluble, and it appeared a priori likely that such a compound would readily penetrate. Although a lower inside concentration than that on the outside had been expected, a discrepancy of 10^4 times between inside and outside seemed difficult to explain satisfactorily in physicochemical terms. In a general way, the observations seemed to point again to the difficulty of interpreting effects of inhibitors on enzymes in the intact structure extrapolated from observations in solution.

In a systematic study of penetration rates of a variety of lipid soluble and insoluble compounds into the axoplasm of squid giant axon, it was found that the former readily penetrate into the interior (HOSKIN and ROSENBERG, 1965). This raised again the question why DFP penetration had been found to be so very poor. The problem was therefore recently reexamined by HOSKIN et al., (1966) with radioactively labeled DFP. Surprisingly, it was found that extruded axoplasm had a high radioactivity. However, as shown by the authors, it was the split product due to the action of a DFPase, first described by MAZUR (1946) and later referred to as phosphorylphosphatase by AUGUSTINSSON and HEIM-BURGER (1954). The enzyme occurs in many tissues and has been extensively studied by a great number of investigators (for a summary, see MOUNTER, 1963). The low concentration of DFP in the axoplasm (less than 0.1 μg/g) reported in the earlier experiments was fully confirmed as well as the irreversible block described previously.

The data still do not indicate how much DFP is actually present *within* the 100 A membrane, but, as discussed, concentration in an organized membrane structure of 100 A has little meaning. In any event, the presence of an active enzyme rapidly hydrolyzing DFP in the preparation is an additional factor which may readily account for the high outside concentration required (1 mg/ml) since only a fraction of this outside concentration may penetrate to the excitable membrane after the passage through the Schwann cell. The high outside concentration was used as the main argument for the assumption of a general toxic effect as the explanation for the irreversible block of electrical activity by DFP, whereas the view that the effect may be due to enzyme inhibition was vigorously rejected. This explanation appears even less satisfactory in view of the finding that the inhibitor acts in a tissue in which it is rapidly hydrolyzed.

Recently, three additional organophosphates were tested by HOSKIN et al., (1969). Isosystox, Tetriso and Selenophos (see Table 1) were applied to the squid giant axons and their effects on electrical activity were reported. Isosystox, which is rapidly hydrolyzed by the tissue, blocks conduction in 5×10^{-3} M within 5 min. After 30 to 60 min, the block becomes progressively irreversible. The effect is quite comparable to that observed on this preparation with DFP and Paraoxon. With the two other compounds, no effects were found in the concentration applied. The compounds are not enzymatically attacked; there is littel spontaneous hydrolysis. Both compounds entered the squid giant axon in concentrations three to five orders of magnitude higher than those which block ACh-esterase in solution. The excess is, therefore, just about as high or even higher than in the case of physostigmine, but still they were inactive. However, these two molecules are quite complex: Tetriso has a tertiary amine with an isopropyl group in the acidic group. Selenophos has a diethylaminoethyl in the acidic group; also the presence of a selenium may strongly modify biological activities and physicochemical properties, as has been demonstrated by the work of Mautner and his associates (1967). It appeared possible that these compounds pass like physostigmine through the membrane into the axoplasm between patches containing the enzyme, but without affecting either receptor or enzyme which may be protected by barriers against this type of structures. To test this possibility, phospholipase A (cottonmouth moccasin venom) was applied and promptly electrical activity was affected in concentrations of 5×10^{-3} M by both these compounds. After 10 min exposure to Tetriso, electrical activity had decreased by 60%. This effect was reversible, as is the case with all other organophosphates applied to this preparation for such a short period. Longer exposure times were not tried. With Selenophos, 5×10^{-3} M applied for 25 min, electrical activity also decreased, but with this exposure period irreversibly, again consistent with the observations with other compounds. The only difference between the effects on electrical activity of these two compounds and of the others is the necessity of pretreatment with venom. Since the concentrations on the inside and outside in excess of those required to block enzyme activity in solution are not very different from those found with eserine, the necessity of using phospholipase A to obtain effects seems to suggest that the barriers for these compounds are present *within* the membrane which may be due to the special properties of these two compounds which prevent them from reacting with the two proteins.

It appeared important to offer further support for the relationship between irreversible block of electrical activity and the irreversible nature of ACh-esterase inhibition by DFP and Paraoxon. The following test seemed to be a promising approach. On exposure to organophosphates, the oxygen of the serine in the active site of ACh-esterase, as in other ester-splitting enzymes, is phosphorylated. It is possible, as will be discussed later, to break the covalent P–O bond, formed in the enzyme reacting with organophosphates, by potent reactivators. Some of them are highly specific for ACh-esterase, such as the pyridine-2-aldoxime methiodide (2–PAM). If a complete and irreversible block of electric activity following exposure to DFP or Paraoxon could be reversed by a reactivator highly specific for ACh-esterase, such a reactivation of electrical activity would be a strong support for the interpendence of electrical activity with the specific chemical reactions involved, formation of the P–O bond blocking electrical activity, break of the P–O bond restoring it. Obviously, this is not an easy experiment to carry out for several reasons. It is difficult to determine the exact moment at which the block of electrical activity becomes irreversible. This usually varies in certain limits from one preparation to the other, even if the same type of fiber is used, in view of many other biological variables. Too long exposure after the irreversible block may result in secondary reactions, which may make the return of the electrical activity impossible even if a large fraction of the enzyme is reactivated.

If the action of one chemical compound on an enzyme in an intact structure is a complex phenomenon, the complexity is potentiated if two different compounds are applied to intact cells even if they act on the same molecule.

The first experiments of this kind were performed on strips of frog sartorius muscle (HINTERBUCHNER and NACHMANSOHN, 1960). Two strips taken from the same sartorius were exposed to Paraoxon until electrical activity was nearly completely and irreversibly abolished; the strips were then returned to a saline solution. One strip was exposed to the reactivator, the other used as control. 2-PAM was not effective, but benzoyl-2- and -4-pyridine alodxime methiodide proved to be active. Both compounds are only slightly less effective reactivators of ACh-esterase in solution than 2-PAM, but both are more lipid-soluble. Eleven experiments were performed: in nine experiments electrical activity returned in the strips kept in solution containing the reactivator; in two experiments electrical activity did not return. As discussed in the paper quoted, the material is for several reasons not particularly suitable. Later experiments were performed on the axons of the lobster walking leg (DETTBARN et al., 1964). The fibers were exposed to 10^{-2} M Paraoxon. Electrical activity was irreversibly abolished in 1 to 3 min. As mentioned before, the excitable membrane in these axons is relatively poorly protected. On exposure to 5×10^{-2} M 2-PAM, electrical activity reappeared within 10 to 15 min. The rapidity of the irreversible block in this preparation supports the assumption of an action on the enzyme; the relatively high concentration is readily explained by the presence of phosphoryl phosphatase in this tissue. In combination with the effect of 2-PAM, these data are strong evidence for the role of ACh-esterase postulated. The possibility of restoring electrical activity by 2-PAM following irreversible block by Paraoxon and DFP was also tested on the squid stellar nerve by ROSENBERG and DETTBARN (1967). Under the experimental conditions used, block of electrical activity was partly reversed even in a few control experiments, and in some fibers electrical activity did not return even after exposure to 2-PAM. Nevertheless, a significant difference was found between the reversibility in presence and absence of 2-PAM. However, the differences became much more impressive and the data appeared quite convincing when they were performed on axons previously treated with cottonmouth moccasin venom. In view of the complexity of this type of experiment, it appears pertinent that restoration of electrical activity by the reactivator has been demonstrated in a significant number of experiments. Still more favorable preparations and experimental conditions may be found. However, even with the data available so far, it appears difficult to interpret the restoration of electrical activity by another chemical reaction than the reactivation of ACh-esterase by the action of 2-PAM displacing the phosphoryl group from the enzyme.

d) Enzyme Activity in Tissues. It would be of interest to follow the decrease of ACh-esterase activity in relation to the modifications of electrical parameters in axons after their exposure to inhibitors, and especially to evaluate the fraction of the enzyme which is the minimum required for axonal conduction. For several reasons this aim is at present unattainable.

The first prerequisite is the availability of procedures which permit precise and reliable quantitative evaluations of the total enzyme activity present in a tissue, such as nerve or muscle fibers. In intact tissue the substrate reaches only a fraction of the enzyme and only this part of the enzyme activity is measurable. The percentage will greatly vary according to the type, the complexity and the chemical composition of the tissue. For several decades it seemed that the use of homogenized suspensions offered the possibility of a quantitative determination of the enzyme concentration in a tissue. Recent observations raise serious questions as to whether simple homogenization is a procedure adequate for obtaining precise values of the total enzyme activity. The values obtained in a homogenized suspension are in many cases markedly higher than in the intact tissue, but the variations are quite large. In unmyelinated nerve fibers, for example, the increase may range from 100—400 percent, even when the same type of fiber is used. These great variations suggest the possibility that the enzyme in the particles of the suspensions is not fully and equally accessible to the substrate. A significant finding was the surprisingly small increase of total activity, only about 25%, in a homogenized suspension of the myelinated fibers of frog sciatic nerve. These fibers offer strong barriers which prevent the substrate from reaching the membrane-bound enzyme. The breakdown of barriers in the suspension would have been expected to result in a very marked increase of measurable activity. Apparently, however, even in the suspension the particles of this fiber are still rich in phospholipids which may prevent or slow down the reaction between enzyme and substrate; therefore, only a fraction of the total activity is measured. The striking effect of Triton 100 X in the tests for the enzyme applied to sections 500—1.000 A thick (Brzin, 1966) is consistent with this assumption.

Brzin et al. (1965) found that by sonic oscillation of small pieces of about 100 µg weight of the envelope of the (unmyelinated) squid giant axons the rate of enzyme activity increased by 250%. It is difficult to know whether the value thus obtained represents 100% of the enzyme present, or perhaps only 80, 90 or 95%. The difference between the values measured and the total activity actually present may vary from one type of tissue to another. Use of organic solvents or detergents may or may not be able to break up the barriers, thereby facilitating the reaction of the substrate with the enzyme. This possibility has not yet been adequately explored. The success of Lundin (1967) and his associates in solubilizing ACh-esterase by the use of bacterial enzyme (or enzymes) offers another illustration of the difficulties involved in solubilizing the enzyme or in making all of it accessible to the substrate simply by homogenization.

Pertinent in this context are the observations of Karlin (1965) on the activity of membrane-bound ACh-esterase. Using electric tissue of *Electrophorus*, he obtained, by differential centrifugation combined with the use of sucrose density gradients, fractions which were rich in large fragments of the cell membrane. Some of the fractions contained 63% of the total enzyme activity, but at four times the specific activity of that of the tissue homogenate. The enzyme associated with the membrane was found to have an increased activity after incubation with sodium deoxycholate which, however, did not solubilize the enzyme. In some fractions, the rates were doubled. Extending the studies on the behavior and properties of membrane-bound ACh-esterase, Silman and Karlin (1968; Silman, 1969) found a fourfold activation of the enzyme activity when the preparation was suspended in 1 M NACl which solubilized the enzyme. These observations make it difficult to consider the values obtained with a homogenized suspension of a tissue, usually formed by structures much more complex than membrane fragments, as representing a precise and reliable evaluation of the total enzyme activity present.

Using single eel electroplax, CHANGEUX et al. (1969) recently separated the part containing the innervated excitable membrane form that containing the non-innervated inexcitable one. The part with the excitable membrane had 10 to 40-fold higher total and specific activity than the other one[1]. They also isolated membrane fragments of the electroplax with procedures similar to those applied by WHITTACKER (1965) and KARLIN (1965). Examination by electron microscopy combined with histochemical staining methods showed clearly the localization of the enzyme in the innervated membrane fragments obtained in the isolation procedures. A strikingly uniform distribution of ACh-esterase in the whole excitable membrane surface was obtained in close agreement with the results of BLOOM and BARRNETT (1966). No enzyme activity was detectable with this method in the non-innervated membrane. The ratio of solubilized enzyme to that in the particulate matter varied from one experiment to the other; this again is an indication of the many variables which prevent reliable quantitative evaluations.

In summary, recent developments have revealed a variety of facts which force us to revise the notion that the enzyme activity measured in homogenized suspensions represents the actual total activity contained in a tissue. Attention should be paid to this aspect in looking at the estimates reported in the literature over the last three decades; they may in many cases supply useful information valuable for a variety of studies. However, where truly quantitative figures of the total enzyme concentration are essential, the procedure seems to be inadequate.

e) ACh-esterase Activity in Tissues after Exposure to Inhibitors. Since no procedures are available at present for a reliable quantitative evaluation of the total enzyme activity in a tissue, it is impossible to answer the question at what level of enzyme activity inhibitors of the enzyme block electrical activity. The difficulties of a precise determination after exposure to inhibitors are compounded by a great variety of complex factors. Some of them were recognized from the beginning, but the number of complicating factors revealed over the years has greatly increased and unexpected new ones are still continuously emerging.

The evaluation of the actual enzyme activity of a tissue after exposure to inhibitors is not only of great theoretical interest, but it is consequential for many problems of practical importance, e. g., for environmental pollution by organophosphate insecticides, as will be discussed later. It appears imperative, therefore, to scrutinize the difficulties encountered in any such evaluation. Even if it is realized that no answer can be given at present, a discussion of the difficulties may stimulate efforts to try new types of approach. Moreover, a critical examination appears necessary in view of the huge number of conflicting, misleading and contradictory reports which have appeared over the

---

[1] These figures are in good agreement with unpublished data obtained in this laboratory by Drs. M. BRZIN and W.-D. DETTBARN.

last three decades, some of them from this writer's laboratory. A critical discussion of the difficulties and the complexities may thus contribute to clarify some of the controversial views prevailing at present. However, the discussion will be essentially limited to some fundamental principles; no attempt will be made to include the vast amount of data which has appeared during the last three decades.

A correlation between the effects of inhibitors on electrical and on enzyme activity cannot be achieved with a reversible type. In order to overcome the barriers preventing the inhibitor from reaching the enzyme in intact tissue, it is necessary to apply a large excess, e. g., of physostigmine, before the first effects on electrical activity appear. If the exposed tissue is homogenized for enzyme determination without removal of the excess inhibitor, all enzyme activity will be inhibited in the homogenized suspension. Some inhibitors will reach the enzyme in the particles much more readily than the substrate: Physostigmine, e. g., is a tertiary amine whose conjugate acid has a dissociation constant of 8×10^{-9}; at pH 6 it is almost exclusively cation, whereas at pH 10 it is predominantly a neutral molecule. In contrast, ACh is charged at every pH. Moreover, many of the inhibitors have a dissociation constant several orders of magnitude lower than the substrate. Low inhibitor concentrations may have, under these conditions, strong effects. Removal of the excess by washing the tissue must, on the other hand, result of necessity in total restoration of the initial activity in view of the reversible nature of this inhibition (see, e. g., NACHMANSOHN, 1959). These factors were not taken into consideration when the function of ACh-esterase was tested with physostigmine on the whole frog sciatic nerve (LOEWI and CANTONI, 1944) and on the whole electric organ of *Electrophorus* (KEYNES and MARTINS-FERREIRA, 1953). The authors did not find any enzyme activity in the homogenized suspension, and the results were taken as evidence that the enzyme has no function in electrical activity.

When organophosphates became available, it was hoped that this type of inhibitor might permit a quantitative evaluation of the enzyme activity in relation to electrical parameters. In this case removal of excess inhibitor, after it was applied to a tissue for a given time, seemed possible; this would permit the determination of the fraction of the enzyme still active at the end of the exposure. At that time, i. e., a quarter of a century ago, very little was known about the complexity of this problem. But the knowledge accumulated in this field since then and particularly in the last few years necessitates drastic revisions of many views, including some expressed by the writer.

It would be difficult, and appears unnecessary for the topic of this article, to discuss in some detail the many complex problems of the chemistry, toxicity, biological and physicochemical properties of organophosphates. Hundreds of organophosphates have been prepared and investigated from a variety of aspects. They differ greatly in their properties and in their reactions with different types of esterases. A bewildering amount of literature has appeared in this field during the last twenty years in view of the theoretical interest and the practical implications. In the early period, virtually nothing was known about the many factors involved and many effects were poorly or not at all understood; it was difficult to avoid pitfalls and sources of error. Even today, there exist wide differences of opinion and interpretations. A huge amount of data on ACh-esterase and its inhibitors may be found in the *Handbook of Pharmacology*, edited by KOELLE (1963). Even if one disagrees with many interpretations presented there — as does the writer — the book is an important source of factual information. Critical and informative discussions about the many attempts, advances and failures of correlating the action of organophosphates with function and toxic symptoms may be found in the monograph of O'BRIEN (1960). Aspects limited more specifically to the particular problem of the interdependence of electrical and ACh-esterase activity, tested with organophosphates, may be found in previous summaries (NACHMANSOHN, 1959, 1963a, 1967c).

For more than two decades, the tissues, after exposure to organophosphates, were washed in saline and the enzyme activity was then tested in a homogenized suspension. As outlined before, this procedure turned out to be unsatisfactory for a quantitative test even in a tissue which has not been exposed to roganophosphates. One additional complication after exposure is the difficulty, or even impossibility, of completely removing, prior to homogenization, the organophosphates accumulated in the tissue, either by prolonged washing or by extraction with organic solvents or by various other procedures tried so far. Some residual organophosphate was found to be present. This fraction may have been located in interstitial tissue, Schwann cells, blood vessels, capillaries, fat deposits, etc. and may have never acted on the excitable membrane nor reached the enzyme in the intact structure. But after homogenization, such a reaction may readily occur. With the abovementioned compounds Tetriso and Selenophos, which readily enter into the axoplasm of squid giant axons but also seem to come out readily, no ACh-esterase activity was detected in the homogenized tissue prepared after prolonged washing. When experiments were performed to test the preparation for residual inhibitor, it was found that more than 99% had been removed by the washing. However, the fraction of less than 1% remaining was in one case 5 and in the other 50 times as high as the concentration required to inhibit all enzyme activity (HOSKIN et al., 1969). It was many years before the real implications of this "homogenization artifact", as it was referred to by VAN ASPEREN (1958), were fully and widely recognized.

Even if methods are developed which will permit a reliable test of the total activity in a tissue, the complication of removing the residue once the tissue was exposed will remain. This additional obstacle must be overcome before the problem of the correlation between enzyme and electrical activity can be solved. In the light of these data, the many reports which have appeared during the last 20 years, claiming that all the enzyme activity was destroyed in nerve fibers exposed to organophosphate whereas electrical activity was still normal, have lost their validity and do not require any discussion. But equally, the determinations in which ACh-esterase activity was found after exposure to organophosphates cannot be considered as a quantitative indication of the total activity actually remaining.

Another factor interfering with a correlation between electrical and enzyme activity and not fully recognized in the early phase of investigations, is the complications introduced by the use of multifiber preparations. In a preparation formed by several thousands axons, e. g., the frog sciatic nerve, the large amounts of interstitial tissue, Schwann cell, blood vessels, myelin, etc. may retain, following exposure, extracellular organophosphate which reacts with the membrane-bound enzyme only during or after homogenization. This source of error may be smaller in a preparation formed by a few axons or even a single one, because they are less surrounded by extracellular structures. The sources of error will be even greater if the inhibitor is injected into the whole animal and intact tissue is taken out, since there is even less control of the concentration of inhibitor to which the tissue was exposed in the intact animal. The effects on electrical activity are also less ambiguous with a single axon than with a multifiber preparation. The inhibitor may reach the membrane of the various fibers at different rates, so the effects may not proceed simultaneously. Under such conditions the relationship between electrical parameters and enzyme activity cannot be adequately evaluated even if procedures will become available for reliable chemical tests. A more detailed discussion of this particular aspect may be found in a paper by DETTBARN and ROSENBERG (1962).

f) **Complexity of Organophosphate Effects.** Several additional complications for the interpretation of organophosphate actions *in vitro* and *in vivo* may be briefly outlined. Whereas the action of organophosphates on electrical activity of axons has been well demonstrated, some of the early unexpected findings require comments in the light of more recent information which has become available. One of the problems encountered was the reversibility of the effect after short periods of exposure. Since the organophosphates inhibit the enzyme irreversibly, how could the block, of electrical activity, at least for a short while, be reversible? Here again, one cannot extrapolate form the solution to the intact structure. In structures there are many components by which modifications of effects may be effected. The first possibility to be considered in evaluating the effects on the electrical activity of axons, as pointed out by the writer more than 20 years ago, is an effect of these compounds on the receptor, but without covalent bond formation, i.e., in a reversible way. Since both components are presumed to be in very close vicinity, the active site of the receptor may become, by the presence of these compounds, less or not at all responsive, to ACh released in very small amounts. This block may take place even before the enzyme is blocked by the organophosphate. Such an interpretation became more likely by the recent advances of our knowledge of membrane structure and organization. In a tightly organized structure of 100 A thickness, competition between chemical compounds for active sites of proteins is not difficult to picture. Allosteric effects (*see* Section III) must also be considered.

The possibility that enzyme inhibitors may also act on the receptor and may even be responsible for some of the primary effects observed, has been apparent for a long time in the case of compounds with the quaternary nitrogen group. Many discussions centered about this problem (*see*, e.g., RIKER, 1953; WERNER and KUPERMAN, 1963). Although an action of organophosphates on the receptor seemed less obvious in the beginning, several observations pointed to such effects and were interpreted in this way. The first observations, in which this possibility was strongly stressed as the most likely one, were experiments on the frog rectus preparation with DFP and Sarin (COHEN and POSTHUMUS, 1955, 1957). However, the complexity of the preparation made a conclusive interpretation difficult.

This applies to other similar reports. More convincing were the observations of FREDERICKSSON and TIBBLING (1959a), who tested the action of several potent methylfluorophosphoryl cholines described by TAMMELIN (1957, 1958) on the frog rectus in the presence of 5×10^{-5} M Sarin. Although some of the interpretations may be open to question, they support indeed the assumption of an action of these compounds on the receptor which, in view of their structure, would be *a priori* likely.

However, in 1956 the monocellular electroplax developed by SCHOFFENIELS (SCHOFFENIELS and NACHMANSOHN, 1957; SCHOFFENIELS, 1957, 1959) proved to be uniquely favorable for studying the reaction of ligands with the receptor protein in the excitable membrane. The results obtained with this preparation form the main topic of Section B III. In regard to the action of organophosphate on the receptor, evidence has now been obtained (BARTELS and NACHMANSOHN, 1969), which clearly shows an action or organophosphates on the receptor protein acting in competition with ACh. In low concentrations Paraoxon, DFP and the tertiary phospholine strongly potentiated the action of ACh. This effect is due to a partial inhibition of ACh-esterase, since it is reversed by 2-PAM. With higher concentrations which by themselves still had no effect, the depolarization of the cell by ACh was rapidly reversed. These effects are readily reversible. The new data are a strong support for direct action on the ACh-receptor by organophosphates competing with ACh even in absence of a methylated amine group; they parallel similar effects mentioned before, obtained on the same preparation by HIGMAN and BARTELS (1961) with physostigmine. This compound, in 5×10^{-5} M, potentiates the action of ACh on the electroplax, more than 100-fold, whereas in 10^{-3} M, and even less (BARTELS, 1968), it antagonizes ACh and acts as a receptor inhibitor. It appears not surprising that compounds such as Tetriso and Selenophos and the choline-like compounds mentioned before will act similarly as receptor inhibitors. All these effects, when studied on the monocellular electroplax preparation may be expected to yield less equivocal and more quantitative results.

Whereas in the light of these data it seems likely that the early reversibility of the block of electrical activity produced by organophosphates is due to their action on the receptor, at least to a large extent if not exclusively, the other question which still remains open is that of the slow rate of the irreversible inhibition of the electrical activitv by organophosphates. However, many observations have been reported which raise the question if the views on the nature of the irreversible inhibition of the enzyme by organophosphates even in solution, but more so in intact structures, do not require considerable revisions.

Initially, the irreversible character of the reaction seemed quite obscure. In the early literature, many statements appeared that the enzyme is "destroyed" by organophosphates. When a highly active and purified ACh-esterase preparation obtained from electric tissue of *Electrophorus* became available (ROTHENBERG and NACHMANSOHN, 1947), investigations on the mechanism of the hydrolytic process were initiated. They have revealed many details about the molecular groups in the active site of ACh-esterase (for summaries, *see* NACHMANSOHN and WILSON, 1951; NACHMANSOHN, 1959; WILSON, 1960; COHEN and OSTERBAAN, 1963). In the physiological process, the enzyme attacks the substrate in a bimolecular nucleophilic substitution (S_N2) reaction. In the first step, an acetylated enzyme is formed with elimination of choline; in a second step the acetyl enzyme reacts rapidly, in a few microseconds, with water, resulting in the formation of acetate and restored enzyme (WILSON et al., 1950). As is well established today from many studies of ester-splitting enzymes, the nucleophilic group in the active site reacting with the carbon of the carbonyl group is the oxygen atom of a serine. An imidazole group participates in the enzymatic process. In the case of organophosphates, the enzyme attacks the electrophilic P atom, eliminating an acidic group. In this $S_N 2$ reaction, the P atom forms a stable covalent bond with the nucleophilic O atom of the serine; a phosphorylated instead of an acetylated enzyme is formed, as was first suggested by BURGEN (1949). The phosphorylated enzyme does not react with water, or only slowly: thus the enzyme is inhibited and in view of ist vital function in the body the effect is fatal.

However, even in solution not all phosphorylated enzymes are really irreversibly inhibited. Dephosphorylation in water, i.e., spontaneous reactivation, may take place. The rates of spontaneous reactivation of different types of phosphorylated ACh-esterase and different types of other esterases vary greatly. Dimethylphosphoryl ACh-esterase (with two methoxy groups) proceeds faster than that of any other type of dialkyphosphoryl ACh-esterase (BURGEN and HOBBIGER, 1951). At 37° C and pH 7.6, the reactivation amounts to 50% in 90 min when rabbit erythrocytes are used as enzyme source (ALDRIDGE, 1953; ALDRIDGE and DAVISON, 1953). With diethylphosphoryl ACh-esterase, obtained from the same source and used under similar conditions, 25 to 40% spontaneous reactivation takes place in 24 hr (BURGEN and HOBBIGER, 1951; HOBBIGER, 1951). No spontaneous reactivation takes place with diisopropylphosphoryl ACh-esterase.

The observations of HESTRIN (1949, 1950) indicating that hydroxylamine attacks the carbon of the carbonyl group of acetylated ACh-esterase to form hydroxamic acid suggested the possibility to attack the phosphorylated enzyme by nucleophilic compounds and possibly to reactivate the enzyme in a displacement reaction. This observation actually opened the way to the development of reactivators of the enzyme and eventually to the development of effective antidotes. Hydroxylamine proved to be a reactivator, but a weak one in high concentrations; it reactivated some but not all types of phosphorylated enzymes – for instance, diethylphosphoryl but not diisopropylphosphoryl enzyme (WILSON, 1951). But in the following years a great number of reactivators were prepared and some of them proved to be very efficient. One of the most effective reactivators was 2-PAM discussed before (WILSON and GINSBURG, 1955; DAVIES and CHILDS, 1955, 1956; CHILDS et al., 1956). This compound was so efficient in reactivating the enzyme in solution that it was tried as an antidote against organophosphate poisoning in animals. It proved to be an efficient antidote against some nerve gases and a spectacularly successful antidote against insecticide poisoning, especially in combination with atropine (KEWITZ et al., 1956; NAMBA and HIRAKI, 1958). Even more effective reactivators in the test tube than the 2-PAM are bisquaternary pyridine aldoximes when tested by the rate of reactivation of phosphoryl enzyme (HOBBIGER et al., 1960; HOBBIGER and SADLER, 1958, 1959; POZIOMEK et al., 1958; WILSON and GINSBURG, 1958; etc.; for a summary, see HOBBIGER, 1963). However, the usefulness of these bis-compounds as antidotes in organophosphate poisoning is complicated by their curarelike properties.

Reactivation experiments with different ester-splitting enzymes and different organophosphates have revealed vast differences of rates of reactivation and of stability of the complexes even in solution. Inhibition of ACh-esterase by phosphoramides, such as the nerve gas Tabun, cannot be reversed even by such a potent reactivator as 2-PAM. The whole notion of reversibility and irreversibility has undergone drastic changes and has become quite fluid. A sharp distinction became even less meaningful when it was found in recent years that the rates of reaction after so-called reversible inhibitions by carbamates may vary from a few minutes to days (WILSON, 1960).

Other observations referred to by HOBBIGER (1955) as "aging" may also have an important bearing on the problem under discussion. HOBBIGER observed that the longer the organophosphate inhibitor and the enzyme were in contact – serum cholinesterase was used – the smaller was the fraction that could be reactivated by appropriate reactivators. He suggested as an explanation that a transphosphorylation takes place transforming the phosphorylated enzyme from one type to another type. For instance, after 30-min contact diethylphosphoryl enzyme was to 90% of a type which could readily be reactivated; after 24-hr contact, only 10% were of this type. The "aging" was faster with diisopropylphosphoryl enzyme. In later studies, HOBBIGER (1959) found aging in vivo with brain and erythrocytes ACh-esterase of mice. DAVIES and GREEN (1956) reported similar observations with erythrocyte esterase inhibited by tetraethylpyrophosphate (TEPP) or Sarin in vitro. They found that the rate of change from the reversible type of phosphorylated enzyme to the irreversible one followed first-order kinetics at 37° C and was pH-dependent. The rate of change was again faster with diisopropyl and dimethylphosphoryl than with dimethylphosphoryl enzyme. Similar observations had been reported by WILSON (1955). The authors suggested that in the beginning the imidazole group of the enzyme might be phosphorylated, and subsequently a transphosphorylation would result in the much more stable phosphoryl serine. A transient phosphorylation of histidine by organophosphate had been previously proposed by WAGNER-JAUREGG and HACKLEY (1953) based on the finding that histidine – and several other compounds tested – promotes DFP hydrolysis in aqueous solution. Another type of aging has been proposed by OOSTERBAAN et al. (1958; see also COHEN and OOSTERBAAN, 1963) in which one of the alkyl groups of a dialkylphosphoryl enzyme is eliminated; thereby the complex becomes more stable. As pointed out by O'BRIEN (1960, p. 108), a process of "aging" may also account for the observation of DAVISON (1955) that spontaneous recovery of rat cholinesterase in vivo was never complete. The enzyme recovered first at a fast rate, but the recovery stopped at 60%.

Another important difference between the reactions and the behavior of organophosphates in tissue and in solution is the much faster rate of reactivation of phosphoryl enzyme in tissue compared to that observed in solution. In the brain of mammals, part of the difference between ACh-esterase concentration found shortly after injection and after successive intervals of recovery may be largely due to removal of interstitial organophosphate (see, for instance, KEWITZ and NACHMANSOHN, 1957; KEWITZ, 1957; ROSENBERG, 1960). However, some of the inhibited enzyme may be actually reactivated by other reactions in the tissue. O'BRIEN (1956) found that cholinesterase of insects inhibited by malathion can readily recover. Similar observations were reported for the in vivo reactivation after exposure of insects to TEPP (SMALLMAN and FISHER, 1958). Further studies of the rapid in vivo recovery of cholinesterase from the inhibition by various organophosphates were carried out by MENGLE and CASIDA (1958). Even after DFP poisoning, an almost complete recovery was found within one day (MENGLE and O'BRIEN, 1960). In isolated electroplax, ACh-esterase, blocked by DFP, Paraoxon and phospholine partially recovered in 20 to 30 min. (DETTBARN, et al. 1970). The differences of recovery rates of inhibited enzyme between in vitro and in vivo are still poorly understood. MENGLE and O'BRIEN (1960) assume that live flies contain a factor capable of inducing recovery of phosphoryl enzyme. The factor seems to be labile, since it disappeared from homogenates within 30 min in the cold. A reactivating factor was also found in mammalian sera by NEUBERT et al. (1958).

Another factor may contribute to the relatively slow rate of irreversible inhibition in tissue. It is well known that organophosphates are hydrolyzable in aqueous solution. Many of them are attacked by OH^- ions in an S_N2 reaction; in some others, e.g., phosphoramidates, the P–N bond is unstable in acid and more readily catalyzed by H^+ ions. Many factors may influence the hydrolysis rates, such as temperature and pH, presence of catalytic agents such as histidine mentioned before, and variations of the substituents on the P atom. Where the acidic group is a thioester, hydrolysis rates may be much faster than for the same type of compound containing an O instead of an S. Without going into details of the various chemical aspects of the many factors which may catalyze organophosphate hydrolysis in aqueous solution or lead to isomerization, oxidation, etc., it is obvious that in a structure such as a membrane several of these factors or a combination of these may play an equal or even more important role. It was pointed out that the notion of concentration in a membrane may be meaningless. The "microscopic" environment (HILLE and KOSHLAND, 1967; KOSHLAND and NEET, 1968) may have significant effects and may even catalyze the hydrolysis of organophosphates in the neighborhood of the active site of the enzyme before the enzyme is able to attack them and become phosphorylated. Catalysts may be efficient in the membrane in amounts ineffective in solution. Hydrolysis of the organophosphate penetrating the structure may slow down the phosphorylation of the enzyme at least for some time, and apparently delay the rate for action until sufficient amounts of inhibitor have accumulated. Furthermore, it is possible to picture nucleophilic agents, such as for instance choline, which are poor reactivators in solution, to be effective within the membrane when located in strategically favorable positions in the neighborhood of the enzyme and thus slow down the rate of phosphorylation of the enzyme. The phosphorylation of the active site of the enzyme may be preceded, in living cells, by phosphorylation of other sites, possibly allosteric sites, before a transphosphorylation leads to the irreversible blocks of the enzyme and thus of electrical activity.

In 1933, A. J. CLARK wrote: "The physical chemist can reasonably hope to simplify his conditions and to reduce the number of variables, until he obtains a system that provides formal proof of the laws which he enunciates, but the pharmacologist is interested in the action of drugs on the living cell and any attempt to simplify this material results in death. Hence he cannot hope to obtain formal proof for his theories and must be content with intelligent guesses. Even in the most favorable cases where quantitative relations have been established for the action of drugs on cells, there probably remain dozens of unknown variables, and there is usually a considerable range of possible alternative explanations." In the 35 years since this statement, the advances of biology have revolutionized our understanding and knowledge of the truly amazing complexities of the biochemical factory that is the cell, of its substructures and organization, and of the great variety of factors by which reactions in the cell are modified and differ from those in solution. A few of them have been analyzed and have been well explained, although we are still at the beginning of the road. But in the light of all the new knowledge accumulated, very few biochemists will be willing to accept simple extrapolations from reactions in solutions to those in intact tissue. The times when knowledge of the properties and behavior of the pure enzyme in solution was the end of the road have long passed. Some of the complex factors which may modify the reactions between organophosphates ane ACh-esterase in the cell compared to those in solutions have been, though brieflyd discussed. It is obvious that marked differences must be expected. The explanation, on a molecular level is still far from complete. Discrepancies such as concentration, initial reversibility or slow rate of action cannot be construed as evidence that the irreversible block of electrical activity in tissue by organophosphates is not due to the irreversible block of ACh-esterase, i.e., to the well-known specific biochemical lesion. With several preparations and enzyme inhibitors, strong evidence for the interdependence of electrical and ACh-esterase activity has been offered. Some of the difficulties encountered with some other preparations and compounds for demonstrating equally well the coincidence of the blocking effects on electrical and on enzyme activities have been explained, others remain unsolved. But they do not justify the assumption of a general toxic effect, repeatedly proposed as being responsible for the block of electrical activity and not related to the inhibition of the enzyme. To deny the relationship between enzyme and electrical activity and to assume a general toxic effect appears all the more unsatisfactory since after 25 years of intensive efforts and research on this particular question not a single reaction has been found and offered as a reasonable alternative explanation.

As an illustration of the reasoning of the writer concerning this particular problem, a well-known example may be mentioned. Once the sliding mechanism and the specific functions of ATP, actin, myosin and Ca^{++} ions in the elementary process of muscular contraction have been demonstrated, the most reasonable assumption is that of their general validity. Difficulties of demonstrating the same mechanism in different types of muscles will not be readily accepted as evidence for a different mechanism; they are most likely due to special features of structure and organization making the preparations less suitable for analysis. Studies on the biochemical and molecular basis of muscular contraction started three decades before those on the chemical basis of nerve impulse conduction since for the latter problem many of the now available methods were not existing. It is noteworthy that MEYERHOF and LOHMANN proposed, in 1934, that ATP provides the energy in the elementary process of muscular contraction; ATP was referred to by MEYERHOF as the 'specific operative substance'. It took more than 30 years until this view was generally accepted and the widespread opposition to it abandoned.

In summary, many of the perplexing problems offered initially by the action of organophosphates on axonal conduction have been explained during the last two decades in physiocochemical and structural terms. A vast amount of evidence has accumulated during the same period in favor of the role of ACh as the specific signal controlling the permeability changes in excitable membranes and the role of ACh-esterase as the biological removal mechanism. Finally, as will be more fully discussed in Section C, the similarity of the role of ACh in the permeability changes of the excitable membranes at the level of axons and that of junctions has been well established and the apparent discrepancies have been elucidated. On the other hand, it has become apparent that there are at present no procedures available for establishing quantitative relationships between electrical parameters and levels of ACh-esterase activity.

g) **Environmental Control and Pollution by Pesticides.** A special aspect of the problem of organophosphate insecticides has suddenly assumed unexpected dimensions because of the widespread interest in the control of the environment and the pollution by insecticides. It appears appropriate to discuss some of the implications resulting from better insight into the complexity of organophosphate action just presented. The new information may have important bearing on the problem of environmental control.

When insecticides, DDT in particular, were first used on a large scale, they were hailed as a great triumph of science in fighting diseases and hunger in underdeveloped countries. The resulting population explosion, however, has created serious problems for man. Few will accept as a solution the discontinuation of the use of insecticides and permit hunger and diseases to exercise again a "natural" control of population growth; there are alternatives preferable for the control of population growth. But a continuation of the use of insecticides requires efforts to eliminate undesirable side effects, chiefly among them the pollution of the environment. There is growing opposition to the use of insecticides of the DDT type in view of the stability of these compounds which may, by accumulation, lead to harmful side effects. Naturally, the interest turns to the development of unstable and harmless insecticides.

Organophosphate insecticides are unstable. They, therefore, offer potentially the possibility of developing a desirable type of compounds. At present, unfortunately, the compounds available are still toxic to man, animals and birds to a degree which is higher then acceptable. The question arises then of whether or not, in view of our present knowledge, the possibility exists to modify the structure of organophosphates and to develop insecticides with the toxicity to man and animals decreased to a negligible degree, whereas they still remain toxic at least to certain insects.

The main difficulty at present is our extremely limited knowledge of the toxicity of organophosphate insecticides. For several decades most of the toxicity studies of organophosphates were carried out on whole animals, or with complex tissue preparation, or on the enzyme in solution. The explosive growth of biology in the last decade has shown how deceiving such type of studies may be unless supplemented by analysis on cellular, subcellular and molecular level. Some of the new factors which have emerged as a result from this approach have been outlined above. For instance, detoxifying abilities due to the presence of enzymes vary from species to species. A study of these detoxifying enzymes, their properties and distribution in insects and animals may lead, by modification of the structure of organophosphates, to the development of insecticides which are specifically fatal to at least some insects, but harmless to man and animals. The incredible variety of the biological, chemical and toxic properties of organophosphates offers a serious hope to develop modifications with the specific properties desired. The substitution of one single atom may drastically modify biological activity of a compound. For example, as will be discussed later (page 69), sulfur and selenium isologs of ACh and congeners have been found to show great variations in biological activity due partly to differences in electron distribution, and partly to small changes in configuration. These studies have opened a new page in the designing of drugs in general, and more specifically, for a completely new approach to the development of new types of insecticides with well defined characteristics.

The development of a more specific type of organophosphate insecticides would, of course, require the combined efforts of many laboratories and teams of ecologists, entomologists, biologists, and chemists. One important aspect would be an intensification of the studies on cellular, subcellular, and molecular level. It was not by accident that the multidiscipline approach, in the writer's laboratory, including studies on molecular level, led to the development of a highly efficient and specific antidote against organophosphate-insecticide poisoning. A systematic and concentrated approach on a broad scale, making use of all the new information, may offer a serious hope to find insecticides with the desirable properties.

h) Molecular Properties. Studies on some of the properties of ACh-esterase were greatly helped when the enzyme was first obtained in a highly active solution by extraction from electric organ (NACHMANSOHN and LEDERER, 1939). In the early 1940's, a more than 400-fold purification of the enzyme was obtained (ROTHENBERG and NACHMANSOHN, 1947). The amounts of material available were, however, extremely limited, and therefore the amounts of homogeneous enzyme protein finally obtained were very small, at best a few milligrams. The partially purified preparations with high specific activity were, however, useful for kinetic studies and for analyzing the reaction mechanisms of many compounds with the enzyme; they provided the information about the molecular groups in the active sites and about the mechanism of the hydrolytic process catalyzed by ACh-esterase, as mentioned before.

The developments of protein chemistry during the last decade which permit an analysis of molecular properties and even of the tridimensional structure made it desirable to obtain large amounts of enzyme protein. Column chromatography was applied for the purification of electric eel tissue by KREMZNER and WILSON (1963). These procedures were modified and a large scale purification was achieved by LEUZINGER and BAKER (1967). Using 10 kg of electric tissue as starting material, about 60—70 mg protein were obtained with the specific activity of about 700 millimoles of ACh split per mg protein per hour. The material is formed by a single component according to disc electrophoresis and analytical ultracentrifugation.

The preparation led to the crystallization of ACh-esterase (LEUZINGER *et al.*, 1968). The crystals have the form of hexagonal prisms (Fig. 6). They have grown so far to 150-μ length and 80-μ width. The molecular weight (M.W.) of the enzyme has been measured by equilibrium centrifugation according to the methods described

Fig. 6. Crystals of ACh-esterase. (A) Regular hexagonal prisms; (B) Possibly pyramidal termination. According to Dr. BARBARA LOW, the most common form of growth observed, A, is compatible with true hexagonal symmetry; B may imply, however, a crystal system of lower order. The crystals exhibit low birefringence so that it is not possible at present to determine whether they are uniaxial or biaxial. (From LEUZINGER *et al.*, 1968)

by VAN HOLDE and BALDWIN (1958) and YPHANTIS (1964) and was found to be 260,000, in good agreement with the previously reported values of 240,000 (LAW- LER, 1961) and 230,000 (KREMZNER and WILSON, 1963). The protein has four subunits of about equal M.W. and two different polypeptide chains, on the basis of the determination of C-terminal groups with two different methods (LEUZINGER *et al.*, 1969).

III. ACh-Receptor

a) **Notion of Receptors. Activators and Inhibitors of the ACh-Receptors.** Since the beginning of this century, it was recognized that effects on cells pro- duced by any chemical compound, drug, toxin, etc., must be due to the reaction with specific molecules, referred to as receptors. Whatever their nature may be, these molecules must have specific sites capable of binding the chemical compound or drug before it produces the characteristic response in the cell. Some of the early formulations of this notion were expressed by EHRLICH, and much of his work was based on the idea that drugs must react with specific receptors in the effector cell before they would be able to exercise their effects. With the advances achieved in the knowledge of cellular mechanisms, it became possible to explain some pharmacological effects by chemical and biochemical reactions. The notion of receptors has been frequently revised and has been formulated in increasingly specific terms. An early attempt at a quantitative approach to the interactions of drugs and receptors is presented in the review of CLARK (1937). He applied the notion of the Langmuir isotherm to the reactions of drug molecules with receptors and tried to relate drug concentrations and the response of the cell to the number of receptors occupied. CLARK himself questioned some of his assumptions and due to the developments of the last few decades about the interaction of drugs and receptors, many of his ideas had to be revised. Some recent notions of receptor mechanisms are discussed in the review of FURCHGOTT (1964) and in the book *Molecular Pharmacology* edited by ARIENS (1964). The most recent evaluation of this rapidly advancing field is the review of MAUTNER (1967) in which classical views of receptors are analyzed and integrated with modern concepts and notions of protein and macromolecular chemistry in general. Several of the facts and problems discussed in Mautner's review are pertinent to the specific topic of this article. The reader interested in this and in the general problem of receptors in chemical terms is referred to this review.

In the early phase of the theory postulating ACh to be a neurohumoral transmitter, the target of this transmitter was not yet discussed. It was simply assumed to stimulate the second cell, nerve and muscle, in a way which was not specified, whereas the inactivation of ACh by hydrolysis was specifically attributed to the action of an enzyme, cholinesterase (*see*, for instance, DALE, 1937). In the last two decades, the notion of a receptor for ACh has emerged and has been widely accepted, particularly after World War II when, due to the widespread investigations with organophosphates, the effects of a great variety of inhibitors of the enzyme were studied. It became apparent to many investigators that it was difficult to explain some of the pharmacological actions and electrophysiological observations simply in terms of enzyme inhibition. WESCOE and RIKER (1951), for instance, clearly recognized that the effects of 3-hydroxyphenyltrimethyl-ammonium, although a strong inhibitor of ACh-esterase, could not be readily attributed to enzyme inhibition (for a more elaborate discussion, *see* RIKER, 1953). In the following decade, an increasing number of observations with inhibitors of ACh-esterase were reported by many investigators, supporting the assumption of an action of some of these inhibitors on the receptor for ACh as distinct from and preceding that on the enzyme (for a review, *see*, for instance, WERNER and KUPERMAN, 1963). It may be mentioned that the notion of the receptor was essentially an operational term, as distinct from ACh-esterase. As to the precise nature, the opinions differed. Some investigators considered the possibility that ACh-esterase and receptor were the same molecule but that there might be a special site on the enzyme; others were inclined to assume two different components. A precise definition of the receptor site was, moreover, complicated by the biochemists' occasional use of the term receptor site as a synonym for the active site of enzymes.

The evidence for the existence of a receptor was based essentially on pharmacological and electrophysiological studies. The complexity of the preparations used (e.g., various types of neuromuscular junctions, frog rectus muscle, guinea pig ileus, etc.) offered serious difficulties for quantitative evaluations and even greater ones for an analysis of the chemical nature and properties of the ACh-receptor. If the ACh-receptor has the function of initiating the permeability changes in the excitable membranes, it must be localized there. In that case, only by the recording of electrical parameters of a suitable membrane in response to chemical compounds and under special, appropriate experimental conditions could it be hoped to obtain the desired information about the nature of the ACh-receptor.

4*

Studies of this kind on the ACh-receptor were greatly helped when techniques were developed in which isolated rows of electroplas were used. Such studies were initiated on the *Torpedo* by AUGER and FESSARD (1939) in the late 1930's (for a summary, *see* FESSARD, 1946). Later, the use of isolated rows of electroplax of *Electrophorus* was introduced simultaneously and independently in the laboratory of Chagas in Rio de Janeiro and in this laboratory (KEYNES and MARTINS-FERREIRA, 1953; ALTAMIRANO *et al.*, 1953). With these techniques, much pertinent information has accumulated about the electrophysiological aspects of electroplax. The reader interested in this angle is referred to the many comprehensive reviews which have appeared in this field (FESSARD, 1958; GRUNDFEST, 1958; CHAGAS and DE CARVALHO, 1961).

The preparation of isolated rows of electroplax of *Electrophorus* on which the electrical parameters were recorded with the use of intracellular electrodes made possible some pertinent advances in the analysis of the ACh-receptor. When ACh and several of its congeners were applied, it became apparent that many of the compounds had strong effects on the electrical parameters at concentrations at which ACh-esterase was not affected to an extent which could explain the action observed; i.e., they acted on the ACh-receptor. Two different types of action were observed. One group of compounds, acting on the receptor only, blocked the electrical activity and simultaneously depolarized the membrane, inducing apparently a change of the receptor comparable to that postulated for the physiological action of ACh. Another group of compounds produced a block of electrical activity, but without any depolarization; they apparently act as antimetabolites preventing the change of the receptor normally induced by ACh. In analogy to enzyme chemistry, in which it is customary to distinguish between substrate and inhibitors, the two types of compounds were referred to as receptor activators and receptor inhibitors (ALTAMIRANO *et al.*, 1955b; SCHLEYER, 1955; ALTAMIRANO *et al.*, 1955a). In addition to ACh, carbamylcholine, neostigmine, decamethonium, and several other quaternary ammonium ions belong to the receptor activators; curare, local anesthetics such as procaine and tetracaine, and several others, as will be seen in the following, belong to the receptor inhibitors.

Although the information obtained was much more quantitative than that previously reported with complex muscle preparations, it was still far from satisfactory. In every multicellular preparation, there are many complicated factors which interfere with the quantitative analysis. Presence of many types of structural barriers, for instance, is bound to interfere with the penetration of the compounds to the excitable membrane. The effects of these barriers may vary greatly for different compounds, depending on their structure and physicochemical properties. The striking difference between the effects of physostigmine and curare on the whole sciatic nerve fiber of frog and a single isolated axon of this fiber has been previously discussed, as well as other aspects of permeability barriers. Another serious complication of multicellular preparations results from the large intercellular spaces. It became increasingly apparent that a real quantitative evaluation of the reactions of compounds applied externally with the ACh-receptor in the membrane, i.e., an analysis on cellular and subcellular level, requires a single cell preparation. Use of a single cell would eliminate many of the most serious complicating factors, although even a cell still remains a highly complex structure, as repeatedly emphasized.

b) Monocellular Electroplax Preparation. A new chapter in the history of the studies on the ACh-receptor was opened when SCHOFFENIELS developed, in 1956, an extraordinary method for testing its properties on the intact cell – the monocellular electroplax preparation (SCHOFFENIELS and NACHMANSOHN, 1957; SCHOFFENIELS, 1957, 1959). One single cell from the posterior part of the electric organ of *Electrophorus*, generally referred to as the Bundle of Sachs, is dissected and mounted between two chambers, so that the cells separate two pools of fluid. The cell is kept between two nylon sheets, one with a window adjusted to the dimensions of the cell, and another with a grid consisting of nylon threads and used for pressing the cell against the window. The great value of this preparation is based on several special features of the cell. 1) It is a cell of extraordinarily large dimensions (6 to 10 mm long, 1 to 2 mm high, and 0.2 mm thick). This size readily permits many types of experiments where smallness might be prohibitive. 2) Only one face is innervated and has a conducting membrane; the other face is not innervated and is nonconducting. Therefore, the fluid of one chamber is bathing one type of membrane, that of the second chamber the other type. Under these conditions, it is possible to test the two membranes separately. 3) The innervated membrane, which faces the window, has a more-or-less rectangular shape. It is, therefore, suitable for many types of studies such as ion movements across the membrane. All compounds moving from one pool across the other must cross the cell. There is no leakage, as has been repeatedly tested with radioactively labeled sulfate. 4) The innervated membrane is formed by many thousands of synaptic junctions (on the average 20,000 to 40,000 per cell), which cover, however, only 5 to 10% of the surface. The remainder, 90% or more, is conducting membrane. Electrical parameters readily permit one to distinguish between the action on the junction (the response to neural stimulation) and that on the conducting membrane (the response to direct stimulation). At the level of the junction, the reactions of the ACh-receptor with the ligand applied externally is measured more or less directly with a minimum of interference by structural barriers, whereas the conducting parts of the membrane are surrounded by barriers which prevent poorly lipid-soluble compounds such as ACh and curare to reach the receptor; the barriers are thus similar to those in axons discussed before.

During the last decade, several refinements and improvements of technique have greatly increased the usefulness of this preparation for the analysis of the ACh-receptor. (For description of the techniques, *see* HIGMAN and BARTELS, 1961, 1962; HIGMAN *et al.*, 1963, 1964; KARLIN and BARTELS, 1966). The use of intracellular electrodes combined with a special switching device permits the precise recording of various parameters, such as simultaneous measurements of the potential difference across the conducting and the nonconducting membranes and across the whole cell. A great improvement for the quantitative evaluation of the potency of the reaction of ligands with the ACh-receptor was the use of recording steady state potentials with increasing concentrations of the compounds used. This procedure permits determination of affinities with a high degree of reliability, using the Michaelis-Menten approach of analysis (HIGMAN *et al.*, 1963).

For the understanding of some aspects of the analyses performed on this preparation, a few comments appear necessary. ACh and other depolarizing quaternary ammonium ions, acting on the level of the synaptic junctions only, nevertheless induce the depolarization of the whole membrane indirectly by the depolarization of the many thousand junctions. But this again immediately raises the question of what processes are responsible for the depolarization, i.e., for increased permeability, in the segments of the membrane which are conducting. Those investigators who consider permeability changes in axonal conduction to be a purely physical process and in synaptic transmission to be effected by neurohumoral transmitters have repeatedly proposed that these two types of bioelectricity exist side by side in the electroplax of *Electrophorus*. The difficulties of such a notion assuming two different mechanisms are the same as in other excitable cells and are difficult to reconcile with the biochemical data. The studies on the electroplax by BLOOM and BARRNETT (1966), using electron microscopy combined with a variety of histochemical staining techniques, have shown that ACh-esterase is equally localized in all membranes, i.e., in the membranes surrounding the nerve fibers, in the pre- and postsynaptic membranes, and in the conducting membranes between the junctions. The presence of the ACh-receptor has been unequivocally demonstrated in the conducting parts of the electroplax membrane, and not only at the junctions, with the use of tertiary, i.e., lipid-soluble, analogues of ACh (*see* p. 33 and 43). Both proteins have been shown to be associated with the electrical activity of the conducting membrane. Thus, we have the same condition as in axons discussed previously; both enzyme and receptor are present and functional, but structural barriers prevent the quaternary ACh applied from the outside to react with the receptor. Once the existence of these barriers surrounding the conducting membrane has been demonstrated, preventing a direct action of quaternary ammonium ions but not of their tertiary analogues on the two proteins and, moreover, once a direct action of ACh on axonal conduction has been shown in suitable preparations or arranging proper conditions, it appears likely that the same explanation holds for the electroplax. In that part of the membrane which is protected, ACh will be released within the membrane by the potential differences developing between the junctions and the rest of the membrane; once released, it will trigger the reactions leading to depolarization of the whole membrane. The presence of 20,000 to 40,000 junctions in the surface of one single cell renders it highly sensitive to the ligands applied; thus, the cell offers an exceptionally favorable sample of excitable membrane for the analysis of the molecular groups reacting in the protein using a great variety of ligands.

Another aspect may be mentioned. Although it is true that ACh and other quaternary compounds act only at the level of synaptic junctions, it is at present an open question what fraction (or fractions) of the synaptic membranes is affected. In fact, it could be a complex effect, including axonal and synaptic membranes and even a limited segment of the conducting membrane of the electroplax in the neighborhood of the junctions, which is not adequately protected, as is known today for muscle membranes near the neuromuscular junction. The presently used techniques do not permit definite distinctions. However, for the analysis of the protein properties it makes little difference in what fraction of the excitable membrane the synaptic junction is located, as long as there exist favorable conditions for such an analysis. The problem whether this protein has a similar function in the whole excitable membrane does not enter here. It will be further discussed in Section C.

A paramount feature of the electroplax from the biochemical point of view remains its extraordinary specialization for generating bioelectricity. It was mentioned earlier that the low protein and high water content of the tissue, on one hand, and the high ACh-esterase activity, on the other, made the tissue an exceptionally favorable choice as a source of extracting and successfully purifying the enzyme. The low rate of respiration and of other metabolic activities further support the assumption that the reactions specifically associated with the excitable membrane will be relatively more dominant than in other cells, and that therefore the proteins and other components of the membrane involved will be more readily identifiable than in a tissue where many other functions go on simultaneously, complicating the chemical machinery. The specialized character of electric tissue greatly helped the study of the sequence of energy transformations and led, in 1943, to the discovery of choline O-acetyl transferase (choline acetylase) and to the evidence that ATP is the energy source for acetylation (NACHMANSOHN and MACHADO, 1943). In combination with the many other remarkable features of the electroplax enumerated before, the specialized function not only makes it suitable for the analysis of the reaction of ligands with the receptor protein, but also offers a serious hope for isolating this protein with the use of affinity labeling to be described later. If such an attempt were successful, it would be the first receptor ever to be isolated and identified. Although this would be a major breakthrough and invaluable for the understanding of the precise molecular events in the excitable membrane, the monocellular electroplax would still continue to be as essential as ever for obtaining the necessary information about the complex chemical events underlying bioelectricity and even for characterizing this protein in view of the many features affecting and modifying protein behavior and properties in the intact structure as compared with the isolated protein.

c) **Active Site of the Receptor; Evidence for its Protein Nature.** It appears a priori likely that the ACh-receptor is a protein. If one accepts the postulated role of ACh, i.e., that it is a specific signal in the excitation process initiating the series of reactions which effect the increased permeability to ions, the most likely assumption for the primary target of the action is that of a protein. Only proteins have the special ability of "recognizing" a specific molecule. Even the transmission of the messages by the genetic code in the formation of proteins requires proteins. Although the writer has postulated the protein nature of the receptor for more than 20 years and has used this hypothesis in many of the notions developed, strong experimental evidence has accumulated during the last few years by the studies with the isolated electroplax. The protein nature of the receptor appears by now well established. Conformational changes of phospholipid micelles may also be induced by ions (LUZZATI et al., 1966), but unlike proteins these cell constituents lack the required specificity.

The single electroplax preparation has made possible quantitative evaluations of the two processes by which the reaction of the ligand with the receptor is characterized: the binding forces and the biological activity. By analyzing the reaction of the receptor with a great number of different types of activators and inhibitors and comparing them with ACh-esterase in solution, a purified preparation obtained from electric tissue of *Electrophorus*, much information has accumulated about the molecular groups in the active site of the receptor which are responsible for binding and for biological activity, and about the features which are different from or similar to those in the active sites of the enzyme. A molecule such as ACh has only a limited number of molecular forces by which it may react with a macromolecule. Thus, it appeared likely that some of these forces, would be used in the reactions with both proteins specifically associated with ACh. For instance, the quaternary ammonium group of ACh should help in the binding of an anionic group of the receptor by Coulombic and van der Waals' forces in a similar way as was shown for ACh-esterase. It may also enhance the biological activity of both proteins in a way to be discussed later. As will be seen, many similarities between the two proteins are based on this particular feature. On the other hand, the two different functions of the two proteins — the one used by the signal to trigger off the permeability changes, the other reversing the effect by the hydrolysis of the ester — must require at least distinctly different molecular groups in the active site, if not necessarily two proteins. As will be seen in the following, the "esteratic" site, i.e., the ester-splitting group, is absent in the receptor protein.

The first two compounds for which dissociation constants with the receptor were determined on the isolated electroplax by recording the increasing degree of depolarization by an activator as a function of concentration and determined

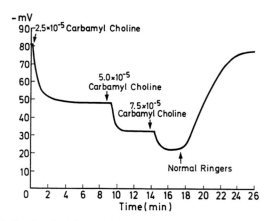

Fig. 7. Strength of the depolarizing action of carbamylcholine tested on the monocellular electroplax preparation (from Higman *et al.*, 1963). The figure presents a typical experiment in which time course and degree of polarization are shown upon addition of increasing concentration, and the repolarization following the return of the cell to Ringer's solution

by the antagonistic effect of an inhibitor, were carbamylcholine and d-tubocurarine (Higman *et al.*, 1963). Fig. 7 illustrates the method of using the steady state potentials obtained with increasing concentrations of carbamylcholine; it also shows the complete repolarization following the removal of the activator. Fig. 8 shows the effects of increasing amounts of curare on the depolarizing action of carbamylcholine. The reader interested in the mathematical analysis and the modifications of the Michaelis-Menten equations used for calculating the cissociation constants is referred to the original paper. The dissociation constant obtained for carbamylcholine is 4.4×10^{-5} M, that for curare 2.4×10^{-7} M. The investigators noticed the deviation of the plot from a simple hyperbola and invoked the similarity of the reaction of hemoglobin with oxygen; however, the implications of the sigmoid shape and the evidence of cooperativity were only later fully realized and discussed by CHANGEUX et al. (1967) and by KARLIN (1967), and analyzed in great detail by CHANGEUX and PODLESKI [(1968) (see pp. 71−78)].

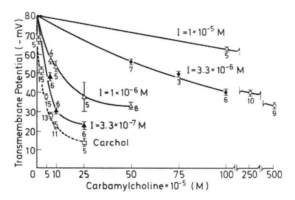

Fig. 8. Strength of d-tubocurarine as a receptor inhibitor tested by the competitive action between carbamylcholine and d-tubocurarine on the electroplax (from HIGMAN et al., 1963). The points of the lowest curve represent the mean values of the constant potential of the membrane for the concentration of carbamylcholine given by the abscissa. The number of experiments for each point is indicated. The degree of depolarization in the presence of carbamylcholine and of receptor inhibitor (I), both at varying concentrations indicated in the figure, is a measure of the antagonistic effect. Vertical bars extend for a distance equal to 1 standard error of the mean on either side of the point

During the last five years, this method of analysis has been applied to the study of a very large number of compounds. Only some groups of the compounds tested will be discussed in so far as they are pertinent for the understanding of the reactions of the ligands with the receptor and provide indications of molecular groups in its active site and of similarities and differences between the two sites. WEBB (1965) determined the dissociation constants of a series of N,N′-bis(diethyl-aminopropyl) quinone (benzoquinonium) (I) and N,N′-bis(diethylaminoethyl)oxa-mide-bisbenzylhalide quaternary salts (ambenonium) (II) derivatives with the ACh-receptor and compared them with those obtained with an active ACh-esterase

Formula I   Formula II

preparation (Formulas I and II). As seen in Table 2, when R in the benzoquinonium is an ethyl group, the dissociation constants of both receptor and esterase are about equal. Substituting the ethyl by a methyl group hardly affects the affinity to the esterase, but decreases significantly that to the receptor. Substitution of the ethyl by a phenyl group increases affinity to both, but stronger to the receptor than to the esterase. Most pronounced is the effect of the addition of a Cl atom to the phenyl group; in the *ortho* position the affinity to the receptor is hardly affected, that to the enzyme is six-fold increased, whereas in the *para* position the affinity to both is decreased, but that to the esterase much more strongly than that to the receptor.

Table 2. *Dissociation constants of benzoquinonium and ambenonium derivatives*

R	$K_{ACh\text{-}R}$	$K_{ACh\text{-}E}$	$\dfrac{K_{ACh\text{-}R}}{K_{ACh\text{-}E}}$
Benzoquinonium derivatives			
C_2H_5	3.8×10^{-7}	2.9×10^{-7}	1
CH_3	1.2×10^{-6}	2.6×10^{-7}	5
(phenyl)—CH_2	1.4×10^{-8}	8.1×10^{-8}	0.2
(o-Cl-phenyl)—CH_2—	1.6×10^{-8}	5×10^{-9}	3
Cl—(phenyl)—CH_2	3.1×10^{-8}	3.2×10^{-7}	0.1
Ambenonium derivatives			
Br	1.2×10^{-6}	2.1×10^{-10}	6000
Cl	1.6×10^{-6}	5.1×10^{-10}	3000
H	3.4×10^{-6}	2.1×10^{-7}	16

Constants with ACh-receptor (ACh-R), tested on the monocellular electroplax preparation, and with ACh-esterase (ACh-E), tested in purified enzyme solution prepared from electric tissue of *Electrophorus*. The last column indicates the strong differences in affinity to the two proteins expressed by the ratio of $K_{ACh\text{-}R}/K_{ACh\text{-}E}$.

Still more striking are the differences observed with the ambenonium derivatives. With either Br or Cl on the benzene ring, the affinity to the enzyme is three orders of magnitude greater than to the receptor. Substitution by a proton brings the two affinities very close to each other. The analysis of these two groups of compounds offers an excellent illustration how similar some of the affinities are and how they may be modified by small substitutions of one group or even one atom, sometimes similarly for the two proteins, sometimes differently, but sometimes in the same and sometimes even in the opposite direction. Since the compounds are receptor and enzyme inhibitors, they are favorable for comparing only the binding forces between them and the two components. The results make it difficult to assume that one of the two components is a protein, and that the other is an entirely different type of molecule.

The depolarizing strength of a series of n-alkyltrimethylammonium ions was measured on the electroplax by PODLESKI (1966, 1969). The potency increased from tetramethylammonium to trimethylbutylammonium, which was found to be the strongest activator. The ability of this compound to depolarize the membrane is as strong as that of ACh. Further increase of the length of the side chain — up to the trimethylhexylammonium — led to a decrease in potency. BERGMANN and SEGAL (1954) had measured the inhibition of ACh-esterase by this series of n-alkyltrimethylammonium ions and had found that increase of the chain length from one carbon through seven resulted in an increase of inhibitory strength. It is true that in the experiments with the enzyme only binding strength of the compounds is compared, whereas in those on the receptor biological activity is recorded. Nevertheless, the marked differences of increased chain length on the reactions with the two components support the assumption that the molecular groups in the active sites are different. In particular, the equal strength of ACh and trimethylbutylammonium suggests that the active site of the receptor does not have an esteratic site. Of special interest is the strong difference of potency between the action of tetramethylammonium and trimethylammonium ions on the receptor found in these studies. The quaternary ion is a thousand times stronger activator than is its tertiary analogue. This remarkable difference in biological activity which is due to the presence or absence of one methyl group strongly supports the assumption that an anionic group is present and functionally important in both proteins, as borne out by many other observations (*see* pp. 65—66).

Tests with the series of aryltrimethylammonium ions also offer pertinent data to compare the reactions with receptor and enzyme (PODLESKI, 1966, 1969). WILSON and QUAN (1958) had found 3-hydroxyphenyltrimethylammonium to be 120 times more effective than phenyltrimethylammonium (PTA) as an inhibitor of ACh-esterase; 3-methoxyphenyltrimethylammonium was only 5 times more effective, whereas benzyltrimethylammonium was about half as effective. The addition of the methyl group in the 3- or 4-position of PTA did not markedly alter the inhibitory strength. The large increase in the inhibitory strength of 3-hydroxyphenyltrimethylammonium is attributed to a hydrogen bond which is oriented so that the bond is probably formed with the esteratic site of the enzyme.

When tested on the isolated electroplax, the compounds are receptor activators, i.e., they depolarize the membrane. 3-hydroxyphenyltrimethylammonium ion was found, as previously observed by BARTELS, to be one-fifth as active an activator as PTA; 3-methoxyphenyltrimethylammonium is a much weaker activator, approximately 65 times less effective than PTA. Also, the maximum depolarization effected is markedly reduced by the presence of the 3-methoxy group, but not by the addition of the hydroxyl group. Again, as with the n-alkyltrimethyl-ammonium series, it is difficult to relate directly the measurements of the activation of the receptor to the inhibition of the esterase, since in the latter case only binding strength is measured. However, one would have expected some similarity in the order of decreasing or increasing strength if the two sides of the receptor and enzyme were identical, whereas there is no similarity at all and no correlation exists in any respect between the inhibitory strength on the enzyme and the potency as activator of the receptor. Another compound may be mentioned which has structural similarities, although it is not a derivative of aryltrimethyl-ammonium. This is N-methylpyridinium; it possesses a quaternary nitrogen, but only one methyl is free; the other carbons on the nitrogen form part of the ring structure. As an inhibitor of ACh-esterase, N-methylpyridinium is about half as strong as PTA. When tested on the electroplax, the compound in 10^{-3} M was found to be without effect either as an activator or as an inhibitor.

The 120-fold increase in inhibitory action of ACh-esterase by the presence of the 3-hydroxyl group in PTA is equivalent to a decrease of about 3 kcal/mole in the free energy of binding. Such a large effect and the much poorer binding of the 3-methoxy derivative and other data strongly support the assumption that a hydrogen bond formation is involved. The phenolic hydroxyl group apparently forms a hydrogen bond with the nucleophilic oxygen of the serine in the active site of the esterase.

In the 3-hydroxyphenyltrimethylammonium, the $(\overset{+}{N} \rightarrow OH)$ distance is about 5 A. As we have seen, tested on the electroplax, the presence of the hydroxyl group in the 3-position in PTA decreased the potency as an activator. It appeared desirable, since inhibition and biological activity are not comparable, to test with a receptor inhibitor having a proper $(N \rightarrow OH)$ distance whether the compound would form a hydrogen bond with an atom in the active site.

The hydrolysis of a series of isomeric 1-methyl-acetoxyquinolinium iodides catalyzed by ACh-esterase has been recently analyzed by Prince (1966) using a newly developed and highly sensitive spectrophotometric method. When isomers were compared with an acetoxy group in the 5-, 6-, 7-, and 8-position, it das found that the K_m of 1-methyl-7-acetoxyquinolinium was the lowest. Estimating the distances of the nitrogen to carbonyl-carbon atoms $(\overset{+}{N} \rightarrow C=O)$ in the 4 isomers with the aid of Fisher-Hirschfelder and Dreiding models, Prince found that the distance in the 7-position was about 4.8 to 5.9 A, which is among the isomers tested closest to the distance of about 5 A separating, according to previous estimates, the anionic binding site and the atom to which the carbonyl-carbon is bound to the enzyme.

The hydrolysis products of 1-methyl-acetoxyquinolinium iodides are receptor inhibitors. They thus offer the possibility of testing the questions of whether the $(\overset{+}{N} \rightarrow OH)$ distance influences the inhibitory strength and whether there is an indication for the formation of a hydrogen bond between the receptor and the inhibitor. A series of isomeric 1-methyl-hydroxyquinolinium were tested as inhibitors of the receptor on the electroplax in the usual way, determing the repolarizing strength after the cell had been depolarized by carbamylcholine. Surprisingly, evidence was obtained for a hydrogen bond formation between the receptor and 1-methyl-7-hydroxyquinolinium. At pH 6.9, the K_I of the 1-methyl-7-hydroxyquinolinium is 11 times lower than the K_I of 1-methylquinolinium; the 6-hydroxyquinolinium has the lowest K_I of the isomers tested. In order to determine whether the ionized or unionized form was the more active inhibitor, the pH of the solution was altered. Changing the pH of the solution from 5.9 to 8.8 reduced the inhibitory strength of 1-methyl-7-hydroxyquinolinium, whereas the pH changes had no effect on the action of 1-methylquinolinium. Since the pK_a of the 7-hydroxyl is 5.9 (Prince, 1966), the most active form is the unionized hydroxyl group.

The increase in inhibitory strength by the presence of the hydroxyl group in the 7-position over the 1-methylquinolinium is 110-fold, comparable to the effect previously observed with the corresponding enzyme inhibitors and equivalent to a decrease in the free energy of binding of about 3 kcal/mole. The distance of the hydroxyl group from the quaternary nitrogen is about 5 A, i.e., about the same $(\overset{+}{N}\rightarrow OH)$ distance as in 3-hydroxyphenyltrimethylammonium. The evidence thus supports the indication that a hydrogen bond is formed in the reaction with both active sites.

In view of the evidence that the active sites of enzyme and receptor are different, except for the presence of an anionic group, the similarity of the hydrogen bond formation in both components was unexpected. It appeared desirable to elucidate this problem and to test whether the hydrogen bond formation between the quinolinium derivative and the receptor was due to a reaction with the esteratic site of the enzyme or with some group in the active site of the receptor. This was tested by PODLESKI (1967) by the following procedure: the electroplax was exposed to methanesulfonyl fluoride (MSF) which forms an irreversible complex with the enzyme, a sulfonyl enzyme comparable to the phosphoryl enzyme formed with organophosphates (MYERS and KEMP, 1954). The advantage of blocking the esteratic site with this compound for the analysis of the problem is the stability of the sulfonyl enzyme complex and the smallness of the sulfonyl group, which reduces the possibility of steric interactions with the anionic group. The action of MSF was tested by the potentiation of the ACh effect in the electroplax. The specificity of the effect was, moreover, tested by addition of tetraethylammonium (TEA), which had been shown by KITZ and WILSON (1963) to accelerate the sulfonylation of the enzyme by MSF approximately 30 times. The presence of TEA plus MSF greatly potentiated the effect of ACh on the electroplax. The strength of the effect remained unchanged 30 and 80 min after removal of TEA.

After the irreversible sulfonylation of the active sites of ACh-esterase, the electroplax was exposed to either tetramethylammonium (TMA) or 3-hydroxyphenyltrimethylammonium ion. The depolarizing strength of both compounds was not affected. These ions had activity identical to that observed on control cells not exposed to MSF plus TEA. When the repolarizing strength of 1-methyl-7-hydroxyquinolinium was now tested, it was found to be exactly the same as in control cells. The K_I of receptor inhibitors had previously been shown to be the same, independent of the depolarizing ions used (PODLESKI, 1966,1969). Since the data show that the pretreatment with MSA plus TEA had no influence whatsoever on the repolarizing activity of 1-methyl-7-hydroxyquinolinium, they appear incompatible with the view that the hydrogen bond is formed with the esteratic site of the enzyme.

Another approach to the study of the protein nature of the ACh-receptor and to the analysis of its active site was applied by KARLIN and BARTELS (1966). If the receptor were a protein, it might react with reagents known to act on side chains of proteins, such as sulfhydryl groups or disulfide bridges. In several observations, it had been described that the inhibition of sulfhydryl groups blocked action potentials in frog and lobster nerves (SMITH, 1958) and in the perfused squid axon (HUNNEUS-COX and SMITH, 1965) and inhibited the action of ACh on the frog heart (NISTRATOVA and TURPAEV, 1959; POHLE and MATTHIES, 1959). KARLIN and BARTELS (1966) tested the effect of the block of sulfhydryl groups by p-chloromercuribenzoate (PCMB) and of the reduction of disulfide bridges by 1,4-dithiothreitol (DTT) on the response of the electroplax to ACh, carbamylcholine and trimethylbutylammonium (TMB). DTT is a reagent designed to reduce disulfides with a minimum formation of the mixed disulfide (CLELAND, 1964); it had not been tested before on excitable membranes.

Both PCMB and DTT were found to inhibit the depolarization of the electroplax by the three receptor activators used. Exposure to 0.5 mM PCMB for 5 min reduced the response by about 40%. The inhibition due to PCMB is not affected by extensive washing with either Ringer's solution (pH 7.0) or Tris-Ringer's solution (pH 8.0). The inhibition is reversed, however, by reducing agents such as 5 mM β-mercaptoethanol and by 5 mM L-cysteine in Tris-Ringer's solution at pH 8.0.

The inhibition of the response of the cell to carbamylcholine due to PCMB is characterized by a shift of the dose response curve toward higher concentrations of activator. The maximum response is not significantly affected. Thus, as described for reactions of PCMB with other proteins, it increases the dissociation constant.

Reduction of the S⁻S bridges by DTT proved to be most effective in inhibiting the response of the electroplax. Exposure to 1 mM DTT (pH 8.0) for 10 min reduces the subsequent response to 50 μM carbamylcholine to more than 30%. The response to ACh or TMB is similarly reduced. The inhibition due to DTT is not reversed by 40-min washing with either Ringer's solution (pH 7.0) or with Tris-Ringer's solution (pH 8.0). It is completely reversed by oxidizing agents such as, for instance, 1 mM potassium ferricyanide (pH 8.0) applied for 10 min. Dilsulfide compound DTNB [5,5'-dithiobis-(2-nitrobenzoic acid)], a compound designed for the assay of sulfhydryl groups (ELLMAN, 1959), also completely reverses the inhibition by DTT. As with PCMB, the inhibition due to DTT is characterized by a shift to the right of the dose response curve of carbamylcholine with little change in the maximum response.

These results offer strong support for the protein nature of the ACh-receptor. No other macromolecules have been described in which sulfhydryl groups or S⁻S bridges play an essential role in activity. Furthermore, when KARLIN (1967) applied PCMB or DTT to ACh-esterase either in solution or bound to the membrane (KARLIN, 1965), he did not find any effect on the K_m or on the V_{max}. Even at twice the concentration and twice the time of exposure, no effect was obtained. This is consistent with the findings of NACHMANSOHN and LEDERER (1939), who did not find an indication of sulfhydryl groups being involved in the activity of ACh-esterase prepared from electric tissue of *Torpedo*; the compounds specific for reactions with SH groups failed to have any effect on the enzyme activity except in very high concentrations at which they cannot be considered as specific for sulfhydryl groups.

The data presented offer strong evidence for the assumption that the active site of the ACh-receptor is distinctly different from that of the enzyme. The negative subsite reacting with the quaternary ammonium group plays an essential role in both active sites; as will be discussed below, the replacement of a quaternary group by the tertiary analog drastically decreases the biological activity of the protomers, sometimes by two to three orders of magnitude. However, the "esteratic", i.e., the ester-splitting group, present in the active site of the enzyme, is absent in the active site of the receptor; the latter seems to form readily hydrophobic bonds with the reacting ligand, thereby increasing its effect. The nature of this group is still unknown. The view that the two active sites are different, but form part of the same protein molecule, as has been repeatedly proposed, is at present not yet substantiated. Studies now in progress on the properties of the subunits of the homogeneous enzyme and on the properties of the receptor may provide a definitive answer.

d) **Effect of Quaternary Groups; Conformational Changes.** In the last 15 years, rapid progress has been achieved in the elucidation of the structure of proteins. The classification, first proposed by LINDERSTROM-LANG (1952), into primary, secondary and tertiary structure is today widely accepted. Primary structure refers to the amino acid sequence of the polypeptide chains. Secondary structure refers to the folded structures, the most important of them being the alpha helix proposed by PAULING et al. (1951), in which 3.7 amino residues are present in each coil, stabilized by hydrogen bonds between the amide linkages; many proteins, however, contain major nonhelical regions. Finally, the tertiary structure refers to the overall conformation of the protein molecule. Subsequently, the term quaternary structure has been introduced for proteins which contain more than single peptide chains (subunits) to describe the arrangements and spatial relationships between different chains (see, e.g., SUND and WEBER, 1966).

One of the most revolutionary and crucial advances in protein and enzyme chemistry during the last decade has been the successful exploration of the tridimensional structure by means of X-ray crystallography in combination with chemical methods. The first proteins thus analyzed were myoglobin (KENDREW, 1962, 1963) and hemoglobin (PERUTZ, 1962, 1963). Subsequently, the tridimensional structure of several enzymes was established; the first of them was lysozyme (BLAKE et al., 1965, 1967). Soon afterwards the tridimensional structure of ribonuclease A by HARKER and coworkers (KARTHA et al., 1967) and of ribonuclease S by RICHARDS and coworkers (WYCKOFF et al., 1967a, b) were established. Since then the analysis of several other enzymes has advanced either to or near completion.

The active site of enzymes and the mechanism of enzymatic catalysis have been explored by many investigators in the last two decades, mainly by a combination of kinetic and chemical studies. An essential tool in these studies has been the use of compounds forming covalent bonds with molecular groups in the active site. Knowledge of the tridimensional structure has added a new dimension to these studies. Some of the predictions were borne out, some others required modifications.

In spite of all these brillant advances, the extraordinary catalytic power of enzymes remains a mystery. The key-lock concept proposed by EMIL FISCHER in the 1890's has long dominated the thinking of enzyme chemists. It assumed a rigid protein which absorbs the substrate to specialized catalytic groups. The resulting formation of an enzyme substrate complex seemed, at least to a certain extent, to account for the specificity of enzymes. However, with the advances of the studies of enzyme mechanisms with the use of refined and sensitive methods, it became apparent that the simple key-lock theory did not provide satisfactory explanations in many cases.

In an attempt to explain some observations for which the theory appeared unsatisfactory, KOSHLAND introduced a notion referred to as induced-fit theory (KOSHLAND, 1960; KOSHLAND et al., 1962; KOSHLAND and NEET, 1968). He postulated on the basis of a vast amount of experimental data that small molecules may induce conformational changes on binding to the enzyme. Thereby, a precise orientation of catalytic groups is produced which makes a reaction possible. The catalytic action takes place only when the proper alignment of the residues is produced by the substrate. A compound related in structure, but unable to induce the fit, would not be affected by the enzyme. Binding alone is not sufficient without the favorable change in the conformation of the protein.

The notions proposed by KOSHLAND have greatly stimulated enzyme chemistry. Much direct evidence has accumulated, by the work of his laboratory as well as that of several others, for conformational changes in proteins induced by ligands. Many implications have resulted from this concept and were pertinent for the interpretation of enzyme action. A strong support came from the work of PERUTZ and his associates (PERUTZ et al., 1964) on hemoglobin; the oxygenated form has a conformation different from that of the nonoxygenated form (for the functional significance of this process, see also pp. 73—75). Conformation and conformational changes of proteins today form an integral part of biochemical thinking. In some reactions, such changes may be quite local and therefore not readily detectable with presently available instrumentation. However, with the continuous improvement of methods and techniques even small changes may become measurable. As an illustration it may be mentioned that it has been possible, with optical rotatory dispersion and circular dichroism, to demonstrate that the removal of the heme from myoglobin induced about 10% of the 154 amino acids of this molecule to change from a helical to a nonhelical form; the process was reversed by the addition of the heme to the apomyoglobin (BRESLOW et al., 1965).

The possibility that a molecule such as a protein may not be a rigid structure and that local changes of conformation may take place and thereby increase the efficiency of the action occurred to the writer some 20 years ago, long before any experimental evidence was available, in connection with certain observations on the reaction of ACh and related compounds with certain proteins. It has been known for nearly a century that certain methylated quaternary ammonium derivatives have a much more potent pharmacological action on cells than their tertiary analogues. In the case of ACh, the difference between the quaternary form and its tertiary analogue is particularly striking. As long as this phenomenon was limited to observations of pharmacological effects, the complexity of the system excluded interpretations on a molecular level. However, when the enzymes associated with the hydrolysis and the formation of ACh became available in purified solutions, it became possible to submit this idea to experimental tests. The results seemed to intimate that some process in the protein may be induced by the extra methyl group. A clue appeared to be the tetrahedral structure of the methylated quaternary nitrogen group. Such a group is more or less spherical. If such a structure is attached in the reaction to a protein surface, a direct contact between all four methyl groups and the protein will not be readily possible, since the fourth methyl group is located in the direction of the solution, i.e., away from the protein. Thus, it seemed possible to picture that a local change of conformation of the protein may take place whereby the macromolecule may have simultaneous contact with all four groups of the ligand by surrounding the quaternary group.

The forces between the ligand and the molecular groups in the protein surface and the effect of the extra methyl group were first analyzed with a highly purified solution of ACh-esterase. When the inhibitory potency of the methyl groups of ammonium and of hydroxyethyl ammonium ions on ACh-esterase was tested, it was found that each methyl group increased the strength of the binding of the inhibitor to the enzyme about sevenfold, except for the fourth methyl group, which hardly increased the binding, or — as was later found — to a markedly smaller extent (WILSON, 1952). It is reasonable to assume that neither the changes in hydration characteristics attending binding nor the entropies of binding differ markedly for any of the four members of the series. Therefore, the additional binding associated with each methyl group may be attributed to VAN DER WAALS' attraction between the methyl group and a hydrocarbon group of the protein. Since the extra methyl group (the fourth one) did not contribute significantly to the binding, the strong difference between the rates of hydrolysis of ACh and its tertiary analogue, dimethylaminoethyl acetate, by ACh-esterase, supported the assumption that a local change of conformation of the protein may take place during activity, whereby the extramethyl group would be enveloped and that this was possibly the explanation of the greater catalytic efficiency.

The possibility of conformational changes of ACh-esterase during the catalytic activity found some further, although still very indirect, support by studies of the enthalpies and entropies of activation, $\Delta H^{\ddagger}$ and $\Delta S^{\ddagger}$, of the ester of ethanolamine and its methylated derivatives with highly purified ACh-esterase from electric tissue (WILSON and CABIB, 1956). Substitution of the first two protons by methyl groups produced little change in the activation energies, but the extra methyl group has a very pronounced effect. The enthalpy of activation, $\Delta H^{\ddagger}$, of the hydrolysis of ACh is about 14,000 cal, as compared with about 8,000 cal for that of the tertiary analogue. On the basis of this finding, one could expect that the hydrolysis of the quaternary ester would be less favored than that of the tertiary. But the entropy of activation is extremely favorable for the quaternary compared to that of the tertiary compound. Whereas it is negative with the tertiary analogue, it is positive with the quaternary group. The difference amounts to about 30 to 40 entropy units. This large favorable entropy of activation and the higher rate of hydrolysis of quaternary esters, when catalyzed by the enzyme, appears all the more significant in view of the recent observations of CHU and MAUTNER (1966) that the nonenzymatic hydrolysis of the tertiary ester is very much faster than that of the quaternary analogue. Thus, the actual difference of efficiency between the enzyme catalyzed hydrolysis of tertiary and quaternary compounds seems to be much greater than is apparent simply on the basis of comparing the two enzyme catalyzed reactions.

A similar and even more striking difference of catalytic efficiency between quaternary and tertiary analogues has been observed with choline acetylase, the enzyme which transfers the acetyl group from acetyl-CoA to choline. The rate of acetylation of choline was found to be about 12 times higher than that of dimethylethanolamine (BERMAN et al., 1953). On the other hand, the difference between the rates of acetylation of dimethyl- and monomethylethanolamine is small.

By far the strongest difference between ACh and its tertiary analogue is, however, observed in the pharmacological action, as mentioned before. In this case, the difference between the two forms is of a different order of magnitude. On the basis of the differences pharmacological actions and of the catalytic efficiencies of the two enzymes associated with ACh formation and hydrolysis, and firmly convinced that the receptor of ACh is also a protein, the writer proposed, in his HARVEY Lecture in 1953, as a possible explanation for the action of ACh on the receptor, that it may induce a conformational local change of the protein and thus initiate the changes leading to increased permeability (NACH-MANSOHN, 1955). The process was referred to as a change of configuration, since at that time conformation was not yet a commonly used term, as it is now. Such a conformational change seemed also to be consistent with the assumption of the shift of charges in the membrane during electrical activity which had been postulated by TEORELL (1953) for many years. A quantitative evaluation of the difference in potency between ACh and its tertiary analogue in their reaction with the receptor has been performed by BARTELS (1962) with the monocellular electroplax preparation. By removal of one methyl group from the quaternary form, the potency of the depolarizing action decreased 200-fold. It would be difficult to attribute such a striking difference in potency simply to attraction by VAN DER WAALS' forces due to the presence of the extra methyl group. At the pH used in the experiments, most of the molecules of the tertiary analogue are present in cationic form. If a second methyl group is removed, the potency of depolarizing action decreases further, but only 20-fold. For such a difference, VAN DER WAALS' forces may be at least partly responsible, although even in this case a change in conformation, although a much weaker one, cannot be excluded, since a 20-fold increase in potency by the addition of one methyl group is still quite high.

At the time the suggestion was made that a conformational change of the receptor may initiate the sequence of reactions resulting in increased permeability, the view appeared quite speculative. Even today, there is still no direct experimental evidence for such processes. However, the indirect evidence mentioned before and other kinetic and chemical observations (e.g. KRUPKA, 1966) and optical data (KITZ and KREMZNER, 1968) have lent support to the view of conformational changes in the case of ACh-esterase. Several additional tests with the now available homogeneous protein are desirable and may eventually provide direct support. Some further data have been recently obtained with the electroplax (see p. 83) which are consistent with the suggestion of conformational changes of the ACh-receptor. With the general acceptance of the notion of conformational changes of proteins the idea of their role in permeability changes of excitable membranes, proposed two decades ago, has greatly gained in probability.

e) Essential Role in Electrical Activity. According the theory, the primary molecular event initiating the permeability changes during electrical activity is the action of ACh on the ACh-receptor inducing a change of conformation. If the specific signal is prevented from acting on the specific protein and from starting the sequence of reactions, electrical activity should be blocked. Thus, a crucial test of the correctness of the theory would be the demonstration that a compound similar in structure to ACh, but unable to induce the conformational change of the protein, i.e., an antimetabolite acting as an inhibitor of the ACh-receptor, would antagonize the action of ACh and in adequate concentrations block electrical activity in excitable membranes in general. Such evidence would parallel and supplement that for the essential role of ACh-esterase shown with the use of potent inhibitors and discussed in Section II.

A series of compounds that are analogues of ACh in structure have long been known as "local anesthetics", such as procaine, tetracaine and others. They differ from ACh in two respects. First, they are receptor inhibitors, i.e., they block electrical activity but do not depolarize the membrane; second, they are able to reach the receptor at the level of junctions as well as of axons, apparently due to their lipid solubility which permits them to penetrate structural barriers protecting axonal membranes (see pp. 31—32). The monocellular electroplax offers the possibility to evaluate quantitatively which substitutions transform the molecular properties of ACh from a molecule that is a potent receptor activator and acts on synaptic junctions only, into one that is a receptor inhibitor and acts equally on synaptic and conducting membranes. The preparation, in addition to its high degree of precision, is particularly suitable for such a study in view of the presence of the two types of excitable membranes.

A systematic analysis by BARTELS, using a great number of compounds in which either the quaternary ammonium or the acyl group of ACh have been substituted, has provided information about the molecular features responsible for the transformation of one type of compound into the other type (BARTELS 1965; BARTELS and NACHMANSOHN, 1965). Table 3 illustrates some of the most essential features of the transformation of the biological action in relation to chemical structure. When, for instance, the methyl group on the carbonyl group is substituted by a saturated benzyl ring, the potency of the compound as an activator is decreased about 200-fold compared to that of ACh. But it is still an activator and acts only at the synaptic junctions; i.e., the compound is unable to reach the receptor in the conducting membrane even if applied in high concentrations. However, the same compound with the small modification that an unsaturated ring is substituted for a saturated one, i.e., benzoylcholine, may be already either an activator or an inhibitor according to experimental conditions. When applied, for instance, to the electroplax together with carbamylcholine, the effect of hexahydrobenzoylcholine is additive; that of benzoylcholine, on the other hand, may be additive or antagonistic, according to the respective concentrations. Moreover, in relatively high concentration, 10^{-3} M, benzoylcholine acts also on the conducting membrane. Thus, benzoylcholine is a typical transitory form in chemical structure as well as in biological action between ACh and the local anesthetics. Addition of an NH_2 group to the phenyl group in *para* position transforms the analog into a receptor inhibitor and increases the penetrating ability; thus, the molecule has acquired the two features characteristic of a local anesthetic. Both inhibitory potency and ability of penetration are greatly increased when the three methyl groups of the quaternary nitrogen are replaced by

5*

Table 3. *Substitutions of ACh transforming the molecule from a receptor activator acting on the synaptic junctions only into a receptor inhibitor acting both on junctions and on the conducting membrane (so-called "local anesthetic"). Benzoylcholine is a transitory form both in structure and in biological activity. For details see the text*

Compound	Synaptic junctions Activator	Inhibitor	Conducting membrane
	M concentration		
Acetylcholine CH_3–$\overset{\oplus}{N}(CH_3)(CH_3)$–$CH_2$–$CH_2$–O–$\overset{(+)}{C}(CH_3)$–$O^{(-)}$	2.5×10^{-6}	0	0
Hexahydrobenzoylcholine CH_3–$\overset{\oplus}{N}(CH_3)(CH_3)$–$CH_2$–$CH_2$–O–$\overset{(+)}{C}$–$O^{(-)}$ (cyclohexyl)	5×10^{-4}	0	0
Benzoylcholine CH_3–$\overset{\oplus}{N}(CH_3)(CH_3)$–$CH_2$–$CH_2$–O–$\overset{(+)}{C}$–$O^{(-)}$ (phenyl)	5×10^{-4}	5×10^{-4}	1×10^{-3}
p-Aminobenzoylcholine CH_3–$\overset{\oplus}{N}(CH_3)(CH_3)$–$CH_2$–$CH_2$–O–$\overset{(+)}{C}$–$O^{(-)}$ (p-NH_2-phenyl)	0	1×10^{-3}	2.5×10^{-3}
Procaine C_2H_5–$\overset{\oplus}{N}(C_2H_5)(H)$–$CH_2$–$CH_2$–O–$\overset{(+)}{C}$–$O^{(-)}$ (p-NH_2-phenyl)	0	2.5×10^{-4}	5×10^{-4}
Tetracainemethiodide CH_3–$\overset{\oplus}{N}(CH_3)(CH_3)$–$CH_2$–$CH_2$–O–$\overset{(+)}{C}$–$O^{(-)}$ (p-C_4H_9–NH-phenyl)	0	2×10^{-5}	1×10^{-5}

the two ethyl groups (procaine), as would be expected from this change of structure. A still greater enhancement of both properties is achieved by substituting one proton of the NH_2 group by a butyl group (tetracaine). Substitution of the methyl groups on the nitrogen also changes both depolarizing potency and ability of penetration. Substitution of one methyl by one ethyl group decreases the depolarizing effect four to five times, substitution of the two methyl by two ethyl groups 100 to 200 times. Both compounds act on the junction only. The triethyl analogue of ACh, however, is already a transitory form like benzoylcholine. Its depolarizing action is insignificant. It antagonizes the depolarizing action of the carbamylcholine at the junction. In high concentrations it blocks the direct response; i.e., it penetrates the structural barriers, although poorly. Both effects, at the junctions and at the conducting membrane, are weaker than those obtained with benzoylcholine.

f) **Sulfur and Selenium Isologs of ACh.** A novel approach to the study of the interaction of ACh with ACh-receptor and ACh-esterase has been introduced by MAUTNER (1967). If one accepts the notion of conformational changes of the protein as an important factor in biological activity, it follows that even small changes of the structure of the ligand may drastically modify the interaction between the small and the macromolecule and thereby greatly affect biological activity. To study these factors, MAUTNER and his associates prepared a great variety of sulfur and selenium isologs of ACh, choline, benzoylcholine, and many of their congeners (see MAUTNER and GÜNTHER, 1961; GÜNTHER and MAUTNER, 1964). The O, S and Se isologs are quite similar in size; however, they may differ in electron distribution and in conformation. The possible role of these two factors in the reaction with ACh-esterase and -receptor has been intensively analyzed in the last few years by MAUTNER and his associates; some of the tests on the receptor protein using the electroplax has been carried out in collaboration with this laboratory.

Evidence has been obtained that the electron distribution in the O, S and Se isologs is different. This was shown by kinetic, spectroscopic and dipole measurements of isologous esters performed by MAUTNER and his associates (MAUTNER et al., 1963; CHU and MAUTNER, 1966; MAUTNER, 1967). These isologs should, therefore, differ in their ability of binding to active sites and to induce conformational changes in the two proteins.

The difference of the reaction of ACh and its S and Se isologs with the receptor was first tested on guinea pig ileus and frog rectus preparations. Addition of enzyme inhibitors increased the pharmacological response to ACh, but reduced it to acetylthio- and acetylselenocholine (SCOTT and MAUTNER, 1964). This was attributed by the authors to a stronger action of the split products, cholinethiol and cholineselenol, compared with choline itself.

A quantitative evaluation of the reactions of the isologs with the ACh-receptor became possible by the use of the monocellular electroplax preparation. One of the most pertinent observations is the remarkable difference between choline and its S isolog cholinethiol. Choline, even at 10^{-1} M, is virtually unable to produce a depolarization. This is about 40,000-fold the concentration at which ACh has a strong depolarizing action. In contrast, the S isolog has only a slightly less depolarizing effect than ACh itself: at 5×10^{-5} M the effect is about equal to that of ACh at 3×10^{-6} M. It is, of course, biologically desirable for an extremely rapid and reversible trigger action that the hydrolytic product should be as inactive as possible, but it is surprising, and at present difficult to explain, what the basis of the extraordinary lack of activity of choline is, although it is apparent that the oxygen atom is responsible. The S atom has a higher polarizability, but it is not the better binding of the anionic form to which the potency can be attributed, since replacement of a proton by a methyl group increases the effect; methylcholinethiol is more than twice as potent than cholinethiol (MAUTNER et al., 1966). Moreover, the preparation is not affected by changes of pH over a wide range; when the depolarizing activity of cholinethiol, which has a pK_a of 7.7 (HEILBRONN, 1958), was tested at 6.2 where only 3% is ionized, and at 9 where 95% is ionized, the compound was found to be more active at the lower pH, i.e., in its mercaptan rather than in its mercaptide form. The methyl group apparently contributes to the binding by hydrophobic bonds, since both methyloxycholine and methylselenocholine are more potent than the corresponding protonated forms. Methoxycholine is still the weakest of the methylated compounds, whereas the seleno isolog is the strongest. The remarkable decrease of potency owing to the presence of the O atom in the position which it has in choline, also becomes apparent from other structural modifications. When, for instance, the 3-carbon analogue is used, $(CH_3)_3N(CH_2)_3OH$, it is much more potent than choline itself, although even in this case the S isolog is more potent than the S isolog of choline; in fact, its depolarizing power is nearly as strong as that of ACh itself (WEBB and MAUTNER, 1966).

Although the choline isologs increase in potency by the substitution of the O atom by S and Se, the opposite is true for the esters. ACh is more effective as receptor activator than acetylthiocholine and very much more than acetylseleno-ocholine. These widely different potencies of the three esters is another support for the difference between the active site of the enzyme and that of the receptor because the hydrolysis rates of the esters catalyzed by the enzyme are similar. Moreover, it is interesting that the biological ester is by far the most active of the three isologs, whereas its hydrolytic product is the most inactive. It may be mentioned that cholinethiol and cholineselenol are readily oxidized and form disulfide and diselenide compounds. The rapid rate of formation of the latter compound makes the quantitative evaluation of the activity of cholineselenol difficult. The oxidation products have a slight blocking effect, but only in concentrations higher than those used in these tests.

More recently, the differences of conformation of the isologs were compared by X-ray analysis in order to separate the effects of this factor on biological activities from that of electron distribution (SHEFTER and MAUTNER, 1967, 1969). The results indicate that the structure in crystals of oxygen and sulfur analogs tend to differ, those of sulfur and selenium isologs are so similar as to make such molecules isotesteric. X-ray analysis of ACh, choline, and a series of related molecules in the solid state shows that in general the $\overset{+}{N}$–C–C–O-grouping is in the *gauche* conformation (CANEPA et al., 1966; SHEFTER et al., in press). In ACh the *gauche* conformation also prevails in solution (D_2O) (CULVENOR and HAM, 1966). In contrast, in acyl-selenocholine crystals (SHEFTER and KENNARD, 1966) the nitrogen and selenium are *trans* to one another (GÜNTHER and MAUTNER, 1964). Acetylthiolcholine was found to be essentially isosteric with acetylselenocholine and different from the *gauche* conformation of ACh (SHEFTER and MAUTNER, 1969). The *trans* conformation of the $-S-C-\overset{+}{C}-N$-group in acetylthiolcholine and the corresponding grouping in acetylselenocholine are rather stable. This was shown by means of nuclear magnetic resonance measurements, indicating that both these compounds retain their *trans* conformation in D_2O solution (R. J. CUSHLEY, *in press*).

Since the conformation of the sulfur and selenium isologs are almost identical, it is reasonable to assume that these compounds will have the same ability to fit the active site of the receptor. Since the biological effects, i.e., the depolarization induced and the hydrolysis by ACh-esterase, are different, these effects may be ascribed to difference in electron distribution brought about by the replacement of the ether oxygen with sulfur and selenium, respectively. Although between ACh and the S and Se isologs there is the *gauche-trans* difference of conformation, the marked differences of electron distribution seem nevertheless to play a role as important as the difference of conformation.

Another important possibility offered by the isomers is a test for similarities, or differences, of their reactions with the ACh-receptor protein in junctional and axonal membranes. If the active sites of the receptors are similar in both types of excitable membranes, the differences of potency between O, S and Se isologs should exhibit comparable differences when applied to the one or to the other type of membrane. This particular test is difficult because of the barriers surrounding axons and preventing quaternary compounds from reacting with the receptor. However, by selecting compounds which in the light of previous information should penetrate and reach the receptor in the conducting membrane, a comparison should be possible.

2-Dimethylaminoethylbenzoate and its sulfur and selenium isologs were selected for the tests (ROSENBERG et al., 1966). The ether oxygen of the compound was replaced by either S or Se. The tertiary analogue of benzoylcholine is, similar to local anesthetics, a receptor inhibitor and capable of reacting with the receptors in the axonal membrane. Tested on the electroplax, the O isolog was found to be the weakest, the S isolog to be stronger, and the Se isolog to be the strongest inhibitor. The differences in potency obtained with these three isologs on the squid giant axon were found to be remarkably similar. When the preparation was first exposed to cottonmouth moccasin venom, the action of all three isologs became much stronger, but the differences in potency between the three isologs remained essentially the same. However, after venom treatment, the effect of the Se isolog became irreversible. Such irreversibility was previously observed with succinoylselenocholine, but not with succinoylthiocholine on junctions (GOODYER and MAUTNER, *in press*). Obviously, irreversible effects may prove useful for affinity labeling of the receptor. The observations have been extended by replacing the carbonyl oxygen with S or Se (ROSENBERG and MAUTNER, 1967). Again, similarly marked differences in potency were observed which were similar in the reactions of the axonal and in the junctional membranes.

g) Allosteric Sites and Cooperativity. Although regulatory and control phenomena of the body as a whole have been known for a long time in biology, extraordinary advances have been made during the last decade in the knowledge of control mechanisms on cellular, subcellular and molecular levels. Control systems exist which involve a series of reactions and interactions determining the rate and direction of many metabolic pathways; there are others essential for the synthesis of proteins and other macromolecules. The investigations of regulatory proteins and enzymes have been particularly aided by the insight obtained into the tridimensional structure and conformation of proteins and the information that they are formed by several subunits.

Certain enzymes seem to act at critical metabolic steps and have specific functions in regulation and coordination (*see* Symposium on: "Cellular Regulatory Mechanisms", 1961). One fundamental characteristic of these *regulatory* enzymes is their susceptibility to activation or inhibition by metabolic effectors that are compounds other than those which are the specific substrates catalyzed by the enzymatically active site. Since there is no structural similarity between the enzyme substrate and the metabolite acting as activator or inhibitor of the catalytic process, the effect on the enzyme activity must be achieved through the binding of the metabolite to a specific site distinctly different from the catalytically active site. This was clearly recognized and emphasized by CHANGEUX (1960, 1961, 1962) when he studied the inhibitory action of L-isoleucine on L-threonine deaminase. Based on a systematic comparative analysis of this and a few other regulatory enzymes, in which a similar structural dissimilarity between substrate and metabolic effectors have been demonstrated, MONOD et al., (1963) formulated certain generalizations concerning the structural and kinetic properties common to the regulatory proteins in spite of the great diversity of these systems. They introduced the notion of *allosteric effectors* and referred to their binding sites as to *allosteric sites*. The proteins possess two, or at least two, stereospecifically different, nonoverlapping receptor sites: 1) the active site which binds the substrate and is responsible for the biological activity of the enzyme; and 2) the allosteric site, complementary to the structure of another metabolite, i.e., the allosteric effector, which it binds specifically and reversibly. The formation of the enzyme-allosteric effector complex does not activate directly a reaction involving the effector, but brings about an alteration of the molecular structure of the protein referred to by the authors as an *allosteric transition*, which modifies the properties of the active site and changes one or several of the kinetic parameters which characterizes the biological activity of the protein. An allosteric effector need not have any particular chemical or metabolic relation with the substrate. The specificity of any allosteric effect in this concept results exclusively from the protein molecule itself, permitting it to undergo a conformational change. This is the fundamental distinction between the action of allosteric effectors from that of coenzymes, secondary substrates or substrate analogues.

Although the conclusion of two different sites and of indirect (allosteric) interactions was first based on structural considerations, it found strong further support by the discovery that after various treatments the enzyme may lose its sensitivity to the metabolic effector, but retain nevertheless its full activity towards the substrate. Such effects referred to as desensitization, may be obtained, for instance, by freezing or heating, by high ionic strength, by the action of urea, by the use of SH blocking agents, and others. A desensitization was first observed with L-threonine deaminase (CHANGEUX, 1961), and aspartate transcarbamylase (GERHART and PARDEE, 1962, 1963), and was soon afterwards also demonstrated with other regulatory proteins and enzymes. A number of observations indicate that this desensitization does not depend on the integrity of the allosteric binding site, but must be attributed to the uncoupling of the interaction between the two sites, i.e., to some disorganization of the protein molecule. The most conclusive evidence for the existence of two different sites came from the studies of aspartate transcarbamylase: the subunits containing the catalytic active site could be separated from those containing the allosteric binding sites GERHART and SCHACH-MAN, 1965). These observations are the strongest support for the assumption that the interaction between the subunits of the protein must be effected by alterations of the protein molecule.

An important characteristic of allosteric transition is the unusual feature of the kinetic properties of allosteric systems: their response to variations of substrate and metabolic effector concentration is atypical and differs from that of most other enzymes. When the reaction velocity of enzymes is plotted against substrate concentration, the curve usually obtained is a hyperbola, corresponding to a Langmuir isotherm, as is to be expected from a simple first-order reaction according to the Michaelis-Menten equation. In contrast, in the case of regulatory enzymes, the curve resulting from the plot of reaction velocity vs. substrate concentration has a sigmoid form. As is well known, the dissociation curve of haemoglobin as a function of oxygen tension is sigmoid. Haemoglobin, although not an enzyme, is a protein whose regulatory functions had long been recognized. Sigmoid curves indicate that a second-order reaction is taking place and that at least two molecules of substrate interact with the enzyme, the binding of one molecule facilitating in some manner the binding of the next: there is a cooperative effect in the binding of more than one substrate molecule to the enzyme. Although in the case of haemoglobin direct interactions between the haem groups could not be excluded as long as the tridimensional structure of haemoglobin was not known, the elucidation of the structure by PERUTZ and his associates (PERUTZ, 1962; PERUTZ et al., 1964) has revealed that the four haem groups lie far apart from each other, the distance being 30 A or more. Thus, the only alternative explanation appears to be that of an indirect interaction presumably due to a conformational change. Such an interpretation is indeed supported by the crystallographic work of PERUTZ and his collaborators. The regulatory effect, i.e., the cooperative binding of oxygen, would then be the result of a reversible, discreet conformational change of the protein, the basis of allosteric transition. Indeed, in contrast to the sigmoid shape of the oxygen dissociation curve of the tetrameric haemoglobin with its four haem groups, that of the monomeric myoglobin with only one haem is a simple hyperbola as obtained with enzymes following a first-order reaction. Whatever the mechanism underlying the sigmoidal response and cooperative action may be, its functional significance is quite apparent: with such a response there exists a concentration below which the protein is relatively insensitive to changes of substrate or effector concentrations, but above which relatively large changes of activity are elicited by slight changes in effector concentrations. Thereby a high sensitivity of protein activity is obtained in a narrow range of substrate, or effector, concentration.

The activity of regulatory enzymes as a function of substrate (S) and allosteric inhibitor (I) concentration may be expressed by the equations:

$$\log \frac{v}{V_{max} - v} = n \log S - \log K \text{ (for substrates)},$$

$$\log \frac{v}{V_0 - v} = \log K' - n' \log I \text{ (for inhibitor)}.$$

These equations are formally identical with Hill's empirical relation for the binding of oxygen to haemoglobin. The coefficient n frequently referred to as Hill coefficient, n_H, for allosteric enzymes and proteins may be 2 or higher, whereas it is 1 for enzymes following the Langmuir isotherm, when no cooperativity is involved.

The explanation of the mechanism of cooperative activity offers a great challenge. MONOD et al., (1965) have proposed a model for explaining the various features and properties of allosteric systems. This model is based essentially on the assumption of the following properties. Most allosteric proteins are polymers, or rather oligomers, involving several identical subunits. Allosteric interactions frequently appear to be correlated with alterations of the quaternary structure of the proteins (i.e., alterations of the bonding between subunits). The protomers of allosteric proteins are associated in equivalent position. This implies that the molecule possesses at least one axis of symmetry. This symmetry will be conserved during changes in protein conformation: the change of one subunit is accompanied by equal changes in all identical subunits. There are two, at least two, states, referred to as R and T states, assumed to be in equilibrium in the absence of a ligand. The conformational change of the protein occurs through an isomerization between the R and T states. If the R state binds the substrate preferentially and is present in small amounts, addition of the substrate will tend to pull the equilibrium in the direction of the R state. According to mathematical analysis, the curve should be sigmoid. The action of inhibitors and activators is readily explained by assuming that the former binds the T state and the latter the R state. The model predicts that the interactions between different ligands, "heterotropic" effects, may be either positive or negative (i.e., cooperative or antagonistuic), interactions between identical ligands, "homotropic" effects, appear to be always cooperative. Cooperative homotropic effects are almost invariably observed with at least one of the two (or more) ligands of the system. Conditions which alter the heterotropic interactions also simultaneously alter the homotropic interactions. This model has received experimental support by the observations with several enzymes, e.g., yeast glyceraldehyde 3-phosphate dehydrogenase (EIGEN et al., 1966), aspartate transcarbamylase (CHANGEUX and RUBIN, 1968), phosphofructokinase (BLANGY, et al., 1968), and others.

The allosteric and cooperative properties of proteins have been approached in recent years by KOSHLAND and his associates on the basis of the induced-fit theory (see p. 64). Their results suggest a flexible interaction between ligand and protein, which may induce a new conformation of the subunit (KIRTLEY and KOSHLAND, 1967; HABER and KOSHLAND, 1957; KOSHLAND and NEET, 1968). When an isolated subunit or a monomeric protein interacts with the ligand and induces a conformational change, this effect may be achieved by three possible pathways: (a) an initial isomerization followed by the binding of substrate to the changed structure;

(b) an initial binding of substrate followed by a conformational isomerization; or (c) a concurrent binding and conformational isomerization. When this monomer is associated with other such monomers, in a polymeric protein, a change in conformation in one subunit may produce varying degrees of change in the shape and stability of adjacent subunits. Each of the changes may increase decrease, or leave unchanged the net attraction between subunits. The changes will occur sequentially with ligand addition. This "sequential" model preocits both positive and negative homotropic and heterotropic effects.

In experiments with several enzymes, negative cooperativity as predicted by theory has been observed, e.g., with rabbit muscle glyceraldehyde 3-phosphate dehydrogenase (CONWAY and KOSHLAND, 1968), with phosphoenolpyruvate carboxylase (CORWIN and FANNING, 1968), and CTP synthetase (LEVITZKI and KOSHLAND, 1969). Although the action of some regulatory enzymes may be well explained with the symmetry model of MONOD et al. (1963, 1965), other enzyme actions fit apparently the more general sequential model. It appears too early to draw final conclusions on the basis of the present evidence (see, e.g., STADTMAN, 1966). Whatever the detailed mechanism may turn out to be, the concept of allosteric proteins and enzymes and cooperativity has become an integral part of biochemical thinking.

Of particular interest to the main topic of this article is the application of the notion of allosteric systems and cooperativity to the reactions in membranes. When the response of the excitable membrane, tested on the monocellular electroplax preparation, to ACh and other congeners was measured with the highly refined techniques developed, a sigmoid shape of the dose-response curve was observed (HIGMAN et al., 1963). This sigmoid shape was interpreted by CHANGEUX and his associates (CHANGEUX et al., 1967) and independently by KARLIN (1967) in terms of allosteric systems and cooperativity. As is today widely accepted, cell membranes are formed by repeating units of lipoprotein complexes (see Introduction). On the premise that biological membranes are an ordered collection of repeating globular lipoprotein units, or "protomers", organized into a two-dimensional crystalline lattice, CHANGEUX et al. (1967) discuss the membrane properties and the possible functional role of conformational states of the proteins and the subunits applying the model proposed by MONOD, WYMAN and, CHANGEUX (1965; see also CHANGEUX and THIÉRY, 1968). In both regulatory enzymes and excitable membranes, biological activity is induced by the binding of specific ligands. In both cases the response exhibits complex cooperative phenomena. This approach permits the interpretation of the classical synergistic or antagonistic interaction between drugs, as observed with various types of excitable membranes, in terms of allosteric sites and cooperativity; a mathematical description is offered which accounts for the phenomena observed. The allosteric model was also independently applied by KARLIN (1967), as just mentioned. In his analysis of the response of the ACh-receptor and electroplax membrane, he found the slope of the Hill plot to yield an n_H close to 2, supporting a second-or higher order reaction. The application of allosteric models to the mechanism by which a compound may induce membrane activity, e.g., the permeability change to ions, and the assumption of an essential role of conformational changes in these processes appear more satisfactory than the classical treatment of dose-response relationship according to simple saturation models (CLARK, 1937) (for a discussion of this particular aspect *see* also MAUTNER, 1967).

Of particular interest is the analysis of the Hill plot by KARLIN before and after treatment with the sulfhydryl blocking and the S–S reducing agents PCMB and DTT. On the basis of the previous data of KARLIN and BARTELS (1966), the analysis shows that the effect of these agents results in a decrease of the Hill coefficient from $\sim$ 2 to $\sim$ 1. These effects are thus in agreement with similar observations on regulatory proteins and enzymes in which the effects of compounds reacting with thiol groups led to a desensitization and were considered to be an indication of the suppression of allosteric interaction.

A detailed and circumstantial analysis of the response of the electroplax membrane to ACh and its congeners was performed by CHANGEUX and PODLESKI (1968) to test how consistent the effects are with the predictions of current models of cooperativity and allosteric interactions. Particular attention was paid to: 1) the sigmoid shape of the dose-response curve of receptor activators; 2) the effect of various receptor activators and inhibitors upon this shape; and 3) the amplitude of the maximal response measured at saturated levels of activators. Titration curves were obtained with several different activators within a large range of concentration, but with particular emphasis on the region close to the origin of the curve where the deviations from the hyperbola were expected to be significant. In all cases the characteristic sigmoid shape was obtained and the Hill coefficient, used as an indication of the cooperative character of the response, was significantly different from one. Fig. 9 shows the response of the membrane of the electroplax to PTA in the upper part; the lower part shows the Hill plot ($n_H = 1.7$ instead of 1.0). The Hill coefficient does not change when the ionic environment of the cell is shifted from the usual Ringer's solution to a high K^+ medium, even when the initial resting potentials are different.

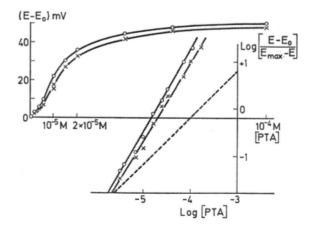

Fig. 9. Response curve of the electroplax to phenyltrimethylammonium. Two different experiments performed on different cells are plotted on the same graph. The resting potential of the cell Eo was in one case 81 mV (o—o), in the other — 80 mV (× — ×). Upper figure: standard plot of response as a function of increasing concentrations of ligand. Lower figure: Hill plot: nH = 1.7 $\pm$ 0.1; Broken line: nH = 1.0 (theroretical). (From CHANGEUX and PODLESKI, 1968)

It has been observed with regulatory proteins that the cooperative (homotropic) interactions for the binding of a given ligand are modified by the presence of a different (heterotropic) ligand. Such conversion of shape is considered as evidence that the interactions between both classes of ligands are indirect or allosteric interactions. CHANGEUX and PODLESKI (1968) studied, therefore, the effects of additional activators or inhibitors on the response of the electroplax in the presence of a given activator. Fig. 10 shows, for instance, the dose-response curve of carbamylcholine as control and the effect of an activator, decamethonium, and

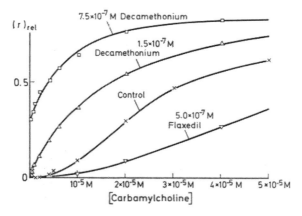

Fig. 10. Effect of a receptor activator, decamethonium, and of a receptor inhibitor, Flaxedil, on the shape and position of the dose-response curve of carbamylcholine (from CHANGEUX and PODLESKI, 1968). Each curve is obtained from a different cell and the maximal response to saturating levels of carbamylcholine is taken as 1. The experiments show the conversion of the sigmoid shape of the response to carbamylcholine to a hyperbola in the presence of the second activator

an inhibitor, Flaxedil. Addition of the activator transforms the sigmoid shape into a hyperbola. No such conversion of shape is observed with the inhibitor; the curve is simply shifted to the right. The observations parallel those on regulatory enzymes; they suggest that decamethonium and carbamylcholine are bound to at least partially different areas and that the same molecular transition accounts for both the process of excitation and the associated cooperative effects.

An important parameter of a dose-response curve is the value of the maximal response at saturating levels of activators. With three activators, PTA, carbamylcholine and decamethonium, different maximal responses are observed. This maximal response, i.e., the degree of change in potential, is thus directly related to the structure of the ligand. It does not constitute an intrinsic electrical parameter of the membrane dependent on the ionic environment. Changes of the ionic environment, either low Na^+ or high K^+ medium, as compared to the usual Ringer's medium, did affect the absolute values of the maximal depolarization, but the relative differences between the different activators remained the same. Thus, the differential amplitude of the maximal response is determined by the elementary interaction of the receptor and the receptor protein in the membrane. This observation would be difficult to interpret in terms of the ionic theory excluding chemical reactions in the membrane and considering the ionic concentration gradients as the only factor determining the electrical potentials. The authors discuss some of the remaining difficulties of definite interpretations of some details. But the parallelism with regulatory enzymes is striking and the observations offer strong evidence for the cooperative character of the response of the proteins of the excitable membrane and for the important role of allosteric effects in this process.

h) Affinity Labeling. Remarkable advances have been made, during the last 20 years, in the knowledge of the molecular groups in the active site of enzymes; the mechanism of the catalytic action of several enzymes has been analyzed. An important tool in these analyses has been the use of compounds which form a covalent bond with a group in the active site of an enzyme and thereby "label" it. Organophosphates were the first such group of compounds used for the study of the active site of ester splitting enzymes, and they are the most widely known example for labeling. With the use of DFP, BALLS, JANSEN and coworkers (JANSEN et al., 1949; BALLS and JANSEN, 1952) showed that the inactivation of chymotrypsin was stoichiometric. Later it was shown (SCHAFFER et al., 1953) that the phosphoryl group in the reaction between organophosphate and esterases is attached to a serine of the active site. Today it is generally accepted that these compounds form a covalent bond with the oxygen of the specific serine located in the active site. The usefulness of organophosphates as a tool in exploring the reactions of ACh-esterase has been discussed before; ACh-esterase was indeed one of the first enzymes with which the catalytic mechanism was explored in terms of reactions with molecular groups in the active site (NACHMANSOHN and WILSON, 1951).

Similarly, alkyl and aryl sulfonyl fluorides shown by MYERS and KEMP (1954) to form a sulfonyl enzyme analogous to the phosphoryl enzyme by organophosphates, were used for labeling of ester splitting enzymes and the analysis of the reaction mechanism. Phenylmethane sulfonyl fluoride is an even more potent irreversible inhibitor of chymotrypsin than DFP and was used for the analysis of the catalytic action of this enzyme (FAHRNEY and GOLD, 1963; GOLD and FAHRNEY, 1964).

The two types of compounds act similarly to the substrate with the difference that the covalent bond is stable in contrast to the extreme lability of the actual substrates; they have, therefore, been referred to as "quasisubstrates" by KOSHLAND (1960). Other types of compounds have been used for analyzing the active sites of proteins by means of covalent bonding: actual substrates and nonspecific reagents, for instance those reacting with S—H groups which may be located in the active site, but also on other places of the molecule, and others. The various types and principles of covalent labeling of active sites have been reviewed by SINGER (1967).

Recently, a new type of covalent bonding referred to as "affinity labeling" has been introduced by WOFSY et al. (1962). The essential principle is as follows. A labeling reagent for the active site is designed in a way that, by virtue of its steric complementarity to the active site, it first combines specifically and reversibly with the site and forms with it a complex. Then, by virtue of a suitably small and reactive group on the reagent, it may react with one or more amino acid residues in the site to form irreversible covalent bonds. The formation of the initial reversible complex increases the local concentration of the labeling reagent in the site as compared to its concentration in free solution; therefore, the reaction with any similar group outside the site. The reactive group need not to be an unusually reactive group. Because of its versatility, the covalent bonding by affinity labeling is a promising method. It has been successfully applied to enzymes (e.g., SCHOELLMAN and SHAW, 1963; SHAW et al., 1965), to antibodies (WOFSY et al., 1962; SINGER and DOOLITTLE, 1966; WOFSY et al., 1967), and has been suggested as a means of labeling and identifying specific proteins of membranes (TRAYLOR and SINGER, 1967).

The first affinity labeling of the ACh-receptor was accomplished by CHANGEUX et al. (1967) with p-(trimethylammonium)benzenediazonium (Tdf) applied to the electroplax. Tdf, being a structural analogue of PTA, a potent activator of the receptor (p. 59), and having in addition the reactive diazonium group, has the two features required for affinity labeling. The quaternary group will be attracted to the anionic group and form a reversible complex, and the diazonium may form a covalent bond with a tyrosine, histidine or lysine side chain in or near the active site. Fig. 11 shows the result of a 20-min exposure of an electroplax to 10^{-4} M Tdf. During this time, the resting potential and the resistance of the

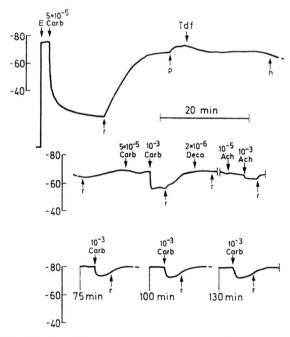

Fig. 11. Irreversible abolition of the response to a receptor activator (carbamylcholine) after a 20-min exposure to Tdf (see text), tested on the electroplax (from CHANGEUX et al., 1967 b). Ordinate: membrane potentials in mV. Abscissa: time. E, impalement; r, Ringer's solution; p, Ringer's solution supplemented with 10^{-2} M Na-phosphate pH 5.9; Tdf, $10^{-4}M$ in p; h, $10^{-2}M$ l-histidine in Ringer's solution to quench unreacted Tdf; $Carb$, carbamylcholine; $Deca$, decamethonium. Lower line: $10^{-3}M$ carbamylcholine is applied 75, 100, and 130 min after the end of the Tdf exposure

innervated membrane remain constant. After removal of Tdf, receptor activators, at concentrations about 10 times larger than their dissociation constants, have no significant effect on the membrane potential: the exposure to Tdf has rendered the cell insensitive to receptor activators. Only at concentrations more than 100 times their apparent dissociation constant may a residual depolarization of about 10 mV still be observed. When this residual response is used to test if the cell recovers its sensitivity, it is, as Fig. 11 shows, exactly the same 20 min after the end of the exposure to Tdf and 130 min later; the action is irreversible.

The irreversible effect of Tdf on the sensitivity to ACh might be due to the irreversible block of the ACh-receptor by Tdf by a covalent bond to an amino acid side chain at or near the active site, or to the block of an intermediate step in the sequence of macromolecular processes mediating the interaction between the ACh-receptor site and the structures which support the membrane potential. The following experiment permits a distinction between the two possibilities. When the electroplax is exposed to PTA until the membrane potential reaches a steady state and Tdf is added at this point without changing the concentration of PTA, a repolarization of the membrane occurs. This repolarization is irreversible. It is thus a consequence of Tdf action, antagonizing PTA. If both compounds are competing for the same binding site in the membrane, then the role of Tdf-induced repolarization should depend upon the concentration of PTA: the rate should decrease with increased concentration of PTA. This was found to be the case.

A competition has also been demonstrated between Tdf and potent receptor inhibitors such as d-tubocurarine, Flaxedil and others. At 10^{-5} M d-tubocurarine protects almost completely against a ten times higher concentration of Tdf (Fig. 12). The possibility of a direct interaction between Tdf and d-tubocurarine was excluded by control experiments. Fig. 12 presents also titration curves with carbamylcholine of electroplax exposed for various periods of time to 10^{-4} M Tdf. An important parameter of such titration curves is the amplitude of the maximal depolarization recorded at saturating levels of the activator; this maximal response is generally considered to be quantitatively related to the total number of ACh-receptor sites occupied by the ligand (CLARK, 1937; ARIENS, 1964). If one assumes that this interpretation is correct and if each receptor site, covalently bound to a Tdf molecule, is inactive, then the maximal response to a receptor activator should decrease progressively as a function of the time of exposure. Fig. 12 shows that this is indeed correct.

The similarities and differences of the active sites of ACh-esterase and ACh-receptor have been repeatedly discussed before. Since the esteratic site is specific for the enzyme and since it has been shown, as described before, that block of this site by a compound forming a covalent bond with the serine oxygen does not at all affect the reaction with the receptor (PODLESKI and NACHMANSOHN, 1966; PODLESKI, 1967), it appeared interesting to block the esteratic site by acylation prior to exposure to Tdf. If the two sites would be identical or only located in close vicinity, block of the esteratic site by a stable acylation should prevent or at least decrease the loss of sensitivity to Tdf. Using DFP and phospholine (see Table 1), CHANGEUX et al. (1967) found that the acylation of the enzyme by these compounds did not interfere with the binding of Tdf to the

ACh-receptor site. However, as was to be expected, the action of ACh was greatly potentiated, whereas the potency of carbamylcholine was not affected.

MEUNIER and CHANGEUX (1969) have tested the reaction of Tdf with homogeneous ACh-esterase in solution. The covalent binding of Tdf to the enzyme results first in a dramatic decrease of the catalytic hydrolysis of both polar and neutral substrates. However, after a few hours of incubation with Tdf, the enzyme,

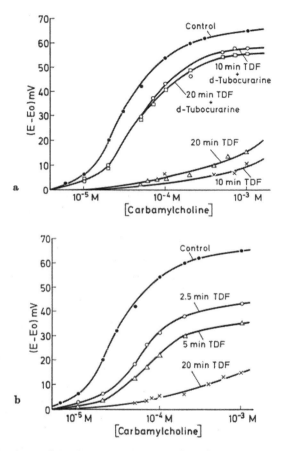

Fig. 12a and b. Response of the electroplax to increased concentrations of carbamylcholine after various Tdf exposures (from CHANGEUX et al., 1967b). Ordinate: E, steady-state potentials recorded in the presence of a given concentration of carbamylcholine; Eo, resting potential in the absence of carbamylcholine (-75 mV $= 10$ mV). Abscissa: log carbamylcholine concentrations. (a) Cells exposed 10 and 20 min to $10^{-4}M$ Tdf in the presence or absence of $10^{-5}M$ d-tubocurarine. (b) Cells exposed to $10^{-4}M$ Tdf from 2.5 to 20 min

while still inactive with the polar substrate, exhibits a selective recovery of its ability to hydrolyze a neutral substrate. These results are interpreted by the authors in terms of the binding of Tdf to several categories of topographically distinct sites.

Recent observations on the reaction of Tdf and some related structures with the ACh-receptor, tested on the electroplax, suggest a more complex mode of action than the first data seemed to indicate (H. G. MAUTNER and E. BARTELS, *personal communication*). In view of the high affinity of PTA to the negative subsite of the receptor, the irreversible block of the membrane response after exposure to Tdf seemed obviously due to the interaction of this group with the anionic site. When, however, p-trimethylammonium was replaced by an uncharged p-nitro group, the effect was just as potent as that of Tdf. Thus, it appears that it is the positively charged diazonium group which is attracted to the negative subsite. When the p-nitro group is replaced by an acetyl group, which attracts electrons less strongly than the p-nitro and the trimethylammonium groups and thereby decreases the positive charge of the diazonium, the blocking potency decreases. These and other experiments still in progress indicate that some modification of the original interpretation will be required.

Another even more efficient, i.e., more potent and more specific, affinity labeling of the ACh-receptor of the electroplax was achieved in a two-step procedure by KARLIN and WINNIK, (1968). As discussed before (p. 62), KARLIN and BARTELS (1966) had shown that the response of the electroplax to ACh and its congeners is blocked by DTT and that this block is completely reversed by subsequent exposure of the cell to the strong oxidizing agent DTNB [5,5'-dithiobis(2-nitrobenzoate)]. Application of N-ethylmaleimide (NEM), after DTT, at al concentration which had no effect before DTT, prevented the reversal by DTNB. The authors concluded that the receptor contains a relatively easily reducible disulfide bond and that one of the sulfhydryl groups formed by DTT may be alkylated by NEM. Since the block of the action on the receptor suggested the possibility that the S–S bridge affected may be near the active site, it appeared possible to increase the specificity of the reaction by making use of the anionic site, as was done repeatedly in the investigations of the active site of ACh-esterase and was also the basis for using Tdf. A maleimide derivative was, therefore, prepared in which the ethyl group was substituted by PTA: 4-(N-maleimido)-phenyltrimethylammonium iodide (MPTA). This compound proved to be an extraordinarily potent and specific affinity label when applied to the receptor reduced by DTT. The rate of the reaction is two to three orders of magnitude faster than that of NEM. The response of the electroplax to carbamylcholine is abolished by MPTA at 10^{-7} M.

Of great interest is the comparison of MPTA with its tertiary analogue (4-maleimido)-N,N-dimethyl aniline (MDA). NEM and MDA are approximately equally active in alkylating the reduced disulfide. Adding the extra methyl group enhances the alkylation of the receptor about 300-fold. This is about the same difference found for the potency as activator between ACh and its tertiary analogue dimethylaminoethyl acetate (BARTELS, 1962) and between tetramethylammonium and its tertiary analogue; in the latter case, the difference is even slightly higher. The ratio of the "rate constants" of the three maleimides measured for the reaction with cysteine, $k(M^{-1} \sec 2^{-1})$, is 1,600 for NEM, 3,920 for MDA, and 8,850 for MPTA. For the alkylation of the receptor, it is 1,660 for MPTA, as compared to 1 for both NEM and MDA. In view of the differences in intrinsic activity with sulfhydryl groups, as measured with cysteine, MDA thus appears even slightly less effective than NEM.

In the MPTA molecule, the distance between the quaternary nitrogen and the ethylenic double bond of the maleimide moiety is more or less fixed and has been estimated by the authors to be approximately 12 A. The distance was measured on Corey-Pauling-Koltun models and was taken from the far side of the reacting carbon atom of the ethylenic double bond to the far side of the methyl groups on the quaternary ammonium group in the maximally extended configuration (KARLIN, 1969).

Two other affinity labeling compounds for the ACh-receptor were designed by SILMAN and tested on the electroplax (SILMAN and KARLIN, 1969; SILMAN, 1970). These two compounds, bromoacetylcholine bromide (BACh) and [(p-nitrophenyl)-p-carboxyphenyl]trimethylammonium iodide, are both receptor activators, i.e., they depolarize the membrane, but both form covalent bonds only after treatment of the membrane with DTT. They presumably react, therefore, like the maleimides, with the sulfhydryl groups. BACh is a good substrate for ACh-esterase. Its depolarizing action, like that of ACh, is obtained only in the presence of physostigmine. After the receptor has been reduced by DTT, the depolarization by BACh is irreversible, indicating a covalent bond formation, although a slight repolarization may be observed.

The p-nitrophenyl ester of (p-carboxyphenyl)trimethylammonium acylates nucleophilic groups with the release of the p-nitrophenyl group. When applied to the electroplax, the ester has no depolarizing action or a barely detectable one. It is a strong competitive inhibitor; this effect is readily reversible. However, when the receptor has been reduced by DTT the compound induces a slight depolarization. After removal of the compound by washing, the depolarization becomes much stronger; this is due to the removal of the excess of the compound which acts as a receptor inhibitor.

In BACh the distance between the α-carbon group reacting with the SH group and the quaternary nitrogen group is 9 A. In the inactive conformational state of the reduced receptor, the distance from the SH groups to the negative subsite seems to be on the basis of the data with MPTA approximately 12 A (or greater). One may then assume that in the active conformation of the reduced receptor the SH group is brought 3 A closer to the negative site than in the inactive form. The flexibility of the active site, especially after reduction by DTT, is also supported by experiments with other maleimide derivatives and quaternary nitrogen derivatives (KARLIN, 1969). In addition to the previously mentioned observations, these data provide further support for a conformational change of the receptor in the reaction with activators, in line with current views of the function of proteins.

Another aspect of general interest resulted from the studies of the ACh-receptor reduced by DTT. Hexamethonium (hexamethylene-bis-tri-methylammonium chloride) is a "curare-like" compound; it is a readily reversible competitive inhibitor of the ACh-receptor in the electroplax: it blocks the neurally evoked action potential and competitively inhibits the depolarization caused by carbamylcholine with an apparent dissociation constant of 3×10^{-5} M (MAUTNER et al., 1966). Hexamethonium partially protects the receptor against alkylation by MPTA; i.e., it slows the reaction of the sulfhydryl group and MPTA, whereas, as would be expected, it does not decrease the rate of the alkylation by NEM. Surprisingly, however, after exposure to DTT, hexamethonium causes a depolarization; i.e., it has become a receptor activator. After reoxidation with DTNB, the compound again becomes a receptor inhibitor. Thus, it must be concluded that the presumably

6*

small change of conformation of the protein by DTT causes the transformation of the compound from an inhibitor into an activator.

Similarly, a drastic change for biological action after exposure to DTT has been observed by PODLESKI et al. (1960). As discussed before, on exposure to TDF in 10^{-4} M, the response of the electroplax to activators is irreversibly blocked. However, after the treatment with DTT, TDF becomes a very potent receptor activator; it depolarizes the electroplax reversibly in 10^{-6} M.

For many years excitation and inhibition have been attributed to "excitatory" or "inhibitory" transmitters, implying that these are different compounds. The conclusion was based simply on the effects on electrical parameters. The present observations illustrate the difficulty of interpreting biological actions on such a basis. Apparently opposite effects may be due to relatively small changes in the system tested e.g., different states of the target protein, of structural organization, or a variety of other factors, but not to a difference of the acting compound. It is well known that ACh may have an inhibitory action on some membranes and an excitatory one on others. The action of ATP may cause a muscular contraction or relaxation depending on the concentration of free Ca^{++} ions. The data are an example for the necessity of explaining the mechanism of the biological action of a given ligand on a molecular level.

i) **Photoregulation of the Membrane Potential by Receptor Inhibitors.** The chief aim of this chapter is the presentation and discussion of our present knowledge of the proteins in excitable membranes which are directly associated with exciation and with the propagation of nerve impulses. The question naturally arises how excitation may be initiated. Experimentally it may be accomplished by electrical currents, chemical compounds, and other means. In the biological process, excitation may be produced by light, sound, smell, touch, etc. Highly specialized organelles are present in receptor cells for transducing the specific stimuli into neural excitation. Although much information is available about the physiology of receptor cells, our knowledge of the chemical reactions involved is extremely limited.

Recently, ERLANGER and his associates prepared a number of diazonium compounds which exist as *cis* and *trans* isomers and are intraconvertible by means of light. Light causes a reversible shift in the *cis-trans* equilibrium about the nitrogen-nitrogen double bond of the compounds. The *trans* isomer predominates at 420 nm length in the light of a photoflood lamp; the *cis* predominates at 320 nm wavelength in ultraviolet light. When these light-sensitive ligands were applied either to chymotrypsin (KAUFMAN et al., 1968) or to ACh-esterase (BIETH et al., 1969), the potency of the two isomers as enzyme inhibitors was different. On exposure to the enzyme systems to different wavelengths, a photoregulation of the activity of the two enzymes was obtained.

The first molecular event in vision is the *cis-trans* isomerization of retinal (WALD, 1968). *Cis*-retinal reacts in the dark with the protein opsin to form rhodopsin. Light-induced isomerization of the *cis*-retinal to the all-*trans* conformation leads to nerve excitation. Absorption of only a few quanta of light is sufficient to produce a measurable response in the retina (HECHT et al., 1942). The mechanism by which these reactions produce neural excitation is unknown. In view of the effects of photochromic substances on ACh-esterase, it appeared of interest to test

Erratum

On p. 84, line 30, and
p. 85, line 5, read

diazo compounds
(instead of diazonium compounds)

whether it is possible to photoregulate the potential across the excitable membrane of the electroplax with ligands which are structural analogs of ACh-receptor activators or inhibitors and with which *cis-trans* isomerization may be obtained by irradiation.

The possibility was studied with two diazonium compounds, N–p-phenyl-azophenyl-N-phenyl-carbamylcholine chloride (azo-CarCh) and p-phenylazophenyl-trimethylammonium chloride (azo-PTA) (DEAL et al., 1969), which are derivatives of potent receptor activators, i.e., strong depolarizing agents. These two derivatives are strong receptor inhibitors; both *trans*-isomers, predominating at 420 nm, are markedly stronger inhibitors than the two *cis* forms predominating at 320 nm. When the electroplax is depolarized by 20 μM carbamylcholine, the two compounds repolarize the membrane in low concentrations. In its *trans* form, azo-CarCh has a strong repolarizing action in 1 μM concentration; the *cis* form has a weak effect. The two isomers of azo-PTA have comparable effects, but at ten times higher concentrations.

When the electroplax is exposed to carbamylcholine in the presence of 1 μM azo-CarCh under ultraviolet irradiation, a marked depolarization takes place. When a steady state is reached and the preparation is then exposed to a photoflood lamp, a repolarization takes place which amounts to 20—30 mV. Similar results were obtained with 20 μM azo-PTA.

These experiments may be considered as a model illustrating how one may link limited changes in conformation of a ligand, such as a light-induced *cis-trans* isomerization, the first step in the initiation of a visual impulse, with substantial changes of the potential difference across an excitable membrane. The changes of 20—30 mV in membrane potential are of the same order of magnitude as those taking place in the visual process leading to a neural impulse. It takes, however, minutes to obtain the effects, whereas the biological response to light occurs within a millisecond. This great difference in efficiency is not surprising. The components of the retina are highly specialized. The light-sensitive molecules form an integral part of the cell. Other events associated with the isomerization of retina are still unexplored; e.g., a conformational change of opsin may be critically involved. Moreover, the efficiency of highly active biological compounds is usually low when applied from the outside. The concentrations may be orders of magnitude higher than those which are released and act in the membrane. In any event, it appears interesting that neural stimulation may be induced by a change of conformation of a ligand of a kind similar to that occuring in the biological process of vision. Obviously, to fully understand the link between the *cis-trans* isomerization of retinal and neural stimulation, an analysis of the specific processes will be necessary.

C. Differences between Axonal Conduction and Synaptic Transmission

In view of the progress achieved in the understanding of chemical reactions in membranes in general and in excitable membranes in particular, a reevaluation of the theory of neurohumoral transmission appears appropriate. One of the criteria for the strength of a new concept, suggested to replace a preceding one, is its ability to provide alternative and more satisfactory explanations for the observations on which the original interpretation had been based. In the particular case of the topic under discussion, the pertinent question to be decided between the two theories is as follows: is ACh the signal released in excitable membranes of axons and junctions and triggering off, by its action on the receptor within the membranes, the sequence of reactions that result in increased ion permeability? Or is it still reasonable to assume, on the basis of the data available now, that it is actually released from the nerve terminals and acts as a neurohumoral transmitter on a second cell, nerve or muscle? A third alternative, occasionally suggested, is the idea that ACh may act in the axonal membranes, as suggested, but at junctions it may still act as a transmitter in the sense originally proposed, i.e., crossing the gap between two cells. All these alternatives center around the question of the soundness and solidity of the assumption that ACh is actually released from the nerve terminal and acts as a transmitter between two cells.

One of the striking facts particularly emphasized as being an indication of a mechanism of synaptic transmission fundamentally different from that of conduction is the powerful pharmacological action of ACh on junctions in contrast to its complete failure to affect axonal conduction. Similarly, more than a century ago, Claude Bernard observed that the activity of curare blocks only the transmission of the impulse from nerve to muscle, but does not affect nerve and muscle fibers. Curare is a receptor inhibitor with a higher affinity to the protein than that of ACh.

The limitation of the action of ACh and curare to the junctional membranes and their failure to affect conduction has been repeatedly discussed in this article and requires, therefore, only a few additional remarks. The specific proteins reacting with ACh during activity are not only present in axonal membranes, but fully functional and associated with electrical activity. The inability of ACh and curare and other quaternary nitrogen derivatives has been fully explained as due to the presence of structural permeability barriers. In several instances, however, direct actions of curare and ACh have been demonstrated where the barriers are either inadequate or may be sufficiently reduced by appropriate treatment such as snake venoms or detergents. The effects on Ranvier nodes of single myelinated fibers and those on the unmyelinated axon of the lobster walking leg have been described. Another preparation on which ACh acts are the unmyelinated axons of the vagus nerve of rabbit (Armett and Ritchie, 1960), although this effect is considered to be a pharmacological curiosity (Ritchie, 1963). The action of ACh and curare on squid giant axons following their exposure to snake venoms has been discussed. Following the exposure of desheathed frog sciatic fibers to cetyltrimethylammonium ions, effects of ACh, neostigmine, curare, etc. were obtained by Walsh and Deal (1959).

On the other hand, even synaptic junctions frequently do not react to ACh or curare, such as, for instance, the neuromuscular junction of lobsters, although the concentrations of ACh-esterase in these junctions are extremely high. They do react, however, to lipid-soluble compounds such as physostigmine and DFP. Thus, the proteins there are present and functional, but apparently protected, as in axons, by barriers against quaternary nitrogen derivatives. Similar barriers were discovered in the nervous system of insects. The total absence of the action of ACh on the insect nervous system was considered as evidence that it is not cholinergic. Therefore, the effect of organophosphate insecticides was attributed to a different factor than the inhibition of ACh-esterase. It took more than a decade of intensive investigation before this situation was clarified. The effect of ACh was demonstrated after mechanical removal of the barriers and the close relationship between the toxic action of organophosphates and enzyme inhibition was established (for a summary, see O'BRIEN, 1960).

Even when a junction reacts to ACh, the discrepancy between the amounts of ACh released and the concentrations which must be applied externally to be effective raises difficult problems. In the squid giant axon, the order of magnitude of ACh released is less than 10^{-15} mole per cm^2/impulse. The amounts of ACh released per unit surface of excitable membrane may vary in different types of axons and species. On the basis of the variations of ACh-esterase activity, such differences may vary by a factor of one to two orders of magnitude. But the amounts would be still very far from the pharmacologically active concentrations at junctions. Applications by electrophoresis may increase the potency by up to two to three orders of magnitude, but there still would remain the fantastic discrepancy of at least six to eight orders of magnitude between the amounts released and those pharmacologically active. A relatively sensitive preparation such as the single electroplax, fails to react to 10^{-5} M ACh unless physostigmine is added. This discrepancy raises the question as to the meaning of the amounts of ACh found in the extracellular perfusion fluid of junctions following stimulation. No trace of ACh appears in the absence of physostigmine, even after prolonged stimulation by many thousands of impulses, as was emphasized repeatedly by Dale and his associates. Since physostigmine blocks the rapid physiological inactivation, the escape of traces of ACh to the outside may be an artifact. ACh appears also in the outside fluid of axons when kept in physostigmine when the surrounding barriers are not entirely impervious to ACh, as is the case of the axons of the walking leg of lobster (DETTBARN and ROSENBERG, 1966). Na^+, K^+, Ca^{++} and Mg^{++} ions have the same effects on the efflux of ACh, increasing or decreasing it, in this preparation, as previously described for junctions.

An alternative explanation for the role of ACh at the junction must then be considered. It is now well established that ACh-receptor and ACh-esterase are localized in the excitable membranes of the nerve terminal and of the postsynaptic membrane. ACh, curare, neostigmine and related compounds act on the membranes on the two sides of the junction; the nerve terminal seems even to be more sensitive than the postsynaptic membrane (MASLAND and WIGTON, 1940; RIKER et al., 1959; for a review, see WERNER and KUPERMAN, 1963). Since both proteins are present and functional in both membranes of the junction, it is reasonable to assume that the amplifier process takes place in the membranes of the junctions in the same way as in those of the axon: ACh, released in the membrane of the nerve terminal, acts on the receptors there and triggers off the sequence of reactions resulting in the influx of Na^+ and efflux of K^+ ions. It is known that K^+ ions release ACh from its storage form. Thus, when the K^+ ions, released from the terminal, reach the postsynaptic membrane, they will release ACh there and thereby trigger off the same amplification process. By this amplification the signal will be much more efficient than if it were directly released and would have to find the specific protein and act on it in a second cell. When a strong efflux of K^+ ions was found subsequent to stimulation at junctions (FELDBERG and VARTIAINEN, 1934) as well as in axons (COWAN, 1934), ECCLES (1935) attributed the transmitter function to K^+ ions. But the release of ions from the nerve terminal requires the amplification process triggered off by ACh. Per 1,000 molecules of ACh released, many millions of K^+ ions would cross the gap and release ACh in the postsynaptic membrane from its storage form. Even if only a fraction would reach the postsynaptic membrane and there release ACh from its storage form, the process would be very much more efficient. This alternative assumption proposed seems all the more reasonable in view of the discrepancy of the order of 10^6 to 10^8 between the amounts of ACh released and those which are pharmacologically effective even in the presence of physostigmine. Chemical reactions in living cells are chemically and thermodynamically coupled; most of them are structurally organized. It seems more plausible that the specific signal in one of the fastest cellular mechanisms known, i.e., nerve impulse propagation, is recognized by a specific protein in the membrane where it is released and acts as a trigger of an amplifier system there in a structurally organized way than to visualize a random process in which a chemical compound released in one cell must cross to another cell and must there find the target protein.

Two other aspects of the problem may be briefly discussed. For many years efforts to detect flow of current from nerve terminals were unsuccessful. This failure was considered as strong evidence for the theory of neurohumoral transmission, since it seemed to exclude electrical transmission across the junction by current flow, in contrast to the propagation along axons. The absence of current flow seemed particularly conspicuous in the giant synapse of the squid (BULLOCK and HAGIWARA, 1957). In this preparation, parts of the pre- and postsynaptic axons are located side by side. It is possible to insert micro pipettes in the two axons, so their tips are approximately 2 mm apart, and to monitor the current flow. In this case the failure of detection of current flow appeared convincing. In another type of preparation, however, electrical transmission across the junction was described (FURSHPAN and POTTER, 1959). Subsequently, flow of current from nerve terminals has been demonstrated by HUBBARD (1963), and his findings were confirmed (KATZ and MILEDI, 1965). Quite recently, flow of current from both axons was found in the squid giant synapse (GAGE and MOORE, 1969). Thus, another objection to the assumption of a fundamentally similar mechanism of conduction and transmission has been removed.

The second aspect is the question of the meaning of the heat production and absorption coinciding with electrical activity. The view of HODGKIN (1951, 1965) that a simple diffusion process accounts for the ion movements has been mentioned before (p. 20). In recent years, he and his associates repeatedly and specifically rejected the assumption of any chemical reactions associated with the permeability changes in conduction (see, e.g. BAKER et al., 1963; KEYNES and AUBERT, 1964). The facts upon which these objections were based have been discussed elsewhere (e.g. NACHMANSOHN, 1965, 1966b) and need no further comment. However, the recent interpretation of HOWARD, RITCHIE, and KEYNES (1968), according to which the heat produced and absorbed is due to ion friction (p. 20), requires some rectification. Discussing the possible source of the net heat, the authors quote the "suggestion (NACHMANSOHN, 1959) that it is the heat of hydrolysis of acetylcholine broken down during the spike. However, since this is of the order of 3,000 cal/mole (NACHMANSOHN, 1959) ...". They use this figure as a target for their discussions. The figure of 3,000 cal/mole, as given in the monograph quoted, represents the ΔF of ACh hydrolysis, whereas only ΔH, the enthalpy change, could be used for their estimates. Neither on p. 150 as quoted by the authors, nor anywhere else in that monograph or in any other publication of the writer has the suggestion quoted by the authors been made. As was stated by the author (NACHMANSOHN, 1966a, p. 269—270), the source of the heat production or absorption cannot be assigned to any specific reaction. "What we definitely know is that there are a whole series of chemical reactions associated with the permeability cycle: the release of ACh; the reaction with the receptor; the hydrolysis of ACh; the release and neutralization of protons; ion mixing; possibly release and binding of Ca; rearrangement of polyelectrolytes and phospholipids, possibly with breaking and formation of hydrogen bonds and hydrophobic bonds. We have very little information about the many chemical reactions involved, and no notion of the associated enthalpies, not even in solution. The recent work of KATCHALSKY has shown how extensively chemical reactions may be affected if they take place in a structure. In view of our ignorance of the reactions and enthalpies involved, your question cannot be answered. For the same reason it seems to me also impossible to draw any conclusions from heat measurements as to the underlying mechanism or to predict whether one should expect heat production or absorption just on the basis of one single reaction out of the many involved."

It seems appropriate to elaborate this brief and casual discussion remark (i) because of the interest of the problem of whether bioelectric currents are generated by chemical reactions or by purely physical events, as implied by the authors, and (ii) because of the developments which have taken place since this remark was made and which are pertinent to the topic under discussion. For evaluating the heat of ligand reactions, the identification of the following thermodynamic parameters is essential: the free energy change (ΔF), the heat or enthalpy change (ΔH), and the entropy change (ΔS). ΔS can be obtained from ΔF and ΔH be the equation $\Delta F = \Delta H - T \Delta S$. In addition to having mixed up two thermodynamic parameters, HOWARTH et al. (1968) did not even consider in their discussion the entropy change which may be the most important in this context. This parameter reflects the change in the degree of randomness of the system or the passage to differently ordered states; it has special significance in macromolecular systems since it has long been recognized that the large values of ΔS in such systems are mainly associated with conformational changes. ΔS may be as high as 40—50 cal/mole, so that $T \Delta S$ may amount to 15,000 cal/mole.

The thermodynamic parameters of many enzymes and protein systems have been evaluated. The complexity of these systems frequently offers experimental and theoretical difficulties which prevent precise separation of individual parameters. An illuminating example of the difficulties is the analysis of these parameters for the reaction of hemoglobin with oxygen, a protein of which the tridimensional structure and the existence of four subunits is known and a reaction in which conformational changes have been demonstrated (ROSSI-FANELLI, ANTONINI, and CAPUTO, 1964; WYMAN, 1964).

If it is difficult to determine and to identify thermodynamic parameters even in isolated and well-defined protein systems, the obstacles to any interpretation of these parameters in a cell or in a subcellular structure such as a membrane are at present insurmountable, since we have here to deal with assemblies of proteins and enzymes. Many of them are probably active, all the time, i.e., in a dynamic state and not in a static condition; equilibrium in cells is reached only at death. In addition to the notions of classical, those of irreversible thermodynamics have gained increasing significance in biological systems (KATCHALSKY and CURRAN, 1965). In view of the incredible complexity of chemical reactions in living cells (their chemical and thermodynamic coupling, their dependence on structure and organization which may change completely their behavior from that in solution, the many control mechanisms, etc.), heat production and absorption, measured on a living cell, must necessarily present overall values resulting from a great number of chemical reactions. Considering the data accumulated about the proteins and enzymes and their activities in biomembranes in general and in excitable membranes in particular, the concept of bioelectricity as a purely physical event and the explanation of the heat during nerve activity as being due to ion friction, appear unsatisfactory.

The specific chemical forces underlying cellular mechanisms, such as motility, vision, energy supply, etc., are remarkably similar throughout the animal kingdom. It has become apparent that this also applies to the specific chemical forces underlying bioelectricity, i.e., the specific proteins controlling the changes of ion permeability in excitable membranes. The great diversity of bioelectrical phenomena and the striking differences of pharmacological effects may be explained by the nearly infinite variations of cellular structure and organization. The multiformity of shape, structure and environment of synaptic junctions is bound to modify the effects of chemical reactions in the membranes or the action of compounds applied externally, as illustrated by several of the observations discussed. Structural differences markedly affect electrical parameters even in axons, in spite of their uniform and relatively simple cylindrical shape. For instance, conduction velocity, in different types of axons, varies from 0.1 to 100 m/sec. Nobody has suggested in this case that different forces may be involved in generating electricity.

A general principle of scientific thinking is not to assume without necessity two different basic mechanisms, even if it is difficult to explain certain variations. As was pointed out by DAVID HUME in his "Treatise on Human Nature", the proposal of two basically different mechanisms frequently serves to cover our ignorance of the truth. In spite of the many variations of pharmacological actions and electrical parameters, there is no necessity, in the writer's view, to assume a basic difference of the trigger function of ACh in the membranes of axons, nerve terminals, postsynaptic membranes and the membranes of muscle fibers.

Knowledge of chemical composition and molecular organization of excitable membranes is extremely limited. The question whether there are variations in this respect between different types of excitable membranes cannot be answered at present. But an elucidation of these questions cannot be supplied by studies of electrical phenomena or pharmacological effects; it requires analysis on a molecular level using possibly still more refined methods and techniques than now available.

Acknowledgments

The author would like to express his appreciation to all those colleagues who permitted the reproduction of the electron micrographs and figures. He is particularly grateful to Dr. VIRGINIA TENNYSON for the authorization to use her electron micrography which she had not published as yet.

References

ABBOTT, B. C., HILL, A. V., HOWARTH, J. V.: The positive and negative heat production associated with a nerve impulse. Proc. roy. Soc. B 148, 149—187 (1958).

ALDRIDGE, W. N.: The inhibition of erythrocyte cholinesterase by tri-esters of phosphoric acid: 3. The nature of the inhibitory process. Biochem. J. 54, 442—448 (1953).

— DAVISON, A. N.: The mechanism of inhibition of cholinesterases by organophosphorus compounds. Biochem. J. 55, 763—766 (1953).

ALTAMIRANO, M., COATES, C. W., GRUNDFEST, H., NACHMANSOHN, D.: Mechanisms of bioelectric activity in electric tissue. I. The response to indirect and direct stimulation of electroplaques of *Electrophorus electricus*. J. gen. Physiol. 37, 91—110 (1953).

— — — — Electrical activity in electric tissue. III. Modifications of electrical activity by acetylcholine and related compounds. Biochim. biophys. Acta (Amst.) 16, 449—463 (1955).

— SCHLEYER, W. L., COATES, C. W., NACHMANSOHN, D.: Electrical activity in electric tissue. I. The difference between tertiary and quaternary nitrogen compounds in relation to their chemical and electrical activities. Biochim. biophys. Acta (Amst.) 16, 268—282 (1955).

ARMETT, C. J., RITCHIE, J. M.: The action of acetylcholine on conduction in mammalian non-myelinated fibers and its prevention by anti-cholinesterase. J. Physiol. (Lond.) 152, 141 to 158 (1960).

AUGER, D., FESSARD, A.: Étude oscillographique des decharges de l'appareil électrique des Raies. Ann. Physiol. Physicochim. biol. 15, 261—270 (1939).

AUGUSTINSSON, K.-B., HEIMBURGER, G.: Enzymatic hydrolysis of organophosphorus compounds. V. Effect of phosphoryl phosphatase on the inactivation of cholinesterases by organophosphorus compounds *in vitro*. Acta chem. scand. 8, 310—318 (1955).

BAKER, P. F., HODGKIN, A. L., SHAW, T. I.: The effects of changes in internal ionic concentrations on the electrical properties of perfused giant axons. J. Physiol. 164, 355—374 (1962).

BALLS, A. K., JANSEN, E. F.: Stoichiometric inhibition of chymotropsin. Adv. Enzymol. 13, 321—343 (1952).

BARRNETT, R. J.: The fine structural localization of acetylcholinesterase at the myoneural junction. J. Cell Biol. 12, 247—262 (1962).

BARTELS, E.: Structure-activity relationship studied on the isolated single electroplax. Biochim. biophys. Acta (Amst.) 63, 365—373 (1962).

— Relationship between acetylcholine and local anesthetics. Biochim. biophys. Acta (Amst.) 109, 194—203 (1965).

— Reactions of ACh-receptor and -esterase studied on the electroplax. Biochem. Pharmacol. 17, 945—956 (1968).

— NACHMANSOHN, D.: Molecular structure determining the action of local anesthetics on the acetylcholine receptor. Biochem. Z. 342, 359—374 (1965).

— — Organophosphate inhibitors of acetylcholine-receptor and -esterase tested on the electroplax. Arch. Biochem. Biophys., 133, 1—10 (1969).

BENSON, A. A.: On the orientation of lipids in chloroplast and cell membranes. J. Am. Oil Chemists' Soc. 43, 265—270 (1966).

BERGMANN, F., SEGAL, R.: The relationship of quaternary ammonium salts to the anionic sites of true and pseudo cholinesterase. Biochem. J. 58, 692—698 (1954).

BERMAN, R., WILSON, I. B., NACHMANSOHN, D.: Choline acetylase specifity in relation to biological function. Biochim. biophys. Acta (Amst.) 12, 315—324 (1953).

BIETH, J., VRATSANOS, S. M., WASSERMAN, N., ERLANGER, B. F.: Photoregulation of biological activity by photocromic reagents. II. Inhibitors of acetylcholinesterase. Proc. nat. Acad. Sci. (Wash.) 64, 1103—1106 (1969).

BLAKE, C. C. F., KOENIG, D. F., MAIR, G. A., NORTH, A. C. T., PHILLIPS, D. C., SARMA, V. R.: Structure of hen egg-white lysozyme. A three dimensional fourier synthesis at 2 A resolution. Nature (Lond.) 206, 757—761 (1965).

— MAIR, G. A., NORTH, A. C. T., PHILLIPS, D. C., SARMA, V. R.: On the conformation of the hen egg-white lysozyme molecule. Proc. roy. Soc. B 167, 365—377 (1967).

BLANGY, H., BUC, H., MONOD, J.: Kinetics of the allosteric interactions of phosphofructokinase from Escherichia coli. J. mol. Biol. 31, 13—35 (1968).

BLOOM, F. E., BARRNETT, R. J.: Fine structural localization of acetylcholinesterase in electroplaque of the electric eel. J. Cell Biol. 29, 475—495 (1966).

BRESLOW, E., BEYCHOK, S., HARDMAN, K., GURD, F. R. N.: Relative conformations of sperm whale metmyoglobin and apomyoglobin in solutuion. J. Biol. Chem. 240, 1639—1646 (1965).

BRINK, F.: The role of calcium ions in neural processes. Pharmacol. Rev. 6, 243—298 (1954).

BRZIN, M.: The localization of acetylcholinesterase in axonal membranes of frog nerve fibers. Proc. nat. Acad. Sci. (Wash.) 56, 1560—1563 (1966).

— DETTBARN, W.-D., ROSENBERG, P., NACHMANSOHN, D.: Acetylcholinesterase activity per unit surface of conducting membranes. J. Cell Biol. 26, 353—364 (1965).

— TENNYSON, V. M., DUFFY, P. E.: Acetylcholinesterase in frog sympathetic and dorsal root ganglia: A study by electron microscope cytochemistry and microgasometric analysis with the magnetic diver. J. Cell Biol. 31, 215—242 (1966).

BULLOCK, T. H., GRUNDFEST, H., NACHMANSOHN, D., ROTHENBERG, M. A.: Generality of the role of acetylcholine in nerve and muscle conduction. J. Neurophysiol. 10, 11—22 (1947).

— HAGIWARA, S.: Intracellular recording from the giant synapse of the squid. J. Gen. Physiol. 40, 565—577 (1957).

— NACHMANSOHN, D., ROTHENBERG, M. A.: Effects of inhibitors of choline esterase on the nerve action potential. J. Neurophysiol. 9, 9—22 (1946).

BURGEN, A. S. V.: The mechanism of action of anticholinesterase drugs. Brit. J. Pharmacol. 4, 219—228 (1949).

— HOBBIGER, F.: The inhibition of cholinesterase by alkylphosphates and alkylphenylphosphates. Brit. J. Pharmacol. 6, 593—605 (1951).

CANEPA, F. G., PAULING, P., SÖRUM, H.: Structure of acetylcholine and other substrates of cholinergic systems. Nature (Lond.) 210, 907—909 (1966).

CANTONI, G. L., LOEWI, O.: Inhibition of cholinesterase activity of nervous tissues by eserine in vivo. J. Pharmacol. exp. Ther. 81, 67—71 (1944).

N. N.: Cellular regulatory mechanisms. Cold Spr. Harb. Symp. quant. Biol. 26, (1961).

CHAGAS, C., CARVALHO, A. P. DE: Bioelectrogenesis. Proc. Symp. Comp. Bioelectrogenesis: A comparative survey of its mechanism with particular emphasis on electric fishes. Amsterdam: Elsevier 1961.

CHANGEUX, J. P.: The feedback control mechanism of biosynthetic L-threomine deaminase leg L-isoleucine. Cold Spr. Harb. Symp. quant. Biol. 26, 313—318 (1961).

— Effet des analogues de la L-threonine et de la L-isoleucine sur la L-threonine desaminase. J. molec. Biol. 4, 220—225 (1962).

— Allosteric interactions on biosynthetic L-threomine deaminase from E. coli K-12. Cold Spr. Harb. Symp. quant. Biol. 28, 497—504 (1963).

— GAUTRON, J., ISRAEL, M., PODLESKI, T. R.: Séparation de membranes excitables à partir de l'organe électrique d'électrophorus électricus. C. R. Acad. Sci. Paris, 269, 1788—1791 (1969).

— GERHART, J. C., SCHACHMAN, H. K.: Allosteric interaction in aspartate transcarbamylase. I. Binding of specific ligands to the nature enzyme and its isolated subunits. Biochemistry 7, 531—538 (1968).

— PODLESKI, T. R.: On the excitability and cooperativity of the electroplax membrane. Proc. nat. Acad. Sci. (Wash.) 59, 944—950 (1968).

— — MEUNIER, J.-C.: On some structural analogies between acetylcholinesterase and the macromolecular receptor of acetylcholine. J. gen. Physiol. 54, 225S—244S (1969).

CHANGEUX, J. P., PODLESKI, T. R., WOFSY, L.: Affinity labeling of the acetylcholine-receptor. Proc. nat. Acad. Sci. (Wash.), 58, 2063—2070 (1967).
— THIÉRY, J.: On the excitability and cooperativity of biological membranes. In: Regulatory Functions of Biological Membranes. J. JÄRNEFELT ed. BBA Library 11, pp. 116—138. Amsterdam: Elsevier 1968.
— — TUNG, Y., KITTEL, C.: On the cooperativity of biological membranes. Proc. nat. Acad. Sci. (Wash.) 57, 335—341 (1967).
CHILDS, A. F., DAVIES, D. R., GREEN, A. L., RUTLAND, I. P.: The reactivation by oximes and hydroxamic acids of cholinesterase inhibited by organophosphorus compounds. Brit. J. Pharmacol. 10, 462—465 (1955).
CHU, S. H., MAUTNER, H. G.: Analogs of neuroeffectors. V. Neighboring effects in the reactions of esters, thiolesters, and selenoesters. The hydrolysis and aminolysis of benzoylcholine, benzoylselenocholine and their dimethylamino analogs. J. Org. Chem. 31, 308—312 (1966).
CLARK, A. J.: The mode of action of drugs on cells. London: Edward Arnold & Co. 1933.
— General pharmacology. In: Handbook of experimental pharmacology, Vol. IV. HEUBNER, W., SCHUELLER, J., Eds. Berlin: Springer 1937.
CLELAND, W. W.: Dithiothreitol. A new protective reagent for SH groups. Biochemistry 3, 480—482 (1964).
COHEN, J. A., OOSTERBAAN, R. A.: The active site of acetylcholinesterase and related esterases and its reactivity towardes substrates and inhibitors. In: Handbook of experimental pharmacology 299—373. Ergw. XV. KOELLE, G. B., (Ed.). Berlin-Göttingen-Heidelberg: Springer 1963.
— POSTHUMUS, C. H.: The mechanism of action of anti-cholinesterases. Acta physiol. pharmacol. neerl. 4, 17—36 (1965).
— — The mechanism of action of anti-cholinesterases. III. The action of anticholinesterases on the phrenic nerve-diaphragm preparation of the rat. Acta physiol. pharmacol. neerl. 5, 385—397 (1957).
COHEN, M.: Concentration of choline acetylase in conducting tissue. Arch. Biochem. 60, 261 to 278 (1955).
CONWAY, A., KOSHLAND, JR., D. E.: Negative cooperativity in enzyme action. The binding of diphosphopyridine nucleotide to glyceraldehyde 3-phosphate dehydrogenase. Biochemistry 7, 4011—4023 (1968).
CORWIN, L. M., FANNING, G. R.: Studies of parameters affecting the allosteric nature of phosphoenolpyruvate carboxylase of Escherichia coli. J. biol. Chem. 243, 3517—3525 (1968).
COWAN, S. L.: The action of potassium and other ions on the injury potential and action current in Maja nerve. Proc. roy. Soc. B. 115, 216—260 (1934).
CULVENOR, C. C. J., HAM, N. S.: The proton magnetic resonance spectrum and conformation of acetylcholine. Chem. Commun. 15, 537—539 (1966).
DAVIES, D. R., GREEN, A. L.: Results quoted under contributions to the general discussion. In: The physical chemistry of enzymes. Discussions Faraday Soc. 20, 269 (1955).
— — The kinetics of reactivation, by oximes, of cholinesterase inhibited by organophosphorus Biochem. J. 63, 529—535 (1956).
DAVISON, A. N.: Return of cholinesterase activity in the rat after inhibition by organophosphorus compounds. A comparative study of true and pseudo cholinesterase. Biochem. J. 60, 339—346 (1955).
DEAL, W. J., ERLANGER, B. F., NACHMANSOHN, D.: Photoregulation of biological activity by photochromic reagents. III. Photoregulation of bioelectricity by acetylcholine receptor inhibitors. Proc. natl. Acad. Sci. (Wash.) 64, 1230—1234 (1969).
DELORENZO, A. J. D., DETTBARN, W.-D., BRZIN, M.: Fine structure and organization of nerve fibers and giant axons in lobster Homarus americanus. J. Ultrastruc. Res. 24, 367—384 (1968).
DETTBARN, W.-D.: New evidence for the role of acetylcholine in conduction. Biochim. biophys. Acta (Amst.) 41, 377—386 (1960a).
— The effect of curare on conduction in myelinated, isolated nerve fibers of the frog. Nature (Lond.) 186, 891—892 (1960b).

DETTBARN, W.-D.: The acetylcholine system in peripheral nerve. In: Symposium on cholinergic mechanism. EHRENPREIS, S. (Ed.). Ann. N. Y. Acad. Sci. 144, 483—503 (1967).
— BARTELS, E., HOSKIN, F. C. G., WELSCH, F.: Spontaneous reactivation of organophosphorus inhibited electroplax cholinesterase in relation to acetylcholine induced depolarization. Biochem. Pharmacol., 1970.
— DAVIS, F. A.: Effets of acetylcholine on axonal conduction of lobster nerve. Biochim. biophys. Acta (Amst.) 66, 397—405 (1963).
— ROSENBERG, P.: Sources of error in relating electrical and acetylcholinesterase activity. Biochem. Pharmacol. 11, 1025—1030 (1962).
— — Effects of ions on the efflux of acetylcholine from peripheral nerve. J. gen. Physiol. 50, 447—460 (1966).
— — NACHMANSOHN, D.: Restoration by a specific chemical reaction of "irreversibly" blocked axonal electrical activity. Life Sci. 3, 55—60 (1964).
ECCLES, J. C.: After-discharge from the superior cervical ganglion. J. Physiol. 84, 50P—52P (1935).
ELBERS, P. F.: The cell membrane: image and interpretation. In: Recent progress of surface science. Vol. 2. 443—503. PANKHURST, K. G. A., RIDDIFORD, A. C., Eds. New York: Academic Press 1964.
ELLMAN, G. L.: Tissue sulfhydryl groups. Arch. Biochem. 82, 70—77 (1959).
ERLANGER, J.: The initiation of impulses in axons. J. Neurophysiol. 2, 370—379 (1939).
FELDBERG, W., VARTIAINEN, A.: Further observations on the physiology and pharmacology of a sympathetic ganglion. J. Physiol. 83, 103—128 (1934).
FESSARD, A.: Some basic aspects of the activity of electric plates. Ann. N. Y. Acad. Sci. 47, 501—514 (1946).
— Les organes électriques. In: Traité de Zoologie, Vol. XIII. GRASSE, P.-P. (Ed.). Paris: Masson 1958.
FREDERIKSSON, T., TIBBLING, G.: Reversal of effects on the rat nerve-diaphragm preparation produced by methylfluorophosphoryl cholines. Biochem. Pharmacol. 2, 63—67 (1959A).
— — Demonstration of direct cholinergic receptor effects of methylfluorophosphorylcholine. Biochem. Pharmacol. 2, 286—289 (1959B).
— — Inhibition of cholinesterase with methylfluorophosphorylcholine and -carbocholine, spontaneous return of activity. Biochem. Pharmacol. 3, 184—189 (1960).
FURCHGOTT, R. F.: Receptor mechanisms. Ann. Rev. Pharmacol. 4, 21—50 (1964).
GERHART, J. C., PARDEE, A. B.: The enzymology of control by feedback inhibition. J. biol. Chem. 237, 891—896 (1962).
— — The effect of the feed-back inhibitor, CTP, on subunit interactions in aspartate transcarbamylase. Cold Spr. Harb. Symp. quant. Biol. 28, 491—496 (1963).
— SCHACHMAN, H. K.: Distinct subunits for the regulation and catalytic activity of aspartate transcarbamylase. Biochemistry 4, 1054—1062 (1965).
GOLD, A. M., FAHRNEY, D.: Sulfonyl fluorides as inhibitors of esterase. II. Formation and reactions of phenylmethanesulfonyl α-chymotrypsin. Biochemistry 3, 783—791 (1964).
GOLDSTEIN, L., LEVIN, Y., KATCHALSKI, E.: A water-insoluble polyanionic derivative of trypsin. II. Effect of the polyelectrolyte carrier on the kinetic behavior of the bound trypsin. Biochemistry 3, 1913—1920 (1964).
GREEN, D. E., MacLENNAN, D. H.: Structure and function of the mitochondrial cristael membrane. Bioscience 19, 213—222 (1969).
— PERDUE, J. F.: Membranes as expressions of repeating units. Proc. nat. Acad. Sci. (Wash.) 55, 1295—1302 (1966).
— SILMAN, I.: Structure of the mitochondrial electron transfer chain. Ann. Rev. Plant Physiol. 18, 147—178 (1967).
GRUNDFEST, H.: The mechanism of discharge of the electric organs in relation to general and comparative elctrophysiology. Progr. Biophys. 7, 1—85 (1957).
— NACHMANSOHN, D., ROTHENBERG, M. A.: Effect of diisopropyl fluorophosphate (DFP) on action potential and cholinesterase of nerve III. J. Neurophysiol. 10, 155—164 (1947).
GÜNTHER, W. H. H., MAUTNER, H. G.: Analogs of parasympathetic neuroeffectors I. Acetylselenocholine, selenocholine and related compounds. J. med. Chem. 7, 229—232 (1964).

HABER, J. E., KOSHLAND, JR., D. E.: Relation of protein subunit interactions to the molecular species observed during cooperative binding of ligands. Proc. natl. Acad. Sci. (Wash.) 58, 2087—2093 (1967).

HECHT, S., SHLAER, S., PIRENNE, M. H.: Energy, quanta, and vision. J. gen. Physiol. 25, 819—840 (1942).

HEILBRONN, E.: Hydrolysis of carboxylic acid esters of thiocholine. Acta chem. scand. 12, 1492—1505 (1958).

HESTRIN, S.: Acylation reactions mediated by purified acetylcholine esterase. J. biol. Chem. 180, 879—881 (1949).

— Acylation reactions mediated by purified acetylcholine esterase II. Biochim. biophys. Acta (Amst.) 4, 310—321 (1950).

HIGMAN, H. B., BARTELS, E.: The competitive nature of the action of acetylcholine and local anestetics. Biochim. biophys. Acta (Amst.) 54, 543—554 (1961).

— — New method for recording electrical characteristics of the monocellular electroplax. Biochim. biophys. Acta (Amst.) 57, 77—82 (1962).

— PODLESKI, T. R., BARTELS, E.: Apparent dissociation constants between carbamylcholine, d-tubocurarine and the receptor. Biochim. biophys. Acta (Amst.) 75, 187—193 (1963).

— — — Correlation of membrane potential and K flux in the electroplax of Electrophorus. Biochim. biophys. Acta (Amst.) 79, 138—150 (1964).

HILL, A. V.: The heat production of muscle. In: Molecular biology. Elementary processes of nerve conduction and muscle contraction. NACHMANSOHN, D. (Ed.). New York: Academic Press 1960.

HILLE, M. B., KOSHLAND, JR., D. E.: The environment of a reported group at the active site of chymotrypsin. J. Amer. Chem. Soc. 89, 5945—5951 (1967).

HINTERBUCHNER, L. P., NACHMANSOHN, D.: Electrical activity evoked by a specific chemical reaction. Biochim. biophys. Acta (Amst.) 44, 554—560 (1960).

HOBBIGER, F. W.: Inhibition of cholinesterases by irreversible inhibitors in vitro and in vivo. Brit. J. Pharmacol. 6, 21—30 (1951).

— Effect of nicotinehydroxamic acid methiodide on human plasma cholinesterase inhibited by organophosphates containing a dialkyl phosphato group. Brit. J. Pharmacol. 10, 356 to 362 (1955).

— Protection against the lethal effects of organophosphates by pyridine-2-aldoxime methiodide. Brit. J. Pharmacol. 12, 438—446 (1957).

— Reactivation of phosphorylated acetylcholinesterase. In: Handb. d. experiment. Pharmakologie. S. 921—988. Ergw. XV. KOELLE, G. B., Hrg. Berlin-Göttingen-Heidelberg: Springer 1963.

— PITMANN, M., SADLER, P. W.: Reactivation of phosphorylated acetocholinesterases by pyridinium aldoximes and related compounds. Biochem. J. 75, 363—372 (1960).

— SADLER, P. W.: Protection by oximes of bis-pyridinium ions against lethal diisopropyl phosphonofluoridate poisoning. Nature (Lond.) 182, 1672 (1958).

— — Protection against lethal organophosphate poisoning by quaternary pyridine aldoximes. Brit. J. Pharmacol. 14, 192—201 (1959).

HODGKIN, A. L.: The ionic basis of electrical activity in nerve and muscle. Biol. Rev. 26, 338—409 (1951).

— The conduction of the nervous impulse. Springfield, Ill.: C. Thomas Publ. 1964.

HOSKIN, F. C. G., KREMZNER, L. T., ROSENBERG, P.: Effects of some cholinesterase inhibitors on the squid giant axon. Biochem. Pharmacol. 18, 1727—1737 (1969).

— ROSENBERG, P.: Penetration of sugars, steroids, amino acids and other organic compouns into the interior of the squid giant axon. J. gen. Physiol. 49, 47—56 (1965).

— — BRZIN, M.: Reexamination of the effect of DFP on electrical and cholinesterase activity of squid giant axons. Proc. nat. Acad. Sci. (Wash.) 55, 1231—1235 (1966).

HOWARTH, J. V., KEYNES, R. D., RITCHIE, J. M.: The origin of the initial heat associated with a single impulse in mammalian non-myelinated nerve fibres. J. Physiol. (Lond.) 194, 745—793 (1968).

HUBBARD, J. I., SCHMIDT, R. F.: An electrophysiological investigation of mammalian motor nerve terminals. J. Physiol. 166, 145—167 (1963).

HUNNEUS-COX, F., SMITH, F. H.: The effects of oxidizing, reducing, and sulfhydryl reagents on the resting and action potentials of the internally perfused axon of *Loligo pealeii*. Biol. Bull. **129**, 408 (1965).

JANSEN, E. F., NUTTING, M. D. E., BALLS, A. K.: Mode of inhibition of chymotrypsin by diisopropylfluorophosphate. I. Introduction of phosphorus. J. biol. Chem. **179**, 201—204 (1949).

KARLIN, A.: The association of acetylcholinesterase and of membrane in subcellular fractionations of the electric tissue of *Electrophorus*. J. Cell Biol. **25**, 159—169 (1965).

— On the application of a plausible model of allosteric proteins to the receptor for acetylcholine. J. theor. Biol. **16**, 306—320 (1967A)).

— Chemical distinctions between acetylcholinesterase and the acetylcholine-receptor. Biochim. biophys. Acta (Amst.) **137**, 358—362 (1967B).

— Chemical modification of the active site of the acetylcholine receptor. J. gen. Physiol. **54**, 245s—264s (1969).

— BARTELS, E.: Effects of blocking sulfhydryl groups and or reducing disulfide bonds on the acetylcholine-activated permeability system of the electroplax. Biochim. biophys. Acta (Amst.) **126**, 525—535 (1966).

— WINNIK, M.: Reduction and specific alkylation of the receptor for acetylcholine. Proc. nat. Acad. Sci. (Wash.) **60**, 668—674 (1968).

KARTHA, G., BELLO, J., HARKER, D.: Tertiary structure of ribonuclease. Nature (Lond.) **213**, 862—965 (1967).

KATCHALSKY, A., CURRAN, P. F.: Nonequilibrium Thermodynamics in Biophysics. Harvard: University Press 1965.

KATZ, B., MILEDI, R.: Propagation of electrical activity in motor nerve terminals. Proc. roy. Soc. B. **161**, 453—482 (1965).

KAUFMAN, H., VRATSANOS, S. M., ERLANGER, B. F.: Photoregulation of an enzymatic process by means of a light-sensitive ligand. Science **162**, 1487—1488 (1968).

KENDREW, J. C.: Side-chain interactions in myoglobin. Brookhaven Symp. Biol. **15**, 216—228 (1962).

— Myoglobin and the structure of the proteins. Science **139**, 1259—1266 (1963).

KENNEDY, E. P.: Some recent developments in the biochemistry of membranes. In: The neurosciences. P. 271—280. QUARTON, G. C., MELNECHUK, T., SCHMITT, F. O., Eds. New York: Rockefeller Univ. Press 1967.

KEWITZ, H.: A specific antidote against lethal alkylphosphate intoxication. III. Repair of chemical lesion. Arch. Biochem. **66**, 263—270 (1957).

— NACHMANSOHN, D.: A specific antidote against lethal alkyl phosphate intoxication. IV. Effects in brain. Arch. Biochem. **66**, 271—283 (1957).

— WILSON, I. B., NACHMANSOHN, D.: A specific antidote against lethal phosphate intoxication. II. Antidotal properties. Arch. Biochem. **64**, 456—465 (1956).

KEYNES, R. D., AUBERT, X.: Energetics of the electric organ. Nature (Lond.) **203**, 261—264 (1964).

— MARTINS-FERREIRA, H.: Membrane potentials in the electroplates of the electric eel. J. Physiol. (Lond.) **119**, 315—351 (1953).

KIRSCHNER, K., EIGEN, M., BITTMAN, R., VOIGT, B.: The binding of nicotinamide-adenine dinucleotide to yeast D-glyceraldehyde-3-phosphate dehydrogenase: Temperature-jump relaxation studies on the mechanism of an allosteric enzyme. Proc. natl. Acad. Sci. (Wash.) **56**, 1661—1667 (1966).

KIRTLEY, M. E., KOSHLAND, JR., D. E.: Models for cooperative effects in proteins containing subunits. Effects of two interacting ligands. J. biol. Chem. **242**, 4192—4205 (1967).

KITZ, R. J., KREMZNER, L. T.: Conformational changes of acetylcholinesterase. Mol. Pharmacol. **4**, 104—107 (1968).

— WILSON, I. B.: Acceleration of the rate of reaction of methanesulfonyl fluoride and acetylcholinesterase by substituted ammonium ions. J. biol. Chem. **238**, 745—748 (1963).

KOELLE, G. B.: Cholinesterase and anticholinesterase agents. In: Handbuch d. experiment. Pharmakologie. KOELLE, G. B. (Hrg.). Berlin-Göttingen-Heidelberg: Springer 1963.

KORN, E. D.: Structure of biological membranes. Science **153**, 1491 (1966).

— Cell membranes: structure and synthesis. Ann. Rev. Biochem. 263—288 (1969).

KOSHLAND, D. E. JR.: The active site and enzyme action. Advanc. Enzymol. 22, 45—97 (1960).
— NEET, K. E.: The catalytic and regulatory properties of enzymes. Ann. Rev. Biochem. 37, 359—410 (1968).
— STRUMEYER, D. H., RAY, W. J., JR.: Amino acids involved in the action of chymotrypsin. In: Enzyme models and enzyme structure. Brookhaven Symp. Biol. 15, 101—133 (1962).
KRUPKA, R. M.: Hydrolysis of neutral substrates by acetylcholinesterase. Biochemistry 5, 1983—1988 (1966).
— Chemical structure and function of the active center of acetylcholinesterase. Biochemistry 5, 1988—1997 (1966).
LAWLER, H. C.: Turnover time of acetylcholinesterase. J. biol. Chem. 236, 2296—2301 (1961).
LEUZINGER, W., BAKER, A. L.: Acetylcholinesterase. I. Large scale purification homogeneity, amino acid analysis. Proc. nat. Acad. Sci. (Wash.) 57, 446—451 (1967).
— — CAUVIN, E.: Acetylcholinesterase. II. Crystallization, absorption spectra, isoionic point. Proc. nat. Acad. Sci. (Wash.) 59, 620—623 (1968).
— GOLDBERG, M., CAUVIN, E.: Molecular properties of acetylcholinesterase. J. molec. Biol. 40, 217—225 (1969).
LEVIN, Y., PECHT, M., GOLDSTEIN, L., KATCHALSKI, E.: A water-insoluble polyanionic derivative of trypsin. I. Preparation and properties. Biochemistry 3, 1905—1913 (1964).
LEVITZKI, A., KOSHLAND, JR., D. E.: Negative cooperativity in regulatory enzymes. Proc. natl. Acad. Sci. (Wash.) 62, 1121—1128 (1969).
LEWIS, P. R., SHUTE, C. C.: Electron microscope distributions of cholinesterase in cholinergic neurones. J. Anat. (Lond.) 99, 941 (1965).
LINDERSTROM-LANG, K. U.: Proteins and enzymes. Lane Medical Lectures, p. 8. Stanford: Stanford Univ. Press 1952.
LOEWENSTEIN, W. R.: Biological membranes: recent progress. Ann. N. Y. Acad. Sci. 137, 403—1048 (1966).
LUNDIN, S. J.: Purification of a cholinesterase from the body muscle of plaice (Pleuronectes platessa). Acta chem. scand. 21, 2663—2668 (1967).
— HELLSTROEM, B.: The ultrastructural localization of a cholinesterase in the body muscle of plaice (Pleuronectes platessa). Z. Zellforsch. 85, 264—270 (1968).
LUZZATI, V., REISS-HUSSON, F., RIVAS, E., BULIK-KRZYWICKI, T.: Structure and polymorphism in lipid-water systems, and their possible biological implications. Ann. N. Y. Acad. Sci. 137, 409—420 (1966).
MARTIN, R., ROSENBERG, P.: Fine structural alterations associated with venom action on squid giant nerve fibers. J. Cell Biol. 36, 341—353 (1968).
MASLAND, R. L., WIGTON, R. S.: Nerve activity accompanying fasciculation produced by prostigmine. J. Neurophysiol. 3, 269—275 (1940).
MAUTNER, H. G.: The molecular basis of drug action. Pharmacol. Rev. 19, 107—144 1(1967).
— BARTELS, E., WEBB, G. D.: Sulfur and selenium isologs of acetylcholine and choline. IV. Activity in the electroplax preparation. Biochem. Pharmacol. 15, 187—193 (1966).
— CHU, S. H., GUNTHER, W. H. H.: The aminolysis of thioacyl and selenoacyl analogs. J. Amer. Chem. Soc. 85, 3458—3492 (1963).
— GÜNTHER, W. H. H.: The relative reactivity of thioacyl and selenoacyl analogs. J. Amer. chem. Soc. 83, 3342—3343 (1961).
MAZUR, A.: An enzyme in animal tissues capable of hydrolyzing the phosphorusfluorine bond of alkyl fluorophosphates. J. biol. Chem. 164, 271—289 (1946).
— Membrane Proteins. Proceedings of a symposium sponsored by the New York Heart Association. Boston: Little, Brown & Co. 1969.
MENGLE, D. C., CASIDA, J. E.: Inhibition and recovery of brain cholinesterase activity in house flies poisoned with organophosphate and carbamate compounds. J. Econ. Entomol. 51, 750—755 (1958).
— O'BRIEN, R. D.: The spontaneous and induced recovery of fly-brain cholinesterase after inhibition by organophosphates. Biochemistry J. 75, 201—207 (1960).
MEUNIER, J.-C., CHANGEUX, J.-P.: On the irreversible binding of p-(trimethylammonium)-benzenediazonium fluoroborate (TDF) to acetylcholinesterase from electrogenic tissue. FEBS Letters 2, 224—226 (1969).

Monod, J., Changeux, J.-P., Jacob, F.: Allosteric proteins and cellular control systems. J. molec. Biol. **6**, 306—329 (1963).
— Wyman, J., Changeux, J.-P.: On the nature of allosteric transitions. A plausible model. J. molec. Biol. **12**, 88—118 (1965).
Mounter, L. A.: Metabolism of organophosphorus anticholinesterase agents. In: Handb. d. experiment. Pharmakologie. Erg. Hrg.: Koelle, G. B. P. 486—504. Berlin-Göttingen-Heidelberg: Springer 1963.
Muirhead, H., Perutz, M. F.: Structure of hemoglobin. Nature (Lond.) **199**, 633—638 (1963) A
— — Structure of reduced human hemoglobin. Cold Spr. Harb. Symp. quant. Biol. **28**, 451—459 (1963 B).
Myers, D. K., Kemp, A., Jr.: Inhibition of esterases by the fluorides of organic acids. Nature (Lond.) **173**, 33 (1954).
Nachmansohn, D.: Metabolism and function of the nerve cell. Harvey Lect. (1953/1954) **49**, 57—99 (1955).
— Chemical and molecular basis of nerve activity. p. 235. New York: Academic Press Inc. 1959.
— Actions on axons and the evidence for the role of acetylcholine in axonal conduction. In: Handb. d. experiment. Pharmakologie (Hrg. Koelle, G.). Erg. Bd. **15**, S. 701—740. Berlin-Göttingen-Heidelberg: Springer 1963.
— Choline acetylase. In: Handbuch d. experiment. Pharmakologie. Ergw. **15**, p. 40—54. Koelle, G., Hrg. Berlin-Göttingen-Heidelberg: Springer 1963.
— Chemical control of ion movements across conducting membranes. In: Symp. on new perspectives in biology. BBA Library 4, 176—204. Ed. by Sela, M. Amsterdam: Elsevier 1964.
— Chemical forces controlling permeability changes of excitable membranes during electrical activity. In: Nerve as a tissue. Ed. by Rodahl, K. New York: McGraw Hill Book Co., Inc. (1966 A).
— Chemical control of the permeability cycle in excitable membranes during electrical activity. In: Biological membranes: Recent progress. Ed. by Loewenstein, W. Ann. N. Y. Acad. Sci. **137**, 877—900 (1966 B).
— Properties of the acetylcholine receptor protein analyzed on the excitable membrane of the monocellular electroplax preparation. In: Current aspects of biochemical energetics. Kaplan, N. O., Kennedy, E. P. (Eds.). New York: Academic Press Inc. (1966 C).
— La membrane excitable. Macromolecules liées à la bioélectrogenèse. Bull. Soc. Chim. biol. (Paris) **10**, 1177—1189 (1967).
— Proteins in bioelectricity. Proc. nat. Acad. Sci. (Wash.) **61**, 1034—1041 (1968).
— Proteins of excitable membranes. J. gen. Physiol. **54**, 187s—224s (1969).
— Lederer, E.: Sur la biochemie de la cholinesterase. Bull. Soc. Chim. biol. (Paris) **21**, 797—808 (1939).
— Machado, A. L.: The formation of acetylcholine. A new enzyme "choline acetylase". J. Neurophysiol. **6**, 397—404 (1943).
— Wilson, I. B.: The enzymic hydrolysis and synthesis of acetylcholine. Advanc. Enzymol. **12**, 259—339 (1951).
Namba, T., Hiraki, K.: PAM (pyridine-2-aldoxime methiodide) therapy for alkylphosphate poisoning. J. Amer. med. Ass. **166**, 1834 (1958).
Neubert, D., Schaefer, J., Kewitz, H.: Reaktivierung der Acetylcholinesterase durch körpereigene Stoffe. Naturwissenschaften **45**, 290 (1958).
Nistratova, S. H., Turpaev, T. M.: The reaction of acetylcholine with choline receptors in tissue homogenates. Biochemistry **24**, 155—160 (1959).
O'Brien, R. D.: The inhibition of cholinesterase and succinoxidase by malathion and its isomer. J. Econ. Entomol. **49**, 484—490 (1956).
— Toxic phosphorus esters. Chemistry, metabolism and biological effects. New York: Academic Press Inc. 1960.
Oosterbaan, R. A., Warringa, M. G. P. J., Jansz, H. S., Berends, F., Cohen, J. A.: The reaction of pseudocholinesterase with diisopropyl-phosphorofluoridate (DFP). In: Proc. Intern. Congr. Biochem., 4th Congress, Vienna 1958, Abstr. 4—12 (1959).

OVERTON, E.: Beiträge zur allgemeinen Muskel- und Nervenphysiologie. Pflügers Arch. ges. Physiol. **92**, 346—386 (1902).

PALADE, G. E.: The organization of living matter. In: The scientific endeavor: Centenn. celebr. of the Nat. Acad. Sci. New York: Rockefeller Inst. Press 1963.

PAULING, L., COREY, B. B., BRANSON, H. R.: The structure of proteins. Two hydrogen-bonded configurations of the polypeptide chain. Proc. Nat. Acad. Sci. (Wash.) **37**, 205—211 (1951).

PERUTZ, M. F.: Proteins and nucleic acids structure and function. Amsterdam: Elsevier 1962.

— X-ray analysis of hemoglobin. Science **140**, 863—869 (1963).

— BOLTON, W., DIAMOND, R., MUIRHEAD, H., WATSON, H. C.: Structure of hemoglobin. Nature (Lond.) **203**, 687—690 (1964).

PODLESKI, T. R.: Effects of quaternary ammonium ions on the membrane potential of electroplax. Ph. D. Thesis. New York: Columbia-University (1966).

— Distinction between the active sites of acetylcholine-receptor and -esterase. Proc. nat. Acad. Sci. (Wash.) **58**, 268—273 (1967).

— Molecular forces acting between ammonium ions and acetylcholine receptor protein. Biochem. Pharmacol. **18**, 211—226 (1969).

— MEUNIER, J. C., CHANGEUX, J. P.: Campared effects of dithiothreitol on the interaction of an affinity-labeling reagent with acetylcholin-esterase and the excitable membrane of the electroplax. Proc. natl. Acad. Sci. (Wash.) **63**, 1239—1246 (1969).

— NACHMANSOHN, D.: Similarities between active sites of acetylcholine receptor and acetylcholinesterase with quinolinium ions. Proc. nat. Acad. Sci. (Wash.) **56**, 1034—1039 (1966).

POHLE, W., MATTHIES, H.: Über den Mechanismus der Acetylcholinwirkung an der Herzmuskulatur. Arch. Exp. Pathol. Pharmakol. **236**, 253 (1959).

POZIOMEK, E. J., HACKLEY, B. E., JR., STEINBERG, G. M.: Pyridinium aldoximes. J. Org. Chem. **23**, 714 (1958).

— KRAMER, D D. N., FROMM, B. W., MOSHER, W. A.: Observation of the geometrical isomerism of formyl-l-methylpyridinium iodide oximes; carbinolamine intermediates. J. Org. Chem. **26**, 423—427 (1961).

— — MOSHER, W. A., MICHEL, H. O.: Configurational analysis of 4-formyl-l-methylpyridinium iodide oximes and its relationship to a molecular complimentary theory on the reactivation of inhibited acetylcholinesterase. J. Amer. chem. Soc. **83**, 3916—3917 (1961).

PRINCE, A. K.: Spectrophotometric study of the acetylcholinesterase-catalyzed hydrolysis of 1-methyl-acetoxyquinolinium iodides. Arch. Biochem. **113**, 195—204 (1964).

— Properties of choline acetyltransferase isolated from squid ganglia. Proc. natl. Acad. Sci. (Wash.) **57**, 1117—1122 (1967).

RACKER, E.: Mechanisms in bioenergetics, p. 259. New York: Academic Press Inc. 1965.

— Resolution and reconstitution of the inner mitochondrial membrane. Federation Proc. **26**, 1335—1340 (1967).

— Resolution and reconstitution of a mammalian membrane. J. gen. Physiol. **54**, 38s—49s (1969).

RIKER, W. F.: Excitatory and anti-curare properties of acetylcholine and related quaternary ammonium compounds at the neuromuscular junction. Pharmaciol. Rev. **5**, 1—86 (1953).

RIKER, W. F., JR., WERNER, G., ROBERTS, J., KUPERMAN, A. S.: The presynaptic element in neuromuscular transmission. Ann. N. Y. Acad. Sci. **81**, 328—344 (1959).

RITCHIE, J. M.: The action of acetylcholine and related drugs on mammalian non-myelinated nerve fibers. Biochem. Pharmacol. **12** (S), 3 (1963).

ROBERTSON, J. D.: The molecular biology of cell membranes. Symp. molec. Biol. Ed. by NACHMANSOHN, D. New York: Academic Press, Inc. 1960.

— The molecular structure and contact relationships of cell membranes. In: Progress in biophysics. Eds.: KATZ, B., BUTLER, J. A. V. New York: Pergamon Press 1960.

ROSENBERG, P.: *In vivo* reactivation by PAM of brain cholinesterase inhibited by paraoxon Biochem. Pharmacol. **3**, 212—219 (1960).

— Effects of venoms on the squid giant axon. Toxicon **3**, 125—131 (1965).

— Use of venoms in studies on nerve excitation. Mem. Inst. Butantan **33**, 477—508 (1966).

— CONDREA, E.: Maintenance of axonal conduction and membrane permeability in presence of extensive phospholipid splitting. Biochem. Pharmacol. **17**, 2033—2044 (1968).

ROSENBERG, P., DETTBARN, W.-D.: Use of venoms in testing for essentiality of cholinesterase in conduction. In: Animal toxins Toxicon 4 (4), 296 (1967).
— HOSKIN, F. C. G.: Demonstration of increased permeability as a factor responsible for the effect of acetylcholine on the electrical activity of venom treated axons. J. gen. Physiol. **46**, 1065—1073 (1963).
— MAUTNER, H. G., NACHMANSOHN, D.: Similarity of effects of oxygen, sulfur, and selenium isologs on theacetylcholine receptor in excitable membranes on junctions and axons. Proc. nat. Acad. Sci. (Wash.) **55**, 835—838 (1966).
— NG, K. Y.: Factors in venoms lesding to block of axonal conduction by curare. Biochim. biophys. Acta (Amst.) **75**, 116—128 (1963).
— PODLESKI, T. R.: Ability of venoms to render squid axons sensitive to curare and acetylcholine. Biochim. biophys. Acta (Amst.) **75**, 104—115 (1963).
ROTHENBERG, M. A., NACHMANSOHN, D.: Studies on cholinesterase. III. Purification of the enzyme from electric tissue by fractional ammonium sulfate precipitation. J. biol. Chem. **168**, 223—231 (1947).
— SPRINSON, D. B., NACHMANSOHN, D.: Site of action of acetylcholine. J. Neurophysiol. **11**, 111—116 (1948).
ROSSI-FANELLI, A., ANTONINI, E., CAPUTO, A.: Hemoglobin and myoglobin. Advan. Protein Chem. **19**, 73—222 (1964).
ROTHFIELD, L., FINKELSTEIN, A.: Membrane biochemistry. Ann. Rev. Biochem. **37**, 463—496 (1968).
SCHAFFER, N. K., MAY, C. S., JR., SUMMERSON, W. H.: Serine phosphoric acid from diisopropylphosphoryl chymotrypsin. J. biol. Chem. **202**, 67—76 (1953).
SCHLAEPFER, W. W., TORACK, R. M.: The ultrastructural localization of cholinesterase activity in the sciatic nerve of the rat. J. Histochem. Cytochem. **14**, 369—378 (1966).
SCHOELLMANN, G., SHAW, E.: Direct evidence for the presence of histidine in the active center of chymotrypsin. Biochemistry **2**, 252—255 (1963).
SCHOFFENIELS, E.: An isolated single electroplax preparation. II. Improved preparation for studying ion flux. Biochim. biophys. Acta (Amst.) **26**, 585—596 (1957).
— Les bases physiques et chimiques des potentiels bioélectriques chez *Electrophorus electricus* L. Thèse d'agrégation. Univ. de Liege, Liege (1959).
— NACHMANSOHN, D.: An isolated single electroplax preparation I. New data on the effect of acetylcholine and related compounds. Biochim. biophys. Acta (Amst.) **26**, 1—15 (1957).
SCOTT, K. A., MAUTNER, H. G.: Analogs of parasymathetic neuroeffectors. II. Comparative pharmacological studies of acetylcholine, its thio and seleno analogs and their hydrolysis products. Biochem. Pharmacol. **13**, 907—920 (1964).
SHAW, E., MARES-GUIA, M., COHEN, W.: Evidence for an active-center histidine in trypsin through use of a specific reagent 1-chloro-3-tosylamido-7-amino-2-heptanone, the chloromethyl-ketone derived from N-tosyl-1-lysine. Biochemistry **4**, 2219 (1965).
SHEFTER, E., KENNARD, O.: Crystal and molecular structure of acetylselenocholine iodide. Science **153**, 1389—1390 (1966).
— MAUTNER, H. G.: The crystal and molecular structure of 2,4-dithiouracil. J. Amer. Chem. Soc. **89**, 1249—1253 (1967).
— — Acetylcholine and its thiolester and selenolester analogs: Conformation, electron distribution, and biological activity. Proc. natl. Acad. Sci. (Wash.) **63**, 1253—1260 (1969).
SILMAN, I.: Covalent attachment of depolarizing groups to the acetylcholine receptor. In: Colloquium E. Molecular Neurology, 6th FEBS Meeting, Madrid 1969. New York: Academic Press 1970.
— KARLIN, A.: Effect of local pH changes caused by substrate hydrolysis on the activity of membrane-bound acetylcholinesterase. Proc. nat. Acad. Sci. (Wash.) **58**, 1664 (1967).
— — Acetylcholine receptor: covalent attachment of depolarizing groups at the active site. Science **164**, 1420—1421 (1969).
SINGER, S. J.: Covalent labeling of active sites. Advanc. Protein Chem. **22**, 1—54 (1967).
— DOOLITTLE, R. F.: Antibody active sites and immunoglobin molecules. Science **153**, 13(1966).
SJÖSTRAND, F. D.: A comparison of plasma membranes, cytomembranes and mitochondrial membrane elements with respect to ultrastructural features. J. Ultrastruct. Res. **9**, 561 to 580 (1963).

SJÖSTRAND, F. D., BARAJAS, L.: Effect of modifications in conformation of protein molecules on structure of mitochondrial membranes. J. Ultrastruct. Res. **25**, 121—155 (1968).

SMALLMAN, B. N., FISHER, R. W.: Effect of anticholinesterases on acetylcholine levels in insects. Canad. J. Biochem. Physiol. **36**, 575—586 (1958).

SMITH, H. M.: Effects on sulfhydryl-blockade on axonal function. J. cell. comp. Physiol. **51** 161—171 (1958).

STADTMAN, E. R.: Allosteric regulation of enzyme activity. Advanc. Enzymol. **28**, 41—154 (1966).

SUND, H., WEBER, K.: The quaternary structure of proteins. Angew. Chem. **5**, 231—245 (1966).

TAMMELIN, L.-E.: Methyl-fluoro-phosphorylcholines. Acta chem. scand. **11**, 859—864 (1957).

— Organophosphorylcholines and cholinesterases. Arkiv Kemi **12**, 287—298 (1958).

TASAKI, I.: Nerve Excitation. Springfield, Ill.: Charles C. Thomas 1968.

TEORELL, T.: Transport prosesses and electrical phenomena in ionic membranes. Progr. Biophys. **3**, 305—369 (1953).

TORACK, R. M., BARRNETT, R. J.: Fine structural localization of cholinesterase activity in the brain stem. Exp. Neurol. **6**, 224—244 (1962).

TRAYLOR, P. S., SINGER, S. J.: The preparation and properties of some tritiated diazonium salts and related compounds. Biochemistry **6**, 881—886 (1967).

VAN ASPEREN, K.: Mode of action of organophosphorus insecticides. Nature (Lond.) **181**, 355—356 (1958).

VAN HOLDE, K. E., BALDWIN, R. L.: Rapid attainment of sedimentation equilibrium. J. physiol. Chem. **62**, 734—743 (1958).

WAGNER-JAUREGG, T., HACKLEY, B. E., JR.: Model reactions of phosphorus-containing enzyme inactivators. III. Interaction of imidazole, pyridine and some of their derivatives with dialkyl halogeno-phosphates. J. Amer. chem. Soc. **75**, 2125 (1953).

WALD, G.: Molecular basis of visual excitation. Science **162**, 230 (1968).

WALSH, R. R., DEAL, S. E.: Reversible conduction block produced by lipid insoluble quaternary ammonium ions in cetyltrimethylammonium bromide treated nerves. Amer. J. Physiol. **197**, 547—550 (1959).

WEBB, G. D.: Affinity of benzoquinonium and ambenonium derivatives for the acetylcholine receptor, tested on the electroplax, and for acetylcholinesterase in solution. Biochim. biophys. Acta (Amst.) **102**, 172—184 (1965).

— MAUTNER, H. G.: Sulfur and selenium compounds related to acetylcholine and choline. VI. Effects of homocholine derivatives on the electroplax preparation. Biochem. Pharmacol. **15**, 2105—2111 (1966).

WERNER, G., KUPERMAN, A. S.: Actions at the neuromuscular junction . In: Handb. d. exp. Pharmac. Ergw. XV. Hrg. KOELLE, G. B.pp. 570—678. Berlin-Göttingen-Heidelberg: Springer 1963.

WESCOE, W. C., RIKER, W. F., JR.: The pharmacology of anti-curare agents. Ann. N. Y. Acad. Sci. **54**, 438—455 (1951).

WHITTACKER, V. P.: The application of subcellular fractionation techniques to the study of brain function. Progr. Biophys. Mol. Biol. **15**, 41—96 (1965).

WILSON, I. B.: Acetylcholinesterase. XI. Reversibility of tetraethyl pyrophosphate inhibition. J. biol. Chem. **190**, 111—117 (1951).

— Acetylcholinesterase. XII. Further studies of binding forces. J. biol. Chem. **197**, 215—225 (1952).

— Promotion of acetylcholinesterase activity by the anionic site. Discuss. Faraday Soc. **20**, 119—125 (1955).

— Acetylcholinesterase. In: The enzymes. Vol. 4, 501—520. BOYER, P. D., LARDY, H., MYRBAECK, K., Eds. New York: Academic Press Inc. 1960.

— BERGMANN, F., NACHMANSOHN, D.: Acetylcholinesterase. X. Mechanism of the catalysis of acylation reactions. J. biol. Chem. **186**, 781—790 (1950).

— CABIB, E.: Acetylcholinesterase. Enthalpies and entropies of activation. J. Amer. chem. Soc. **78**, 202—207 (1956).

— GINSBURG, S.: A powerful reactivator of alkylphosphate-inhibited acetylcholinesterase. Biochim. biophys. Acta (Amst.) **18**, 168—170 (1955).

WILSON, I. B., GINSBURG, S.: Reactivation of alkylphosphate inhibited acetylcholinesterase by bis quaternary derivatives of 2-PAM and 4-PAM. Biochem. Pharmacol. 1, 200—206 (1958).
— HARRISON, M. A.: Turnover number of acetylcholinesterase. J. biol. Chem. 236, 2292—2295 (1961).
— QUAN, C.: Acetylcholinesterase studies on molecular complementariness. Arch. Biochem. 73, 131—143 (1958).
WOFSY, L., BING, D. H., KUMURA, J., PARKER, D. C.: Affinity labeling of rabbit antisaccharide antibodies. Biochemistry 6, 1981—1988 (1967).
— METZGER, H., SINGER, S. J.: Affinity labeling: A general method for labeling sites of antibody and enzyme molecules. Biochemistry 1, 1031—1038 (1962).
WYCKOFF, H. W., HARDMAN, K. D., ALLEWELL, N. M., INAGAMI, T., JOHNSON, L. N., RICHARDS, F. M.: The structure of ribonuclease-S at 3.5 A resolution. J. biol. Chem. 242, 3984—3988 (1967).
— — — — TSERNOGLOU, D., JOHNSON, L. N., RICHARDS, F. M.: The structure of ribonuclease-S at 6 A resolution. J. biol. Chem. 242, 3749—3753 (1967).
WYMAN, JR., J.: Linked function and reciprocal effects in hemoglobin: A second look. Advan. Protein Chem. 19, 233—286 (1964).
YPHANTIS, D. A.: Equilibrium ultracentrifugation of dilute solutions. Biochemistry 3, 297—317 (1964).

Chapter 3

Transmission Action on
Synaptic Neuronal Receptor Membranes

By

L. Tauc, Paris (France)

With 15 Figures

Contents

General Considerations

Following the historical work on the inhibitory synapse of the motoneurone in the vertebrate spinal cord (Brock et al., 1952), the neurohumoral theory of synaptic transmission was generally accepted as a satisfactory description of the mechanism by which informations is transmitted to a neuron. It is now known that some neurons can also interact by a purely electrical process, but the electrically working junction offers less functional flexibility than the chemical synapse.

The short-term action of each of these mechanisms is to act at the spike trigger zone of the postsynaptic neuron, facilitating or inhibiting the initiation of a spike. This final action is accomplished in both cases through a similar electrical mechanism; in both cases, the extrajunctional membrane represents a return pathway for an electrical current the origin of which, however, is different. In the case of electrical transmission, the sink is presynaptically formed at the side of discharge of the presynaptic spike, the flow of current between the presynaptic and postsynaptic neuron being facilitated by a special low-resistance contact which is felt by the postsynaptic neuron as the virtual sink. The duration depends on the duration of the spike and on biophysical membrane constants of the involved membranes.

In chemically transmitting synapses, the origin of current flow is represented by the subsynaptic membrane in which the chemical or transmitter liberated by the presynaptic neuron has combined with a specific receptor molecule. This combination of the two molecules, transmitter and receptor, is followed by a change in membrane permeability specific to given ions, thus creating a current flow between the subsynaptic and extrasynaptic membranes of the postsynaptic neuron. Depending on the ions involved and the differences in their intra- and extracellular concentration, the current goes in one or the other direction, thus depolarizing and exciting or hyperpolarizing and inhibiting the extrasynaptic membrane of the postsynaptic cell (Fig. 1). The efficacy of the synaptic action will depend on the quantity of the transmitter which will reach the postsynaptic membrane, on the intensity of the inactivating mechanism, on the quantity and sensitivity of the specific postsynaptic receptors, on the ionic content of the post-

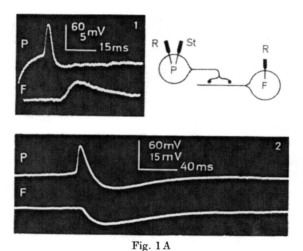

Fig. 1 A

Fig. 1. Excitatory and inhibitory activity and schema of the mechanism. A) Recording from presynaptic (P) and postsynaptic (F) neurons showing, respectively, the spikes and the excitatory and inhibitory (2) postsynaptic potentials. The delay between the spike and PSP is mainly due to conduction time (from Tauc, 1959). B) Schema of the synaptic action and of the equivalent electrical circuitry. The spike arriving at the nerve ending (pre) liberates the transmitter (black circles) which crosses the synaptic cleft and some of it combines with receptor molecules situated on the postsynaptic membrane. The liberation of the transmitter cannot occur in the absence of Ca^{++}. The transmitter is inactivated by diffusion, by enzymatic action (En) or/and taken up by the presynaptic structure. The combination receptor-transmitter acts on the permeability of the postsynaptic membrane and is followed by a current generation inbetween the subsynaptic and extrasynaptic membranes. Depending on which of the ions are involved in the permeability change, the event can be recorded across the membrane between the interior of the cell (B) and the medium (A) as a excitatory (EPSP) or inhibitory (IPSP) postsynaptic potential. The duration of the synaptic current (Is) might be shorter (upper left insert) than the potential difference measured (Vs) but it is not necessarily so.

In the equivalent electrical circuit presented, the extrasynaptic membrane made up by the resistance Rm and capacity Cm and generator of the resting potential Em, forms a return circuitry for the subsynaptic membrane which, when activated by the combination transmitter-receptor (closing the switch) introduces a generator Es in series with a small resistor Rs. The value of Es represents the equilibrium potential for the ions involved in the transient permeability change.

synaptic neuron, and on the biophysical constants of the postsynaptic neuronal membrane. The geometry of the postsynaptic neuron has also to be taken into account and especially the respective locations of the activated synapse and of the spike trigger zone. In close proximity the synaptic action will be felt more intensely, but the conductance increase on subsynaptic membrane might then represent an additional charge on the spike generating mechanism and in these conditions even

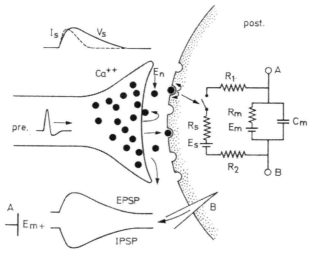

Fig. 1 B

a depolarizing action might have an inhibitory effect. The synaptic action will be transient because the liberation of the transmitter occurs in definite amounts following arrival of the presynaptic spike at the terminal and it is rapidly removed from the extracellular space through diffusion, enzymatic destruction and/or is taken back by the presynaptic terminal ("uptake"). The combination of the transmitter with the receptor is reversible.

The potential changes produced by the synaptic action can be recorded in the postsynaptic neuron as the so-called postsynaptic potential (PSP), either excitatory (EPSP) or inhibitory (IPSP) (Fig. 1). The use of capillary microelectrodes which can be introduced into the neuron without affecting its properties permits intracellular recording of the PSP and offers to neurophysiology and pharmacology a precious quantitative tool for studying the synaptic mechanism. Indeed, the modifications in form, size and duration of a PSP, when the neuronal environment is modified by the addition of drugs and by changes in ionic composition, are highly significant, if one has taken the necessary experimental precautions. The experiment should be done in a neuronal monosynaptic system with the exclusion of interneurons; preferentially on a unitary synapse and under the control of secondary modifications of membrane properties of the postsynaptic neuron. Such conditions in fact could only be realized in a few neuronal structures.

In most cases it was shown that the general pattern of the chemical transmission mechanism in between neurons is analogous to that which occurs at vertebrate neuromuscular juntions, with variations of the chemical nature of transmitters, receptors, enzymes and inactivating mechanisms. Also, the anatomical picture which electron microscopy attributes to a neuro-neuronal synapse is simpler: it reveals an active synaptic zone with thickening of postsynaptic and presynaptic membranes, separated by an enlarged synaptic cleft, rather than the more complex structure observed on skeletal muscle endplate. The presynaptic vesicles, however, which, are probable transmitter storage sites, are generally present in both endplate and synapse.

The functional situation in most synapses is also similar; as in neuromuscular junction, no regenerative reaction follows the transmitter action on the membrane receptors (Katz, 1962). The ionic conductance change is considered to be solely responsible for the generation of PSP; the ions present which can contribute significantly to this conductance change are sodium, potassium and chloride. Since the conductance is independent of the level of membrane potential, a possible way to gain insight into the ionic mechanism involved in generation of PSP is the measure of the equilibrium potential, i.e. the membrane potential at which the PSP gives zero potential change. This can be achieved by passing a steady current through an intracellular electrode, thus artificially displacing the membrane potential to any desired level. The equilibrium potential is virtually at zero potential for the EPSP in mammalian spinal motoneuron (Coombs et al., 1955c) and in the squid giant synapse; this suggests that more than one ion is involved in the transmission process. The equilibrium potential for IPSP in the spinal motoneuron is −80 mV, it is a value which falls between the equilibrium potentials for chloride and potassium ions. The potential for the cholinergic EPSP in gastropods is − 60 mV i.e., below that of the potassium equilibrium potential (Fig. 2 A).

More information can be obtained when the displacement of the equilibrium potential is studied after electrophoretical injection of main ions involved or when ionic concentration in the fluid surrounding the neuron is changed selectively. For example, injection of different ions in the cat motoneuron and the consequent displacement of the equilibrium potential for the IPSP (Eccles, 1964) pointed to chloride and in lesser degree to potassium as the responsible ions. In gastropod cells, some IPSP are due to chloride permeability changes, others are potassium dependent (Fig. 2B).

Owing to the capacitative component of the extrasynaptic membrane, the duration of these PSP may be longer than the duration of the underlying conductance change. This is well known for the motor end plate of the striated muscle fiber (Takeuchi and Takeuchi, 1959) and was also demonstrated on several synapses, especially on the excitatory input to the spinal motoneuron (Curtis and Eccles, 1959) and in the frog sympathetic ganglion cell (Nishi and Koketsu, 1960). In some gastropod neurons, however, the EPSP have much longer duration than the membrane time constant, pointing to a prolonged action of the transmitter (Tauc, 1958).

The most common synaptic action is represented by one transmitter acting on a homogenous group of receptors; the induced permeability change involves only one ion, or, if several ions are involved, their respective permeabilities are affected simultaneously or nearly simultaneously. In such a case, the recorded PSP is monophasic, either purely excitatory or purely inhibitory. Recently, in some neuronal structures more complex transmissions processes have been found, to which analogy with events on neuromuscular endplate cannot be applied. These processes involve phenomena which are produced on the postsynaptic membrane by the combination of a single transmitter substance with two different receptors. Such events, mostly conveyed by acetylcholine (ACh) have been described in sympathetic ganglion cells, in Renshaw cells and in ganglionic cells of gastropod

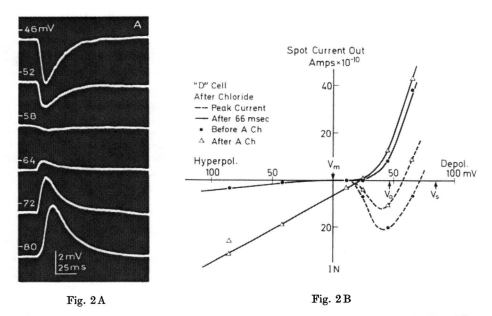

Fig. 2 A   Fig. 2 B

Fig. 2. A) Amplitude and polarity of the inhibitory postsynaptic potential in an H cell at different levels of the membrane potential. Reversal potential for the IPSP is close to —60 mV (From TAUC, 1958). B) Voltage current relationship as studied under voltage clamp (i.e. measuring the current requirement to bring the membrane potential to different but steady potential levels) in a piece of somatic membrane, before and after local electrophoretical injection of ACh. *Helix* ganglion cell. Peak and late currents are represented. Previous injection of chloride has displaced the zero current under ACh about 10 mV to the left (from FRANK and TAUC, 1964)

mollusca. The biphasic effects are usually of much longer duration than classical synaptic effects; the duration of unitary potential is expressed in seconds or even in minutes. Although these events are observed in a limited number of neurons, they might represent an important central mechanism and to this effect they are described in more detail.

Sympathetic Ganglion Cells

The relative simplicity of vertebrate sympathetic ganglia has permitted to neurophysiology and pharmacology an appreciable gain in the knowledge of inside events even before intracellular penetration of composing nerve cells was possible. Frog and rabbit sympathetic ganglia have been studied most.

In the frog sympathetic ganglion the external collagen matrix contains, along with nerve cells, other connective tissue cells and pigment-forming or chromaffin cells. The sympathetic nerve cell is usually unipolar, giving rise to a single, thick postganglionic axon. By the so-called "pericellular nest" the preganglionic fiber surrounds the postganglionic cell body in a sort of spiral formation, forming synaptic contact through widening or varicosities reminiscent of "boutons terminaux" (Fedorov and Matwejewa, 1935; Huber, 1900; Majorow, 1957, 1960; Sjöqvist, 1963). Electron microscopy shows that the postganglionic neuron is covered by a Schwann cell, except in the region of synaptic contacts where the presynaptic fiber contains several types of vesicles; some one clear, 300 to 500 A in diameter, possibly containing ACh (Pipa et al., 1959; Watanabe and Bullock, 1960), and others with dark cores, 800 to 1,500 A in diameter, have been considered as catecholamine storage sites by some (Bloom et al., 1961; Pick, 1963; Uchizono, 1964), but not by others (Watanabe and Bullock, 1960).

Mammalian sympathetic ganglion cells, unlike those of the frog, are multipolar with extensively branching dendrites and axons (Castro, 1932). Chromaffin cells are also present. The ganglion is supplied by preganglionic B and C fibers. Each cell usually receives contacts from several presynaptic fibers; in the area of the synapses, electron microscopy has revealed the widenings of preganglionic fibers, forming the well-known varicosities (Elfvin, 1963). The contacts are usually axodendritic, but axosomatic contacts also have been described (Elfvin, 1963; Taxi, 1965). Dendro-dendritic and dendro-somatic contacts also have been observed (Elfvin, 1963); possibly they can insure a direct functional interaction between ganglion cells, but this proposition lacks physiological evidence. As in the frog, mammalian presynaptic fibers contain clear vesicles of 300 to 500 A and larger vesicles af 700 to 1,000 A with a dense central region, possible storage sites for ACh and a catecholamine.

Most of the information concerning the synaptic transmission in sympathetic ganglia has been obtained by recording ganglionic potentials from the surface of the ganglia (with one electrode on the surface of the ganglion and another on a distal portion of the isolated postganglionic nerve trunk). The sucrose gap method was also used with success (Kosterlitz and al., 1968). In the presence of a curarizing drug such as d-tubocurarine or dihydro-β-erythroidine which prevents orthodromic initiation of ganglionic spikes, a characteristic sequence in potential changes is observed when a synchronized preganglionic volley reaches the ganglion. The sequence involves a fast surfaced negative "N-potential", a slow-surfaced positive "P-potential", and a still slower-surfaced negative wave, the "LN-potential"; N for negative, P for positive and L for late [(Eccles and Libet, 1961; Libet, 1964, 1967; (Fig. 3A)]. The P and LN waves are distinctly larger after a tetanic train of preganglionic volleys than after a single volley. The durations of the waves are different — about 100 msec for the N wave, several hundred milliseconds for the P wave, and several seconds for the LN wave. They all seem to be electrical expressions of synaptic events: all the waves decrease rapidly with distance along the postsynaptic nerve and are absent on the preganglionic site; all are depressed or abolished when the ganglion is bathed in Ringer solution with low Ca^{++}/Mg^{++} ratio; and all of them show post-tetanic potentiation. This evidence also indicates that this synaptic activity involves the presynaptic release of chemical transmitters. The N wave certainly represents the classical depolarizing EPSP and was designed as fast EPSP. The P wave denotes a hyperpolarizing membrane event, a type of slow IPSP. The LN wave expresses a long-lasting depolarizing membrane potential which has been designated as a slow EPSP (Libet, 1965). The N wave has the usual short latency of a classical EPSP. Selective depression of the fast EPSP by strong curarization has permitted an estimate of 35 to 100 msec for the P wave latency; that of the slow EPSP was observed to be about 200 to 300 msec.

The three waves are differently affected by different drugs. The N wave is selectively depressed by curare and curarelike drugs, whereas the P and LN waves are dperessed by atropine, which suggests the participation of ACh in their formation. In addition, dibenamine depresses the P wave more than the LN and N potentials, and this effect is different in the presence of reserpine (LIBET, 1962), pointing to a

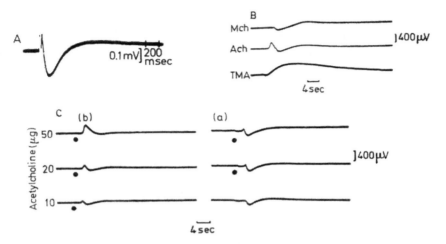

Fig. 3. A) Ganglionic potentials recorded from the rabbit sympathetic ganglion as responses to a single preganglionic volley. Fast and slow EPSP and the slow IPSP are clearly visible (from LIBET, 1967). B) A comparison of ganglion potentials evoked by methacholine, ACh and tetramethylammonium (superior cervical ganglion of the cat). Only ACh gives triphasic response analogous to the response of nervous stimulation (from TAKESHIGE and VOLLE, 1964). C. The effects of hexamethonium (HMT) on the ganglion potentials produced by 10, 20 and 50 µg of ACh. The responses of the untreated ganglion to ACh are shown in the left column, and those recorded after HMT in the right column. HMT selectively blocks initial fast response (better seen with lower doses of ACh), disengaging the slow hyperpolarizing and depolarizing responses (from TAKESHIGE and VOLLE, 1964)

possible intervention of a monoamine in the process. However, all three waves are abolished in the presence of botulinum toxin (which is generally considered to paralyze specifically the release of ACh from the presynaptic structures without affecting the postsynaptic membranes (AMBACHE, 1949, 1951; BROOKS, 1956; CALHOUN, 1958) indicating that there was a cholinergic link involved in the three.

ECCLES and LIBET (1961) have given the following explanation of ganglionic transmission: ACh released by presynaptic fibers causes depolarization by acting on curare-sensitive sites (which give the N potential or the fast EPSP) and on atropine-sensitive sites (which give the LN wave or slow EPSP). It also affects atropine-sensitive sites on some other structures (at first chromaffin cells were tentativelly proposed but other candidates are now known for this role), which in turn release catecholamine that acts on ganglion cells and then produces a hyper-polarization (the P wave or slow IPSP).

This theory received large support from other pharmacological observations. The biphasic potential produced by nerve activity in partially curarized ganglion can be strikingly reproduced by administration of ACh (TAKESHIGE and VOLLE,

1964; Fig. 3). The components of the ACh-induced triphasic wave show the same sensitivity to curare and atropine (TAKESHIGE and VOLLE, 1964; VOLLE, 1965). The curare- and atropine-sensitive receptors have been further dissociated by taking advantage of the higher sensitivity of atropine-sensitive receptors; using low ACh concentration, only hyperpolarization and late depolarization were produced (Fig. 3C). Moreover it has been found that only ACh is able to activate both types of receptors (Fig. 3B). Nicotinic drugs like tetramethylammonium produce only an early depolarization blocked by curare, whereas muscarinic drugs such as methacholine or pilocarpine induce a hyperpolarization followed by late depolarization, both blocked by atropine. Noradrenaline has a depressive effect on the ganglionic transmission and was considered to be a possible candidate for the transmission of inhibition. An extensive picture of the action of drugs on synaptic mechanisms in autonomic ganglia can be found in an excellent review by VOLLE (1966).

The presence in the sympathetic ganglion of the two types of ACh receptors and one type of another receptor involved in inhibition being thus definitely established, it remained to study their distribution in the neuronal structures, that is, whether they are distributed on the same or different neurons.

Electrophysiological studies using intracellular recordings have shown two types of postganglionic cells, B and C, in frog sympathetic ganglia. The B cells have a myelinated axon and are innervated by a single, preganglionic myelinated B fiber (NISHI and SOEDA, 1962; NISHI et al., 1965); the C cells have a nonmyelinated axon and are innervated by more than one presynaptic nonmyelinated C fiber of multisegmental origin. The C cells are smaller than the B cells and so are the afferent and efferent fibers. Hence, the conduction in the B system is much faster than in the C system.

In earlier studies, only fast EPSP were observed in both B and C cells; they were found to have the same duration and the same reversal potential (about $-10\,\mathrm{mV}$) and to result consequently from the same ionic mechanisms. Recently, LIBET and TOSAKA (1966) and TOSAKA et al., (1968) have succeeded in making intracellular recordings of both slow hyperpolarizing IPSP and depolarizing EPSP in the posterior paravertebral ganglia of the frog (Fig. 4). Slow IPSP ranging from 5 to 25 mV have been detected in C neurons upon the selective stimulation of C and B fibers. Although the same C fiber elicits both initial fast EPSP and slow IPSP, the stimulation of preganglionic B fibers elicits only slow IPSP (in smaller increments) in these cells. The slow IPSP is abolished by atropine; this is in accordance with the above theory which proposes that the slow IPSP would be mediated by an intercell which is itself cholinergically activated by preganglionic fibers via atropine-sensitive receptors in view of its atropine sensitivity. Because of the observed hyperpolarizing action of noradrenaline on the postganglionic cell, the intercell is supposed to be adrenergic in nature. It does not need to be the chromaffin cells as earlier proposed, since it now appears that they are rare and not properly distributed for this purpose (LIBET, personal communication). This function might be assumed by other types of neurons now known to be present in the sympathetic ganglia, for instance, by a peculiar type of interneuron described by WILLIAMS (1967). In addition there is evidence of adrenergic-type endings on sympathetic ganglion cells (HAMBERGER et al., 1963a, b, 1965; NORBERG and HAMBERGER, 1964; NORBERG and SJÖQVIST, 1966). It is then possible that the adrenergic intercell affects the ganglionic neuron through a classical synaptic mechanism, but a loose release of the catecholamine in the immediate vicinity of the ganglion cell could also explain the slow IPSP. Because the stimulation of the B fibers also produces slow IPSP in the C cells (but not the classical fast EPSP), it is necessary to admit that the postulated adrenergic intercell is also excited by B fibers (Fig. 4). It seems that the slow IPSP is physiologically active; in fact, the inhibition of the postganglionic discharge by the P potential has been reported (KOKETSU and NISHI, 1967).

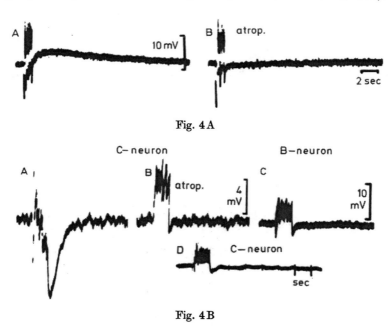

Fig. 4 A

Fig. 4 B

Fig. 4. Slow PSP recorded from frog sympathetic ganglion cells. *Upper records*: slow EPSP recorded intracellularly in uncurarized B neuron responding to orthodromic trains: (*A*) before atropine, and (*B*) after atropine which blocks the slow EPSP. *Lower records*: slow IPSP recorded intracellularly in a curarized C neuron to orthodromic train: (*A*) before atropine, and (*B*) after atropine which blocks the slow IPSP. (*C*) and (*D*), respectively, represent recordings from a B neuron and a C neuron which show no IPSP for the same type of stimulation. Time calibration of 1 sec in *D* applies to all. Voltage calibration in *B* applies to *A—B*; in *C* to *C—D* (from LIBET and TOSAKA, 1968)

The slow IPSP is absent in B cells; but a train of B volleys elicits in a part of B-type cells, besides the fast EPSP, a small but very long (30 sec) slow EPSP which is wiped out by atropine. One preganglionic B fiber apparently elicits both the initial curare-sensitive fast EPSP and all of the atropine-sensitive slow EPSP in a given B neuron. It is thus clear that nicotinic postsynaptic receptor sites (for fast EPSP) and muscarinic ones (for slow EPSP) coexist on the same cell.

A still more interesting situation is found in the mammalian sympathetic ganglion cells; intracellular recordings in the superior cervical ganglion of the rabbit have shown cells which exhibited both slow IPSP and slow EPSP; every ganglion cell always exhibited the initial of fast EPSP in addition (LIBET and

Tosaka, 1969, (Fig. 5). The fast EPSP was selectively blocked by curare, the slow IPSP and EPSP by atropine. Thus, in mammalian sympathetic ganglion the two types of nicotinic and muscarinic receptors are present on the same cell, in addition to receptors mediating inhibition.

There is no information about the spatial distribution of the two types of cholinergic receptors. The long latency or the slow (200 to 300 msec) could be explained by a distant localization of atropine-sensitive receptors from the place of ACh release, but the long delay in this case is now considered to be due more to the nature of the transmitter-receptor response on the postsynaptic membrane than to the long diffusion pathway for acetylcholine (Libet, *personal communication*).

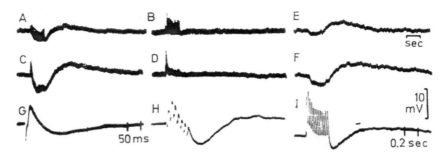

Fig. 5. Examples of fast and slow PSP all in the same cells, in the cervical sympathetic ganglia of the cat, with effects of atropine and strong curarization. Slow IPSP and EPSP can both be seen in curarized cells (A, C, E, F, G, H, I) and can be completely separated from fast EPSP in strongly curarized ganglia (A, F). Both slow IPSP and slow EPSP are blocked in the presence of atropine (B, D). Sweep speed at 1 sec per division in A to F, 0,2 in H and I, and 50 msec in G (from Libet and Tosaka, 1968)

In addition to the slow EPSP, Nishi and Koketsu (1968) recently described in the frog a so-called late slow EPSP, which can be intracellularly recorded from the B as well as from the C cells and is a prolongation of the slow EPSP. It is not blocked by nicotine, by d-tubocararine, or by atropine. It was postulated that the so-called late slow EPSP was generated by a noncholinergic postsynaptic mechanism in response to a specific transmitter substance which is released when the preganglionic stimulation is strong enough to stimulate the C fibers.

In contrast to the general scheme of humoral synaptic transmission valid for the fast EPSP, Libet and Kobayashi (1968) consider that the slow EPSP and IPSP are not generated by increases in ionic conductance in the postsynaptic membrane. In both frog and mammalian sympathetic ganglia, the amplitude of the slow EPSP was increased by moderate depolarization of membrane potential, and was reduced by hyperpolarization. This is the reverse of effects on the initial fast EPSP. No equilibrium potential could be found for the slow EPSP, nor was any decrease in membrane resistance observed. The slow IPSP also appeared to behave in a manner not in accord with ionic conductance changes (Nishi and Koketsu, 1968).

The responses of applied ACh and norepinephrine, supposed transmitters of slow EPSP and IPSP, also did not show any measurable increase in membrane conductance of ganglionic neurons (Libet and Kobayashi, 1969). It was concluded that these neurotransmitters can activate electrogenic mechanisms which do not involve movements of ions down their electrochemical gradient.

A hypothesis has been proposed explaining the slow IPSP as resulting from the activation by the transmitters of an electrogenic sodium-potassium pump (NISHI and KOKETSU, 1968). However, the slow IPSP as well as the response to norepinephrine can still be observed in the presence of ouabain which is a specific inhibitor of such an active transport mechanism for sodium coupled to potassium intake (LIBET and KOBAYASHI, 1969). Further work is required to assess the hypothesis which would disclose a new interesting mechanism of neuronal inter-action. At present it does not seem that a conclusive evidence has been found for the exclusion of a classical permeability change in the initiation of slow EPSP and IPSP. A possibility still remains that the absence of measurable changes in conductance is simply due to a distant location of the responsible receptors in respect to the soma where recording electrodes were inserted.

Renshaw Cells

Two main types of cholinergic receptors have been equally described in the vertebrate central nervous system on the so-called Renshaw cells, first described by RENSHAW (1946). Although anatomists have so far failed in the search for these cells, there is strong electrophysiological evidence for their existence in the spinal cord. They are excited by axon collaterals of spinal motoneurons, and their excitation then produces recurrent inhibition in the motoneurons. It is thus possible to activate the Renshaw cells by a monosynaptic pathway through the antidromic stimulation of the motoneurons, using the ventral root. Although such a stimulation is quite unphysiological, it gives in the Renshaw cells a peculiar sequence of spike activities which, besides aiding in the identification of the cells, has been used for the analysis of the pharmacology of the junction between the motoneuron and the Renshaw cell. This latter work was originally inspired by the principle of DALE (1934, 1952) who postulated that a given neuron releases the same transmitter on all synapses which it establishes; thus it was supposed that the terminal of motoneurons would release ACh as they do on the neuromuscular junction.

Intracellular recordings from the Renshaw cells are very rare (ECCLES et al., 1961), but the spike activity is quite a good indication of polarization changes which can occur in the cell under different treatments. The activity of a single cell can be recorded by a microelectrode placed in the immediate proximity of the cell. When this microelectrode is included in a multibarrelled micropipette system, filled with different drugs which can be injected electrophoretically close to the individual neuron, the recorded spike activity gives a good indication of the action of a given drug (CURTIS and ECCLES, 1958a; CURTIS et al., 1960; CURTIS and WATKINS, 1963; ANDERSEN and CURTIS, 1964; CURTIS and RYALL, 1966). This method has the advantage of bringing the drug on the other side of the blood brain barrier, which is otherwise impermeable to many substances. The most complete study of the receptors on the Renshaw cell can be found in a series of papers by CURTIS and RYALL (1966a, b, c).

The typical response of a Renshaw cell to a ventral root volley consists of an early high-frequency discharge and a late low-frequency firing separated by a period of silence (Fig. 7A). The initial high fast response was selectively depressed by dihydro-β-erythroidine (DHβE), a powerful nicotinic blocking agent, whereas atropine specifically depressed the late firing (Fig. 6). Along the same lines, it has been found that ACh and different cholinomimetics administered electrophoretically from multibarrelled glass micropipettes exerted an excitatory action on the

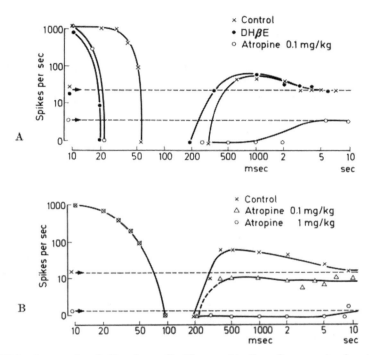

Fig. 6. A) Firing frequencies of a Renshaw cell with respect to time after a maximal root stimulus, under normal conditions, and after successive addition of DHβE and atropine, which have a depressive action, respectively, on the early and late firing. In B, in a different cell, only atropine was added to show that it has no effect on the early firing. In each case the mean spontaneous firing rate is indicated by an arrow and the upper horizontal broken line (from Curtis and Ryall, 1966)

Renshaw cells. The action of ACh, nicotine and carbamylcholine was predominantly due to interaction with nicotinic receptors and was blocked by DHβE. Acetyl-β-methylcholine and DL-muscarine interact with muscarinic receptors which were blocked specifically by intravenously administered atropine (crossing, contrary to curare, the blood brain barrier). A nonspecific activation of the cells by DL-homocysteic acid has been used as test for the absence of excitability modification of the treated cells. Moreover, the observation that hemicholinium-3, which is known to reduce ACh synthesis, readily suppresses the transmission represents additional evidence for the transmitter role of ACh on this synapse (Quastel and Curtis, 1965).

It is therefore clear that both nicotinic and muscarinic receptors are present on the Renshaw cell and that their responses have some analogies with those on the sympathetic ganglion cells, especially in the differences of delay and duration of action. It must be emphasized, however, that in both cases the observed type of responses is completely unphysiological and is seen only when the majority or all of the cholinergic synapses on a neuron are releasing ACh simultaneously. Concerning the Renshaw cell, it was proposed that the late atropine-sensitive response may have no functional role, but muscarinic receptors seem to be involved

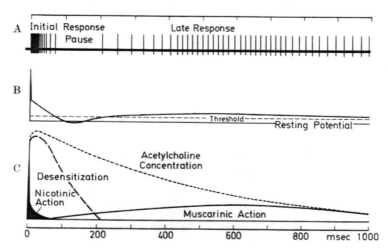

Fig. 7. Diagrammatic representation of the response of a Renshaw cell to a maximal ventral root volley. (A) Firing patterns. (B) Intracellularly recorded postsynaptic potential. (C) Analysis of nicotinic and muscarinic actions of ACh, showing desensitization as a possible factor of the silent period (from CURTIS and RYALL, 1966)

in the spontaneous activity which is suppressed by atropine (CURTIS and RYALL, 1966). In sympathetic ganglia, the sensibility of atropine-sensitive receptors was found to be many times greater than that of curare-sensitive receptors, and the physiological role of muscarinic receptors in prolonged activity was established (LIBET, 1964).

When the Renshaw cell is under the influence of both DHβE and atropine, and when its spontaneous firing is restored by administration of DL-homocystoic acid, cholinomimetics may depress and subsequently excite the cell. The presence of a third type of ACh receptor has been proposed to account for this biphasic response; such receptors have also been found on other cells (CURTIS et al., 1966) where no physiological action of ACh was expected.

When the Renshaw cell is excited by a pathway different than that which involves the intervention of cholinergic receptors, the resulting excitation can be blocked neither by DHβE nor by atropine. Thus, another noncholinoceptive type of excitatory receptor must be present on the Renshaw cell sensitive to an unknown excitatory transmitter. Finally the depression observed after a synchronous antidromic volley (Fig. 7A) can be explained by the desensitization of nicotinic receptors prior to the activation of the muscarinic ones (Fig. 7C), but the possibility of interaction of the synaptic noncholinergic inhibitory mechanism with adequate inhibitory receptors is equally possible.

8*

Gastropods Central Ganglion Cells

The number of ganglion cells in the central nervous system of *Aplysia* was estimated to be of the order of 10,000. Given the larger size of some of these cells accessible to microelectrode penetration on the direct visual controls, it can be understood why neurophysiologists currently favor the ganglionic preparations of *Aplysia* and other gastropods such as *Helix*. The relative simplicity has made possible the unitary analysis of a large spectrum of properties and of transmission in particular. It must, however, not be forgotten that we face a structure homologous to a "brain" which has a complex function; consequently, there is in the gastropod ganglion no trace of this structural homogeneity observed in the sympathetic vertebrate ganglion. Morphological and functional properties of the identified neurons in abdominal ganglion have recently been reviewed (FRAZIER *et al.*, 1967; TAUC, 1966, 1967).

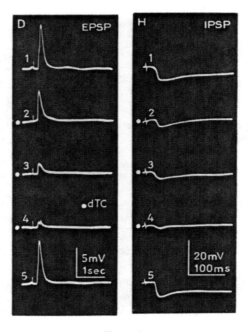

Fig. 8 A

Fig. 8. A) Depression of EPSP in a D cell and IPSP in an H cell of *Aplysia* by d-tubocurarine in 10⁴ g/ml concentration (white dots, *2—4*). The removal of curare restores the initial amplitude of PSP. B) *1* Effect of brief electrophoretical injections of ACh (arrows) on the somatic membrane of a D cell (left column) and an H cell (right column) firing spontaneously. The drug depolarizes the D cell and accelerates its firing, whereas it hyperpolarizes the H cell and inhibits firing. 2. Same amount of ACh as in (*1*) is injected on the *D*-cell (left) and the *H* cell (right), but in the presence in the bathing fluid of 10⁴ g/ml of d-tubocurarine, which clearly inactivates both D and H cell membranes to the action of ACh. 3. The same electrophoretical injection of ACh on D and H cells as in (*1*), but in the presence in the bathing medium of 10⁴ g/ml of hexamethonium bromide. This latter drug inactivates the D cell membrane to the action of ACh (left) but has no effect on the response of the H cell membrane to ACh (from TAUC and GERSCHENFELD, 1961)

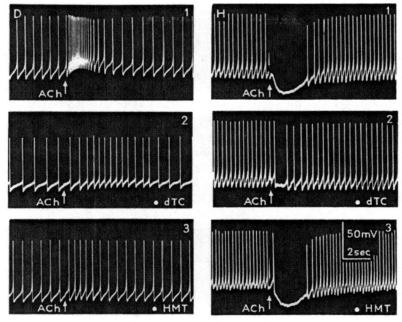

Fig. 8 B

Gastropod ganglion cells are integrative neurons, and most of them receive both excitatory and inhibitory impulses from fellow interneurons or receptors. Besides classical EPSP and IPSP, other transmission processes have been described. With respect to the input which they receive and the pharmacological properties of the postsynaptic membrane, the ganglion cells can be divided into several groups.

Transmission in D and H Cells

The two major divisions represent the so-called D and H cells. The application of ACh by perfusion, or ever better by local electrophoretical injection on these cells, produces either of the following reactions (TAUC and GERSCHENFELD, 1961, 1962): (1) Cells of one type, conventionally designated as D neurons, are depolarized and excited; or (2) Cells of another type, the H neurons, are hyperpolarized and inhibited (Fig. 8 B).

In both D and H cells, threshold effects are observed by ACh concentrations less than 10^{-10} g/ml, and the membrane conductance is considerably increased. When the membrane potential is artificially changed, the reversal potential of ACh action is identical to that of the EPSP in D cells and that of the IPSP in H cells. The d-tubocurarine abolishes the effect of ACh on both types of cells, and in a similar fashion the EPSP in D cells and IPSP in H cells are rapidly depressed by this drug (Fig. 8 A). This inactivating effect of curare on ACh injections and on the PSP constitutes an important argument in favor of the cholinergic nature of excitatory synapses in D cells and inhibitory synapses in H cells. The same drug is able to produce both excitation and inhibition; thus, the character of the response does not depend on the nature of the chemical transmitter but on physicochemical differences in the postsynaptic membranes. Cholinoceptive receptors are different on D and H cells, as is shown by the action of hexamethonium which blocks the ACh action on D cells and has no influence on the ACh action on H cells (TAUC and GERSCHENFELD, 1961). In a similar way, hexamethonium depresses the EPSP in the D cells, but has no effects on IPSP in H cells.

A proposition has been made that synaptic endings of the same interneuron might have an excitatory action on some D cells and an inhibitory action on some H cells, both actions being mediated by ACh (TAUC and GERSCHENFELD, 1961) (Fig. 9A). This situation was found by STRUMWASSER (1962) in the buccal ganglion of *Aplysia* and more recently by KANDEL *et al.* (1967) in the visceral ganglion.

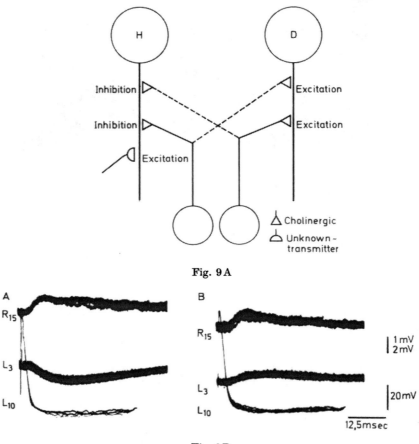

Fig. 9A

Fig. 9B

Fig. 9. A) Diagram illustrating the nature of synaptic contacts on D and H cells. The inhibitory input to the H neuron and the excitatory one to the D cell are cholinergic. A possible contact of a single presynaptic neuron on both D and H cells is indicated by the dotted lines. The excitatory input on the H cell is conveyed by an unknown transmitter (from TAUC and GERSCHENFELD, 1961). B) Demonstration of a single interneuron producing excitation in one cell (upper trace showing EPSP) and inhibition in another one (lower trace showing IPSP in the left column). Displacement of the membrane potential in both follower cells slightly affects the EPSP, but causes the IPSP to be inverted to a depolarizing potential (right column) (from KANDEL *et. al.*, 1967)

KANDEL et al. (1967) have been able to demonstrate, in identified cells, that a single interneuron can produce excitation in one type of neuron and inhibition in other types of neurons, both of these effects being conveyed by ACh (Fig. 9 B). The EPSP could be blocked hlocked by hexamethonium, whereas IPSP have been shown to be refractory to hexamethonium. These are important observations as they indicate that not only is the same interneuron able to produce excitation and inhibition in different cells, but also that these differences of action are due to the differences in receptors present on the cells as well as to the ionic content of the cells.

The EPSP in D cells are mostly due to chloride permeability change, possibly combined with that of sodium (CHIARANDINI et al., 1967; FRANK and TAUC, 1964; KERKUT and THOMAS, 1964; KERKUT and MEECH, 1966). In H cells, the chloride permeability change seems to be exclusively responsible for the IPSP (KERKUT and THOMAS, 1963; FRANK and TAUC, 1964).

When chloride is injected electrophoretically into H cells by an intracellular electrode, the equilibrium potential for chloride diminishes and the inhibitory postsynaptic potential first decreases in size and then reverses, giving a depolarizing response (FRANK and TAUC, 1964; KERKUT and THOMAS, 1963). It is possible to inject in place of chloride some other anions which, like chloride, change the reversal potential of the EPSP; apparently they are able to cross the permeability channels normally opened for chloride (KERKUT and THOMAS, 1963b). These anions are Br^-, I^-, NO_2^-, NO_3^-, N_3^-, ClO_4^-, SCN^-, BF_4^-, OCN^-, ClO_3^-, HCO_2^-; whereas F^-, IO_3^-, SO_4^-, SO_3^- and many others have no effect on the IPSP reversal potential. The comparative study of such anions revealed, in fact, that the size of the anion in aqueous solution is an important factor, the nonpermeability point being obtained with anions which in their hydrated form are larger than HCO_2^-. The general picture was the same as that obtained by ARAKI et al. (1961) on the inhibitory synapse on the spinal motoneuron. This has given rise to the notion of pores having a precisely standardized size, opened in the membrane by synaptic action.

An attempt has been made to study the arrangement of cholinoreceptors on the neuronal membranes of D cells in the gastropod mollusca Planorbis corneus and Limnea stagnalis. The method used consisted of comparing the sensitivity of the postsynaptic membrane to different bisquarternary compounds which have an internitrogen chain containing 9 or 10 atoms (about 14 A) such as decamethonium, hexamethonium, d-tubocurarine, and others containing 16 atoms (about 22 A) between nitrogen atoms, such as hexadecamethonium and dicholinic esters of suberic and sebacinic acids (VULFIUS et al., 1967). Such a study has previously been made on mammalian skeletal muscles where, besides the C–16 structures, the C–10 structures have also been found. The C–16 structure is also present in the mammalian autonomic ganglia (BARLOW and ZOLLER, 1964; KHROMOV-BORISOV and MICHELSON, 1966). Planorbis D cells are very sensitive to substances with internitrogen distances of 22 A. This suggests the existence of C–16 structures on the neuronal membrane of Planorbis with a mutual disposition of receptors with about 26 A between the anionic points of two neighboring receptors. The results obtained on L. stagnalis are not so clear.

In a group of cells called DINHI cells, noncholinergic types of IPSP have been observed which result from a permeability change primarily conveyed by potassium (GERSCHENFELD and CHIARANDINI, 1965). DINHI cells are depolarized by ACh and have been shown to have a cholinergic excitatory input.

Nonclassical Transmission Processes

A number of neurons in the Gastropod ganglia have been designated as CILDA cells (GERSCHENFELD and TAUC, 1964). The common property of these cells is the presence of a peculiar biphasic response called inhibition of long duration [(ILD) (TAUC, 1958, 1959, 1967)]. Two types of CILDA cells have been described: DILDA cells in which the initial fast phase of ILD is represented by a depolarizing wave of a few hundred milliseconds (Fig. 10), and HILDA cells in which the fast initial phase is formed by a hyperpolarizing potential identical to an IPSP (Fig. 12). DILDA cells are depolarized by ACh, HILDA cells are hyperpolarized. The second component of ILD in both types of CILDA cells is a long-lasting hyperpolarizing wave, the duration of which varies from several seconds

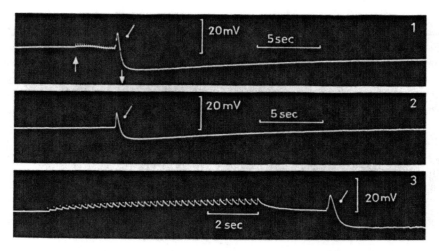

Fig. 10. ILD in an identifiable *Aplysia* neuron "Oberon". (*1*) Unitary ILD following a train of orthodromic stimuli which give no direct effect in the postsynaptic cell. (*2*) Spontaneous unitary ILD, appearing after several discharges of the responsible interneuron occuring several minutes before this record. Note that the amplitude of the E- phase in *1* and *2* is identical, whereas the I-phase in *2* is smaller. (*3*) Unitary EPSP produced by the liminal stimulation of a connective (see clear activation potentiation) followed by a unitary ILD (from TAUC, 1968)

to 30 min (Fig. 11). In both HILDA and DILDA cells, ILD was observed on a unitary level resulting from the activity of a single interneuron; the duration of isolated unitary response is shorter but still clearly surpasses the classical IPSP. In any case, the biphasic nature is maintained even on this unitary level and can never be separated under normal conditions.

There is no doubt that ILD is due to the action of a chemical transmission mechanism. In both types of cells, the study of the reversal potential for the hyperpolarizing long-lasting phase has disclosed the participation of potassium ions (TAUC, 1967; KEHOE, 1967). The chemical nature of the transmitter in the DILDA type is unknown although there is a suggestive evidence that dopamine might be the substance responsible for the hyperpolarizing phase. Indeed this phase can be imitated by the injection of dopamine on the axonal synaptic zone of DILDA cells (ASCHER *et al.*, 1967).

The second type of inhibiton of long duration which appears in HILDA cells is better known. Identifiable HILDA cells have been found in the pleural ganglion of *Aplysia*, and it was possible to analyze the mechanisms of this type of double

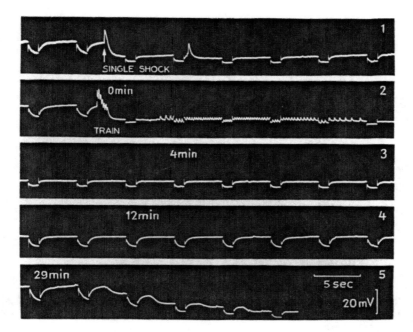

Fig. 11. Conductance changes during ILD, as measured by amplitude modification of the electrotonic potential produced by square current injection pulses from a second intracellular microelectrode. *1*. After a single shock. *2*. to *5*. After a train of eight shocks to an afferent nerve. The ILD interneuron was active for about 20 sec. The total duration of ILD in this cell was 3 min after a single shock and 29 min after the train. In *5*, the polarization of the cell membrane was artificially altered by the applied current. The modifications of the electrotonic potentials point to the presence of anomalous rectification which must be considered in the interpretation of the above results (from TAUC, 1968)

inhibition on a unitary level (KEHOE, 1967, 1968). The reversal potentials of the fast initial inhibitory phase and for the later long-lasting phase have been found to be different (Fig. 12). Consequently, the two phases resulted from different ionic permeability changes; in fact, the initial phase is now known to represent chloride permeability change, whereas the second inhibitory phase results from potassium permeability change (KEHOE, 1967). A single electrophoretical injection of ACh reproduces both phases. ACh is thus able to produce chloride and potassium permeability changes which occur nearly simultaneously but are of different duration; the difference in duration explains the biphasic form of the phenomenon.

Pharmacologically, the two phases are differently affected. D-tubocurarine selectively blocks the initial fast phase but leaves the slow phase unaffected

(Fig. 13); strychnine has the same effect. Atropine has no action either on the initial or the later phase. The second phase can be selectively blocked by tetraethylammonium (Kehoe, 1968). The physiological inputs are similarly affected. It seems necessary to admit that ACh, liberated on HILDA cells by a responsible interneuron, combines with two different receptors – one acting

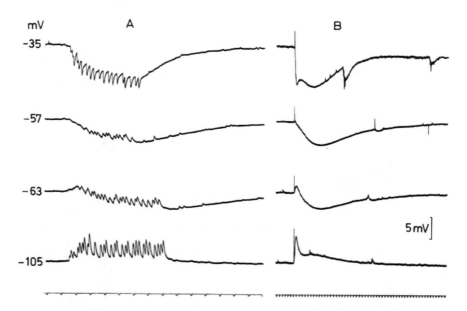

Fig. 12. (A) Inhibition of long duration produced in a HILDA-type cell upon physiological stimulation of the head of the animal. One can clearly distinguish the first fast inhibitory phase (several responses), which was shown to be Cl^-—dependent, and the following K^+—dependent slow inhibitory phase. The two phases can be reversed at different polarity levels, adjusted by injected transmembrane current. (B) The two ILD phases can be imitated by electrophoretical injection of ACh onto somatic membrane; the reversal potentials are then identical as for the physiological effect. The depolarizing-hyperpolarizing response at —63 mV perfectly illustrates the differences in ionic mechanisms for the first and second phase. Time in seconds (from Kehoe, 1967)

on chloride, the other on potassium permeability mechanism. One of the receptors not affected by curare of atropine represents a new and peculiar cholinoceptive structure. The existence of such a receptor, which without any doubt is physiologically active, indicates that in a general pharmacological study it is now impossible to discard the presence of a cholinergic mechanism when the application of classical cholinolytic agents is without effect.

The second long-lasting inhibitory phase of any type of ILD shows peculiar cumulative effects. Repetitive activation of the responsible interneuron produces, besides the usual spatio-temporal summation, a considerable increase in the duration of this phase which is in direct relation to the quantity of the initial input (Fig. 11; Tauc, 1967). In this respect, it is significant that, whereas ACh is injected on HILDA cells, the duration of potassium permeability change is also proportionate to the quantity of ACh initially injected and it reaches several minutes in duration. In the presence of large amounts of acetylcholinesterase which has been found in the ganglion, this long lasting effect of ACh is remarkable and can be explained either by admitting a considerable stability of the combination of ACh-receptor responsible for potassium permeability change or by the long duration of effects which the ACh-receptor combination produces in a secondary mechanism reponsible for potassium permeability.

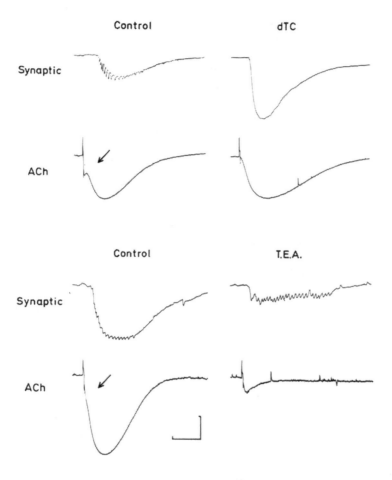

Fig. 13. Action of d-tubocurarine (*dTC*) and tétraethylammonium (*T.E.A.*) on ACh potential and the physiological response in HILDA-type cells. *Upper half:* *dTC* depresses the fast inhibitory phase of ACh potential (arrow) and the fast IPSP. *Lower half:* *T.E.A.* selectively depresses the slow inhibition phase of ACh potential and the slow IPSP. It must be remembered that atropine is uneffective on any of these potentials. Time in 10 sec for ACh injection and 2 sec for synaptic stimulation amplitude 5 mV for all records (KEHOE, 1968; *unpublished*)

In addition to ILD, CILDA cells have other classical excitatory inputs. In DILDA cells cholinergic excitatory input has been found, but in these cells ACh does not produce a double effect. Sodium permeability change is involved in this cholinergic EPSP (CHIARANDINI et al., 1967). Besides this cholinergic input, another classical excitatory input was described mediated by an unknown transmitter but producing the same permeability changes as the cholinergic input (GERSCHEN-FELD et al., 1967).

A peculiar type of transmission has been observed in identifiable cells of Aplysia (WACHTEL and KANDEL, 1967). In these cells, an interneuron produces excitatory postsynaptic potential which at high firing rates gradually diminish in size and inverts to inhibitory postsynaptic potentials. Electrophoretical injection of ACh onto these cells produces depolarization for a small quantity of ACh and hyperpolarization for a large quantity; biphasic effect (depolarization followed by hyperpolarization) was observed for intermediate concentrations. Both postsynaptic potentials and ACh effects are blocked by curare. It was thus proposed that a single transmitter, ACh, mediates both excitation and inhibition. ACh would combine with two different receptors on the postsynaptic membrane, the excitatory one having a low threshold and opening permeability channel for sodium ions as in DILDA-type cells, and the other one having a higher threshold and opening permeability channel to chloride as in H-type cells. This unique cell in Aplysia appears thus to combine the receptor properties of cells receiving either purely excitatory or purely inhibitory branches from the same cholinergic presynaptic interneuron.

Localization of Receptors

In contrast to the neuromuscular junction where receptors are strictly localized in the region of the motor end-plate, in most molluscan ganglion cells ACh receptors can be found also on the cell body which is completely deprived of synapses located on the axon at some distance from the soma. Apparently the concentration of receptors at the level of the soma is similar to that of the axonic region. In addition, the properties of cholinergic receptors are comparable in both localizations. This is very convenient for pharmacological study because it is possible to inject the drug on the visible somatic membrane which is easily accessible.

It seems that this somatic localization of receptors is restricted to cholinergic receptors. Receptors to dopamine or to serotonine which are supposed transmitters on some cells, are strictly localized in the axonal synaptic zone (GERSCHENFELD and STEFANI, 1965; ASCHER et al., 1967). It is evident that such a problem escapes direct analysis in structures like the Renshaw cells or sympathetic ganglion cells where access for unitary study is more difficult.

An exception to the rule of the somatic localization of ACh receptors was found on five identified giant cells in the buccal ganglion of a marine mollusc Navanax (LEVITAN and TAUC, 1969, unpublished). Injection of ACh on the soma of these cells shows no response, but when injected on different parts of the axon, depolarizing PSP-like potentials were observed. When the postsynaptic cell was artificially depolarized close to the firing level, the ACh response obtained in the most proximal part of the axon was reversed to a hyperpolarizing and inhibitory potential, whereas activation by ACh of the more distal part of the axon always produced depolarizing and excitatory responses. Injected inbetween these two zones ACh produced a biphasic, depolarizing-hyperpolarizing potential. It has been found that the proximal hyperpolarizing response was due to chloride permeability change, the distant depolarization to sodium permeability change. The sodium dependent response could be blocked by curare, atropine, and other cholinolitics and a number of other drugs, but the chloride dependent response was found to be resistant to all these drugs.

It is clear that these cells have two different ACh receptors which are localized on different parts of the axon with a possible overlap in the central zone. Activation of one of them has an inhibitory effect on the spike generation, the other has excitatory properties. In physiological terms, such distribution of receptors to the same type of interneurons releasing ACh would mean that the same type of interneurons releasing ACh would exert an action the character of which, excitatory or inhibitory, would be determined solely by the anatomical localization of the synapses.

Episynaptic Transmission Processes

Another type of contact between neurons which produce long-lasting effects can be demonstrated indirectly. Modifications in synaptic efficacy have been observed when a third neuron is stimulated and affects by an unknown process the size of the test EPSP. In *Aplysia*, both increase and decrease of synaptic efficacy produced in such way has been found, designated as presynaptic inhibition (TAUC, 1964, 1965) and as presynaptic facilitation (KANDEL and TAUC, 1964; KANDEL and TAUC, 1965a, 1965b; BRUNER and TAUC, 1965, 1966; TAUC and EPSTEIN, 1967). Presynaptic inhibition has already been observed in the spinal cord producing a decrease in synaptic efficacy of a duration of about 400 msec when a heterosynaptic conditioning pathway is stimulated prior to the test pathway (ECCLES, 1964). In *Aplysia*, the duration of presynaptic inhibition may reach several tens of minutes.

Presynaptic facilitation has so far been described only in *Aplysia;* heterosynaptic stimulation produces an increase of efficacy of the test pathway which can reach a duration of several tens of minutes (Fig. 14). These phenomena can be explained by admitting a physiological contact between the conditioning and the test pathway which will somehow modify the quantity of the transmitter released on the synapse. This contact has been designated under the name of episynapse (TAUC, 1965), but the mechanism of function of this structure is completely unknown. Because of the long-lasting effect, however, it is necessary to admit some sort of chemical interaction. It is not impossible that the episynapse in fact works on the same basis as inhibition of long duration in CILDA cells.

The difference between presynaptic facilitation and presynaptic inhibition seems to consist only in polarity of the effect, the basic mechanism being identical. It appears that the dynamics of presynaptic inhibition and presynaptic facilitation are very similar. It is to be underlined that presynaptic effects are observed in excitatory or inhibitory inputs to only some cells and, when present, affect the majority of the same type of presynaptic fibers. However, inputs to a given cell will show only presynaptic inhibition or presynaptic facilitation (TAUC, 1965). The diffuse character of this type of interaction has lead to the supposition that the episynaptic transmitter is liberated by the conditioning pathway in a diffuse way rather than at places of specific contacts which would have determined anatomical configuration (TAUC, 1966, 1967). The transmitter could then reach by diffusion proximal and less proximal receptive fields of several neurons without being limited in its action to only one postsynaptic neuron. Only structures which possess the adequate receptor would be affected. The observed accumulation of effects and differences in duration of presynaptic effect in a neighboring cell seem to support this hypothesis.

Another possible mechanism of heterosynaptic effects may consist in the action of the episynaptic transmitter not on the presynaptic but on the postsynaptic neuron, modifying the sensitivity of receptor molecules to the transmitter normally released on the synapse in unchanged quantity.

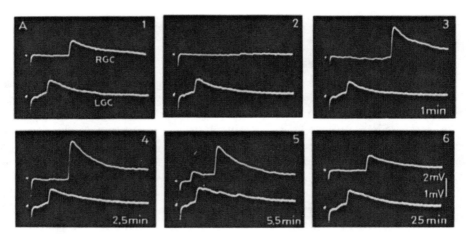

Fig. 14. Heterosynaptic facilitation of a unitary EPSP in the right giant cell of *Aplysia*. The upper and lower recordings represent unitary EPSP from the same interneuron recorded respectively in the right giant cell (*RGC*) and in the left giant cell (*LGC*), the two cells being situated in different ganglia. During the conditioning stimulation (*2*), the left and right cells have been temporarily separated by a sucrose block, so that the conditioning stimulation could affect only the axonal ending on the right giant cell; the presence of synaptic relays between the conditioning pathway and the test pathway was excluded. When the sucrose block was removed, the EPSP in the right giant cell increased considerably as shown in *3* through *6*. The peak was reached after $2^{1}/_{2}$ min after addition of the priming stimulation, and the total duration of heterosynaptic facilitation was 25 min (from Tauc and Epstein, 1967)

The effects produced by presynaptic action might be very powerful. Facilitations observed might increase the amplitude of the test EPSP to several hundred per cent of its initial value; on the other hand, presynaptic inhibition practically wipes out the test potential. All phenomena have been observed in physiological conditions when physiological stimuli have been applied to produce presynaptic effect; their physiological role is thus perfectly clear.

Desensitization of Synaptic Receptors

The diminution of chemosensitivity to a given drug in the presence of this drug is found in a number of structures. The "desensitization" of physiological transmitters was first described in the vertebrate motor end-plate and concerns ACh (Axelsson and Thesleff, 1958; Katz and Thesleff, 1957). A similar phenomenon was observed in vertebrate sympathetic ganglia (Blackman et al., 1962, 1963) and in Renshaw cells (Curtis and Ryall, 1966a). In molluscan ganglia, the effectiveness of injected ACh on the cell membranes of some D and H cells is progressively diminished if the application of ACh is prolonged or repeated (TAUC and BRUNER, 1963). After the removal of ACh the effectiveness of ACh action is progressively recovered. Presumably, the afinity of ACh with receptor molecules is progressively and reversibly changed. In muscle and synaptic neurons,

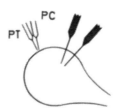

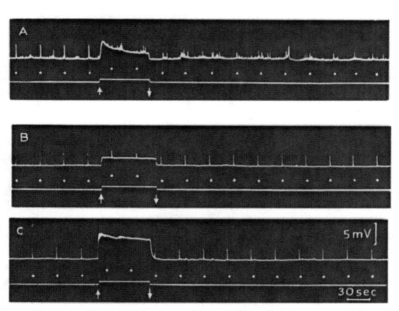

Fig. 15. Effects of injection of ACh on the chemosensitivity of its receptors in *Aplysia* D neurons. Dots show the time of injection of the test dose of ACh, lower record represents the current ejecting the conditioning dose of ACh from another ACh-filled microelectrode (*see* schema). (A) A cell which shows an intense desensitization followed by long recovery time for the test potential (the small potentials represent spontaneous EPSP). (B) A cell with pratically no desensitization; the decrease of amplitude of the test ACh potential during the conditioning of ACh is solely due to conductance change of the membrane. (C) Same cell as in B, still with practically no desensitization in spite of the increased conditioning dose (TAUC and BRUNER 1963, *unpublished*)

the recovery time is relatively rapid; in some *Aplysia* cells, however, it may reach several minutes and even tens of minutes. Surprisingly enough, cholinergic receptors in different cells belonging to the same pharmacological type have different aptitudes for desensitization, and the recovery times are different (Fig. 15). Also, recovery time depends closely on the quantity of ACh previously administered. Moreover, cells have been found in the same ganglion whose cholinergic receptors cannot be desensitized at all. Desensitization thus seems to be bound to some specific cellular properties as yet unknown. Desensitization of other types than ACh receptors has been described for both 5-HT (Gerschenfeld and Stefani, 1965) and dopamine receptors (Ascher et al., 1967).

The physiological role of desensitization has never really been demonstrated. Even on the neuromuscular end-plate, such a role is doubtful and attempts made on molluscan ganglia to demonstrate such a role have so far been fruitless. The decrease of the amplitude of the cholinergic EPSP in CILDA cells which has been called habituation (Bruner and Tauc, 1965) is apparently not related to desensitization (Bruner, 1967, unpublished results). The extrapolation of effects seen on using applied ACh to the effects of natural transmitter released by the nerve is difficult, mainly because the concentration and the surface affected by the action of drugs are different in the two cases.

Conclusions

It now seems evident that excitation and inhibition are two facets of the same synaptic mechanism. Findings in gastropod cells have clearly shown in analogous structures that excitation and inhibition can be mediated by the same transmitter and that the polarity effect depends on postsynaptic receptors and on the ionic content of the postsynaptic cells. Studies that attempt to ascribe a general presynaptic structural correlate to inhibitory synapses, such as those which attribute to inhibitory synapses an oval form of presynaptic vesicles (Uchizono, 1964, 1965, 1968), are devoid of any rational basis.

In fact, the same transmitter is able to produce excitation or inhibition, or both, on the same neuron (Kandel and Watchel, 1967; Tauc, 1967). In Renshaw cells and in the sympathetic ganglion cells, the complex effects of ACh have been explained by the presence of two different cholinoceptive receptors, one nicotinic and one muscarinic. A nicotinic receptor and a peculiar type of receptor not blocked by most classical cholinergic drugs have been considered responsible for the double inhibition in gastropod HILDA cells. The two ACh receptors which coexist on the same neurons in Navanax have still other pharmacological specifications (Levitan and Tauc, 1969, unpublished). Excitation and inhibition observed by Watchel and Kandel (1967) are somehow different in the sense that curaresensitive receptors are able to produce both excitation and inhibition.

When describing different cholinergic receptors situated on the same cell, it is not absolutely necessary that this concern different molecules. It is perfectly possible that the same receptor molecule has two different ACh-sensitive sites which trigger different permeability channels. In the case of ILD in HILDA cells, it is clear that two such different sites must exist on the same or different molecules, as it was shown that the two permeability channels are opened nearly simultaneously. In the case of the biphasic effect of ACh which is entirely blocked by curare, the double phenomenon could be explained by a modification of molecular properties of a unique receptor or of the mechanism which it affects.

The mechanism through which the activated receptor is able to modify the permeability for specific ions of the postsynaptic membrane is still obscure. The change in configuration of the receptor molecule may affect the permeability in a more or less direct manner. The concept of pores being more or less charged, which would permit ionic transfer, represents only a convenient and imaginative term to describe observed events.

Slow synaptic events represent an excitability modification which would otherwise be induced only by maintaining sustained repetitive activity of a fast cynaptic input. The interesting feature which concerns the slow excitatory post-synaptic potential in sympathetic ganglion, the late response in Renshaw cells, the inhibitory phase of ILD and the episynaptic action in *Aplysia* ganglion cells is the dependence of their duration on the quantity of transmitters received in a given time at the beginning of the action. Long duration of action can be obtained by accumulation of effects which can occur in a short period of time. In this property, their mechanism differs considerably from that of the classical cholinergic synapses. On these latter structures the duration of synaptic action is independent of the quantity of transmitter previously released, mainly because this duration depends on the time of inactivation of ACh by acetylcholinesterase. As the destructive enzyme is present in excess, the time of destruction is constant. In the case of slow postsynaptic potentials, especially in the case of ILD, once the transmitter combines with the receptor, it appears to be protected from the destructive enzyme or it triggers a modification of the postsynaptic membrane which evolves independently of the enzymatic action. Even if one considers only diffusion as the limiting factor in the absence of enzyme, the evolution should be nearly as fast as the duration of classical EPSP.

Perhaps there is some analogy between the dependence of duration of the slow postsynaptic events on the quantity of transmitter released and of the duration of the recovery time after receptor desensitization. Indeed, this recovery time is not constant and can be greatly prolonged when larger doses of ACh have been put in contact with the receptors. Of course, the finality of desensitization might be opposite to slow postsynaptic action, but in all cases the receptor molecules undergo modifications of comparable duration dependent upon the quantity of the drug put in their presence.

The discovery, on the same neurons, of several receptors to the same transmitter substance has permitted understanding of the complex synaptic events described here. It is perhaps important, however, not to be tempted to explain a priori any unitary double effect by a double action of a unique transmitter. Although it has never really been demonstrated, the possibility cannot be discarded that a single presynaptic neuron can release more than one transmitter. Such a mechanism would be equally effective, but in addition it would introduce new variables consisting in the separate presynaptic control of release through temporal differences in intensities of the metabolisms of the two transmitters and of their inactivating mechanisms.

The question can be posed whether the action of a transmitter consists only in immediate permeability changes or whether more permanent modifications could be impressed by this action in the postsynaptic cells. Such a concept can be encouraged by observations of the modification of metabolism of phosphatidyl inositol in the mammalian superior cervical ganglion (LARRABEE and KLINGMAN, 1963; LARRABEE and LEICHT, 1965). In this ganglion incubated in vitro in inorganic 32P phosphate, the label is incorporated more or less in different lipids. On orthodromic stimulation via the activation of presynaptic fibers, the incorporation of 32P increases, but only in phosphatidyl inositol. This increase is blocked by curare and is absent when the ganglion is stimulated antidromically; that is, it is independent of the formation of spikes. It seems probable, therefore, that the increase in labelling of phosphatidyl inositol is an effect of the synaptic transmitter on the postsynaptic cell via ACh receptors and does not result from impulse generation or any associated ion transport. The role of this increased metabolism of phosphatidyl inositol in a nervous activity is unknown but, in spite of our ignorance, this important observation represents the first chemical evidence for a "trace" in synaptic transmission.

References

AMBACHE, N.: The nicotinic action of substances supposed to the purely smooth-muscle stimulating (B) Effect of Ba Cl₂ and pilocarpine on the superior cervical ganglion.J.Physiol. (Lond.) 110, 164—172 (1949).
— A further survey of the action of *clostridium botulinum* toxin upon different types of autonomic nerve fibres. J. Physiol. (Lond.) 113, 1—17 (1951).
ANDERSEN, P., CURTIS, D. R.: The pharmacology of the synaptic and acetyl-choline-induced excitation of ventrobasal thalamic neurones. Acta pyhsiol. scand. 61, 100—120 (1964).
ARAKI, T., OSCARSSON, O.: Anion permeability of the synaptic and non-synaptic motoneurone membrane. J. Physiol. (Lond.) 159, 410—435 (1961).
ARVANITAKI, A., CHALAZONITIS, N.: Configurations modales de l'activité, propres à différents neurones d'un même centre. J. Physiol. (Paris) 50, 122—125 (1958).
ASCHER, P., KEHOE, J. S., TAUC, L.: Effets d'injections électrophorétiques de dopamine sur les neurones d'Aplysie. J. Physiol. (Paris) 59, 331—332 (1967).
AXELSSON, J., THESLEFF, S.: The "desensitizing" effect of ACh on the mammalian motor end-plate. Acta physiol. scand. 43, 15—26 (1958).
BARLOW, R. B., ZOLLER, A.: Some effects of long chain polymethylene bisonium salts on junctional transmission in the peripheral nervous system. Brit. J. Pharmacol. 23, 131—150 (1964).
BLACKMAN, J. G., GINSBORG, B. L., RAY, C.: The release of acetylcholine at a ganglionic synapse. J. Physiol. (Lond.) 162, 58 P—59 P (1962).
— — — Synaptic transmission in the sympathetic ganglion of the frog. J. Physiol. (Lond.) 167, 355—373 (1963).
BLOOM, G., ÖSTLUND, E., EULER, U. S. VON, LISHAJKO, F., RITZEN, M., ADAMS-RAY, J.: Studies on catecholamine-containing granules of specific cells in cyclostome hearts. Acta physiol. scand. 53, Suppl. 185 (1961).
BROCK, L. G., COOMBS, J. S., ECCLES, J. C.: The recording of potentials from motoneurones with an intracellular electrode. J. Physiol. (Lond.) 117, 431—460 (1952).
BROOKS, V. B.: An intracellular study of the action of repetitive nerve volleys and of Botulinum toxin on miniature end-plate potentials. J. Physiol. (Lond.) 134, 264—277 (1956).
BRUNER, J., TAUC, L.: Long lasting phenomena in the molluscan nervous system. In: Neural and humoral mechanisms of nervous integration. Exp. Biol. Symp. St.-Andrews 1965.
— — Habituation at the synaptic level in *Aplysia*. Nature (Lond.) 210, 37—39 (1966).
BURGEN, A. S. V., DICKENS, F., ZATMAN, L. J.: The action of Botulinum toxin on the neuromuscular junction. J. Physiol. (Lond.) 109, 10—24 (1949).
CASTRO, F. DE: Sympathetic ganglia, normal and pathological. In: Cytology and cellular pathology of the nervous system. Ed. by HOEBER. I, p. 317—379. New York: Penfield.
CHIARANDINI, D. J., GERSCHENFELD, H. M.: Ionic mechanism of cholinergic inhibition in molluscan neurons. Science 156, 1595—1596 (1967).
— STEFANI, E., GERSCHENFELD, H. M.: Ionic mechanisms of cholinergic excitation in molluscan neurons. Science 156, 1597—1599 (1967).

COOMBS, J. S., ECCLES, J. C., FATT, P.: Excitatory synaptic action in motoneurones. J. Physiol. (Lond.) **130**, 374—395 (1955).

CURTIS, D. R., ECCLES, R. M.: The excitation of renshaw cells by pharmacological agents applied electrophoretically. J. Physiol. (Lond.) **141**, 435—445 (1958).

— — The time course of excitatory and inhibitory synaptic actions. J. Physiol. (Lond.) **145**, 529—546 (1959).

— PHILLIS, J. W., WATKINS, J. C.: The chemical excitation of spinal neurones by certain acidic amino acids. J. Physiol. (Lond.) **150**, 656—682 (1960).

— RYALL, R. W.: The excitation of renshaw cells by cholinomimetics. Exp. Brain Res. **2**, 49—65 (1966a).

— — The acetylcholine receptors of renshaw cells. Exp. Brain Res. **2**, 66—80 (1966b).

— — The synaptic excitation of renshaw cells. Exp. Brain Res. **2**, 81—96 (1966c).

— — WATKINS, J. C.: The action of cholinomimetics on spinal interneurones. Expl. Brain Res. **2**, 97—106 (1966).

— WATKINS, J. C.: Acidic amino acids with strong excitatory actions on mammalian neurons. J. Physiol. (Lond.) **166**, 1—14 (1963).

DALE, H. H.: Pharmacology and nerve endings. Proc. roy. Soc. Med. **28**, 319—332 (1935).

— Transmission of effects from nerve endings. London: Oxford University Press 1952.

ECCLES, J. C.: The physiology of synapses. Berlin-Göttingen-Heidelberg-New York: Springer 1964.

— IGGO, A., LUNDBERG, A.: Electrophysiological investigations on renshaw cells. J. Physiol. (Lond.) **159**, 461—478 (1961).

— LIBET, B.: Origin and blockade of the synaptic responses of curarized sympathetic ganglia. J. Physiol. (Lond.) **157**, 484—503 (1961).

ELFVIN, L. G.: The ultrastructure of the superior cervical ganglion of the cat. II. The structure of the preganglionic end fibers and the synapses as studied by serial sections. J. Ultrastruct. Res. **8**, 441—476 (1963).

FEDOROV, B., MATWEJEWA, S. J.: La structure des connexions interneuronales dans le système autonome de la Grenouille. Trab. Lab. Invest. Biol. Univ. Madr. **30**, 379—401 (1935).

FRANK, K., TAUC, L.: Voltage-clamp studies of molluscan neuron membrane properties. In: The cellular functions of membrane transport. HOFFMAN, J. F., Ed. Englewood Cliffs, N. J.: Prentice-Hall 1964.

FRAZIER, W. T., KANDEL, E. R., KUPFERMAN, I., WAZIRI, R., COGGESHALL, R. E.: Morphological and functional properties of identified neurons in the abdominal ganglion of *aplysia californica*. J. Neurophysiol. **30**, 1288—1351 (1967).

GERSCHENFELD, H. M., ASCHER, P., TAUC, L.: Two different excitatory transmitters acting on a single molluscan neurone. Nature (Lond.) **213**, 358—359 (1967).

— CHIARANDINI, D. J.: Ionic mechanism associated with non-cholinergic synaptic inhibition in molluscan neurons. J. Neurophysiol. **28**, 710—723 (1965).

— STEFANI, E.: 5-Hydroxytryptamine receptors and synaptic transmission in molluscan neurones. Nature (Lond.) **189**, 924—925 (1965).

— TAUC, L.: Différents aspects de la pharmalogie des synapses dans le système nerveux central des mollusques. J. Physiol. (Paris) **56**, 360—361 (1964).

HAMBERGER, B., NORBERG, K.-A., SJOQVIST, F.: Evidence for adrenergic nerve terminals and synapses in sympathetic ganglia. Int. J. Neuropharmacol. **2**, 279—282 (1963a).

— — — Cellular localization of mono-amines in sympathetic ganglia of the cat. A preliminary report. Life Sci. **9** 659—661 (1963b).

— — — Correlated studies of mono-amines and acetylcholinesterase in sympathetic ganglia, illustrating the distribution of adrenergic and cholinergic neurons. Proc. 2. Int. Pharmacol. Meeting 1963. In: Pharmacology of cholinergic and adrenergic transmission. Praha: Czechoslovak Medical Press 1965, 41—53.

HUBER, G. C.: A contribution on the minute anatomy of the sympathetic ganglia of the different classes of vertebrates. J. Morphol. **16**, 27—90 (1900).

KANDEL, E. R., FRAZIER, W. T., WAZIRI, R., COGGESHALL, R. E.: Direct and common connections among identified neurons in *aplysia*. J. Neurophysiol. **30**, 1352—1376 (1967).

— TAUC, L.: Mechanism of prolonged heterosynaptic facilitation. Nature (Lond.) **202**, 145—147 (1964).

KANDEL, E. R., TAUC, L.: Heterosynaptic facilitation in neurones of the abdominal ganglion of *aplysia depilans* J. Physiol. (Lond.) 181, 1—27 (1965a).
— — Mechanism of heterosynaptic facilitation in the giant cell of the abdominal ganglion of *aplysia depilans*. J. Physiol. (Lond.) 181, 28—47 (1965b).
— WACHTEL, H.: The functional organization of neural aggregate in *aplysia*. In: Physiological and biochemical aspects of nervous integration. CARLSON, F. D., (Ed.), Englwood Cliffs, N. J.: Prentice Hall 1968.
KATZ, B.: The transmission of impulses from nerve to muscle, and the subcellular unit of synaptic action. Proc. roy. Soc. B. 155, 455—479 (1962).
— THESLEFF, S.: A study of the "desensitization" produced by acetyl-choline at the motor end-plate. J. Physiol. (Lond.) 138, 63—80 (1957).
KEHOE, J. S.: Pharmacological characteristics and ionic bases of a two component postsynaptic inhibition. Nature (Lond.) 215, 1503—1505 (1967).
— Double inhibition de certains neurones d'Aplysie. J. Physiol. (Paris) 60, 266 (1968).
KERKUT, G. A., MEECH, R. W.: The internal chloride concentration of H and D cells in the snail brain. Comp. Biochem. Physiol. 19, 819—832 (1966a).
— — Microelectrode determination of the intracellular chloride concentration in nerve cells. Life Sci. 5, 453—456 (1966b).
— THOMAS, R. C.: Acetylcholine and the spontaneous inhibitory postsynaptic potentials in the snail neurone. Comp. Biochem. Physiol. 8, 39—45 (1963a).
— — Anion permeability of the inhibitory postsynaptic membrane of *Helix* neurones. J. Physiol. (Lond.) 168, 23—24 P (1963b).
— — The effect of anion injection and changes in the external potassium of the IPSP and acetylcholine. Comp. Biochem. Physiol. 11, 199—213 (1964).
KHROMOV-BORISOV, N. V., MICHELSON, M. J.: The mutual disposition of cholino receptors of locomotor muscles, and the changes in their disposition in the course of evolution. Pharmacol. Rev. 18, 1051—1090 (1966).
KOKETSU, K., NISHI, S.: Characteristics of the slow inhibitory postsynaptic potential of bull-frog sympathetic ganglion cells. Life Sci. 6, 1827—1836 (1967).
KOSTERLITZ, H. W., LEES, G. M., WALLIS, D. I.: Resting and action potentials recorded by the sucrose gap method in the superior cervical ganglion of the rabbit. J. Physiool. (Lond.) 195, 39—53 (1968).
LARRABEE, M. G., KLINGMAN, J. D., LEICHT, W. S.: Effects of temperature, calcium and activity on phospholipid metabolism in a sympathetic ganglion. J. Neurochem. 10, 549—570 (1963).
— LEICHT, W. S.: Metabolism of phosphatidyl inositil and other lipids in active neurones of sympathetic ganglia and other peripheral nervous tissues. The site of the inositide effect. J. Neurochem. 12, 1—13 (1965).
LIBET, B.: Slow synaptic responses in sympathetic ganglia. Fed. Proc. 21, 345 (1962).
— Slow synaptic responses and excitory changes in sympathetic ganglia. J. Physiol. (Lond.) 174, 1—25 (1964).
— Postsynaptic nature and long synaptic delays of the slow responses in sympathetic ganglia. Physiologist 8, 219 (1965).
— Slow synaptic responses in autonomic ganglia. In: Studies in physiology. Berlin-Heidelberg-New York: Springer 1965.
— Long latent periods and further analysis of slow synaptic responses in sympathetic ganglia. J. Neurophysiol. 30, 494—514 (1967).
— KOBAYASHI, H.: Generation of adrenergic and cholinergic potentials in sympathetic ganglion cells. Science 164, 1530—1532 (1969).
— KOBAYASHI, M.: Electrogenesis of slow postsynaptic potentials in sympathetic ganglion cells. Fed. Proc. 27, 750 (1968).
— TOSAKA, T.: Slow postsynaptic potentials recorded intracellularly in sympathetic ganglia. Fed. Proc. 25, 270 (1966).
— — Slow inhibitory and excitatory postsynaptic responses in single cells of mammalian sympathetic ganglia. J. Neurophysiol. 32, 43—50 (1969).
MAJOROW, V. N.: Über den Ausschau der lebenden perizellulären Apparate. Z. mikr.-anat. Forsch. 65, 547—556 (1959).

MAJOROW, V. N.: Neue Angaben über den Aufbau der lebenden und absterbenden Neurone sowie der interneuronalen Verbindungen. Z. mikr.-anat. Forsch. 66, 225—235 (1960).

NISHI, S., KOKETSU, K.: Electrical properties and activities of single sympathetic neurons in frogs. J. cell. comp. Physiol. 55, 15—30 (1960).

— — Early and late afterdischarges of Amphibian sympathetic ganglion cells. J. Neurophysiol. 31, 109—121 (1968).

— KOKETSU, J.: Analysis of slow inhibitory postsynaptic potential of bullfrog sympathetic ganglion. J. Neurophysiol. 31, 717—728 (1968).

— SOEDA, H.: The electrical activities of sympathetic B and C neurons and the mode of presynaptic innervation. Kurume med. J. 9, 178—192 (1962).

— — KOKETSU, K.: Studies on sympathetic B and C neurons and patterns of preganglionic innervation. J. cell. comp. Physiol. 66, 19—32 (1965).

NORBERG, K.-A., HAMBERGER, B.: The sympathetic adrenergic neuron. Acta physiol. scand. 63, Suppl. 238, (1964).

— SJOQVIST, F.: New possibilities for adrenergic modulation of ganglionic transmission. II. catecholamine Symp. Milan (Italy) 1965. Pharmacol. Rev. 18, 743—751 (1966).

PICK, J.: The submicroscope organisation of the sympathetic ganglion in the frog (rana pipiens). J. comp. Neurol. 120, 409—462 (1963).

QUASTEL, D. M. J., CURTIS, D. R.: A central action of hemicholinium. Nature (Lond.) 208, 192—194 (1965).

RENSHAW, B.: Central effects of centripetal impulses in axons of spinal ventral roots. J. Neurophysiol. 9, 191—204 (1946).

ROBERTIS, E. DE, BENNETT, H. S.: Some features of the submicroscopic morphology of synapses in frog and earthworm. J. biophys. biochem. Cytol. 1, 47—58 (1955).

SMIRNOW, A.: Zur Kenntnis der Morphologie der sympathischen Ganglienzellen beim Frosch. Anat. Hefte, Abt. I, 14, 409—432 (1900).

STRUMWASSER, F.: Postsynaptic inhibition and excitation produced by different branches of a single neuron and the common transmitter involved Proc. 22. Int. Congr. Physiol. Sci., Leiden 2 (1962).

TAKESHIGE, C., VOLLE, R. L.: Modification of ganglionic responses to cholinometic drugs following preganglionic stimulation, anticholinesterase agents and pilocarpine. J. Pharmacol. exp. Ther. 146, 335—343 (1964b).

— — A comparison of the ganglion potentials and block produced by acetylcholine and tetramethylammonium. Brit. J. Pharmacol. 23, 80—89 (1964c).

TAKEUCHI, A., TAKEUCHI, N.: Active phase of frog's end-plate potential. J. Neurophysiol. 22, 395—411 (1959).

TAUC, L.: Processus postsynaptique d'excitation et d'inhibition dans le soma neuronique de l'Aplysie et de l'Escargot. Arch. ital. Biol. 96, 78—110 (1958).

— Preuve expérimentale de l'existence de neurones intermédiaires dans le ganglion abdominal de l'Aplysie. C. R. Acad. Sci. (Paris) 248, 853—856 (1959a).

— Sur la nature de l'onde de surpolarisation de longue durée observée parfois après l'excitation synaptique de certaines cellules ganglionnaires de Mollusques. C. R. Acad. Sci. (Paris) 249, 318—320 (1959b).

— Inhibition présynaptique dans les neurones centraux de l'Aplysie. C. R. Acad. Sci. (Paris) 259, 885—888 (1964).

— Presynaptic inhibition in the abdominal ganglion of Aplysia. J. Physiol. (Lond.) 181, 282—307 (1965).

— Physiology of the nervous system. In: Physiology of Mollusca WILBUR, K. M., YONGE, C. M. (Eds.). New York: Academic Press Inc. 1966.

— Physiology of the nervous system. In: Physiology of Mollusca WILBUR, K. M., YONGE, C. M. (Eds.). New York: Academic Press Inc. 1966.

— Transmission in invertebrate and vertebrate ganglia. Physiol. Rev. 47, 522—593 (1967).

— Some aspects of the postsynaptic inhibition in Aplysia. In: Structure and functions of inhibitory neuronal mechanisms. Proc. 4. Int. Meeting Neurobiol., 1966, Stockholm. Oxford: Pergamon Press 1968.

— BRUNER, J.: "Desensitization" of cholinergic receptors by acetyl choline in molluscan central neurones. Nature (Lond.) 198, 33—34 (1963).

Tauc, L., Epstein, R.: Heterosynaptic facilitation as a distinct mechanism in *Aplysia*. Nature (Lond) **214**, 724—725 (1967).
— Gerschenfeld, H. M.: Cholinergic transmission mechanisms for both excitation and inhibition in molluscan central synapses. Nature (Lond.) **192**, 366—367 (1961).
— — A cholinergic mechanism of inhibitory synaptic transmission in a molluscan system. J. Neurophysiol. **25**, 236—262 (1962).
Taxi, J.: Contribution à l'étude des connexions des neurones moteurs du système nerveux autonome. Thèse Sciences, Paris, 1964. Ann. Sci. Nat. Zool. **7**, 413—674 (1965).
Tosaka, T., Chichibu, S., Libet, B.: Intracellular analysis of slow inhibitory and excitatory postsynaptic potentials in sympathetic ganglia of the frog. J. Neurophysiol. **31**, 396—409 (1968).
Uchizono, K.: On different types of synaptic vesicles in the sympathetic ganglia of amphibia. Jap. J. Physiol. **14**, 210—219 (1964).
— Characteristics of excitatory and inhibitory synapses in the central nervous system of the cat. Nature (Lond.) **207**, 642—643 (1965).
— Inhibitory and excitatory synapses in vertebrate and invertebrate animals. In: Structure and function of neuronal inhibitory mechanisms. Proc. 4th Int. Meeting Neurobiol. 1966. Stockholm, Oxford: Pergamon Press 1968.
Volle, R. L.: Interactions of cholinometic and cholinergic blocking drugs at sympathetic ganglia. In: Pharmacology of cholinergic and adrenergic transmission. Proc. 2. Int. Pharmacol. Meeting. Prague: Czech. Med. Press 1965.
— Modification by drugs of synaptic mechanisms in autonomic ganglia. Pharmacol. Rev. **18**, 839—869 (1966).
— Muscarinic and nicotinic stimulant actions at autonomic ganglia. In: Int. Encyclop. Pharmacol. Therap., Sect. 12. Vol. I. Oxford: Pergamon Press 1966.
Vulfius, E. A., Veprintzev, B. N., Zeimal, E. V., Michelson, M. J.: Arrangement of cholinoreceptors on the neuronal membrane of two pulmonate gastropods. Nature (Lond.) **216**, 400—401 (1967).
Wachtel, H., Kandel, E. R.: A direct synaptic connection mediating both excitation and inhibition. Science **158**, 1206—1208 (1967).
Williams, T. H. W.: Electron microscopic evidence for an autonomic interneuron. Nature (Lond.) **214**, 309—310 (1967).

Chapter 4

The General Electrophysiology
of Input Membrane in Electrogenic Excitable Cells

By

Harry Grundfest, New York (USA)

With 17 Figures

Contents

Introduction

The general electrophysiology of sensory receptors is concerned basically with the mechanisms by which changes in the environment are transformed (*transduced*) into a language that is intelligible to the nervous system. To satisfy this function, the signals transported by the sensory neuron must carry a reasonable facsimile of the information that was received (*sensed*) at the input. They must also be capable of propagation, including the capacity for transmission from one cell to another in the synaptic and/or ephaptic chains of the nervous system. In this context, the principles of the general electrophysiology of sensory receptors can be encompassed within the varieties of activity that bring the sensory data into the nervous system via the individual sensory neuron. The various requirements for specificity of the information and for its modulation with respect to stimulus conditions or states of the organism generally involve large assemblies of neurons. Although these functions are vital to the total behavioral reactions of the animal to the primary stimulus, they are of secondary importance in the present context.

The recognition that electrogenic membranes may include a number of components with fundamentally different excitability properties (GRUNDFEST, 1957a, b, c, d, 1959a, 1961a, b, 1966a, b, 1967a) has led to a clarification of the functional characteristics of various receptor systems. Different combinations of the various electrogenic components of the cell membrane give rise to different phenomenological manifestations of the electrogenesis that is associated with reception of the sensory stimulus, with its transduction into the signal code of the organism, and with delivery to the nervous system (GRUNDFEST, 1958a, b, 1959b, 1961a, d, 1964a, 1965).

I. Classification of Sensory Systems

The sensory neuron is the element that delivers the sensed message to the higher order neuronal assemblies of the nervous system; thus, it is essential to all sensory systems. The sensory neuron, however, may be the primary, secondary, or tertiary target of the specific stimulus, depending upon the stimulus-response linkages of various receptor systems.

Only primary sensory neurons are excited directly by the specific stimulus, e.g., the mechanosensory crayfish and lobster stretch receptors (*see* TERZUOLO, this volume), chordotonal organs of arthropods (COHEN, 1963), the vertebrate mechanosensory Pacinian corpuscles and muscle proprioceptors (*see* LOEWENSTEIN, this volume; OTTOSON and SHEPHERD, this volume), the chemosensory olfactory epithelium (MOULTON and BEIDLER, 1967) and carotid body of vertebrates (EYZAGUIRRE and KOYANO, 1965), and the sensillary and other sensory neurons of insects (BOECKH, *et al.* 1965; DETHIER, 1963; WOLBARSHT, 1965; HODGSON, 1964).

Systems that include secondary sensory neurons are exemplified by the vertebrate chemosensitive taste receptors (BEIDLER, 1965; OAKLEY and BENJAMIN, 1966), electro- and mechano-receptors of the lateral line, and the auditory and vestibular receptors (DIJKGRAAF, 1967; BENNETT, this volume; FLOCK, this volume; FURUKAWA and ISHII, 1967a, b), and by the photosensory visual systems of arthropods (*see* FUORTES, this volume). A specialized taste bud, hair cell, electroreceptor cell, and retinular cell, are the primary targets of the stimulus. In turn, these *receptor cells* (GRUNDFEST, 1958a, 1959b, 1965; DAVIS, 1961, 1965) excite the sensory neuron.

The only presently known case where the sensory neuron is a tertiary target is that of the vertebrate retina. The primary targets are the rods and cones. They, in turn, act upon intermediate cells (bipolar, horizontal, and/or amacrine cells) which then relay the message to the sensory neurons (ganglion cells). The anatomical and presumably also the functional interrelations among these various components are very complex.

In all cases the sensory neuron must be regarded as the final common path that carries a train of spikes into the nervous system. The interposition of one or more receptor cells in the stimulus-response linkage presumably introduces more degrees of freedom for the functional performance of the sensory system (GRUNDFEST, 1964a, 1965). However, it also gives rise to some complexities in the electrophysiological manifestations of the sensory activity.

II. The Sensory Neuron

a) **Essentiality of Spike Electrogenesis.** The requirement for speed in delivering a message ruled out diffusional, and in most cases also humoral circulatory mechanisms and called for electrical propagation. The neuron is not a generator of strong electric power. Furthermore, its axonal portion is a poor electrical conductor, since it is a strand of very small diameter, filled with a medium of relatively high specific resistivity and is subject to large resistive and capacitive losses (HODGKIN and RUSHTON, 1946; DAVIS and LORENTE DE NÓ, 1947). The invention of spike electrogenesis thus was essential to overcome the cable losses of the axon. Since conduction of information is its prime function, the neuron with a long axon is the only cell for which spike electrogenesis is a sine qua non.

The means for spike electrogenesis are severely limited by the characteristics of cells and their environment (GRUNDFEST, 1963, 1966a). Under most conditions, the only variety of autogenetic bioelectrogenesis (i.e., one that is capable of generating a current) is a shift of the membrane potential from a resting inside-negative state, toward or into inside-positivity, an effect that is achieved by an increased membrane permeability for an appropriate ion (HODGKIN and HUXLEY, 1952; GRUNDFEST, 1966a). Thus, in order to permit the regenerative performance that results in an all-or-none spike, the trigger for the spike electrogenesis must also be a depolarization.

b) **The Generator Potential.** This requirement, therefore, sets the condition for the electrogenesis of the receptive or input section of the sensory neuron (Fig. 1). The input depolarization, or *generator potential* (GRANIT, 1955), is evoked by the activity of a specialized component of the cell membrane. As a rule, this input component responds only to specific stimuli and, except in electroreceptors (GRUNDFEST, 1959b, 1964a) it is *electrically inexcitable* (GRUNDFEST, 1957a, b), a property that permits the input transducer to have an approximately linear transfer function (GRUNDFEST, 1959b, 1965). The generator depolarization is graded in proportion to the stimulus intensity and usually it can last as long as the stimulus is applied.

The depolarization of the input membrane component, in turn, affects the *electrically excitable* spike generating and conductile component that predominantly, or exclusively, constitutes the reactive membrane of the axon. The interaction of the generator potential upon the conductile component leads to one

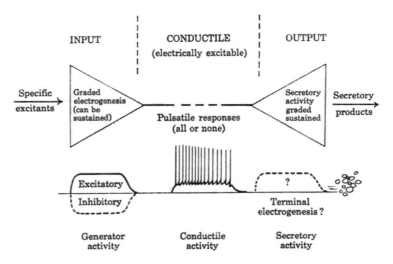

Fig. 1. Diagrammatic representation of functional components and electrical responses of a receptor or correlational neuron. The electrically inexitable input produces electrogenesis, which is graded in proportion to its specific stimulus and is usually sustained as long as the stimulus is applied. The possibility of hyperpolarizing electrogenesis, which may be produced by inhibitory synaptic membrane, is shown but is not further considered. The depolarization at the input, operating upon the conductile electrically excitable component, can evoke spikes. The spikes are encoded by number and frequency in proportion to the generator depolarization. These pulsatile signals, propagated to the output, command secretory activity there, roughly proportional to the information encoded in the message of the pulses and sustained as long as the message demands. The transmitter which is released at the output can initiate a synaptic transfer by operating upon the input of another cell. The possibility of a special output electrogenesis is indicated, but is not further considered. (From Grundfest, 1957a.)

or more spikes of the latter. Since the spikes are digital messages, information regarding the amplitude and duration of the specific stimulus to the input must be encoded into a frequency-number pattern of spikes.

c) **Inhibitory Input.** In many and perhaps in the majority of cases, the sensory neuron also receives inhibitory innervation. The activity of these nerve fibers evokes activity of an inhibitory postsynaptic component of the sensory neuron, so that the amplitude and effectiveness of the generator potential are thereby depressed. The interrelations between the excitatory and inhibitory mechanisms may be such that they modulate the excitatory activity. In a number of cases, for example, in the eccentric cell of *Limulus*, the inhibitory activity is so organized as to restrict the number of sensory neurons that report on a stimulus to the receptor system. It is believed that this process of *lateral* *inhibition* (HARTLINE and RATLIFF, 1957) plays an important role in sensory perception. A collateral of the axon of an eccentric cell also activates its own inhibitory synapses (*self-inhibition*; see PURPLE and DODGE, 1965). The inhibitory synapses of sensory neurons have the same properties as do synapses in general; i.e., their membrane is electrically inexcitable.

d) Output Activity. The information encoded in a frequency-number message of spikes is carried without decrement to the output terminals of the neuron, where as a rule there is conversion of the electrical message into a secretory activity. The transmitter agent released by that activity in turn acts upon the input of the next (postsynaptic) cell which may be a neuron or an effector cell.

e) Presynaptic Modulation. The presynaptic terminals may receive synaptic inputs from other neurons, by which the output activity of the terminals may be modulated. The modulation may be excitatory or inhibitory in effect. Such connections have been found on the dorsal root afferents to the spinal cord (WALL, 1964; FETZ, 1968).

f) Transmission from the Sensory Neuron. In general, the sensory neuron forms synaptic contacts with the next line of neurons. The postsynaptic membrane component that receives the incoming information is also electrically inexcitable and reacts only to a rather restricted group of agents that are chemically and structurally related to the natural transmitter. The activity of the postsynaptic input may be an electrogenesis that is either excitatory or inhibitory, depending upon the specific characteristics of the input membrane. In vertebrate neurons and arthropod muscle fibers, both the excitatory and inhibitory membrane components are intermingled and may be interspersed among electrically excitable components.

It is possible that the transmission may be ephaptic (GRUNDFEST, 1959a, 1967b), i.e., by electrotonic spread of depolarization from the pre-fibers to the post-cell. This type of transmission is usually, but not necessarily, mediated by "tight junctions" (BENNETT, 1966). Such sites of membrane fusion have been observed, e.g., in the retina (YAMADA and ISHIKAWA, 1965; DOWLING and BOYCOTT, 1965). However, as will be described below, other conditions are also necessary for ephaptic transmission and probably do not exist in the retinal system.

g) Identification of the Membrane Components. There is no method presently available for morphological or chemical characterization of the different components of a sensory neuron. However, the electrophysiological and pharmacological differences that distinguish these functionally different components are clear-cut (GRUNDFEST, 1957c, 1961a, b, 1966a; GRUNDFEST and BENNETT, 1961). For example, in some cells there are large discrete regions of the surface that possess the properties only of various electrically inexcitable or only of various electrically excitable components. Functional isolation of one or another component may also be achieved by the use of pharmacological agents that act specifically upon one or another component. Thus, it has been possible by such means to demonstrate that the secretory activity of presynaptic terminals, as well as the electrogenesis of input components, are independent of spike electrogenesis (GRUNDFEST, 1961b, 1964b; REUBEN and GRUNDFEST, 1960; LOEWEN-

STEIN *et al.*, 1963; NAKAJIMA, 1964; KATZ and MILEDI, 1967a, b; OZEKI *et al.*, 1966; BLOEDEL *et al.*, 1966; KUSANO *et al.*, 1967).

The region of transition from the input component to the spike generator is not precisely delineated in most sensory neurons. For example, there appear to be "patches" of electrically excitable membrane along the dendrites of the crustacean stretch receptor cell (Fig. 2). These sites are too sparse to evoke action

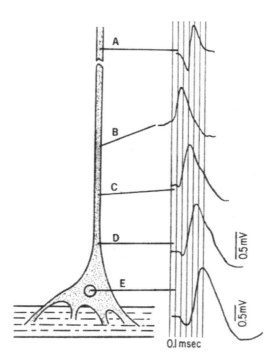

Fig. 2. The probable density of electrically excitable membrane components (dots) in different portions of the lobster stretch receptor (left) and the site of the spike trigger zone (right). The dots are sparse in the dendrite portion and become more dense along the axon, and particularly at the region of the trigger zone. (Modified from GRAMPP, 1966.) The extracellular recordings were of responses to a stretch stimulus. The spikes occurred first at *B*, and propagated antidromically (*C-E*) to invade the cell body, as well as orthodromically (*A*) along the axon. Site *B* was about 0.5 mm from the cell body; site *A* was about 0.8 mm from *B*. Long vertical lines represent 100-μsec intervals. Short vertical lines represent 0.5 mV. *A-D* were recorded at the same amplification. (Modified from EDWARDS and OTTOSON, 1958)

potentials of all-or-none characteristics, but under appropriate conditions can lead to graded electrically excitable responses (GRUNDFEST, 1957b, c). The receptor region of the Pacinian corpuscle may also possess some electrically excitable membrane, the electrogenesis of which can be eliminated by tetrodotoxin, whereas that of the receptor component remains (LOEWENSTEIN *et al.*, 1963; HUNT and TAKEUCHI, 1963; SATO and OZEKI, 1963; OZEKI and SATO, 1965). In some receptor neurons, the presence of spike electrogenesis at the dendrite would appear to be obligatory. For example, the sensillary dendrites of insects are relatively long and thin and are accessible to natural stimulation at only a small sensillary pore (see Boeckh et al., 1965). There would be little likelihood that the generator potential localized at the tip region could excite spike electrogenesis in the cell body or axon[1]. Thus, it is not surprising that the dendrite also generates a spike (Wolbarsht and Hanson, 1965). Indeed, even the morphological distinction between "dendrite" and "axon" is largely lost in the spinal afferent neurons of vertebrates. The receptor portion is confined to a small region, which is particularly well delineated in the Pacinian corpuscle. The major portion of the neurite is an axon and the cell body is a mere appendage .It is unnecessary for the conductile performance of the neuron although, of course, it is vital for maintaining the integrity of the cell.

III. Secondary Sensory Systems

a) **Receptor-effector Cells.** Until recently, secondary systems were known only in vertebrates where they are represented in the taste, vestibulo-acoustic, and lateral-line sensory mechanisms. The primary target of the stimulus is a specialized receptor cell. This primary target cell as a rule is linked to the innervating afferent nerve fiber by synaptic transmission (Fig. 3). Thus, the primary cell is a receptor-effector unit which responds to the stimulus with secretory activity, releasing a specific transmitter that acts upon the input terminals of the nerve fiber (Grundfest, 1958a, b, 1959b, 1964a; see also Davis, 1961; Furukawa and Ishii, 1967a, b; Flock, 1967, and this volume; Bennett, 1967). However, at least in some electroreceptors, there is probably an electrotonic (ephaptic) linkage between the cell and the secondary neuron (Bennett, 1967). An ephaptic linkage apparently also occurs between the retinular cells of the *Limulus* eye, which are primary receptors, and the eccentric cell, which is the sensory neuron (Smith et al., 1965). In some cases, the receptor cell receives a synaptic input from an efferent fiber (Fig. 3). Presumably this innervation has a modulating effect on the output activity of the receptor cell.

b) **Linkage by Secretory Activity.** The presence of a primary receptor-effector cell in synaptic linkage with the afferent axon provides an extra degree of freedom with respect to the electrogenesis of the system. The primary cell is essentially neurosecretory in function (Grundfest, 1958a, b). It need not generate spikes, and it seems likely that many primary receptor cells are electrically inexcitable (Grundfest, 1964a, 1965). Indeed, an overt electrogenesis seems unnecessary for secretory activity; if an electrogenesis is present, it may be hyperpolarizing or depolarizing in sign. The sign and its presence or absence would be only epiphenomena, mere reflections of the ionic processes that might accompany the secretory activity (Grundfest, 1965). They might, in fact, be due to electrogenic pumping processes (Grundfest, 1955; Grundfest et al., 1954) rather than to specific changes in the permeability of the membrane.

---

[1] The dendrites of many neurons are predominantly or entirely electrically inexcitable, yet they are the loci of dense synaptic innervation. The function of such inputs is primarily to integrate information from many sources and to modulate upon the all-or-none outputs of the neurons (Grundfest, 1958c).

c) **Ephaptic Linkage.** In the case of ephaptic transmission from the primary receptor cell to the secondary neuron, however, the two cells are essentially in continuity and coupled through a low resistance (GRUNDFEST, 1959b; WATANABE and GRUNDFEST, 1961; BENNETT, 1966). The effective stimulus for the neuron

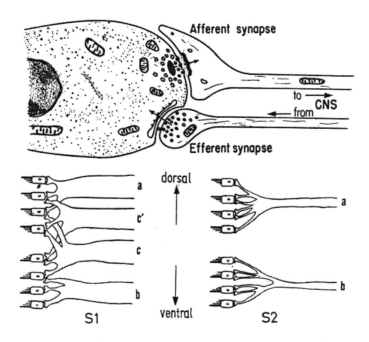

Fig. 3 Synaptic relations of hair cells. *Upper:* Lateral-line hair cell receives innervation from an efferent axon as well as sending information out by the afferent fiber. The efferent innervation presumably modulates the output of the hair cell. (Modified from FLOCK, 1967.) *Lower:* Probable connections in the goldfish of auditory hair cells with afferent nerve fibers. Note the different orientations of the kinocilia (thick lines) in the different hair cells. They determine the direction of the mechanoresponsiveness of the cells. S1 and S2 axons, which are connected to only one type of hair cell (*a*, *b*), respond only to the rarefaction or compression phase of the auditory stimulus, whereas S1 axons that are connected to both types of cells (*c*, *c'*) respond to both phases of stimulation. Responses of S1 and S2 axons are shown in Fig. 16. (Modified from FURUKAWA and ISHII, 1967b)

must then be a depolarization of the receptor cell. Ephaptic transmission appears to occur in the case of the *Limulus* lateral eye (see SMITH et al., 1965; FUORTES, this volume). Upon illumination, the retinular cells generate a depolarization which is impressed upon the eccentric cell; i.e., the generator potential is evoked in a photosensory receptor and is transmitted ephaptically to an electrically excitable component in the eccentric cell. It is not clear at present if the membrane of the eccentric cell also has an electrically inexcitable input component of depolarizing electrogenesis. The inhibitory synapses of the cells do, however, constitute an electrically inexcitable electrogenic input component (PURPLE and DODGE, 1965). The details of the electrogenic input in insect photoreceptors is also unclear at this time.

In addition to compound lateral eyes, *Limulus* also possesses photosensitive neurons in the so-called "olfactory nerves" (ventral eyes) and in the endoparietal and rudimentary lateral eyes (MILLECCHIA et al., 1966). These cells, like the retinular cells (BENOLKEN and RUSSELL, 1966; GRUNDFEST, 1967c), may generate graded electrically excitable responses, but although they possess axons, these seem to be nonfunctional since they do not generate spikes. There are no ephaptic contacts between adjacent cells (CLARK et al., 1969; MILLECCHIA et al., 1966; MILLECCHIA and MAURO, 1969a, 1969b).

IV. The Tertiary System of the Vertebrate Retina

a) **The Cellular Components.** Still more degrees of freedom are made possible by the complexities introduced by the structure-function relations of the vertebrate retina. The primary photoreceptors are dual in nature (rods and cones), and at least the cones are subdivided into several different varieties among those

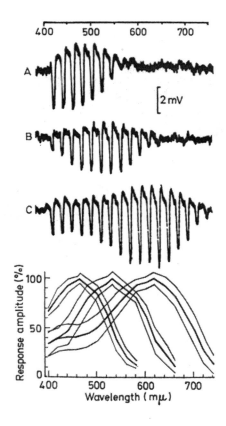

Fig. 4. Responses of single cones of carp to flashes of monochromatic lights of constant energy. *Above:* Electrogenesis in three different cones is mainly hyperpolarizing and lasts as long as does the stimulus. Adjacent flashes differed by steps of 20 mμ. The peak effectiveness was in the blue, green and red, respectively, for the three cells. *Below:* Average response curves (with standard deviation) for the three types of cones. (Modified from TOMITA et al., 1967.)

vertebrates that possess color vision. These complexities give rise to the various phenomena of scotopic and photopic vision and to some other aspects of the physiology of vision. The activity of these primary receptors thus seem to be transmitted to intermediate cells (horizontal, bipolar, and/or amacrine cells) which in turn are in connection with the tertiary neurons, or ganglion cells.

b) **Nature of the Intercellular Linkages.** It seems likely that all the interconnections between the rods and cones and the intermediate receptor cells are synaptic rather than ephaptic in character. As noted in the previous section,

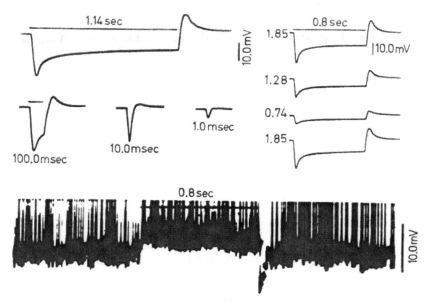

Fig. 5. Different kinds of responses of intermediate cells and of a ganglion cell of the cat retina. *Upper left:* Hyperpolarization that was graded in duration and amplitude was produced in the intermediate cell by flashes of constant intensity which varied in duration from 1 msec to 1.14 sec. *Upper right:* Gradation of the hyperpolarization by changing the intensity of a light flash of 0.8-sec duration. *Below:* Intracellular recording from a retinal ganglion neuron, showing depolarizing "generator" potential and augmented spike activity during a light flash of 0.8 sec. Since the depolarizing response of the ganglion cell was evoked in conjunction with hyperpolarization of the preceding element, electrical activation of the ganglion cells by the intermediate cells appears to be ruled out, and excitation probably occured by chemical mediators. (From GRUNDFEST, 1961d, modified from BROWN and WIESEL, 1959.)

ephaptic transmission requires that the coupling resistance between pre- and post-cells be small, and that the electrogenesis of the pre-cell be depolarizing and sufficiently large so as to elicit spikes in the post-cell. Only one group of investigators (TOMITA et al., 1967) has thus far succeeded in recording from single cones of the carp retina (Fig. 4). There are three varieties of cones, distributed rather as proposed in HECHT's trichromatic theory (HECHT, 1934; see also RUSHTON, 1965; WALD, 1964). From the standpoint of electrophysiology it is noteworthy that the recorded potentials are small (+10 mV maximally) and that the electrogenesis is predominantly hyperpolarizing and lasts as long as does the light stimulus (Fig. 4). Thus, it is unlikely that the secondary receptor cells, which are those that probably generate the S-potentials, would be excited ephaptically. This inference is supported by the findings (Figs. 5, 6) that some of the latter cells generate only hyperpolarizing long-lasting potentials (luminosity, or L-responses), whereas others respond with diphasic electrogenesis, the sign of the

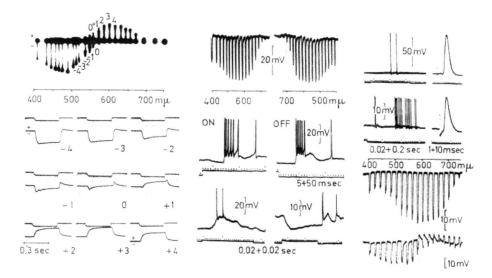

Fig. 6. Retinal activity in cold-blooded vertebrates. *Left column.* Fish retina. *Above:* Action spectrum by scanning method (wavelength scale immediately below). Stimuli of different wavelengths but of constant energy evoked hyperpolarizing electrogenesis for shorter wavelengths, and depolarizing activity at longer wavelengths ("chromaticity" response). *Below:* The responses numbered on the scanning trace above are swept out on a time base in the lower trace of each set. The upper traces monitor the light stimulus. Note the absence of spikes. The potentials last as long as does the stimulus. At the record marked 0 the hyperpolarizing potentials appear to be nearly balanced. (From MacNichol and Svaetichin, 1958.) *Center column.* Frog retina. *Top:* Only hyperpolarizations ("luminosity" responses) are observed in the action spectrum. *Middle:* Intracellular record from a ganglion cell which responded to both "on" and "off" of a stimulus. The latter and a time scale are shown on the trace below. The initial response was sustained hyperpolarization on which was superimposed a later depolarization accompanied by spikes. Toward the end of the illumination, the cell was hyperpolarized, but it depolarized and produced spikes when the stimulus was terminated. *Bottom:* Records from two ganglion cells. One neuron responded "on" with depolarization leading to spike. The other hyperpolarized during the stimulus and developed spikes only after the light was turned off. *Right column.* Two upper sets show intracellular records of responses in axons of frog optic nerve. Slow sweep traces (left) show a pair of spikes during illumination (above) and a train of spikes in another axon which responded only when the light was turned off. Fast sweep traces (right) show the spikes of the two axons evoked by electrical stimuli. (From Tomita et al, 1961.) Two lower sets of records, action spectra from turtle retina, one of "luminosity" responses, the other "chromaticity" responses, which change in sign during illumination with different wavelengths. (From Tomita, 1963.)

potential depending upon the wavelength of the stimulus (chromatic or C-potentials). The large, predominantly hyperpolarizing S-potentials in carp develop on a background of a resting potential of -20 to -60 mV and they may be some

30 to 50 mV in amplitude (Witkovsky, 1967). Thus, the large amplitudes as well as the signs of the potentials indicate that the electrogenesis of the S-potential is intrinsic to the intermediate cells[2].

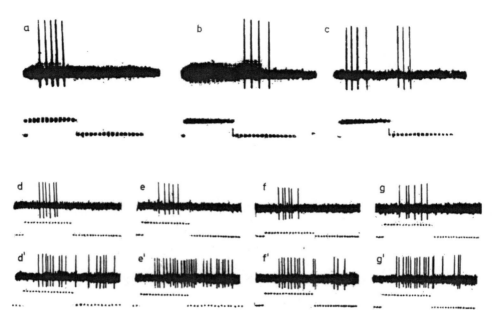

Fig. 7. Patterns of responses evoked in cat retinal neurons by different stimulating conditions. Extracellular recording from a single ganglion cell. a, "on" response during the light flash (signaled on lower trace) applied as 0.2-mm spot at central region of the receptive field for this neuron, at the site of the recording microelectrode. b, "off" response developed with spot moved 0.5 mm peripherally. c, at an intermediate site of stimulation "on-off" responses occurred. d, d'-g, g', another neuron subjected to changes in four stimulus parameters. In each case, the "on" response was converted to "on-off" activity. d, d', retinal area of the stimulating spot was increased from 0.2 to 3 mm. e, e', background illumination was decreased from 19 to 4 mc. f, f', the intensity of the light source was increased. g, g', the site of stimulation was shifted from the center of the receptive field to a more peripheral part. Time scale appears as 20-msec intervals on lower traces of all records. (Modified from Kuffler, 1953.)

The fact that the S-potentials are hyperpolarizing except for C-responses at some wavelengths (Fig. 6) immediately leads to the conclusion that the intermediate cells are also neurosecretory in function (Grundfest, 1958a, b), exciting the ganglion cells by release of transmitter agents. The responses of the ganglion cells may be depolarizations that generate spikes, or hyperpolarizations which tend to suppress spike electrogenesis (Figs. 5 and 6). There is also evidence for lateral inhibition and other functional complexities (Fig. 7). The interconnections among the various components are quite complex (Fig. 8). Thus, the intermediate

---

[2] Dethier (1953) has called attention to the general rule that the "resting potential" recorded between the cornea and the receptive surface of the retina is cornea-negative in invertebrates, but cornea-positive in vertebrates. The difference is in line with and might be related to the fact that the receptor electrogenesis is depolarizing in the arthropods whereas it is predominantly hyperpolarizing in the vertebrates (Grundfest, 1959b).

cells might have inhibitory and/or excitatory effects upon one another and upon the ganglion cells; there may also be feedback from the ganglion to the intermediate cells. However, these possible interactions cannot be characterized further at present.

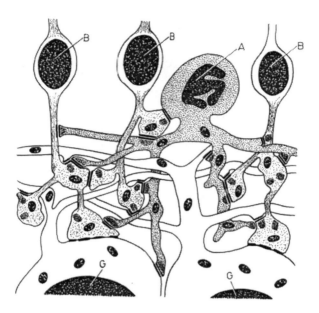

Fig. 8. The linkages of cells in the inner plexiform layer of the primate retina. Diagram based on electron micrographs. An amacrine cell (*A*) and three bipolar cells (*B*) form synaptic linkages with one another and with two ganglion cells (*G*). Two of the bipolar cells also form tight junctions on the cell bodies of the ganglion cells (From DOWLING and BOYCOTT, 1965.)

V. Nature of the Electrogenesis

a) Receptor and Intermediate Cells. It is not possible at this time to detail the mechanisms by which the primary or intermediate receptor cells carry on their transducer activity, or what are the relations between the latter and the potentials they may generate. Presumably the potentials might result from diffusional ion fluxes like the mechanisms that generate postsynaptic potentials (GRUNDFEST, 1961a, b, 1966a; ECCLES, 1964), from electrogenic pumps (GRUND-FEST, 1955; GRUNDFEST et al., 1954) such as produce the long lasting hyperpolarizing afterpotential of the slowly adapting crayfish stretch receptor (NAKA-JIMA and TAKAHASHI, 1966), or from combinations of these. It is noteworthy that the hyperpolarizing potential which accompanies activity of salivary glands and is apparently due to Cl movement does not appear to have a reversal potential (LUNDBERG, 1956), resembling in this respect the hyperpolarizing afterpotential of the crayfish stretch receptor, which results from the activity of an electrogenic Na pump (NAKAJIMA and TAKAHASHI, 1966).

b) Generator Potentials. The generator potentials of the sensory neuron appear to be due to the same ionic mechanism as the depolarizing excitatory postsynaptic potentials (EPSP) of postsynaptic membrane. The stimulus initiates

10*

an increased permeability of the receptive membrane for the monovalent cations Na and K. Since E_{Na} (the emf of the ionic battery for Na) is inside-positive and E_K (the emf of the K battery) is inside-negative, the maximum generator potential is a depolarization toward some intermediate value (GRUNDFEST, 1961 c). The ionic requirements of the generator electrogenesis have been studied most completely in the stretch receptors (EDWARDS *et al.*, 1963; OBARA and GRUNDFEST,

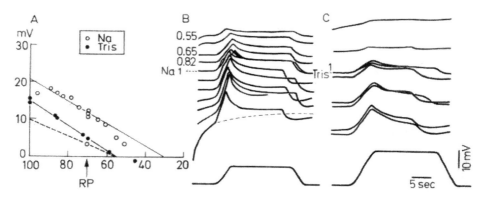

Fig. 9. The Na dependence of the reversal potential of the generator depolarization in crayfish stretch receptor. The amplitude of the generator depolarization in response to a constant stretch was determined while the membrane potential was changed by an intracellularly applied current. The data obtained in the presence of Na are plotted (*A*) as open circles and the records are shown in *B*. The membrane potential at which the generator potential may be expected to disappear (reversal potential) is —30 mV. The filled circles and the records in *C* show the data with Na replaced by Tris. The reversal potential is about —60 mV or about 10 mV positive to the resting potential which is shown by the arrow in *A*, and by "Na" and "Tris" in *B* and *C*. The four numbers in *B* indicate the changes in slope resistance of the cell during depolarizations relative to the resting resistance. The decrease in resistance accounts for the deviations from a linear relation in *A*. The lowest trace in the recordings represents the stretch stimulus. Since the stimuli were larger in *C* than in *B*, the broken line of *A* shows normalized values. (From OBARA, 1968.)

1968; OBARA, 1968). The reversal potential, which normally is some 30 mV inside-negative, becomes much more negative when Na is removed from the bathing medium being substituted with a less permeant cation (Fig. 9), but the generator potential is not affected when Cl is substituted with an impermeant anion. These data thus indicate that the generator potential, like the EPSP, is due to a simultaneous increase in diffusional flux for both Na and K.

In theory, of course, removal of Na should tend to abolish the generator potential, since the only remaining emf would be that of the K battery, which is at or close to the resting potential. This result has been observed at the excitatory synapses of crayfish muscle (OZEKI and GRUNDFEST, 1967), but the electrically inexcitable membrane of the stretch receptor does not discriminate strongly against various ionic substitutes for Na (Fig. 10). The "simple" photosensitive neurons of *Limulus* exhibit an interesting variation (Fig. 11). When Li or Tris is substituted for Na, the generator potential at first disappears, indicating that the membrane discriminates between Na and the substitute cations. However, after some time (5 to 15 min, typically) the neuron can develop a response to light. The generator potential is smaller than that evoked in the presence of Na. Thus, the receptive membrane appears to become modified after some time so as to lose its specificity for Na. (MILLECHIA and MAURO, 1969a).

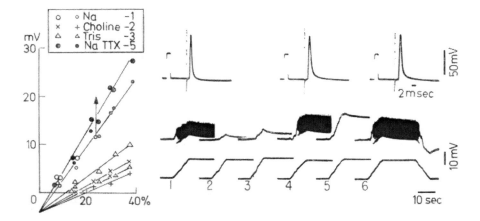

Fig. 10. Generator potentials elicited in slowly adapting crayfish stretch receptor neuron in the absence of Na. *Right:* Records of generator potential (middle trace) evoked by a constant stimulus (lower trace) in various media. (1) Preparation in the standard Na-containing saline. The stretch stimulus evoked a generator depolarization and a train of spikes. Upper trace shows a spike recorded in the cell following stimulation of the axon with a brief pulse. (2) Spike electrogenesis was abolished on replacing all the Na with choline. The stretch evoked a small generator potential. (3) The latter was larger after Tris was substituted for the choline, but spike electrogenesis was still absent. (4) Spike electrogenesis returned after Na was reintroduced. (5) Tetrodotoxin (TTX) was added. Spikes were abolished but the generator potential remained and its large amplitude was no longer masked by the spike electrogenesis. (6) Effects of TTX were reversible. *Left:* The amplitude of the generator potentials (ordinate) for various degrees of stretch (abscissa). Large symbols denote the peak of the generator potential, and the smaller symbols indicate the steady state values. (From OBARA, 1968.)

Generator potentials usually exhibit a high initial depolarization which subsides to a smaller steady value (Figs. 10 and 11). Some portion of the initial elevation may be due to electrically excitable depolarization. Even when there is no clear-cut spike electrogenesis the peak is sometimes accentuated by the occurrence of graded responses that can be eliminated by tetrodotoxin (GRUNDFEST, 1964a, 1965, 1967c). However, a time variant change in amplitude is seen in crayfish stretch receptors even after spike electrogenesis is eliminated by removal of Na (Fig. 10) or by applying TTX (cf. Fig. 14). This "generator adaptation" of mechanoreceptors appears to be due in large measure to visco-elastic deformations of the tissue. For example, the decrease in generator potential of the crayfish stretch receptor can be greatly diminished when the muscle fibers in which the dendrites are embedded are clamped to a constant length (NAKAJIMA and ONODERA, 1969b). On the other hand, mechanical deformation is probably not a factor in the generator adaptation observed in visual receptors (Figs. 5, 6 and 11). The presence of an electrically excitable conductance increase for K or Cl might induce a decline from the peak of the generator potential (GRUNDFEST, 1964a)[3]. However, it is very likely that other factors, such as a change in the characteristics of the electrically inexcitable transducer membrane, also participate. This matter and also the nature of spontaneous "miniature" depolarizations (see DOWLING, 1968) are at present far from clear.

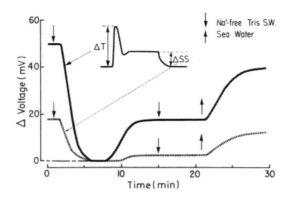

Fig. 11. Modification of the Na-dependence of generator membrane. Photosensitive neuron from the rudimentary eye of *Limulus*. The typical generator response to a light stimulus is shown diagrammatically in the inset. The peak (ΔT) may be contaminated with a graded response of electrically excitable depolarizing electrogenesis. The undershoot below the steady state (ΔSS) also indicates that the large initial depolarization had caused an electrically excitable conductance increase, probably for K. The first arrow of the graph denotes a change from the standard seawater to a Na-free (Tris) saline. The generator potential was abolished within about 4 min, but began to return again a few minutes later. Both phases of the responses were reduced in amplitude, however, They were not affected by exposure to a fresh Tris saline (second set of downward arrows), but the response increased again soon after restoration of the Na (upward arrows). (Unpublished data kindly provided by Drs. MILLECHIA and MAURO.)

c) Inhibitory Postsynaptic Potentials. Inhibitory electrogenesis of sensory neurons is similar to that of other neurons and of effector cells (GRUNDFEST, 1961b). The transmitter that is released from the inhibitory axon or mimetics of the transmitter that are applied to the neuron activate a specialized electrically inexcitable membrane component. The activation is manifested by an increased permeability for K or Cl. The conductance increase (Fig. 12) thus emphasizes the contribution of the relevant ionic battery to the membrane potential (HODGKIN and HUXLEY, 1952). Since E_K and E_{Cl} are close to or at the resting membrane potential, the activation of the inhibitory membrane tends to "clamp" or "poise" the average membrane potential to the region of the resting level. A concurrent tendency toward depolarization through the activity of the excitatory postsynaptic membrane is thereby diminished, whence the inhibitory action of the inhibitory postsynaptic potentials (IPSP).

However, if the emf of the ionic battery for the IPSP can be made depolarizing relative to the resting potential, the activation of the „inhibitory" synapse may cause sufficient depolarization so as to evoke a spike of the electrically excitable

---

[3] The presence of this conductance increase is usually evidenced by an undershoot on the generator potential. It may even cause the latter to appear to be oscillatory.

membrane. Most IPSPs, including those of the crayfish stretch receptor (HAGI-WARA et al., 1960; OBARA and GRUNDFEST, 1968), are due to a Cl battery. A shift of E_{Cl} toward positivity is induced rather easily by reducing the Cl in the medium

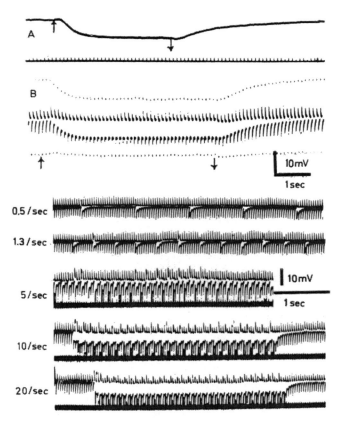

Fig. 12. Inhibitory postsynaptic activity induced in *Limulus* eccentric cell by antidromic stimulation. *Above:* The axon of the cell was stimulated at 30/sec (arrows). *A.* The antidromic invasion of the spike gave rise to a sustained hyperpolarizing IPSP. *B.* The cell was in a bridge configuration, with the antidromic stimulation delivered while brief pulses of current were applied intracellularly. Both sets of traces mirror the IPSP. The increased separation of the traces during the hyperpolarization denotes an increased conductance which is responsible for the synaptic electrogenesis. *Below:* A series of recordings with different frequencies of nerve stimulation. As the frequency increased, the IPSP fused and increased in amplitude. The change was associated with larger conductance increase. (Modified from PURPLE and DODGE, 1965.)

bathing the isolated preparation. The replacement of Cl with an impermeant anion causes the IPSP to become sufficiently depolarizing so as to evoke a spike of the sensory neuron (Fig. 13).

d) **The Ionic Basis of Spike Electrogenesis in the Sensory Neuron.** The spike-generating mechanism appears to be the same in all axons that have been studied to date. Spikes are abolished in crayfish stretch receptors, Pacinian corpuscles, eccentric cells of *Limulus* and honey bee, and insect sensillary receptors by applications of tetrodotoxin at about the same concentrations of the drug that are effective in blocking spikes of squid, lobster, crayfish and frog axons (GRUNDFEST, 1964b, 1965). In the crayfish stretch receptor, spikes are produced when all the Na is replaced with Li, but are blocked when Na is replaced with larger cations (Tris, choline) or with sucrose (OBARA and GRUNDFEST, 1968; see Fig. 10).[4]

The kinetics of ionic processes in soma and axon apparently may differ considerably since the threshold for the soma spike of the stretch receptor is higher than that of the axon (Fig. 2), and the spike electrogenesis of the soma is blocked

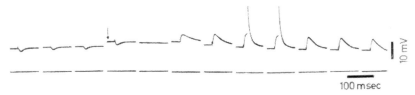

Fig. 13. Excitatory effects produced by depolarizing electrogenesis of the inhibitory synapses of crayfish stretch receptor. Stimulation of the inhibitor axon caused hyperpolarization (first three records) when the sensory neuron was in the standard bathing medium. At the arrow, the latter was replaced with a saline in which a large anion (glutamate) substituted for the Cl. The hyperpolarizing IPSP now became depolarizing to such an extent that they evoked spikes in the electrically excitable component. The lower line is a reference potential. (Modified from HAGIWARA et al., 1960.)

more readily by various conditions than is that of the axon (OBARA and GRUNDFEST, 1968). The soma may have very different ionic properties from the axon. In the sensory neurons of the frog dorsal root ganglion, the soma can generate spikes in the absence of Na when this cation is replaced by a variety of organic cations (see KOKETSU, 1961). It seems likely that the profound differences between the membrane of the soma and the axon are related to an anatomical feature (GRUNDFEST, 1967c). The cell body of these unipolar dorsal ganglion neurons lies on a branch of the afferent axon out of the direct line of progression of a spike on its way to the spinal cord. The elimination of the cell body of the unipolar neurons in the crab does not prevent the passage of the impulse along the axon (BETHE, 1897). Thus, it seems that the position of the cell body in a cul-de-sac made possible evolutionary experimentation that resulted in spike electrogenesis of a different type from that of the axonal membrane. Another possible evolutionary quirk has already been noted. The photosensitive cells of the accessory eyes of Limulus and most, if not all, the retinular cells of the compound eyes possess axons that do not generate spikes. The cell bodies appear to have morphological characteristics of neurons, but seem to have lost their conductile mechanism as a specialization, perhaps for neurosecretory activity.

e) **Electrogenesis at the Nerve Terminals.** Data on the activity of presynaptic terminals derive from only a few favorable preparations, and particularly

---

[4] Under various experimental conditions the squid giant axon can generate spikes utilizing emf's other than that of the Na battery (TASAKI, 1969; TASAKI et al., 1969; YAMAGISHI, 1969). However, although the spikes may be of large amplitude the currents that are associated with the electrogenesis are very small and in general would not cause the response to propagate.

from the squid giant axon synaptic system. Spike electrogenesis is not essential for transmitter release, since spontaneous miniature potentials remain in muscle fibers of arthropods (REUBEN and GRUNDFEST, 1960; GRUNDFEST, 1961b; OZEKI et al., 1966), and of vertebrates (ELMQVIST and FELDMAN, 1965) after the spikes of the axons are eliminated by saxitoxin or tetrodotoxin. EPSP are elicited in the squid giant axon (BLOEDEL et al., 1966; KATZ and MILEDI, 1967b; KUSANO et al., 1967), frog (KATZ and MILEDI, 1967a), and crayfish (OZEKI et al., 1966) muscle when the presynaptic axon is depolarized with applied current. The EPSP is graded, increasing to a maximum and then decreasing with increasing depolarization. Thus, this and other data indicate that the relation between membrane potential of the pre-fiber and the EPSP of the post-fiber in squid is complex. These findings and the persistence of spontaneous miniature EPSP in the neuromuscular preparations after axonal spikes are abolished suggest that membrane processes other than those of conductile electrogenesis are associated with transmitter release. Furthermore, in the receptor cells of the vertebrate retina, as well as in salivary gland cells, secretory activity is accompanied by hyperpolarization. It is likely, therefore, that the depolarization which is required for the transmitter release from axonal terminals initiates other events that are more directly responsible for this secretory activity (see KATZ, 1969).

The electrogenic characteristics associated with the modulating (presynaptic) effects on the terminals are also unclear at present. Presynaptic inhibition in arthropod neuromuscular systems appears to be due to an effect of the inhibitory transmitter on the terminals of the exciter axon. This effect is also induced by applications of γ-aminobutyric acid and is blocked by picrotoxin. It is believed, therefore, that the presynaptic inhibition is due to an increased conductance of the terminals for Cl (GRUNDFEST and REUBEN, 1961; DUDEL, 1965; EPSTEIN and GRUNDFEST, 1968).

f) Early Receptor Potentials. The recently discovered early receptor potentials (ERP) of photosensitive cells are discussed in the chapter by CONE and PAK (this volume). It may be noted that this electrical manifestation of photochemical activity occurs in specialized receptor cells of vertebrates and invertebrates, as well as in primary sensory neurons and in cells that seem to have become degenerated neurons in the accessory eye of *Limulus*. ERP occur in conjunction with hyperpolarizing electrogenesis in vertebrates and with depolarizing electrogenesis in invertebrates. This ubiquity of the ERP reinforces other evidence that the ERP has a different origin from the subsequent electrogenesis which is mainly or entirely due to diffusional emf.

VI. Interplays of Different Components of the Sensory Neuron

a) The Coding Problem. HODGKIN (1948) noted and classified the different responses of sensory axons which tend to fire repetitive trains of spikes. Although the depolarizing generator potential is, indeed, the stimulus for the spike electrogenesis, the patterns of the resulting spike trains are not predictable from the application of the Hodgkin-Huxley analysis (1952) in the form applied to the

squid axon (A. MAURO, personal communication). The ranges of the latencies
with which the first spike is elicited, of the frequency variation, and of the
durations of the trains are far greater in the sensory axons than is predicted by
computer analysis of the responses to excitation by depolarizations that simulate
the generator potential.

b) Electrogenic Specializations. A striking example of the role that secondary factors may play is the difference between the slowly and rapidly adapting

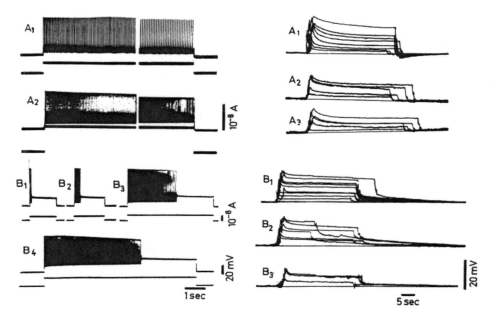

Fig. 14. Nature of the differences in adaption of crayfish stretch receptors. *A*, slowly and *B*,
rapidly adapting sensory neurons. *Right:* Three cells of each type, each treated with tetrodotoxin (10^{-7} gm/ml) to eliminate spike electrogenesis. Intracellular recordings. The generator
potentials which were induced by various degrees of stretch were similar in both types of
neurons. The depolarizations were graded depending on the stretch and, except for a small
early decline, the generator potentials were maintained in all the cells as long as the stimulus
was applied. Irregularities at the beginning and end of the stretches are artifacts due to
irregularities in applying the mechanical stimuli. *Left:* Oscillographic records of responses
(upper traces) to intracellularly applied depolarizing currents (monitored on lower traces).
The slowly adapting neuron continued to generate spikes as long as the depolarizing stimulus
was applied, 140 sec in A_1 and 26 sec in A_2. The rapidly adapting cell produced spikes for
only brief periods at the beginning of each stimulation, the frequency and duration of the
spikes increasing with increasing currents, B_1 to B_4. Still stronger current, however, curtailed
the duration of firing. (Modified from NAKAJIMA, 1964.)

stretch receptors (NAKAJIMA, 1964). These cannot be ascribed to differences in
the time courses of the generator potentials, since the latter are nearly identical
for both types of cells (Fig. 14, right); nor can they be ascribed to differences
in the mechanical coupling of the receptor dendrites to the muscle fiber, since
the differences in response are evidenced when the two types of neurons are
depolarized by intracellularly applied currents (Fig. 14, left), thereby eliminating

possible artifacts due to mechanical coupling. The slowly adapting receptors continue to develop spikes during long-lasting depolarizations, the frequency increasing with increasing depolarization. In the rapidly adapting cells, however, spike electrogenesis is quickly terminated.

It is obvious from the foregoing that the difference between the rapidly and slowly adapting cells resides in some differences in the properties of their electrically excitable components. Furthermore, there are also regional differences in the electrically excitable membrane (Fig. 2). In the slowly adapting crayfish stretch receptors, the region of axon, about 0.5 mm from the cell body, appears to be the site of the persistent spike electrogenesis (NAKAJIMA and ONODERA, 1969a). A possible factor in the electrogenesis of the slowly adapting cells that may account for the differences is the presence of a strongly electrogenic Na-pump (NAKAJIMA and TAKAHASHI, 1966). The latter activity causes hyperpolarization, although by an as-yet-unknown mechanism, and presumably it contributes to the overall membrane potential (GRUNDFEST, 1955). This hyperpolarization, which is smaller or absent in the rapidly adapting cells, would tend to counteract the Na inactivation induced by a steady depolarization, particularly during the post-spike period when the persisting K activation of spike electrogenesis has reduced the generator depolarization. The role of the electrogenic pump would then be similar to that which has been ascribed to it in the activity of pacemaker cells (GRUNDFEST, 1966b). Obviously, however, much further work remains to be done at the level of the single sensory neuron before the problem of encoding the input electrogenesis into a train of spikes is solved.

VII. Specificity of Transducer Membrane

The question of the specificity of transducer membrane has received relatively little attention. It is likely that the transducer component may have only relative specificity, being much more sensitive to activation by the adequate stimulus, but also responsive to various other stimuli. For example, it is reported that the crayfish stretch receptor is depolarized by acetylcholine as well as by mechanical stimuli (WIERSMA et al., 1953). A counterpart of this finding may be the classical observations (see SCHAEFFER, 1940) that noninjurious mechanical stimuli can excite nerve fibers. This mechanotransducer property has been restudied recently (GOLDMAN, 1965).

Taste receptors, however, appear to be rather nonselective, since intracellularly recorded depolarizations of rather large magnitudes are elicited by applying different modalities of stimulation (Fig. 15). Some cells responded to all four of the tested modalities (salt, sweet, bitter, acid), although to different degrees, and most cells responded to at least three modalities. These findings must lead to the conclusion that the recognition of any modality of taste is the resultant of a complex pattern of messages to the central nervous system. The principles of the discriminatory process, however, are not yet clear.

A high degree of specificity as to color sensitivity is also lacking in cones (Fig. 4). Each cell responds to a broad spectrum, so that, at a given wavelength and intensity of illumination, a number of cones of different psychophysical properties are brought into activity (Hecht, 1934; Wald, 1964; Rushton, 1965). Nevertheless, the organism can decode this welter of information to permit exquisite hue discrimination.

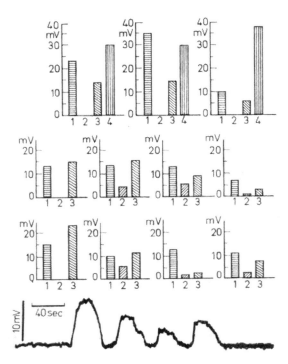

Fig. 15. Electrical activity in single taste buds produced by four standardized stimuli (1) salt, (2) sweet, (3) bitter, and (4) acid. *Below:* Intracellular records of potential evoked by the four substances acting on a single taste bud. Note long-lasting potential and absence of spikes. *Above:* Different patterns of responses in 11 different cells. (From original data by Kimura and Beidler, 1961, as published in Grundfest, 1964a.)

Single auditory nerve fibers of goldfish respond to a wide range of sound frequencies (Fig. 16). If the same situation pertains in terrestrial animals (as is likely), then pitch discrimination and other psychophysical properties of audition must also depend upon central analysis of the complex pattern. The complexity of the latter is increased further by the possibility that the characteristics of the transmissional process between hair cells and axons may have a wide range of differences (Furukawa and Ishii, 1967a, b). For example, the S1 nerve fibers of the goldfish adapt rapidly. Although the hair cells continue to generate microphonic potentials during continuous stimulation, the generator (postsynaptic) potential of the nerve fiber (sensory neuron) diminishes and is almost abolished, and there is cessation of spike electrogenesis. Other fibers (S2, L) continue discharging spikes, but the patterns of the responses may differ considerably in these two fiber types.

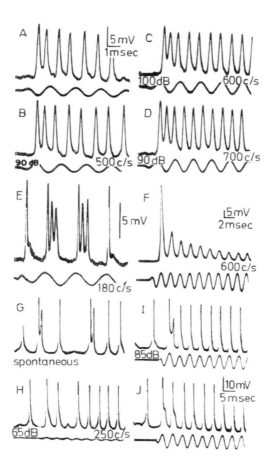

Fig. 16. Discharges recorded from S1 and S2 axons from the saccular nerve of goldfish. Action potentials on upper, sound stimuli on the lower traces, with intensities and frequencies of the latter indicated. *A-D:* Extracellular recordings from S1 fiber that responded to rarefaction as well as compression, and that was presumably connected to oppositely directed hair cells, as in *c, c'* of Fig. 3. Each cycle of the stimulus thus resulted in two spikes in the axon. The stimulus was in reversed phase in *A* and *B*, but the latency of the responses was not changed by the reversal. *E-J:* Intracellular records. *E*, this S1 fiber responded only to compression, sometimes with multiple discharges. *F*, another S1 fiber that responded only to rarefaction, initially with a spike, and then with graded decrementing activity. *G-J:* S2 fiber that discharged spontaneously (*G*). It responded only to rarefaction phase of the stimulus. *H:* The discharge locked in during a weak stimulus. *I, J:* The stimulus was stronger. When the initial phase was one of rarefaction (*J*), the latency of the first evoked response was shortened. (Modified from FURUKAWA and ISHII, 1967b.)

VIII. Cell-Cell Interactions

a) Influences on the Primary Sensory Neuron. Even when the sensory neuron is the primary target of the specific stimulus, it may be subject to various modulating influences from other cells. Among the best-studied examples of such influences are the inhibitory innervations of the crayfish stretch receptor (KUFFLER and Eyzaguirre, 1955) and the γ-neuron control of the performance of vertebrate muscle spindles (Kuffler and Hunt, 1952). As already noted, the modulation of the crayfish stretch receptor by the IPSP reduces the effectiveness of the generator potential for initiation of spikes. The γ-neuron modulation acts by changing the mechanical condition of the intrafusal muscle fibers, and thereby applies a bias that increases or decreases the sensitivity of the system to a mechanical stimulus. The precise effect on the mechanoreceptors is not yet clear, but the net effect presumably is to change the amplitude of generator depolarization for a given mechanical stimulus.

b) **Ephaptic Connections of Secondary Receptor Systems.** Conceptually, the simplest form of interaction between cells is that which is due to their electrical coupling (Grundfest, 1967c). As already noted, this type of coupling depends on a low-resistance (tight electrical coupling; Watanabe and Grundfest, 1961) pathway between the cells. In neuronal assemblies, tight electrical coupling is correlated with fusion of regions of the unit membranes of the two cells (Bennett et al., 1963; Bennett et al., 1967a—d) to form "tight junctions" (Farquhar and Palade, 1963). The electrical activity of the prejunctional cell must excite the postjunctional unit. Thus, the latter must have an electrically excitable membrane, and the former must develop a depolarizing electrogenesis. The depolarization, furthermore, must be sufficiently large so as to overcome the losses in the coupling system, as well as the losses in the transmissional cables of the two cells.

Since taste receptors can generate large depolarizations (Fig. 15), ephaptic transmission is feasible in this system. However, neither morphological nor electrophysiological evidence for electrotonic coupling is as yet available (De Lorenzo, 1963; Farbman, 1965). The question is of considerable theoretical interest in view of the finding that there is a very active turnover of taste buds (Beidler and Smallman, 1965), and of the effects of denervation and reinnervation (Zelena, 1964; Robbins, 1967).

Ephaptic transmission is believed to be the mode of action in the coupling of the retinular to eccentric cells of *Limulus* (Smith et al., 1965; Borsellino et al., 1965). It might also be operative in the visual system of other arthropods, where the primary receptors have a depolarizing electrogenesis. It is likely that the coupling of a number of retinular cells to a single sensory neuron, as in the arthropod compound eye, is a specialization that permits a larger amount of current flow from the receptor cells to the neuron, thereby enhancing the effectiveness of the light stimulus on the neuron. It is not known if the numerous foldings of the rhabdomeres are specializations to decrease the resistance. It may be noted that electric organs of the strongly electric Torpedinidae and *Electrophorus* have developed a quasi-ephaptic specialization without the need for tight junction contacts (Grundfest, 1959b; Grundfest and Bennett, 1961).

The performance of the *Limulus* retinular-eccentric cell combination is, however, complicated by a number of factors (Fig. 17). As already noted, the retinular cell may have an electrically excitable depolarizing electrogenesis which is eliminated by tetrodotoxin. The cells can also develop a high membrane resistance when hyperpolarized (Smith et al., 1965) and may then generate hyperpolarizing responses like those in many other cell types (Reuben et al., 1961; Grundfest, 1961c, 1966a, 1967a). The ephaptic linkage is indicated in the equivalent circuit

(Fig. 17) by a T-network which represents respectively the membrane resistances of the pre- and post-components of the tight junction and the leak resistance of the open channels between the cells.

The dendrite and soma of the eccentric cell, as well as a portion of the neurite proximal to the soma, appear to be electrogenically unreactive, but the latter gives off a number of thin axonal branches of as-yet-unknown function. This part of the cell is represented by a "leak" resistance across which some of the generator

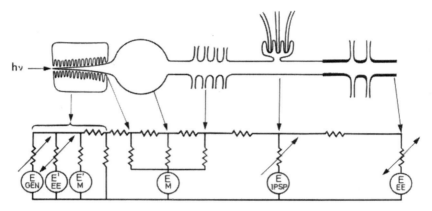

Fig. 17. The anatomical and electrical relations in *Limulus* photoreceptors. Two retinular cells are shown, their rhabdomes in contact with the dendrite of the eccentric cell. In the equivalent circuit, the retinular cells are represented by a generator component, E_{GEN}, in series with a resistance that decreases when the membrane is excited by light. An electrically excitable component is represented by E'_{EE}, with a series resistance that can be increased as well as decreased in response to electrical stimuli. Unreactive membrane is represented by E'_M (the resting potential) and a fixed resistance. The ephaptic junction with the eccentric cell is represented by three resistances in T configuration. The membrane of the dendrite, soma and proximal portion of the neurite are also represented as electrically inexcitable with the emf of the resting potential (E_M). The properties of the proximal axon collaterals are discussed in the text. The electrically inexitable reactive membrane of the inhibitory synapses is represented by E_{IPSP} and a variable resistance which is decreased by transmitter action. The electrically excitable component of the axon is represented as an emf (E_{EE}) in series with a resistance that can increase or decrease under appropriate stimulation. (Modified from PURPLE and DODGE, 1965.)

current must be dissipated. Interposed between this region and that of the spike generator is the zone of inhibitory synaptic innervation. The location of the inhibitory synapses is strategic (GRUNDFEST, 1959a), since the conductance increase of the synaptic membrane tends to shunt the ephaptic generator current as well as to poise the membrane potential below the critical firing level for spike electrogenesis in the main axon.

The axon collaterals originate in a region of electrogenically inert membrane, and thus far no spikes have been observed that might be attributed to these collaterals (PURPLE and DODGE, 1965). However, even if the collaterals do not generate spikes, the spikes that originate at the main axon must cause an electrotonic change in their membrane potential since large spikes can be recorded in the soma (FUORTES, this volume). These collaterals may form additional pathways for lateral inhibition of adjacent neurons as well as providing a pathway for self-inhibitory feedback.

c) **Synaptic Coupling.** When the cell-cell coupling is synaptic, the system possesses more degrees of freedom. Furthermore, the relations between transmission and electrogenesis are radically altered. The receptor cell conceivably might release several varieties of secretory products which might have various effects on different postsynaptic components, or a single product might have different effects on different varieties of postsynaptic membrane. When another synaptic linkage is included, as in the case of the intermediate cells of the vertebrate retina, many more possibilities of interactions must be expected, particularly if these cells make complex interconnections as they appear to do (Fig. 8).

A synaptic linkage couples a secretory output to an electrically inexcitable postsynaptic membrane component. The electrogenesis of the pre-cell therefore does not participate directly in the transmissional activity; it is only the sign that secretory activity occurs. Depending upon as-yet-unknown factors of secretory activity, the electrogenesis may be depolarizing or hyperpolarizing in sign and, conceivably, there might be no change in potential despite the secretory activity.

d) **Significance of Receptor Potential.** These considerations have led to the view that "receptor potentials" are usually of no significance to the transmissional chain of activity (Grundfest, 1958a, 1961a, 1964a, 1965). The view is perhaps best illustrated by the "receptor potentials" of the vertebrate retina, which are predominantly, or sometimes exclusively, hyperpolarizing in sign. Hence, they cannot be directly involved in the excitatory processes of spike electrogenesis[5]. Likewise, if transmission from hair cell to auditory nerve fibers is synaptic (Grundfest, 1958a, 1964a), as is now generally believed (Davis, 1965; Furukawa and Ishii, 1967b), then the auditory microphonics which originate in the hair cells are also likely to be epiphenomena without direct function (Grundfest, 1965).

e) **Significance of Tight Junctions.** Tight junctions have been described between horizontal cells (Yamada and Ishikawa, 1965) and between bipolar and ganglion cells (Dowling and Boycott, 1965) of the vertebrate retina (Fig. 8). Since tight junctions are common between electrogenically unreactive epithelial cells (Loewenstein, 1966), their presence in sensory systems does not necessarily signify that there is functional ephaptic coupling, but reflects rather the epithelial origin of the nervous system (Grundfest, 1966a, 1967b). The horizontal cells do not generate spikes and must be regarded as electrically inexcitable (Grundfest, 1958a). Furthermore, they as well as the other varieties of intermediate cells generate mainly hyperpolarizations (Figs. 5, 6). If there is electrotonic transmission across the tight junctions between the bipolar and ganglion cells, it is likely that the effect is an inhibitory action on the ganglion cells. The potency of such an effect cannot be evaluated at present. It is obvious, even from a schematic illustration like that of Fig. 8, that there are vast possibilities for cell-cell interactions at all stages of the transmissional process within the retina.

---

[5] When these hyperpolarizing potentials were first described, they were regarded as artifacts (*see* Granit, 1955), because their sign did not accord with the prevalent ideas of transmission. Svaetichin (1956) himself, noted the apparent contradiction with neurophysiological data.

Acknowledgement: Work in the author's laboratory is supported in part by grants from the Muscular Dystrophy Associations of America; by U.S. Public Health Service Research Grants (NB 03728, NB 03270 and Training Grant NB 5328 from the National Institute of Neurological Diseases and Blindness, and by National Science Foundation grant GB 6988 X.

References

BEIDLER, L. M.: Comparison of gustatory receptors, olfactory receptors, and free nerve endings. Cold Spr. Harb. Symp. quant. Biol. **30**, 191—200 (1965).

— SMALLMAN, R.: Renewal of cells within taste buds. J. Cell Biol. **27**, 263—272 (1965).

BENNETT, M. V. L.: Physiology of electronic junctions. Ann. N.Y. Acad. Sci. **37**, 509—539 (1966).

— Mechanisms of electroreception. In: Lateral line detectors. Bloomington: Indiana Univ. Press 1967.

— ALJURE, E., NAKAJIMA, Y., PAPPAS, G. D.: Electrotonic junctions between teleost spinal neurons: electrophysiology and ultrastructure. Science **141**, 262—264 (1963).

— NAKAJIMA, Y., PAPPAS, G. D.: Physiology and ultrastructure of electrotonic junctions. I. The supramedullary neurons. J. Neurophysiol. **30**, 161—179 (1967a).

— — — Physiology and ultrastructure of electrotonic junctions. III. The giant electromotor neurons of *Malapterurus electricus*. J. Neurophysiol. **30**, 209—235 (1967b).

— PAPPAS, G. D., ALJURE, E., NAKAJIMA, Y.: Physiology and ultrastructure of electrotonic junctions. II. Spinal and medullary electromotor nuclei in mormyrid fish. J. Neurophysiol. **30**, 180—208 (1967c).

— — GIMENEZ, M., NAKAJIMA, Y.: Physiology and ultrastructure of electrotonic junctions. IV. Medullary pacemaker and relay nuclei of the electromotor system of the gymnotid fish. J. Neurophysiol. **30**, 236—300 (1967c).

BENOLKEN, R. M., RUSSELL, C. J.: Dissection of a graded visual response with tetrodotoxin. In: The functional organization of the compound eye. London: Pergamon Press 1966.

BETHE, A.: Das Zentralnervensystem von *Carcinus maenas*. Arch. f. Mikr. Anat. **50**, 589—634 (1897).

BLOEDEL, J., GAGE, P. W., LLINAS, R., QUASTEL, D. M. J.: Transmitter release at the squid giant synapse in the presence of tetrodotoxin. Nature (Lond.) **212**, 49—50 (1966).

BOECKH, J., KAISSLING, K. E., SCHNEIDER, D.: Insect olfactory receptors. Cold Spr. Harb. Symp. quant. Biol. **30**, 263—280 (1965).

BORSELLINO, A., FUORTES, M. G. F., SMITH, T. G.: Visual responses in Limulus. Cold Spr. Harb. Symp. quant. Biol., **30**, 429—443 (1965).

BROWN, K. T., WIESEL, T. N.: Intraretinal recording with micropipette electrodes in the intact cat eye. J. Physiol. **149**, 537—562 (1959).

CLARK, A. W., MILLECHIA, R., MAURO, A.: The ventral photoreceptor cells of *Limulus*. I. The microanatomy. J. gen. Physiol. **54**, 289—309 (1969).

COHEN, M. J.: The crustacean myochordotonal organ as a proprioceptive system. Comp. Biochem. Physiol. **8**, 223—243 (1963).

CONE, R. A., PAK, W. L.: The early receptor potential. In: Handbook of sensory physiology, vol. I, pp. 345—365. Berlin-Heidelberg-New York: Springer 1971.

DAVIS, H.: Some principles of sensory receptor action. Physiol. Rev. **41**, 391—416 (1961).

— A model for transducer action in the Cochlea. Cold Spr. Harb. Symp. quant. Biol. **30**, 181—190 (1965).

DAVIS, L., Jr., LORENTE DE Nó, R.: Contribution to the mathematical theory of the electrotonus. In: A study of nerve physiology. New York: Studies from the Rockefeller Inst. Vol. 131, 1947.

DE LORENZO, A. J. D.: Studies on the ultrastructure and histophysiology of cell membranes, nerve fibers and synaptic junctions in chemoreceptors. In: Olfaction and taste. New York: Pergamon Press 1963.

DETHIER, V. G.: Vision. In: Insect physiology. New York: John Wiley & Sons. Inc. 1953.

— The physiology of insect senses. London: Methner 1963.

Dijkgraaf, S.: Biological significance of the lateral line organs. In: Lateral line detectors. Bloomington: Indian Univ. Press 1967.

Dowling, J. E.: Discrete potentials in the dark-adapted eye of the crab *Limulus*. Nature (Lond.) **217**, 28—31 (1968).

— Boycott, B. B.: Neural connections of the retina: fine structure of the inner plexiform layer. Cold Spr. Harb. Symp. quant. Biol. **30**, 393—402 (1965).

Dudel, J.: Presynaptic and postsynaptic effects of inhibitory drugs on the crayfish neuromuscular junction. Pflügers Arch. ges. Physiol. **283**, 104—118 (1965).

Eccles, J. C.: The physiology of synapses. Berlin Göttingen-Heidelberg-New York: Springer 1964.

Edwards, C., Ottoson, D.: The site of impulse initiation in a nerve cell of a crustacean stretch receptor. J. Physiol. (Lond.) **143**, 138—148 (1958).

— Terzuolo, C. A., Washizu, Y.: The effect of changes of the ionic environment upon an isolated crustacean sensory neuron. J. Neurophysiol. **26**, 948—957 (1963).

Elmquist, D., Feldman, D. S.: Spontaneous activity at a mammalian neuromuscular junction in tetrodotoxin. Acta physiol. scand. **64**, 475—477 (1965).

Epstein, R., Grundfest, H.: Desensitization of Gamma Aminobutyric Acid (GABA) Receptors in Muscle Fibers of the Crab, *Cancer borealis*. J. gen. Physiol. **56**, 33—45 (1970).

Eyzaguirre, C., Koyano, H.: Origin of sensory discharges in carotid body chemoreceptors. Cold Spr. Harb. Symp. quant. Biol. **30**, 227—231 (1965).

Farbman, A. I.: Fine structure of the taste bud. J. Ultrastruct. Res. **12**, 328—350 (1965).

Farquhar, M. G., Palade, G. E.: Junctional complexes in various epithelia. J. Cell Biol. **17**, 375—412 (1963).

Fetz, E. E.: Pyramidal tract effects on interneurons in the cat lumbar dorsal horn. J. Neurophysiol. **31**, 69—80 (1968).

Flock, Å.: Ultrastructure and function in the lateral line organs. In: Lateral line detectors. Bloomington: Indiana Univ. Press 1967.

— Sensory transduction in hair cells. In: Handbook of sensory physiology, vol. I, pp. 396—441. Berlin-Heidelberg-New York: Springer 1971.

Fuortes, M. G. F.: Generation of responses in receptor. In: Handbook of sensory physiology, vol. I, pp. 243—268. Berlin-Heidelberg-New York: Springer 1971.

Furukawa, T., Ishii, Y.: Effects of static bending of sensory hairs on sound reception in the goldfish. Japan. J. Physiol. **17**, 572—588 (1967a).

— — Neurophysiological studies on hearing in goldfish. J. Neurophysiol. **30**, 1377—1403 (1967b).

Goldman, D. E.: The transducer action of mechanoreceptor membranes. Cold Spr. Harb. Symp. quant. Biol. **30**, 59—68 (1965).

Grampp, W.: The impulse activity in different parts of the slowly adapting stretch receptor neuron of the lobster. Acta physiol. scand. **66**, Suppl. 262, 1—36 (1966).

Granit, R.: Receptors and sensory perception. New Haven: Yale Univ. Press 1955.

Grundfest, H.: The nature of the electrochemical potentials of bioelectric tissues. In: Electrochemistry in biology and medicine. New York: John Wiley Sons, Inc. 1955.

— Electrical inexcitability of synapses and some of its consequences in the central nervous system. Physiol. Rev. **37**, 337—361, (1957a).

— Excitation triggers in post-junctional cells. In: Physiological triggers. Washington, D. C.: Amer. Physiol. Society 1957b.

— The mechnisms of discharge of the electric organs in relation to general and comparative electrophysiology. Progr. Biophys. **7**, 1—85 (1957c).

— General problems of drug action on bioelectric phenomena. Ann. N. Y. Acad. Sci. **66**, 537—591 (1957d).

— An electrophysiological basis for cone vision in fish. Arch. ital. Biol. **96**, 135—144 (1958a).

— Discussion. In: Electrophysiology of the visual system. Amer. J. Ophthal. **46**, II, 43—46 (1958b).

— Electrophysiology and pharmacology of dendrites. Electroenceph. Clin. Neurophysiol. Suppl. **10**, 22—41 (1958c).

— Synaptic and ephaptic transmission. In: Handbook of Physiology: Neurophysiology I. Washington, D. C.: Amer. Physiol. Soc. 1959a.

GRUNDFEST, H.: Evolution of conduction in the nervous system. In: Evolution of nervous control. Washington, D. C.: Amer. Ass. Advanc. Sci. 1959 b.
— Functional specifications for membranes in excitable cells. In: Regional neurochemistry. London: Pergamon Press 1961 a.
— General physiology and pharmacology of junctional transmission. In: Biophysics of physiological and pharmacological actions. Washington, D. C.: Amer. Ass. Advanc. Sci. 1961 b.
— Ionic mechanisms in electrogenesis. Amnn. N. Y. Acad. Sci. **94**, 405—457 (1961 c).
— Excitation by hyperpolarizing potentials. A general theory of receptor activities. In: Nervous inhibition. London: Pergamon Press 1961 d.
— Impulse conducting properties of cells. In: The general physiology of cell specialization. New York: McGraw-Hill Book Co., Inc. 1963.
— Evolution of electrophysiological varieties among sensory receptor systems. In: Essays on physiological evolution. London: Pergamon Press 1964 a.
— Effects of drugs on the central nervous system. Ann. Rev. Pharmacol. **4**, 341—364 (1964 b).
— Electrophysiology and pharmacology of different components of bioelectric transducers. Cold Spr. Harb. Symp. quant. Biol. **30**, 1—14 (1965).
— Comparative electrobiology of excitable membranes. In: Advances in comparative physiology and biochemistry, Vol. 2. New York: Academic Press, Inc. 1966 a.
— Heterogeneity of excitable membrane: electrophysiological and pharmacological evidence and some consequences. Ann. N. Y. Acad. Sci. **137**, 901—949 (1966 b).
— Some comparative biological aspects of membrane permeability control. Fed. Proc. **27**, 1613—1626 (1967 a).
— Synaptic and ephaptic transmission. In: The neurosciences. A study program. New York: Rockefeller Univ. Press 1967 b.
— Tetrodotoxin: action on graded responses. Science **156**, 1771 (1967 c).
— BENNETT, M. V. L.: Studies on morphology and electrophysiology of electric organs. I. Electrophysiology of marine fishes. In: Bioelectrogenesis. Amsterdam: Elsevier 1961.
— KAO, C.-Y., ALTAMIRANO, M.: Bioelectric effects of ions microinjected into the giant axon of *Loligo*. J. gen. Physiol. **38**, 245—282 (1954).
— REUBEN, J. P.: Neuromuscular synaptic activity in lobster. In: Nervous inhibition. London: Pergamon Press 1961.
HAGIWARA, S., KUSANO, K., SAITO, N.: Membrane changes in crayfish stretch receptor neuron during inhibition and under action of gamma-aminobutyric acid. J. Neurophysiol. **23**, 505—515 (1960).
HARTLINE, H. K., RATLIFF, F.: Inhibitory interaction of receptor units in the eye of *Limulus*. J. gen. Physiol. **40**, 357—376 (1957).
HECHT, S.: Vision. II. The nature of the photoreceptor process. In: A handbook of general experimental psychology. Worcester, Mass.: Clark Univ. Press 1934.
HODGKIN, A. L.: The local electric changes associated with repetitive action in a non-medullated axon. J. Physiol. (Lond.) **107**, 165—181 (1948).
— HUXLEY, A. F.: A quantitative description of membrane current and its application to conduction and excitation in nerve. J. Physiol. (Lond.) **117**, 500—544 (1952).
— RUSHTON, W. A. H.: The electrical constants of a crustacean nerve fibre. Proc. Roy. Soc. (London) B. **133**, 444—479 (1946).
HODGSON, E. S.: Chemoreception. In: The physiology of insecta, Vol. 1. New York: Academic Press Inc. 1964.
HUNT, C. C., TAKEUCHI, A.: Responses of the nerve terminal of the Pacinian corpuscle. J. Physiol. **160**, 1—21 (1962).
KATZ, B.: The release of neural transmitter substances. Liverpool: Liverpool Univ. Press 1969.
— MILEDI, R.: Tetrodotoxin and neuromuscular transmission. Proc. roy. Soc. (London) B. **167**, 8—22 (1967 a).
— — A study of synaptic transmission in the absence of nerve impulses. J. Physiol. (Lond.) **192**, 407—436 (1967 b).
KIMURA, K., BEIDLER, L. M.: Microelectrode study of taste receptors of rat and hamster. J. cell. comp. Physiol. **58**, 131—140 (1961).
KOKETSU, K.: Mechanism of active depolarization. Dispensability of sodium. In: Biophysics of physiological and pharmacological actions. Washington, D. C.: Amer. Ass. Advanc. Sci. 1961.

Kuffler, S. W.: Discharge patterns and functional organization of mammalian retina. J. Neurophysiol. **16**, 37—68 (1953).
— Eyzaguirre, C.: Synaptic inhibitions in an isolated nerve cell. J. gen. Physiol. **39**, 155—184 (1955).
— Hunt, C. C.: The mammalian small-nerve fibers: a system for efferent nervous regulation of muscle spindle discharge. Res. Publ. Ass. nerv. ment. Dis. **30**, 24—47 (1952).
Kusano, K., Livengood, D. R., Werman, R.: Correlation of transmitter release with membrane properties of presynaptic fibers of the squid giant synapse. J. gen. Physiol. **50**, 2579—2597 (1967).
Loewenstein, W. R.: Permeability of membrane junctions. Ann. N. Y. Acad. Sci. **137**, 441—472 (1966).
— Mechano-electric transduction in the Pacinian Corpuscle. Initiation of the sensory impulses in the mechanoreceptors. In: Handbook of sensory physiology, vol. I, pp. 269—290. Berlin-Heidelberg-New York: Springer 1971.
— Terzuolo, C. A., Washizu, Y.: Separation of transducer and impulse generating processes in sensory receptors. Science **142**, 1180—1181 (1963).
Lundberg, A.: Secretory potentials and secretion in the sublingual gland of the cat. Nature (Lond.) **177**, 1080—1081 (1956).
MacNichol, E. F., Jr., Svaetichin, G.: Electric responses from isolated retinas of fishes. Amer. J. Ophthal. **46**, 26—46 (1958).
Millecchia, R., Bradbury, J., Mauro, A.: Simple photoreceptors in *Limulus polyphemus*. Science **154**, 1199—1201 (1966).
— Mauro, A.: The ventral photoreceptor cells of *Limulus*. II. The basic photoresponse. J. gen. Physiol.**54**,310—330(1969a).III. A voltage clamp study. J. gen. Physiol.**54**,331—351 (1969b).
Moulton, D. G., Beidler, L. M.: Structure and function in the peripheral olfactory system. Physiol. Rev. **47**, 1—52 (1967).
Nakajima, S.: Adaptation in stretch receptor neurons of crayfish. Science **146**, 1168—1170 (1964).
— Onodera, K.: Membrane properties of the stretch receptor neurones of a crayfish with particular reference to mechanisms of sensory adaptatron. J. Physiol. (Lond.) **200**, 161—185 (1969a).
— — Adaptation of the generator potential in the crayfish stretch receptors under constant length and constant tension. J. Physiol. (Lond.) **200**, 187—204 (1969b).
— Takahashi, K.: Post-tetanic hyperpolarization and electrogenic Na-pump in stretch receptor neuron of crayfish. J. Physiol. (Lond.) **187**, 105—127 (1966).
Oakley, B., Benjamin, R. M.: Neural mechanisms of taste. Physiol. Rev. **46**, 173—211 (1966).
Obara, S.: Effects of some organic cations on generator potential of stretch receptor of crayfish. J. gen. Physiol. **52**, 22—45 (1968).
— Grundfest, H.: Effects of lithium on different membrane components of crayfish stretch receptor neurons. J. gen. Physiol. **51**, 635—654 (1968).
Ottoson, D., Shepherd, G. M.: Transducer properties and integrative mechanisms in the frog's muscle spindle. In: Handbook of sensory physiology, vol. I, pp. 442—499. Berlin-Heidelberg-New York: Springer 1971.
Ozeki, M., Freeman, A. R., Grundfest, H.: The membrane components of crustacean neuromuscular systems. I. Immunity of different electrogenic components to tetrodotoxin and saxitoxin. J. gen. Physiol. **49**, 1319—1334 (1966).
— Grundfest, H.: Crayfish muscle fiber: ionic requirements for depolarizing synaptic electrogenesis. Science **155**, 478—481 (1967).
— Sato, M.: Changes in the membrane potential and the membrane conductance associated with sustained compression of the nonmyelinated nerve terminal in Pacinian corpuscles. J. Physiol. (Lond.) **180**, 186—208 (1965).
Purple, R. L., Dodge, F. A.: Interaction of excitation and inhibition in the eccentric cell in the eye of *Limulus*. Cold Spr. Harb. Symp. quant. Biol. **30**, 529—537 (1965).
Reuben, J. P., Grundfest, H.: Inhibitory and excitatory miniature postsynaptic potentials in lobster muscle fibers. Biol. Bull. **119**, 335 (1960).
— Werman, R., Grundfest, H.: The ionic mechanisms of hyperpolarizing responses in lobster muscle fibers. J. gen. Physiol. **45**, 243—265 (1961).
Robbins, N.: The role of the nerve in maintenance of frog taste buds. Exp. Neurol. **17**, 364—380 (1967).

RUSHTON, W. A. H.: Chemical basis of colour vision and colour blindness. Nature (Lond.) **206**, 1087—1091 (1965).

SATO, M., OZEKI, M.: Response of the non-myelinated nerve terminal in Pacinian corpuscles to mechanical and antidromic stimulation and the effect of procaine, choline and cooling. Jap. J. Physiol. **13**, 565—582 (1963).

SCHAEFFER, H.: Elektrophysiologie. Bd. I: Allgemeine Elektrophysiologie. Wien: Franz Deuticke 1940.

SMITH, T. G., BAUMANN, F., FUORTES, M. G. F.: Electrical connections between visual cells in the ommatidium of *Limulus*. Science **147**, 1446—1447 (1965).

SVAETICHEN, G.: Spectral response curves from single cones. Acta physiol. scand. **39**, Suppl. 134, 17—112 (1956).

TASAKI, I.: Nerve excitation: A macromolecular approach. Springfield, Ill. C. C. Chomas 1968.

— LERMAN, L., WATANABE, A.: Analysis of excitation process in squid giant axons under bi-ionic conditions. Amer. J. Physiol. **216**, 130—138 (1969).

TERZUOLO, C. A., Knox, C. K.: Static and dynamic behavior of the stretch receptor organ of *crustacea*. In: Handbook of sensory physiology vol. I, pp. 500—522. Berlin-Heidelberg-New York: Springer 1971.

TOMITA, T.: Electrical activity in the vertebrate retina. J. Ophthal. Soc. Am. **53**, 49—57 (1963).

— KANEKO, A., MURAKAMI, M. PAUTLER, E. L.: Spectral response curves of single cones in the carp. Vision Res. **7**, 519—531 (1967).

— MURAKAMI, M., HASHIMOTO, Y., SASAKI, Y.: Electrical activity of single neurons in frog's retina. In: The visual system: neuropharmacology and psychophysics. Berlin-Göttingen-Heidelberg: Springer 1961.

WALD, G.: The receptor of human color vision, Science **145**, 1007—1016 (1964).

WALL, P. D.: Presynaptic control of impulses at the first central synapse in the cutaneous pathway. In: Physiology of spinal neurons. Amsterdam: Elsevier 1964.

WATANABE, A., GRUNDFEST, H.: Impulse propagation at the septal and commissural junctions of crayfish lateral giant axons. J. gen. Physiol. **45**, 267—308 (1961).

WIERSMA, C. A. G., FURSHPAN, E. J., FLOREY, E.: Physiological and pharmacological observations on the muscle receptor organs of the crayfish, *Cambarus clarkii* Girard. J. exp. Biol. **30**, 136—150 (1953).

WITKOVSKY, P.: A comparison of ganglion cell and S-potential response properties in carp retina. J. Neurophysiol. **30**, 546—561 (1967).

WOLBARSHT, M. L.: Receptor sites in insect chemoreceptors. Cold Spr. Harb. Symp. quant Biol. **30**, 281—288 (1965).

— HANSON, F. E.: Electrical activity in the chemoreceptors of the blowfly. III. Dendritic action potentials. J. gen. Physiol. **48**, 673—683 (1965).

YAMADA, E., ISHIKAWA, T.: The fine structure of the horizontal cells in some vertebrate retinae. Cold Spr. Harb. Symp. quant. Biol. **30**, 383—392 (1965).

YAMAGISHI, S.: Ionic mechanism of the prolonged spike of squint axons perfused with pertease. Soc. gen. Physiol. Woods Hole (Abstr.) 1969).

ZELENA, J.: Development, degeneration and regeneration of receptor organs. In: Progress in brain research, Vol. 13: Mechanism of neural maturation. Amsterdam: Elsevier 1964.

Addendum

Work on vertebrate and invertebrate visual systems that has appeared since the manuscript was completed support the viewpoint expressed in the text. Pertinent references will be found in two recent papers (McReynods & Gorman, 1970a, b) and the review by Tomita (1970).

References

MCREYNOLDS, J. S., GORMAN, A. L. F.: Photoreceptor potentials of opposite polarity in the eye of the scallop, *Pecten irradians*. J. gen. Physiol. **56**, 376—391 (1970a)

— — Membrane conductance and spectral sensitivities of *Pecten* photoreceptors. J. gen. Physiol. **56**, 392—406 (1970b).

TOMITA, T.; Electrical activity of vertebrate photoreceptors. Quart. Rev. Biophysics, **3**, 179—222 (1970).

Chapter 5

Formation of Neuronal Connections in Sensory Systems

By

Marcus Jacobson, Baltimore, Maryland (USA)

With 6 Figures

Contents

The development of peripheral sensory innervation and the formation of synaptic connections in sensory systems have attracted relatively little attention from neurophysiologists. This neglect may be due to the dominating preoccupation with impulse traffic and coding which few sensory physiologists have succeeded in escaping. The purpose of this review is to draw attention to the developmental mechanisms which result in the formation of connections in the nervous system, and which continue to maintain these connections throughout life. The main emphasis will be on the development of connections in sensory systems, but it is not always possible or desirable to remain within this narrow frame of reference. On the other hand, the whole problem of the development of neuronal connections is too large to deal with here. Other aspects of that problem are considered in recent reviews (Székely, 1966; Gaze, 1967; Jacobson, 1967a).

1. Control of the Density of Sensory Innervation

There are differences in the density of sensory innervation of different parts of the body which result in differences in regional sensitivity. The most outspoken differences are to be found in the density of innervation to different regions of skin, and the differences in the numbers of stretch receptors present in different muscles and tendons. It is not known how these differences develop. Very little experimental work has been devoted to this problem, which is still largely in the realm of pure speculation.

RAMÓN Y CAJAL, (1919) suggested that each region of the skin attracts a
certain quantity of nerves by virtue of the amount of neurotropic influence which
it exerts. He further postulated that the ingrowing nerves inactivate or utilize the
neurotropic substance, so that ingrowing sensory nerves regulate their own density by depleting the material which attracts them to the skin. The theory of neurotropism is based entirely on circumstantial evidence. Although there is direct
evidence that mechanical guidance of growing nerve fibers occurs in tissue culture
(WEISS, 1934) and in vivo (WEISS and TAYLOR, 1944), no direct evidence has been
found in favor of neurotropism. Nevertheless, there are some phenomena, such as
the collateral sprouting of sensory nerves into denervated areas of skin (RAMÓN Y
CAJAL, 1959; SPEIDEL, 1941; WEDELL et al., 1941; LIVINGSTON, 1947), and the
collateral sprouting of motor nerves which re-innervate partially denervated
muscle (EDDS, 1950; 1953; HOFFMAN, 1950; HOFFMAN and SPRINGELL, 1951),
which are best explained as a form of biochemical action of denervated tissue on
neighboring nerve fibers. FITZGERALD (1961, 1962) showed that collateral nerve
sprouting is the normal method of innervation of the skin during growth of the
pig's snout. It is not known what regulates the collateral sprouting of dermal nerves
and the growth of new nerves into the epidermis. However, some control must
exist, because nerve sprouting and growth keep pace with the increase in area of
the skin. No regions of skin remain devoid of nerves neither do other regions acquire
excessive innervation.

Regulation of the density of cutaneous innervation apparently continues
throughout life, as there is a continual destruction and replacement of cutaneous
nerves and sense organs (TELLO, 1932; COWDRY, 1932; FITZGERALD, 1962). The
skin remains in a steady state in spite of a continual turnover of epidermal cells,
nerves and sense organs. Proliferation of epidermal cells in the germinal layer of
the skin is regulated by chalone (BULLOUGH and LAURENCE, 1959, 1964; BULLOUGH
et al., 1967). The normal concentration of chalone in the skin is sufficient to inhibit
mitosis, but the concentration falls as a result of loss of chalone after injuries. It
seems most probable that a similar chemical mechanism regulates the growth of
nerves into the skin during development and during repair of normal wear and
tear.

2. Induction of Receptors

Sensory nerve terminals always arrive before sense organs start to develop. The
literature describing this time sequence in the ontogeny of sensory nerves and their
sense organs has been reviewed recently by ZELENÁ (1964). She concluded that
development of cutaneous sense organs, taste buds, lateral-line organs, muscle
spindles and tendon organs all depend in an unknown way on their sensory nerves.

There have been no physiological experiments on the normal development of
sense organs, probably because of the technical difficulty of working with embryos
at the stages when their sensory nerves reach the periphery. One way of avoiding
some of these difficulties is to study the delayed development of sensory nerves
into an amphibian limb which has developed without sensory innervation (PIATT,
1942; TAYLOR, 1944). Sensory nerves may be excluded from the developing limb
by extirpating the spinal ganglia supplying the limb. Presumably the sense organs

would not develop in such a limb. Re-innervation of the limb with neighboring sensory nerves might result in the formation of sense organs. The functional characteristics of growing sensory nerve terminals and of developing sense organs might be studied in this preparation.

I have described this experiment as an illustration of the value of combining neuroembryological and neurophysiological techniques. Experiments of this type may provide evidence regarding the functional specificity of sensory nerves during development. It may then become possible to answer the question of whether the sensory nerve confers its specificity on the sense organ or vice versa, or whether there is a reciprocal interaction between nerve terminal and sense organ which determines their functional specificity. At present we have to rely mainly on morphological evidence from studies of denervation and re-innervation of sense organs.

3. Effects of Denervation and Re-Innervation on Receptors

The dependence of the taste buds on their innervation is the classic example of the trophic action of a sensory nerve on its sense organ. Since this phenomenon was first studied nearly a century ago (Von Vintschgau and Honigsmied, 1876; Von Vintschgau, 1880), a succession of investigators have shown that taste buds degenerate after denervation and reform after regeneration of their nerves (reviews by Olmsted, 1920a; May, 1925; Guth, 1958; Zelená, 1964).

The rate of degeneration after denervation depends on the sense organ as well as on the species. In the rat and rabbit, reduction of the size of taste buds is evident 8 hr after denervation, and the taste buds degenerate completely in 4 to 7 days (Beidler, 1963). The taste buds on the barbels of the catfish show atrophic changes within a few days of denervation, and disappear in 11 to 19 days (Olmsted 1920a, b; May, 1925). In the frog, complete degeneration of taste buds occurs 40 weeks after denervation (Robbins, 1967a). Encapsulated sense organs in the skin, for example, the corpuscles of Herbst in the duck's beak (Quillian and Armstrong, 1961) or the muscle spindles (Zelená, 1964), persist for weeks or months after denervation, but they eventually degenerate.

The effects of denervation on sense organs may be reversed after reinnervation if the sense organ has not degenerated. It is not known if the nerves reconnect with the same sense organs with which they were originally connected. In cases where the sense organ has degenerated completely before the return of their nerves, epithelial cells in contact with the sensory nerve terminals differentiate into new receptor cells.

There is some evidence that the nerves are attracted to the positions occupied by degenerating sense organs. For example, May (1925) observed that regenerated nerve terminals returned to the positions previously occupied by taste buds in the catfish. After cutting the dorsal lateral-line nerve and cutaneous nerves in frog tadpoles, Speidel (1964) observed a marked preference for re-innervation of lateral-line organs by lateral-line nerve fibers rather than cutaneous nerve fibers.

There is a continual turnover of all the epithelial cells of the tongue, including taste cells. The time between administration of H^3-thymidine and disappearance of all labelled tongue epithelial cells in the mouse and rat is between 4 and 8 days

(WALKER, 1960; TOTO and OJHA, 1962; BEIDLER, 1963). Normally the rate of cell death in the gustatory epithelium is precisely balanced by production of new cells. In the taste buds of the catfish (MAY, 1925), the frog (ROBBINS, 1967), and the rat (BEIDLER, 1963; BEIDLER and SMALLMAN, 1965), the cells move from the periphery of the bud where mitosis occurs to the center of the bud where the cells degenerate.

There has been some speculation about the way in which nerves control the number of receptor cells in taste buds. GUTH (1963) proposed that the nerves merely induce a number of epithelial cells to differentiate into recepter cells, whereas epithelial proliferation continues independently at a rate which compensates for cell death. This seems very likely in view of the evidence that the control of epithelial proliferation by chalone appears to be independent of the influence of nerves (BULLOUGH and LAURENCE, 1959, 1964; BULLOUGH et al., 1967). However, ROBBINS (1967a) has presented some preliminary results on two frogs which show that DNA synthesis in taste buds is almost completely abolished after the lingual nerve is cut. This evidence that the nerve controls cell proliferation is unconvincing because the rates of DNA synthesis and cell proliferation in the frog's taste buds are extremely slow.

What happens to the nervous connections of taste cells during their migration from the periphery to the center of the taste bud? It is very probable that the newly formed cells may be innervated at random by branches of any nearby gustatory nerve. The functional specificity might then be determined either by the nerve or by the receptor cell; at present there is no evidence one way or the other. There are two possible mechanisms whereby the sensory nerve could maintain connections with a taste cell during its movement from the periphery to the center of the taste bud. Either the cells move from one nerve terminal to another or the nerve moves with the cell until it dies. BEIDLER and SMALLMAN (1965) favor the former hypothesis, and suggest that taste cells change their specificity as they move from one nerve to another. However, the rate of growth of axon terminals is usually at least 1 mm per day, and this would be sufficient to allow the nerve terminal to maintain contact with the taste cell. which moves at the rate of a few micrometers per day in the rat (BEIDLER, 1963). At present there is insufficient evidence with which to choose between these hypothetical alternatives.

An important question is whether the nerve fiber imposes its specific response pattern on the epithelium in which the receptor cells develop, or whether the receptor cells impose their response properties on the nerve. IGGO and BROWN (personal communication; BROWN and IGGO, 1962) have found that regenerating cutaneous nerve fibers give relatively nonspecific responses to mechanical and thermal stimulation of the skin. When the fibers re-innervate cutaneous sense organs, they regain the responses typical of the receptors they innervate. For example, rapidly adapting responses typical of hair follicle receptors can be recorded. Although not conclusive, these results, together with results of re-innervation of receptors after skin grafting and nerve-crossing (which will be described below), suggest that receptor cells impose their response properties on the nerves. However, the evidence does not rule out a two-way traffic between sense organ and sensory nerve.

4. Effects of Heterotopic Grafting and of Nerve-crossing on Sense Organs

The question whether the specificity is inherent in the epithelium or whether it is conferred on the receptors by the nerves can, at present, be dealt with experimentally by heterotopic grafting of sense organs and by crossing sensory nerves. The simplest method of re-innervating sense organs with foreign nerves is by grafting to a new position. For example, DIJKSTRA (1933) transferred skin containing Herbst and Grandry corpuscles from the ducks beak to the foot which normally lacks these sense organs. Many of the corpuscles degenerated after transplantation but reformed when the transplanted skin was re-innervated by sensory nerves of the foot. This suggested that the skin nerves could exert the same inductive or trophic effect as the trigeminal nerve, which normally innervates Herbst and Grandry corpuscles on the beak. This evidence is not entirely convincing because Herbst corpusles degenerate very slowly after denervation, and because the functional specificity of the re-innervated corpuscles was not examined.

Another experiment to determine whether innervation pattern depends on the graft or on the re-innervating nerves was performed by KADANOFF (1925), who exchanged skin between hairy skin of the snout and the hairless skin of the soles of the feet in mice and guinea pigs. Re-innervation of both types of grafts occurred. In some cases the innervation pattern of the graft resembled that which was normal for the grafted skin and not for the skin in which the nerves originated. This was most clearly seen in the case of hair follicles which were innervated normally in snout skin transplanted to the hairless skin of the sole.

Observations on transplanted tongue also support the hypothesis that nongustatory sensory nerves can induce the formation of taste buds in gustatory epithelium. Transplantation of tongue primordia from salamander larvae to the body wall of donor salamanders resulted in innervation of the tongue by cutaneous nerves and normal development of taste buds (STONE, 1940). Taste buds degenerated within 2 to 3 weeks after transplantation of the tongue to the orbit of the newt, but the taste buds gradually reappeared as the graft became innervated by orbital nerves (PORITSKY and SINGER, 1963). Before any final conclusions can be drawn from these observations, it is necessary to have some evidence of the physiological activity of sense organs in skin and tongue grafts innervated by foreign nerves. However, the evidence from heterotopic grafting experiments suggests that outaneous sense organs or taste buds on the tongue may be innervated and maintained by foreign sensory nerves. This is also in agreement with the results of the cross-innervation experiments described below.

Although the evidence suggests that sensory nerves are equivalent in their trophic effect on sense organs, it is very doubtful that motor nerves can exert any trophic effect on sense organs or that sensory nerves can exert a trophic effect on muscle. It has definitely been shown that motor nerves do not innervate the skin of a frog's leg, even when sensory nerves are excluded from the limb (TAYLOR, 1944). Although sensory nerves grafted into muscles come into close apposition with muscle fibers, they never form functional connections with them (WEISS and EDDS, 1945; GUTMANN, 1945). It is, therefore, difficult to accept the observation of BOEKE (1917) that motor nerve fibers of the hypoglossal nerve can induce differentiation of taste buds in the tongue after suturing the proximal end of the hypoglossal to the distal stump of the lingual nerve. Moreover, GUTH (1958) was unable to find any re-innervation of taste buds in the circumvallate papillae of the cat after the proximal end of the hypoglossal nerve was sutured to the distal stump of the glossopharyngeal. GUTH (1958) did find that vagus nerve fibers could re-innervate circumvallate papillae, but the functional effect of this foreign innervation is not known.

A start has been made in using neurophysiological techniques to obtain information about the functional activity of cross-innervated taste buds. For example, OAKLEY (1967) has recorded electrical responses in the chorda tympani and glossopharyngeal nerves of the rat before and after cross-innervation. In the rat, the chorda tympani normally innervates the front and the glossopharyngeal innervates the back of the tongue. The front and back differ in their responses to chemicals, and the front is normally much less responsive to cooling than is the back of the tongue. After the chorda tympani and glossopharyngeal nerves were crossed, the massed responses from the whole nerves tended to conform to their new terminations (Fig. 1). This indicates that the functional specificity of the taste cells may be an inherent property of the epithelium. The nerves appear to exert a nonspecific morphogenetic action on the epithelium resulting in the formation of taste cells. It might be argued that OAKLEY's results were due to the fact that the chorda tympani and glossopharyngeal nerves both contain many nerve fibers with similar functional specificity, and that reinnervation occurred only when the functional specifities of taste cells and gustatory nerve fibers matched. This possibility can be ruled out only by the demonstration that taste buds can be functionally reinnervated by nongustatory sensory nerves. It was shown by ROBBINS (1967b) that denervation of the frog tongue resulted in gradual reduction in the size of the taste buds, until they disappeared after about 40 weeks. Re-innervation by their own nerves, or by a cutaneous branch of the trigeminal nerve, resulted in recovery of the taste buds to about half their normal size. After cross-innervation, chemical stimulation of the tongue resulted in a discharge of action potentials in the cutaneous nerves innervating the tongue. Unfortunately, this result is not completely unambiguous because the cutaneous branches of the trigeminal nerve may innervate chemoreceptors in the skin of the frog head. It is conceivable that only the latter fibers re-innervate taste buds on the frog tongue.

Because of these ambiguities in the results of OAKLEY (1967) and ROBBINS (1967b), it is not yet possible to say whether the functional specificity of receptors is determined by the sensory nerve or is inherent in the receptor cells. If the receptor cells impose their response properties on the nerve, how is information from cross-innervated sense organs dealt with centrally? Perhaps the central connections of the afferent nerve fibers also change so as to conform to the peripheral rearrangement. This possibility will be discussed later at some length.

At present it is not even possible to decide how many different processes are involved in the formation and maintenance of connections between sensory nerves and receptor cells. It seems that there are at least three: (1) the inductive effect of nerve on "competent" cells which causes them to differentiate into receptor cells; (2) determination of the response properties of the receptor cells and sensory nerve; and (3) the trophic effect of the nerve in maintaining the vitality of the receptor cells. The nature of this trophic effect is still a subject of speculation and will be discussed below.

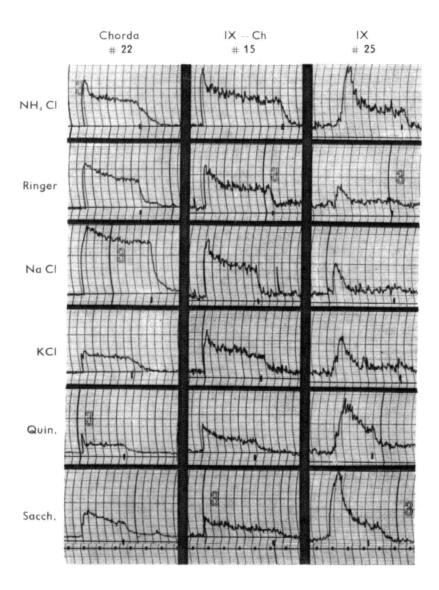

Fig. 1. Original summated records of action potential activity from three nerves: the normal chorda of rat no. 22 in the left column, the cross-regenerated IXth of rat no. 15 in the center, and the normal IXth of rat no. 25 in the right column. Each row represents the response of these three nerves to a specific taste stimulus. The small vertical dash beneath each record is a marker which gives a crude indication of the onset of the distilled water rinse. Time base in 10-sec intervals. (From Oakley, 1967, with permission of the Editors of The Journal of Physiology, London)

5. Nature of the Trophic Action of Sensory Nerves on Sense Organs

OLMSTED (1920b) postulated that the trophic effect of the gustatory nerve on taste buds is due to some substance "of the nature of a hormone" which is continuously released by the sensory nerve terminals. This hypothesis was elaborated by MAY (1925) who was the first to postulate the "flow of a hormone-like substance from the cell body of the neurone to its terminations". Since then an impressive body of evidence has been discovered (reviewed by LUBINSKA, 1964) which shows that substances move in the axon towards its terminals as well as towards the perikaryon. The existence of proximodistal movement of substances in the axon is no longer in question. This seems to have increased the plausibility of the hypothesis that trophic substances are released from both sensory and motor nerve terminals. However, there is no direct evidence of any kind to support this hypothesis, and other alternatives should also be considered.

It is interesting to compare the induction of receptors by nerves with embryonic induction. In both cases the inductive stimulus seems to be nonspecific and the result of the induction seems to depend only on the competence of the tissue in which the induction occurs. A variety of chemical and physical agents can act as embryonic inductors, and no specific inductor substance has been found after nearly 50 years of intensive research on embryonic induction (SAXÉN and TOIVONEN, 1962). The same may apply to the induction and maintenance of receptors by sensory nerves. It is even possible that no molecules pass from the nerve to the receptor cells. Mere physical contact between them may be sufficient inductive and trophic stimulus. Having said this, it is only fair to point out that there are some observations which can be interpreted most easily in terms of proximodistal movement of some trophic substance in the axon.

PARKER (1932) was the first to show that the latent period between cutting a nerve and the onset of atrophic changes in sense organs or muscles is proportional to the length of the distal stump of the nerve. After the lateral-line nerve of the catfish is cut, the degeneration of lateral-line organs spreads from the cut proximodistally at a rate of about 2 cm per day (PARKER, 1932; PARKER and PAINE, 1934). A similar correlation has been found between the length of the peripheral stump of a motor nerve and the time of onset of atrophic changes in muscle. LUCO and EYZAGUIRRE (1955) suggested that the latent period was due to the gradual depletion of a reservoir of trophic substance in the peripheral stump of the nerve. After cutting the motor nerve to the tenuissimus muscle of the cat, they showed an apparent movement of trophic substance at the rate of 2 mm/hr for the first 73 hr, and from 1.4 to 0.5 mm/hr thereafter.

From this evidence, it seems as if a trophic substance is transported proximodistally in sensory and motor nerves at about the same rate: on the order of 1 to 2 cm/day.

6. Neuronal Specificity and the Development of Neuronal Circuits

Orderly mapping of sensory receptive fields into the corresponding sensory centers is one of the most outstanding features of connectivity of sensory neurons. It is not known how the sensory input to the central nervous system becomes organized during ontogeny. The existence of rival theories which attempt to deal with this controversial problem is really an indication of a grave lack of the right kind of experimental evidence. At present there is insufficient data about the functional activities of developing sensory neurons. Equally little is known about their ultrastructure and cytochemistry or about the chronology and spatial order of formation of their central and peripheral connections.

It is essential to have more information of this kind before the developing neuron may be considered from the point of view of contemporary cell biology. Eventually it may become possible to describe the ontogeny of neuronal connectivity in terms of genetic and epigenetic control of molecular synthesis. Present evidence falls far short of permitting such a description. There is general agreement that some kind of specificity is concerned in the formation of orderly central-peripheral relations in the nervous system. However, the nature of the specificity and its mode of operation are in question.

Apart from transducer specificity of receptors, which is beyond the scope of this discussion, the specificity may be in the order of connections between neurons or in their functional properties, or in both. Accordingly, the development of functional neuronal circuits may depend primarily on the selective formation of connections, or it may depend mainly on the selective responses to different rhythms of nerve impulses.

The antithesis between the theories of impulse specificity and specific connectivity originated in neuroembryology with the observation that supernumerary grafted limbs in salamanders may move synchronously with normal limbs (Detwiler, 1920, 1925, 1936; Weiss, 1924, 1928, 1931; Verzár and Weiss, 1930). This phenomenon was called homologous or myotypic response. It only occurs when the grafted limb is close to a normal limb and is innervated by some limb nerves. Limbs which are innervated by trunk nerves or by cranial nerves have incoordinated movements (Braus, 1905; Nicholas, 1933; Platt, 1957; Székely, 1959; Hibbard, 1965). Homologous response may occur when the grafted limb receives a fraction of the number of nerve fibers supplying a normal limb, and under these circumstances a single motor axon may branch to supply several muscles, including antagonists. Weiss (1928, 1931) therfore concluded that the peripheral connections are made at random but the muscles respond selectively to different patterns of nerve impulses. Although this was soon disproved by Wiersma (1931), the possibility that temporal patterning of impulses might allow selective responses without specific connection has continued to intrigue some neuroembryologists (Székely, 1959, 1966; Székely and Szentágothai, 1962; Weiss, 1966). The evidence in favor of this theory is insignificant. However, it might be asserted that single unit recording has excluded such evidence and that only simultaneous recording from many units can determine if the rhythm of impulse traffic has any functional significance.

There are two serious criticisms of the observations of homologous response. First, the observed movements result from the combined activity of many neurons and muscles and include an unknown quantity of feedback control. Second, the urodeles (salamanders and newts) on which the observations were made "so differ from other tetrapod vertebrates respecting nerve muscle relationships that they form a group unto themselves" (Straus, 1946). In urodeles the somatotopic arangement of sensory and motor neurons is very blurred, even when compared with the frog (SZÉKELY and CZEH, 1967). There are great variations in the pattern of innervation of specific limb muscles in salamanders (PIATT 1939, 1942). Multiple innervation of many muscles by branches of a single motor axon and termination of several different motor axons on a single muscle fiber are normal in urodeles (TIEGS, 1953; MARK et al., 1966). For these reasons, there is a much greater probability in urodeles than in other vertebrates that a grafted limb will receive sufficient innervation from which functionally adequate connections may be selected. Whether the selection is made peripherally or centrally, or both, is not clear. There is some evidence that peripheral selection of muscles by motor axons occurs in a fish, which has multiple muscle innervation similar to that in urodeles. MARK (1965) found that normal function was restored after regeneration of nerves to the muscles moving the fin, but permanent incoordination of movement occurred after cross-union of nerves to antagonistic fin muscles. From these observations it was concluded that multiple innervation of muscles greatly increases the probability of selective reconnection of motor nerves to specific muscles. In urodeles there is also an increased probability of selective central connection and functional recovery, because the central organization of the spinal cord in urodeles is based on functional rather than somatotopic localization (SZÉKELY and CZEH, 1967). Because of these peculiarities of neuronal organization, evidence obtained from experiments on the development of neuronal circuits in urodeles may have limited general validity.

In the past 20 years, the evidence has been growing in favor of the hypothesis that synapses form as a result of selective affinity between pre- and postsynaptic membranes. Most of this evidence has been obtained by SPERRY (1951 a, b, 1965) who observed the changes in behavior following a variety of ingenious surgical changes of neuronal connections. His experiments have shown that the affinity between neurons is so great that selective reconnection occurs after mechanical derangement of the normal connections. This conclusion leads to a reinterpretation of the phenomenon of homologous response, for if the peripheral connections to muscles or sense organs are switched, their central connections break down and reconnect to restore the status quo ante (SPERRY, 1966). The assumption (which will be discussed at length in the following section) is made that the central connections react to a change in the peripheral contacts. This capacity to readjust the central connections decreases ontogenetically and phylogenetically. For this reason, sensory nerve crosses in adult mammals always result in false referral of the origin of the sensation, whereas motor nerve crosses result in permanent maladaptive movements. A critical review of the evidence shows that the maladaptive effect of nerve crosses in adult mammals may be inhibited or disguised, but even in man readaptation does not occur (SPERRY, 1945a).

Of all the neuronal circuits which have been studied, the formation of connections between the retinal ganglion cells and optic tectum of fishes and frogs has been analyzed most rigorously. SPERRY (1944, 1945b, 1951a, b, 1963) has shown that the retinal ganglion cells and the tectal neurons with which they connnect have such great affinity that they reconnect selectively after the optic axons have been chaotically scrambled by cutting the optic nerve. Recovery of point-to-point retinotectal connections after section and regeneration of the optic nerve of the frog and goldfish has been confirmed by recording action potentials in the tectum evoked by a small visual stimulus (Gaze, 1960; Gaze and Jacobson, 1963; Jacobson and Gaze, 1965). Single unit analysis of visual responses in the frog tectum has demonstrated that optic axons of different functional classes regenerate to specific depths in the tectum as well as to specific places in the retinotectal map (Maturana et al., 1959, 1960). This evidence shows that there must be refined differences in the specificity extending down to the single unit level.

Assuming that the final selection of a specific region of postsynaptic membrane by the axon tip is a matter of chemoaffinity, as Sperry (1963) has proposed, how does the axon terminal reach its proper destination? The probability of error is too great for mechanical guidance to play an important part in the selection of pathway. All the evidence points to a highly refined chemotaxis of some kind. Even when optic axons were surgically deflected from their usual course, they were observed histologically to have grown directly back to their correct tectal terminals bypassing all other available tectal neurons on the way (Attardi and Sperry₂ 1960; Arora and Sperry, 1962). When optic axons from only half the retina were permitted to regenerate, they connected only with their normal loci, leaving half the tectum without optic connections (Jacobson and Gaze, 1965). Optic nerve fibers grew toward their appropriate tectal loci from small pieces of retina grafted on the pial surface of the tectum in chick embryos, although it was not known if they formed functional connections (De Long and Coulombre, 1968). Visual function was demonstrated in eyes whose optic nerve fibers had been surgically deflected into the oculomotor root and had then grown along an aberrant pathway into the tectum (Hibbard, 1967). These observations that optic nerve fibers can reach their tectal loci along abnormal routes also suggests that the fibers grow down a chemical gradient rather than along preformed mechanical pathways.

7. Development of Neuronal Specificity

There is evidence that neuronal specificity may develop in two different ways. Some neurons, and perhaps all peripheral sensory and motor neurons, have their central synaptic associations specified through their peripheral connections (Weiss, 1936, 1952; Sperry, 1950; Sperry and Miner, 1949; Miner, 1951, 1956; Eccles et al., 1960, 1962a, b; Jacobson and Baker, 1968). Other neurons are specified independently before they form connections. This probably occurs in many neurons which synapse only within the central nervous system, as, for example, in the retinal ganglion cells which develop specificity during the early neuroblast stage (Jacobson, 1967, 1968a, b, c).

Evidence for specification of sensory neurons through their peripheral connections has come from many experiments in which changing the peripheral loci of receptive fields of sensory neurons resulted in corresponding changes in their central connections.

In salamanders, stimulation of the cornea of an eye transplanted to the nose or ear region (Weiss, 1942) or to the gill region (Kollros, 1943) resulted in lid-closure of the ipsilateral normal eye. These observations indicated that the trigeminal or vagus nerve fibers innervating the cornea of the grafted eye had made central connections with the oculomotor neurons concerned with the corneal reflex of the normal eye. Székely (1959) has objected to this interpretation because he found that neither corneal reflexes nor limb reflexes could be elicited by stimulating a limb grafted to the gill region of the newt. However, corneal reflexes did occur on stimulating the regeneration blastema which grew after amputating the terminal part of the grafted limb. As an alternative explanation, SzÉKELY proposed that the lid-closure reflex was a selective response to similar patterns of nerve impulses which originated in similar kinds of naked nerve endings in the regeneration blastema and in the cornea.

Additional evidence in favor of peripheral specification of cutaneous nerves was obtained by MINER (1951, 1956). She removed a piece of skin on one side of the trunk of a frog tadpole from the midline of the back to the midline of the belly, and replaced the skin in a dorsoventrally inverted position. The skin underwent self-differentation, so that after metamorphosis dark dorsal skin developed on the belly, and light ventral skin on the back. Mechanical stimulation of the graft elicited limb reflexes aimed at the original site of the skin which was stimulated, and not at the point of stimulation. These reflexes could have resulted from three alternative mechanisms. First, the cutaneous nerve might have reconnected with their former receptive fields; MINER partially excluded this by anatomical dissection. The second alternative, favored by MINER, is that respecification of the cutaneous nerves might have occured. The third alternative, extremely difficult to eliminate at present, is that specific patterns of impulse traffic in the cutaneous nerves originated from stimulating different places on the skin.

In order to distinguish further between these alternatives, we have repeated MINER'S experiments and extended them by recording impulses in cutaneous nerve fibers and mapping their receptive fields in normal and grafted skin (JACOBSON and BAKER, 1968). Two types of skin grafts were made in frog tadpoles at different stages of development (stages I to XIV; TAYLOR and KOLLROS, 1946). The stages at which the grafts were made did not affect the results. In the first type of graft, the skin was excised from the trunk on one side from the mid-dorsal to midventral line and from the tympanic membrane to the base of the tail. The skin was replaced in anteroposterior and dorsoventral inverted position. These were called DV-interted grafts. In the second type of graft, the skin on the back was excised from between the eyes to the base of the tail, and replaced after anteroposterior and left-right inversion. These were called AP-inverted grafts. All nervous connections of the grafted skin were broken during these operations. Cutaneous reflexes appeared in these frogs at the normal time after metamorphosis. Stimulation of the skin with a nylon bristle (0.4-mm diameter, exerting a pressure of 200 mg) resulted in limb movements directed at the point of stimulation on the normal skin (BAGLIONI, 1913; FRANZISKET, 1963).

Stimulation of DV-inverted grafts frequently resulted in reflex movements of a limb which were misdirected at the original site of the grafted skin rather than at the point of stimulation (Fig. 2). This conformed with MINER'S observations. However, stimulation of AP-inverted grafts always resulted in normal reflexes directed at the stimulated point on the skin. These reflexes were observed at regular intervals for more than a year. Then the receptive fields of cutaneous nerves were mapped by recording action potentials evoked in cutaneous nerve fibers by stimulating the skin as before.

In all cases, irrespective of the original position of the grafted skin, the cutaneous nerve entered the skin within its own receptive field (Fig. 3). There were
some individual variations in the distribution of receptive fields of dorsal cutaneous
nerves, but the receptive fields of ventral cutaneous nerves were always arranged
in regular segmental order. There was considerable overlap of receptive fields, in

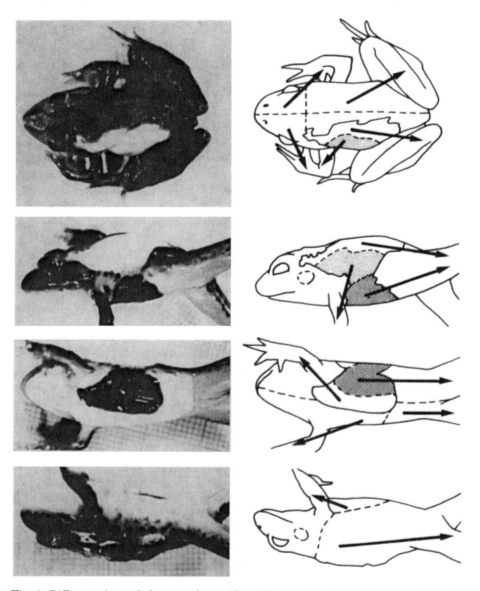

Fig. 2. Different views of the same frog with a DV-inverted skin graft prepared 280 days
previously at larval stage XIV. Each arrow points from the reflexogenic zone of the skin to
the limb which responded when the skin-zone was stimulated. Misdirected limb movements
resulted from stimulating the stippled zones. When the lightly stippled zone was stimulated,
the forelimb wiped the darkly stippled zone; when the darkly stippled zone was stimulated,
the hindlimb wiped the lightly stippled zone

agreement with the observations of ADRIAN *et al.* (1931). It is remarkable that in DV-inverted grafts the receptive fields of normal and grafted skin did not overlap but always ended precisely at the edge of the graft. This did not occur in AP-inverted grafts, where the receptive fields of normal and grafted skin overlapped as in

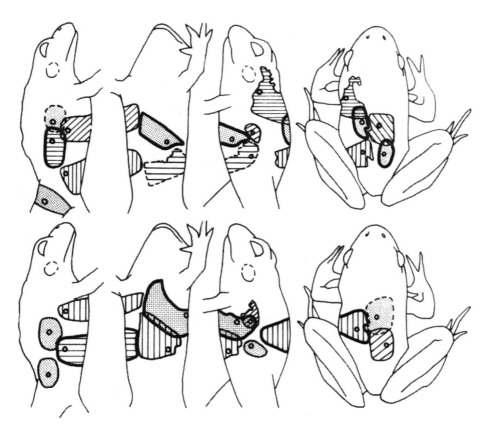

Fig. 3. Different views of the same frog as in Fig. 2 with DV-inverted graft. Receptive fields of cutaneous nerves were mapped by recording action potentials in the nerve which entered the skin at the position shown by a circle within each receptive field

normal dorsal skin. Receptive fields of dorsal cutaneous nerves were usually roughly circular or oval with the AP axis longer than the DV axis, whereas the receptive fields of ventral cutaneous nerves were always much longer in the DV axis than in the AP axis (Fig. 2). It is noteworthy that these distinct differences between the normal shapes of dorsal and ventral receptive fields were maintained in DV-inverted grafts. Therefore, the shape of the receptive field is mainly determined by the nerve rather than by the skin into which the nerve grows.

These observations tend to eliminate the possibility that misdirected reflexes were due to selective growth of sensory nerves back to their original places in the skin grafts. The second alternative, that specific patterns of impulses originate from different regions of skin, is considered unlikely, although it is difficult to exclude it directly. In these experiments, activity was usually recorded simultaneously in several fibers and frequently from a single unit in each cutaneous nerve. It would be very difficult to exclude the possibility, remote as it seems, that there are specific differences in the patterns of impulses in fibers in different cutaneous nerves. There is no evidence of this type of local specificity in the discharge from mechanoreceptors in frog skin: the pattern of impulses appears to be determined by the mode of application of the stimulus (Maruhashi et al., 1952; Loewenstein, 1956; Catton, 1958; Lindblom, 1962, 1963; Högland and Lindblom, 1961).

The most satisfactory explanation of the misdirected reflexes was that respecification of cutaneous nerves had occurred through their terminal connections, resulting in relocation of central synapses and restoration of the original reflex circuits from grafted skin to limb muscles. Most probably the extent of synaptic relocation is limited and occurs only between neurons which are fairly close together. This would account for the normal reflexes elicited from AP-inverted grafts. It seems unlikely that a complete anteroposterior switch of reflex circuits in the spinal cord could occur, but it is more probable that a switch of connection within a short segment of the cord resulted in the reversal of reflexes observed in DV-inverted grafts.

Specification of sensory nerves through their peripheral contacts is not limited to the lower vertebrates. In kittens, new reflex connections of Group Ia afferents to alpha motoneurons may form which partially reverse the effects of crossing peripheral nerves to antagonistic muscles (Eccles et al., 1960, 1962a, b). The Group Ia input to alpha motoneurons is a most favorable system in which to detect changes in specific connectivity. Aberrant connections can be detected because of the high degree of specificity with which Group Ia afferents from muscle spindles make monosynaptic connections with alpha motoneurons of the same or synergic muscles (Eccles et al., 1957; Eccles and Lundberg, 1958). Cross-union of the nerves to antagonistic muscles of the hindlimb of the neonatal kitten (M. medial gastrocnemius and M. peroneus, or M. lat. gastrocnemius and M. plantaris) has the effect of switching both the motor axons and the sensory axons from muscle stretch receptors. Some months after such nerve crosses, the Group Ia monosynaptic input from muscle spindles was recorded intracellularly from alpha motoneurons (Eccles et al., 1960, 1962b). As a result of the nerve crosses the peroneal motoneurons acquired a statistically significant increase in Group Ia input from the synergic muscles, lateral gastrocnemius and plantaris. There was also a significant decrease in the Group Ia inputs which had become functionally inappropriate as a result of crossing the nerves. The results showed that to a limited extent new monosynaptic reflex connections formed which tended to reverse the effects of switching the peripheral connections. No such rearrangements were found after cutting and self-union of the nerves to medial gastrocnemius and peroneal muscles (Eccles et al., 1962a). The best explanation of the observed changes in the monosynaptic input from muscle spindles to alpha motoneurons after nerve crosses was that they resulted from respecification of the neurons through their new peripheral connections.

Although each of the experiments described above is in itself not fully conclusive, the evidence taken as a whole leaves little doubt that respecification of the central connections of sensory neurons can occur through their peripheral connections. It should now be possible to devise experiments to uncover the mechanism of this type of neuronal specificity. Presumably it involves some type of message passing from the peripheral tissue to the nerve terminals, and a reaction which then affects their central terminals. According to SPERRY (1950, 1951a, b), the respecification causes a change in the selective affinity between the neurons and its association neurons. Respecification is assumed to cause a breakdown of the original synapses and the formation of new ones on the basis of the new selective affinities.

Extensive formation of new synaptic associations as a result of peripheral specification of sensory and motor neurons may only be possible for a short time after peripheral connections are first formed during development. The purpose of this type of reorganization could be to reduce the numbers of aberrant connections which may result from accidents of development and to insure that peripheral and central connections are congruent. The extent of these readjustments undoubtedly becomes progressively more limited during ontogeny. Behavioral tests show that reorganization is absent in adult mammals (SPERRY, 1945), and that none occurs after crossing nerves to antagonistic muscles as early as 10 days after birth in the rat (SPERRY, 1941). However, these conclusions can only be accepted with reservations because observations of behavior or tests of the reflexes of the whole limb could not have detected the changes which were shown by microelectrode recording to have occurred in kittens after peripheral nerve crosses (ECCLES et al., 1960, 1962b). Even in adult mammals, some sensory neurons may retain the capacity for respecification and relocation of central synapses, perhaps limited to adjacent neurons. This could be a means of compensating for the small changes in the periphery, for example, in the relation of gustatory nerves to taste cells (BEIDLER, 1963; BEIDLER and SMALLMAN, 1965) or of cutaneous nerves to the normal turnover of cutaneous sense organs which have been demonstrated by FITZ-GERALD (1962).

Specification of some neurons which have their connections entirely within the central nervous system probably occurs during the early neuroblast stage, before the outgrowth of axon and dendrites. The evidence for this is indirect in the case of some neurons in the chick spinal cord (WENGER, 1950) or the chick medulla (HARKMARK, 1954) where it has been found that the type of neuron is already determined before the neuroblasts migrate from the germinal zones. Precise determination of the time of neuronal specification is possible, in principle, by moving the neuroblast to a new position at progressively later stages of development, and by determining the last stage at which it regulates its synaptic associations to its new position. An experiment of this type has been done to determine the time of specification of retinal ganglion cells in the clawed toad, Xenopus (JACOBSON, 1967, 1968a). The advantages of working with the retinal ganglion cells are that they form retinotopically ordered connections in all vertebrates, and the methods of mapping the topography of these connections have sufficient accuracy to detect aberrant connections (JACOBSON, 1962; GAZE and JACOBSON, 1963). Moreover, it is possible to alter the positions of the retinal ganglion cells relative to the tectal cells with which they connect by rotating the eyecup in embryo amphibians before any neuronal connections develop between the eye and the brain.

Behavioral tests of vision in salamanders and newts in which the eyecup had been rotated 180 degrees at various stages of development showed that normal vision developed if the operation was done at a sufficiently early stage, but inverted visuomotor behavior resulted from eye rotation at later stages (STONE, 1944, 1948, 1960; SZÉKELY, 1954). These results lead to the inference that during "functional polarization" (STONE, 1944) or "functional specification" (SZÉKELY, 1954) of the retina, the retinal ganglion cells acquired an unknown kind of local property which enabled them to connect with the correct places in the optic tectum. This inference was tested by 180° rotation of one eyecup in a series of *Xenopus* embryos at various stages of developnent before the formation of retinotectal connections. After metamorphosis, the projection from the rotated eye to the optic tectum was mapped electrophysiologically and compared with the retinotectal projection from the normal eye (JACOBSON, 1967, 1968a).

These experiments showed that the central connections of the retinal ganglion cells formed in the same order after eyecup rotation at stages 28-29 as they would

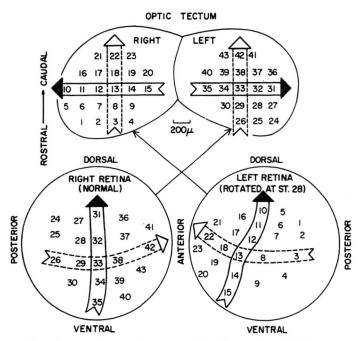

Fig. 4. Map of the retinotectal projection in an adult *Xenopus* in which the left eye had been DV- and AP-inverted (rotated 180°) at embryonic stage 28-29. The projection from the inverted eye is normal. Each number on the tectum represents the position at which a micro-electrode recorded action potentials in response to a small spot of light at the position shown by the same number on the retina. The following figures conform to the same conventions

normally have formed, showing that the ganglion cells were unspecified at those stages of development (Fig. 4). Eye rotation at stage 30 resulted in inversion of the retinotectal map in the AP axis, but not in the DV axis of retina (Fig. 5). Therefore, specification of the ganglion cells in the AP axis of the retina occurred before inversion of the eye, but DV specification only occurred later in accordance with the inverted position of the eye. After stage 30, inversion of the eye resulted in complete inversion of the retinotectal projection (Fig. 6).

These results showed that *specification of retinal ganglion cells* occurred over a period of about 10 hr, in two steps, related to the relative positions of the ganglion cells in the AP and DV axes of the retina. At stage 30 when the ganglion cells

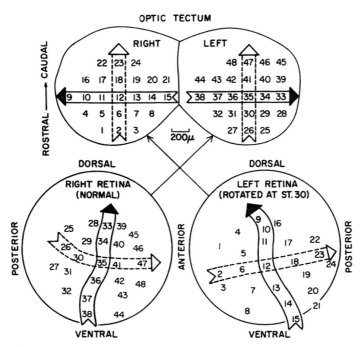

Fig. 5. The retinotectal projection in adult *Xenopus* from the normal right retina and from the left retina which had been inverted at stage 30. The projection from the left retina is normal in the DV axis (black arrow) but is inverted in the AP axis (white arrow)

were specified, they were in the early neuroblast stage of development. Their central connections were already determined about 20 hr before the initial growth of optic axons from the ganglion cells, and about 80 to 100 hr before the axons reached their terminal loci in the tectum. Presumably, the tectal cells became specified independently, in parallel with the retinal ganglion cells with which they formed connections.

The change from an unspecified to a specified state, whatever its nature may be, occurs within about 10 hr (JACOBSON, 1968b). By labelling *Xenopus* embryos with thymidine-H³, it was shown that the precursors of the retinal ganglion cells ceased DNA synthesis and therefore completed their final mitotic division at stage 29 shortly before specification of their central connections occurred. It was suggested that DNA synthesis and specification of retinal ganglion cells are mutually exclusive, perhaps because specification involves synthesis of specific macromolecules. DNA synthesis and specific macromolecular synthesis have been shown to be mutually exclusive in embryonic muscle cells (STOCKDALE and HOLTZER, 1961; OKASAKI and HOLTZER. 1965), in single antibody-forming cells (MÄKELÄ and

NOSSAL, 1962), in pancreatic acinar cells (WESSELS, 1963, 1964), and in the formation of lens-specific proteins (EISENBERG and YAMADA, 1966; YAMADA, 1966). There is also evidence that in some kinds of cells new transcription may *depend upon* an immediately preceding replication of DNA (EBERT and KAIGHN, 1966), and perhaps neurons are of this kind. The hypothesis was proposed that specification of ganglion cells, which occurs immediately after DNA replication ceases,

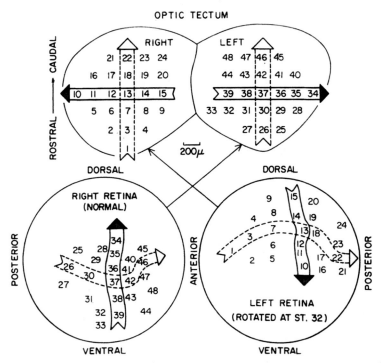

Fig. 6. The retinotectal projection in adult *Xenopus* to the left tectum from the normal right eye, and to the right tectum from the left retina which had been inverted at stage 32. The projection from the left retina is totally inverted

involves the synthesis of specific macromolecules which uniquely label each ganglion cell according to its position in the retina (JACOBSON, 1968b). The mechanism of neuronal specification may either be selective or instructive, in the sense that these terms are used in connection with antibody synthesis. A selective mechanism in which all the information for specifying the connections of each neuron is coded in DNA is implausable because of the limited information capacity of the genome. In an instructive mechanism, a bi-axial gradient system in the retina might interfere with transcription of the genetic code into the nucleotide sequence of RNA, or it might affect the translation of the latter into amino acid sequence, or it might affect the binding of subunits to form macromolecules. Whatever the mechanism of neuronal specification in retinal ganglion cells, it occurs rapidly in two all-or-none steps. The absence of any other intermediate states is striking, and so is the fact that all the ganglion cells (or rather, a good sample of them, as shown in Figs. 4, 5, and 6) were specified in one axis or both, and there were no cases in which ganglion cells had different degrees of specification within the same retina. This suggests that the cellular mechanism of neuronal specification is quantal, and that it occurs in two all-or-none steps.

We are now in the exciting stage of progress in this field where several neuronal mechanisms have been recognized which can be interpreted in terms of contemporary cell biology. The trophic activities of nerves, the induction of receptors and the specification of neuronal connections all involve cellular interactions which result in cellular differentiation. As a result of these interactions, specific molecular synthesis in neurons produces the diversity of form and of connectivity which has been the main preoccupation of neurobiologists for nearly a century. It now seems fair to say that it may only be possible to understand the circuitry of the nervous system when much more has been discovered about the regulation of molecular synthesis in neurons.

Acknowledgement:

The author's work has been supported since 1966 by grants National Science Foundation grants GB 4622 and GB 6564.

References

ADRIAN, E. D., CATTELL, M., HOAGLAND, H.: Sensory discharges in single cutaneous nerve fibers. J. Physiol. (Lond.) **72**, 377—391 (1931).

ARORA, H. L., SPERRY, R. W.: Optic nerve regeneration after surgical cross-union of medial and lateral tracts. Amer. Zoologist **2**, 610 (1962).

ATTARDI, D. G., SPERRY, R. W.: Preferential selection of central pathways by regenerating optic fibers. Exp. Neurol. **7**, 46—64 (1963).

BAGLIONI, S.: Die Hautreflexe der Amphibien. Ergebn. Physiol. **13**, 454—546 (1913).

BEIDLER, L. M.: Dynamics of taste cells. In: Olfaction and Taste. ZOTTERMAN, Y. (ed.). Oxford: Pergamon 1963.

— SMALLMAN, R. L.: Renewal of cells within taste buds. J. Cell Biol. **27**, 263—272 (1965).

BOEKE, J.: Studien zur Nervenregeneration. II. Die Regeneration nach Vereinigung ungleichartiger Nervenstücke (heterogene Regeneration), und die Funktion der Augenmuskel- und und Zungennerven, Die Allgemeinen Gesetze der Nervenregeneration. Verhandel. Koninkl. Akad. Wetenschap. Amsterdam (Afdeel. Naturk.) **19**, 1—71 (1917).

BRAUS, H.: Experimentelle Beiträge zur Frage nach der Entwicklung peripherer Nerven. Anat. Anz. **26**, 433—479 (1905).

BROWN, A. G., IGGO, A.: The structure and function of cutaneous touch corpuscles after nerve crush. J. Physiol. (Lond.) **165**, 28—29 (1962).

BULLOUGH, W. S., LAURENCE, E. B.: The control of epidermal mitotic activity in the mouse. Proc. roy. Soc. Lond. B. **151**, 517—536 (1959).

— — Mitotic control by internal secretion: The role of the chalone-adrenalin complex. Exp. Cell Res. **33**, 176—194 (1964).

— — IVERSEN, O. H., ELGJO, K.: The vertebrate epidermal chalone. Nature (Lond.) **214**, 578—580 (1967).

CATTON, W. T.: Some properties of frog skin mechanoreceptors. J. Physiol. **141**, 305—322 (1958).

COWDRY, E. V.: The skin. In: Special cytology. 2. ed. New York: Hoeber 1932.

DE LONG, R. G., COULOMBRE, A. J.: The specificity of retino-tectal connections studied by retinal grafts onto the optic tectum in chick embryos. Develop. Biol. **16**, 513—531 (1968).

DETWILER, S. R.: Experiments on the transplantation of limbs in Amblystoma. The formation of nerve plexuses and the function of the limbs. J. Exp. Zool. **31**, 117—169 (1920).

— Coordinated movements in supernumerary transplanted limbs. J. comp. Neurol. **38**, 461 (1925).

— Neuroembryology. An Experimental Study. New York: Macmillan 1936.

Dijkstra, C.: Die De- und Regeneration der sensiblen Endkörperchen des Entenschnabels (Grandry- und Herbst-Körperchen) nach Durchschneidung des Nerven, nach Fortnahme der ganzen Haut und nach Transplantation des Hautstückchens. Z. mikr.-anat. Forsch. **34**, 75—158 (1933).

Ebert, J. D., Kaighn, M. E.: The keys to change: Factors regulating differentiation. In: Major problems in developmental dibiology. Locke, M. (Ed.). New York: Academic Press Inc. 1966.

Eccles, J. C., Eccles, R. M., Lundberg, A.: Types of neurones in and around the intermediate nucleus of the lumbo-sacral cord. J. Physiol. (Lond.) **154**, 89—114 (1960).

— — Magni, F.: Monosynaptic excitatory action on motoneurons regenerated to antagonistic muscles. J. Physiol. (Lond.) **154**, 68—88 (1960).

— — Shealy, C. M.: An investigation into the effect of degenerating primary afferent fibers on the monosynaptic innervation of motoneurons. J. Neurophysiol. **25**, 544—558 (1962a).

— — — Willis, W. D.: Experiments utilizing monosynaptic excitatory action on motoneurons for testing hypothesis relating to specificity of neuronal connections. J. Neurophysiol. **25**, 559—580 (1962b).

— Lundberg, A.: Integrative pattern of Ia synaptic actions on motoneurons of hip and knee muscles. J. Physiol. (Lond.) **144**, 271—298 (1958).

Edds, M. V.: Collateral regeneration of residual motor axons in partially denervated muscles. J. Exp. Zool. **113**, 517—552 (1950).

— Collateral nerve regeneration. Quart. Rev. Biol. **28**, 260—276 (1953).

Eisenberg, S., Yamada, T.: A study of DNA synthesis during the transformation of the iris into lens in the lentectomized newt. J. Exp. Zool. **162**, 353—368 (1966).

Fitzgerald, M. J. T.: Developmental changes in epidermal innervation. J. Anat. (Lond.) **95**, 495—514 (1961).

— On the structure and life history of bulbous corpuscles (corpuscula nervorum terminalia bulboidea). J. Anat. (Lond.) **96**, 189—208 (1962).

Franzisket, L.: Characteristics of instinctive behavior and learning reflex activity of the frog. Anim. Behav. **11**, 318—324 (1963).

Gaze, R. M.: Regeneration of the optic nerve in amphibia. Int. Rev. Neurobiol. **2**, 1—40 (1960).

— Growth and differentiation. Ann. Rev. Physiol. **29**, 59—86 (1967).

— Jacobson, M.: A study of the retino-tectal projection during regeneration of the optic nerve in the frog. Proc. roy. Soc. B. **157**, 420—448 (1963).

Guth, L.: Taste buds on the cat's circumvallate papilla afer reinnervation by glosso-pharyngeal, vagus, and hypoglossal nerves. Anat. Rec. **130**, 25—38 (1958).

— Histological changes following denervation of the circumvallate papilla of the rat. Exp. Neurol. **8**, 336—349 (1963).

Gutman, E.: Reinnervation of muscle by sensory nerve fibers. J. Anat. (Lond.) **79**, 1—7 (1945).

Harkmark, W.: Cell migrations from the rhombic lip to the inferior olive, the nucleus raphe and the pons. A morphological and experimental investigation on chick embryos. J. comp. Neurol. **100**, 115—209 (1954).

Hibbard, E.: Innervation of intrinsic limb musculature by cranial nerves in *Pleurodeles waltlii*. Anat. Rec. **151**, 360—361 (1965).

— Visual recovery following regeneration of the optic nerve through oculomotor nerve root in *Xenopus*. Exp. Neurol. **19**, 350—356 (1967).

Hoffman, H.: Local re-innervation in partially denervated muscle: A histo-physiological study. Austr. J. exp. Biol. med. Sci. **28**, 383—397 (1950).

— Springell, P. H.: An attempt at the chemical identification of "neurocletin" (the substance evoking axon-sprouting). Aust. J. exp. Biol. med. Sci. **29**, 417—424 (1951).

Högland, D., Lindblom, U.: The discharge in single touch receptors elicited by defined mechanical stimuli. Acta physiol. scand. **52**, 108—119 (1961).

Jacobson, M.: The representation of the retina on the optic tectum of the frog. Correlation between retinotectal magnification factor and retinal ganglion cell count. Quart. J.Exp. Physiol. **47**, 170—178 (1962).

— Starting points for research in the ontogeny of behavior. In: Major problems in developmental biology Locke, M., ed. New York: Academic Press Inc. 1967a.

JACOBSON, M.: Retinal ganglion cells: specification of central connections in larval *Xenopus laevis*. Science **155**, 1106—1108 (1967 b).
— Development of neuronal specificity in retinal ganglion cells of *Xenopus*. Develop. Biol. **17**, 202—208 (1968 a).
— Cessation of DNA-synthesis in retinal ganglion cells correlated with the time of specification of their central connections. Develop. Biol. **17**, 219—232 (1968 b).
— Specification of neuronal connections during development. In: Physiological and biochemical aspects of nervous integration. CARLSON, F. C. Ed. Englewood Cliffs, N. J.: Prentice Hall 1968 c.
— BAKER, R. E.: Neuronal specification of cutaneous nerves through connection with skin in the frog. Science **160**, 543—545 (1968).
— GAZE, R. M.: Selection of appropriate tectal connections by regenerating optic nerve fibers in adult goldfish. Exp. Neurol. **13**, 418—430 (1965).
KADANOFF, D.: Untersuchungen über die Regeneration der sensiblen Nervenendigungen nach Vertauschung verschieden innervierter Hautstücke. Arch. Entwickl.-Mech. Org. **106**, 249—278 (1925).
KOLLROS, J. J.: Experimental studies on the development of the corneal reflex in amphibia. III. The influence of the periphery upon the reflex center. J. Exp. Zool. **92**, 121—142 (1943).
LINDBLOM, U.: The relation between stimulus and discharge in a rapidly adapting touch receptor. Acta physiol. scand. **56**, 349—361 (1962).
— Phasic and static excitability of touch receptors in toad skin. Acta physiol. scand. **59**, 410—423 (1963).
LIVINGSTON, W. K.: Evidence of active invasion of denervated areas by sensory fibers from neighboring nerves in man. J. Neurosurg. **4**, 140—145 (1947).
LOEWENSTEIN, W. R.: Excitation and changes in adaptation by stretch of mechanoreceptors. J. Physiol. (Lond.) **133**, 588—602 (1956).
LUBINSKA, L.: Axoplasmic streaming in regenerating and in normal nerve fibres. In: Progress in brain research. Vol. 13: Mechanisms of Neural Regeneration. SINGER, M., SCHADÉ, J. P., (Eds.). Amsterdam: Elsevier 1964.
LUCO, J. V., EYZAGUIRRE, C.: Fibrillation and hypersensitivity to ACh in denervated muscle: Effect of length of degenerating nerve fibres. J. Neurophysiol. **18**, 65—73 (1955).
MÄKELÄ, D., NOSSAL, G. J. V.: Autoradiographic studies of the immune response, II. DNA-synthesis among single antibody-producing cells. J. Exp. Med. **115**, 231—243 (1962).
MARK, R. F.: Fin movements after regeneration of neuromuscular connections: an investigation of myotypic specificity. Exp. Neurol. **12**, 292—320 (1965).
— CAMPENHAUSEN, G. VON, LISCHINSKY, D. J.: Nerve-muscle relations in salamander: possible relevance to nerve regeneration and muscle specificity. Exp. Neurol. **16**, 438—449 (1966).
MARUHASHI, J., MIZUGUCHI, K., TASAKI, T.: Action currents in single afferent nerve fibers elicited by stimulation of the skin of the toad and the cat. J. Physiol. (Lond.) **117**, 129—151 (1952).
MATURANA, H. R., LETTVIN, J. Y., McCULLOCH, W. S., PITTS, W. H.: Evidence that cut optic nerve fibers in a frog regenerate to their proper places in the tectum. Science **130**, 1709 to 1710 (1959).
— — — — Anatomy and physiology of vision in the frog (Rana pipiens). J. gen. Physiol. Suppl. **43**, 129—175 (1960).
MAY, R. M.: The relation of nerves to degenerating and regenerating taste buds. J. Exp. Zool. **42**, 371—410 (1925).
MINER, N.: Cutaneous localization following 180° rotation of skin grafts. Anat. Rec. **109**, 326—327 (1951).
— Integumental specification of sensory fibers in the development of cutaneous local sign. J. comp. Neurol. **105**, 161—170 (1956).
NICHOLAS, J. S.: The correlation of movement and nerve supply in transplanted limbs of Amblystoma. J. comp. Neurol. **57**, 253—283 (1933).
OAKLEY, B.: Altered temperature and taste responses from cross-regenerated sensory nerves in the rats' tongue. J. Physiol. (Lond.) **188**, 353—371 (1967).

OKASAKI, K., HOLTZER, H.: Analysis of myogenesis *in vitro* using fluorescein-labeled anti-myosin. J. Histochem. Cytochem. **13**, 726—739 (1965).

OLMSTED, J. M. D.: The nerve as a formative influence in the development of taste-buds. J. comp. Neurol. **31**, 465—468 (1920a).

— The results of cutting the seventh cranial nerve in Ameiurus nebulosus (Lesueur). J. Exp. Zool. **31**, 369—401 (1920b).

PARKER, G. H.: On the trophic impulse so-called, its rate and nature. Am. Naturalist **66**, 147—158 (1932).

— PAINE, V. L.: Progressive nerve degeneration and its rate in the lateral-line nerve of the catfish. Amer. J. Anat. **54**, 1—25 (1934).

PIATT, J.: A study of nerve-muscle specificity in the forelimb of Triturus pyrrhogaster. J. Morph. **65**, 155—185 (1939).

— Transplantation of aneurogenic forelimbs in Amblystoma punctatum. J. Exp. Zool. **91**, 79—101 (1942).

— Studies on the problem of nerve pattern. II. Innervation of the intact forelimb by different parts of the central nervous system in Amblystoma. J. Exp. Zool. **134**, 103—125 (1957).

PORITSKY, R. L., SINGER, M.: The fate of taste buds in tongue transplants to the orbit in urodele. Triturus. J. Exp. Zool. **153**, 211—218.

QUILLIAM, T. A., ARMSTRONG, J.: Structural and denervation studies of the Herbst corpuscle. In: Cytology of nervous tissue. Proc. Anat. Soc. Great Britain 1961.

RAMÓN Y CAJAL, S.: Acción neurotrópica de los epitelios (Algunas datalles sobre el mecanismo genetico de las ramificaciones nerviosas intra-epiteliales, sensitivas y sensoriales)). Trab. Lab. Invest. Biol. Univ. Madrid **17**, 181—228 (1919).

— Degeneration and Regeneration of the Nervous System. Transl. by MAY, R. M. New York: Hafner Publication Co. 1959.

ROBBINS, N.: The role of the nerve in maintenance of frog taste buds. Exp. Neurol. **17**, 364—38 (1967a).

— Peripheral modification of sensory nerve responses after cross-regeneration. J. Physiol. (Lond.) **192**, 493—504 (1967b).

SAXÉN, L., TOIVONEN, S.: Primary embryonic induction. London: Logos Press 1962.

SPEIDEL, C. C.: Adjustments of nerve endings Harvey Lect. **36**, 126—158 (1941).

— Correlated studies of sense organs and nerves of the lateral-line im living frog tadpoles. IV. Patterns of vagus nerve regeneration after single and multiple operations. Amer. J. Anat. **114**, 133—160 (1964).

SPERRY, R. W.: The effect of crossing nerves to antagonistic muscles in the hindlimb of the rat. J. comp. Neurol. **75**, 1—19 (1941).

— Optic nerve regeneration with return of vision in anurans. J. Neurophysiol. **7**, 57—69 (1944).

— The problem of central nervous reorganization after nerve regeneration and muscle transposition. Quart. Rev. Biol. **20**, 311—369 (1945a).

— Restoration of vision after crossing of optic nerves and after contralateral transplantation of the eye. J. Neurophysiol. **8**, 15—28 (1945b).

— Neuronal specificity. In: Genetic Neurology. Ed. by WEISS, P. Chicago: Univ. Chicago Press 1950.

— Mechanisms of neural maturation. In: Handbook of experimental psychology. Ed. by STEVENS, S. s. New York: John Wiley & Sons, Inc. 1951a

— Regulative factors in the orderly growth of neural circuits. Growth Symp. **10**, 63—87 (1951b).

— Chemoaffinity in the orderly growth of nerve fiber patterns. Proc. nat. Acad. Sci. (Wash.) **50**, 703—710 (1963).

— Embryogenesis of behavioral nerve nets. pp. 161—186. In: Organogenesis. DEHAAN, R. L., URSPRUNG, H., Eds. New York: Holt, Rinehart & Winston 1965.

— Selective communication in nerve nets: impulse specificity vs. connection specificity. In: Neuroscience Res. Symp. Summaries, SCHMITT, F. O., MELNECHUK, T., Eds. Cambridge, Mass.: M. I. T. Press 1966.

— MINER, N.: Formation within sensory nucleus V of synaptic associations mediating cutaneous localization. J. comp. Neurol. **90**, 403—423 (1949).

STOCKDALE, F. E., HOLTZER, H.: DNA-synthesis and myogenesis. Exp. Cell Res. **24**, 508—526 (1961).

STONE, L. S.: The origin and development of taste organs in salamander observed in the living condition. J. Exp. Zool. **83**, 481—506 (1940).

— Functional polarization in retinal development and its reestablishment in regenerated retinae of rotated eyes. Proc. Soc. exp. Biol. Med. **57**, 13—14 (1944).

— Functional polarization in the developing retinae of transplanted eyes. Ann. N. Y. Acad. Sci. **49**, 856—865 (1948).

— Polarization of the retina and development of vision. J. Exp. Zool. **145**, 85—93 (1960).

STRAUS, W. L.: The concept of nerve-muscle specificity. Biol. Rev. **21**, 75—91 (1946).

SZÉKELY, G.: Zur Ausbildung der lokalen funktionellen Spezifität der Retina. Acta biol. Acad. Sci. Hung. **5**, 157—167 (1954).

— The apparent "corneal specificity" of sensory neurons. J. Embryol. exp. Morph. **7**, 375—379 (1959).

— Embryonic determination of neural connections. Advanc. Morphogenes **5**, 181—219 (1966).

— CZEH, G.: Localization of motoneurons in the limb moving spinal cord segments of Ambystoma. Acta physiol. Acad. Sci. hung. **32**, 3—18 (1967).

— SZENTÁGOTHAI, J.: Reflex and behavior patterns elicited from implanted supernumerary limbs in the chick. J. Embryol. exp. Morph. **10**, 140—151 (1962).

TAYLOR, A. C.: Selectivity of nerve fibers from the dorsal and ventral roots in the development of the frog limb. J. Exp. Zool. **96**, 159—185 (1944).

— KOLLROS, J. J.: Stages in the normal development of Rana pipiens larvae. Anat. Rec. **94**, 7—23 (1946).

TELLO, J. F.: Contribution à la connaissance des terminaisons sensitives dans les organes génitaux externes et de leur developpement. Trab. Lab. Invest. Biol. Univ. Madrid **28**, 1—58 (1932).

TIEGS, O. W.: Innervation of voluntary muscle. Physiol. Rev. **3**, 90—144 (1953).

TOTO, P. D., OJHA, G.: Generation cycle of oral epithelium in mice. J. Dent. Res. **41**, 388 (1962).

VERZÁR, F., WEISS, P. A.: Untersuchungen über das Phänomen der identischen Bewegungsfunktion mehrfacher benachbarter Extremitäten. Pflügers Arch. ges. Physiol. **223**, 671—684 (1930).

VINTSCHGAU, M. VON: Beobachtungen über die Veränderungen der Schmeckbecher nach Durchschneidung des N. glossopharyngeus. Pflügers Arch. ges. Physiol. **23**, 1—13 (1880).

— HONIGSCHMIED, J.: Nervus glossopharyngeus und Schmeckbecher. Pflügers Arch. ges. Physiol. **14**, 443—448 (1876).

WALKER, B. E.: Renewal of cell populations in the female mouse. Amer. J. Anat. **107**, 95—105 (1960).

WEDELL, G., GUTTMANN, L., GUTMANN, E.: The local extension of nerve fibers into denervated areas of skin. J. Neurol. Pschiat. **4** (N. S.), 206—225 (1941).

WEISS, P.: Die Funktion transplantierter Amphibienextremitäten. Aufstellung einer Resonanztheorie der Motorischen Nerventätigkeit auf Grund abgestimmter Endorgane. Arch. Entwickl.-Mech. Org. **102**, 635—672 (1924).

— Erregungspecifität und Erregungsresonanz. Ergebn. Biol. **3**, 1—151 (1928).

— Das Resonanzprinzip der Nerventätigkeit. Wien. klin. Wschr. **39**, 1—17 (1931).

— In vitro experiments on the factors determining the course of the outgrowing nerve fiber. J. Exp. Zool. **68**, 393—448 (1934).

— Selectivity controlling the central -perpheral relations in the nervous system. Biol. Rev. **11**, 494—531 (1936).

— Lid-closure reflex from eyes transplanted to atypical locations in Triturus torosus: Evidence of a peripheral origin of sensory specificity. J. comp. Neurol. **77**, 131—169 (1942).

— Central versus peripheral factors in the development of coordination. Res. Publ. Ass. nerv. ment. Dis. **30**, 3—23 (1952).

— EDDS, M. V.: Sensory-motor nerve crosses in the rat. J. Neurophysiol. **8**, 173—193 (1945).

— TAYLOR, A. C.: Further experimental evidence against "neurotropism" in nerve regeneration. J. Exp. Zool. **95**, 233—257 (1944).

WEISS, P. A.: Specificity in the neurosciences. In: Neuroscience Res. Symp. Summaries. Eds.: SCHMITT, F. O., MELNECHUK, T. Cambridge, Mass.: M. I. T. Press 1966.

WENGER, E. L.: An experimental analysis of the relations between parts of the brachial spinal cord of the embryonic chick. J. Exp. Zool. 114, 51—85 (1950).

WESSELS, N. K.: DNA synthesis, mitosis, and differentiation in pancreatic acinar cells *in vitro*. J. Cell Biol. 20, 415—434 (1963).

— Tissue interactions and cytodifferentiation. J. Exp. Zool. 157, 139—152 (1964).

WIERSMA, C. A. G.: An experiment on the "resonance theory" of muscular activity. Arch. neer. Physiol. 16, 337—345 (1931).

YAMADA, T.: Control of tissue specificity: the pattern of cellular synthetic activities in tissue transformation. Am. Zoologist 6, 21—31 (1966).

ZELENÁ, J.: Development, degeneration and regeneration of receptor organs. Progr. Brain Res. 13, 175—211 (1964).

Chapter 6

The Relation of Physiological
and Psychological Aspects of Sensory Intensity

By

Leo E. Lipetz, Columbus, Ohio (USA)

With 18 Figures

Contents

A. Introduction

The magnitude of a perceived sensation is naturally dependent on the magnitude of stimulation to the corresponding sensory system. The mathematical form of that dependence can be determined experimentally. It will be shown that much of such experimental data can be explained in terms of certain elementary neurophysiological models.

One model relates the amplitude of potential change at a receptor membrane to the stimulus strength. This model is shown to fit a variety of receptors. Further models are suggested which would relate this receptor potential to the generation of propagating impulses more centrally along the sensory pathways.

A model originally proposed by MacKay (1963) is used to relate the magnitude of sensation to the frequency of nerve impulses. Relations between physiological functions and psychophysical data are then derived and shown to be consistent with the available physiological data from several sensory systems.

All of these models and data are limited to steady state cases, and no attempt is made to account for any except the simplest transients in sensory systems.

B. The "Self-shunting" Model of a Receptor Membrane

1. Conductance of a Membrane

It is convenient (Kornacker, 1969) to define a membrane's conductance, g_γ, to an ion species γ as

$$g_\gamma = \frac{I_\gamma - J_\gamma}{V - E_\gamma} \tag{1}$$

where V is the voltage between inside and outside of the membrane, I_γ is the current carried from inside to outside by ion species γ, E_γ is the chemical equilibrium potential for the γ species, and J_γ is the active transport current for the γ species. Since V is the same for all ion species, Eq. (1) can be summed over γ to yield

$$V = \frac{\Sigma I_\gamma - \Sigma J_\gamma + \Sigma g_\gamma E_\gamma}{\Sigma g_\gamma}. \tag{2}$$

During the normal stimulation of a receptor membrane, there is usually no significant change in E_γ for any ion. Therefore, a change of V for given total current ΣI_γ is produced during stimulation either by: (a) changing some or all g_γ; or (b) changing ΣJ_γ.

2. Directly Induced Change of Conductance

In case (a) above, stimulation directly affects the ionic conductances of the receptor membrane. This case has been considered by LOEWENSTEIN (1961) for the Pacinian corpuscle and an electrical analog for its response has been developed. The receptor potential, ΔV_m, appears as a change in the membrane potential, V_m. From this analog it was calculated that the measured receptor potential, normalized in terms of the maximum value of the momentary peak amplitude of ΔV_m, is given approximately by the equation

$$\Delta V_m = kEbx/(1 + bx). \tag{3}$$

E is a measure of the maximum value of ΔV_m, k and b are constants, and x is a measure of the stimulus strength, that is, of the fraction of unit areas of the nerve

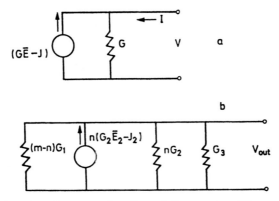

Fig. 1 a and b. Circuit analogs for current flow through the receptor cell's membrane. a Generalized circuit. b Circuit for the inexcitable portion and the partially excited excitable membrane. For definitions of the terms see Appendix 1

ending which are excited (LOEWENSTEIN, 1959, 1961). (See Appendix 1 and Fig. 1 for an alternative derivation which obtains the equivalent expression, Eq. (8a), from Eq. (2), and which clarifies the meaning and limitations of Eq. (3).)

The curve of Eq. (3) is plotted as the solid lines in Fig. 2a and b. These lines can be seen to fit the data (solid circles) for the normalized receptor potential amplitude as a function of stimulus strength (mechanical deformation) for the Pacinian corpuscle.

The experimental data of Fig. 2 are also plotted (open circles) as receptor potential versus logarithm of stimulus strength. It is shown in Appendix 2 that Eq. (8a) when transformed to logarithmic units, w, of the stimulus strength (S) becomes the hyperbolic tangent function of Eq. (14a), $V_0 = {}^1/_2 + {}^1/_2 \tanh (w - w_0)$.

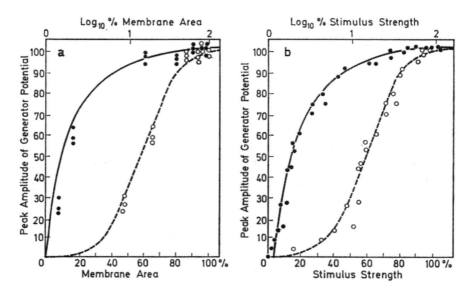

Fig. 2a and b. Filled circles represent experimental data of the relation between receptor potential and stimulus strength. The open circles show the same data replotted on a logarithmic scale of stimulus intensity as the abscissa. The ordinates are linear values of the amplitude of the receptor potential normalized to make the maximum amplitude equal 100%. The solid curves are the theoretical curve from Eq. (3). The broken curves are the hyperbolic tangent curve which best fits the logarithmic plots of the data. a The receptor potential-excited membrane area relation. Varying membrane areas of the ending of a decapsulated Pacinian corpuscle were stimulated (mechanically compressed) with equal-amplitude mechanical pulses delivered by styli of varying diameters. b A typical generator potential-stimulus amplitude relation obtained by varying the amplitude of compression applied by a fixed diameter stylus to an intact Pacinian corpuscle. The solid curve has been displaced to the right in order to best fit the data. This implies that a compression amplitude of over 3% must be applied to the intact corpuscle before the receptor membrane deforms in proportion to the additional amplitude of applied compression. Adapted from LOEWENSTEIN (1959)

For convenience, this function will hereafter be referred to as the tanh log function. This function is plotted as the broken lines of Fig. 2a and b, and can be seen to fit the logarithmically plotted data.

Note that the tanh log function is nearly a straight line over a range of stimulus strength of about one logarithmic unit. Only over this limited range is the transfer function of this receptor a logarithmic transform. The tanh log function is a better description of the transfer function because it fits the data for a larger range of stimulus strength.

If the data of any stimulus-response relationship are plotted with the normalized response amplitude on a linear ordinate scale and the stimulus strength on a logarithmic abscissa scale, and if from the data the asymptotic maximum and minimum of the response amplitude can be estimated, one can readily test the fit of the curve of Eq. (8a) to the data. The curve will be the tanh log "template" curve, and it can be fitted to any data by the procedures described in Appendix 3.

3. Indirectly Induced Change of Conductance

The light receptor cell (retinula cell) of the *Limulus* lateral eye has been found to show an increased conductivity from the inside of the cell to the outside of the ommatidium upon illumination (BORSELLINO et al., 1965; STIEVE, 1965).

The potential and conductance changes in the postsynaptic, spike-generating eccentric cell as a function of light intensity have been extensively measured (MACNICHOL, 1956; PURPLE, 1964; FUORTES, 1959). It now appears (BORSELLINO, et al., 1965) that these changes passively follow the change in the membranes of the retinula cells surrounding the eccentric cell and do not indicate a primary change in the membranes of the eccentric cell. The significant conductance change is at the retinular membranes contacting the epithelial sheath which forms the outside of the ommatidium, yet this is on the other side of the cell from the retinular membrane (rhabdomere) which contains light-sensitive pigment.

On analyzing these data plus the data of FUORTES (1959) as analyzed by RUSHTON (1959) I find that the resistance change measured in the eccentric cell of the lateral eye for various illuminations can be adequately described by the formalism given at the end of Appendix 1.

SMITH et al. (1968a, b) have shown that in the *ventral eye* of *Limulus* the effect of illumination is to reduce the active transport current ΣJ_y across some membrane of the photoreceptor cell. As can be seen from Eq. (2), this would lead to a change of V for the cell; indeed, a depolarization was observed upon illumination. SMITH et al. have also shown that the cell membrane conductance is voltage-dependent, increasing as the cell is depolarized. (This voltage-dependent mechanism might explain how, in the *lateral eye*, light acting at the rhabdomere of the retinula cell evokes the conductance increase at the opposite side of the cell with a much shorter latency than could be accomplished by diffusion across the retinula cell.)

4. Other Examples of "Self-shunting" Receptors

A number of other receptors show upon stimulation the full tanh log relation of the self-shunting transducer mechanism.

a) Photoreceptor of Carp. TOMITA et al. (1967) have succeeded in recording an increased negativity of membrane potential from within single retinal cones upon illumination. Typical data for the relation between logarithm of light intensity and potential change in a single cone of the carp retina are given in Fig. 3. The good fit of the hyperbolic tangent curve to this data suggests that the light-to-cone potential transducer involves a self-shunting membrane.

13*

b) Giant Synapse of Squid. The postsynaptic membrane of a chemical transmission type of synapse acts as a receptor for that chemical. Katz and Miledi (1966) studied this stimulus-response relationship (transfer function) in the giant synapse of the squid. The changes in membrane potential (depolarization) at the

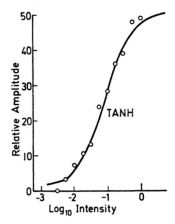

Fig. 3. Relative amplitudes of potential change recorded from a single carp cone as a function of logarithm of light intensity. The curve is a tanh function fitted to the data (dots). This cone had a maximal sensitivity at 610 nm. Adapted from Tomita et al. (1967)

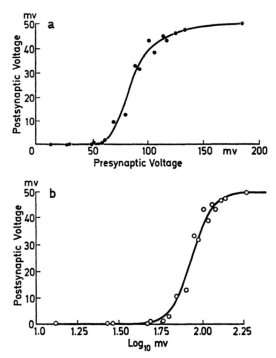

Fig. 4a and b. Transfer function of a single synapse, the "giant synapse" of the squid. Abscissae: presynaptic depolarization. Ordinates: postsynaptic response. a Data plotted on a linear scale of abscissa. b Data replotted on a logarithmic scale of abscissa and fitted by a tanh function as the solid line. Data from Katz and Miledi (1966)

presynaptic terminal (stimulus amplitude) and at the postsynaptic terminal of the synapse (response amplitude) were measured simultaneously. This was made easier by poisoning the synapse with a drug (tetrodotoxin) which prevented the initiation of spike potentials. As can be seen in Fig. 4, the transfer function of the synapse fits the tanh log curve. Normally, the action potential of the presynaptic neuron penetrates into the presynaptic terminal and depolarizes it sufficiently to cause a large depolarization to be produced in the postsynaptic terminal. This postsynaptic depolarization is more than enough to cause the initiation of a nerve impulse. Thus this particular synapse serves as a one-to-one transmitter of nerve impulses from the input to the output neuron.

c) **Taste Receptors.** The receptor potential evoked in a taste bud of the tongue of rat or hamster has been measured by KIMURA and BEIDLER (1961) as a function of the concentration of the stimulant, any of a number of salts. These authors found the data to be fitted by the equation derived by BEIDLER (1954),

$$\frac{C}{R} = \frac{C}{R_s} + \frac{1}{KR_s} \tag{4}$$

where C is the concentration of the stimulus, R is the amplitude of the receptor potential, R_s is the maximum amplitude evoked by high stimulus values, and K is a constant. The equation can be rewritten as

$$\frac{R}{R_s} = \frac{KC}{1 + KC} . \tag{5}$$

It can be seen that the left-hand term is the normalized receptor potential. The equation is of the same form and meaning of terms as Eq. (8a) and thus is the relation for a self-shunting transducer mechanism. It seems reasonable to interpret the results in terms of such a mechanism instead of the chemical reaction mechanism suggested by BEIDLER (1954, 1961). However, since the constant K then becomes a measure of the stimulating effectiveness of a given salt concentration, it might be a measure of the total number of receptor sites available to the stimulant molecules.

The success of the self-shunting transducer mechanism in fitting the behavior of the diverse types of receptors described above suggests that it may be a very common mechanism for producing nonlinear stimulus-response characteristics.

C. Relation between Receptor Potential and the Frequency of Axonal Firing

The receptor potential, when evoked, usually leads to the firing of nerve impulses by an axon. This may be the axon of the receptor cell or, if it has none, the axon of a neuron with which it synapses.

1. The Limulus Lateral Eye

The receptor potential of the retinula cell is transmitted passively to the dendrite of a neuron, the eccentric cell, and passively along the dendrite and soma to the axon of that cell. There, at a particular region of the axon, the resulting depolarization causes the initiation of spike-type action potentials which are propagated along the axon [for a review, *see* PURPLE and DODGE (1965)]. The frequency of firing of the axon is linearly proportional to the voltage change in the eccentric cell (Fig. 5a and b), and therefore is linearly proportional to the receptor potential. This has also been shown by FUORTES (1959) and RUSHTON (1959).

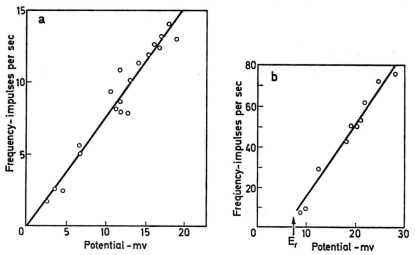

Fig. 5a and b. The relationship between frequency of discharge and the potential change within an eccentric cell of the *Limulus* lateral eye in response to prolonged illumination. a From MACNICHOL (1956). b From PURPLE (1964)

2. Muscle Spindle

A linear relation between the amplitude of the receptor potential and the frequency of firing was found for the muscle spindle of frog (KATZ, 1950).

3. Olfactory Neurons

Most of the single neurons measured in the olfactory bulb of the frog (DÖVING, 1964) discharged at a rate linearly proportional to the olfactory receptor potential (electro-olfactogram) over the greater part of the stimulus strength range. These neurons appeared to be those which received the terminations of the olfactory receptors.

4. Gustatory Nerve

The massed discharge recorded from the entire gustatory nerve (chorda tympani) of rats or hamsters had the same relation to stimulus intensity as did the summed receptor potentials of a number of taste bud receptors (KIMURA and BEIDLER, 1961). This implies a linear relation between the summed firing rate of all the gustatory axons and the summed receptor potentials.

D. Relation between Stimulus Strength and Firing Frequency

1. Tanh Log Function

In those cases for which the firing frequency is linearly proportional to the receptor potential and the receptor has the self-shunting transducer mechanism, the firing frequency should be a tanh log function of the stimulus strength. This was shown to be true for gustatory second-order neurons (KIMURA and BEIDLER, 1961) and for certain ganglion cells, third-order neurons, of the vertebrate retina (see Fig. 14).

2. Linear Function

If the sensitivity of the receptor membrane is low [small b in Eq. (3)] and the gain of the impulse firing mechanism is high, the rate of firing will be linearly related to stimulus strength. These conditions restrict the firing rate to a small range near the origin of the tanh log function (see solid curves, Fig. 2). Such a linear relation has been observed for nerve fibers from the Iggo touch corpuscles of the cat by DARIAN-SMITH et al. (1968).

A linear relation holds over the entire range of stimulus strength if the receptor membrane has no change in total conductance when excited but does have a change in its active transport [see Eq. $(8a_1)$ of Appendix 1].

If in this linear range, the receptor transducer mechanism is not affected by stimulus strength, S, below some threshold value, S_0, then the firing frequency, f, becomes

$$f = k(S - S_0) \tag{6}$$

where k is a constant. This relation has been observed for touch receptor nerve axons in the palmar branch of the medial nerve of monkey by MOUNTCASTLE et al. (1966).

If the axon has a spontaneous firing rate, f_0, then it is the change in firing rate, $f - f_0$, which is controlled by the receptor potential. Therefore, the relation becomes

$$f - f_0 = k(S - S_0) \ . \tag{7}$$

This relation was found by WERNER and MOUNTCASTLE (1965) to fit the data for the massed firing of axons of Iggo corpuscles of cat and monkey when the stimulus strength was expressed as the force applied to the corpuscle.

3. Function of a Power Less than One

If, for the impulse firing mechanism, there is a large threshold amplitude of receptor potential, S_0, below which it is not activated, this is equivalent to shifting its operating range more to the right along the tanh log function. This results in the firing frequency approximating a power law relation of stimulus strength

$$f_1 - f_0 = k(S - S_0)^n \tag{8}$$

where n is a constant and can vary from 1 to 0 as the operating range is shifted to the right. Such a relation has been observed for the human gustatory nerve by BORG et al. (1967), and for the Iggo touch corpuscle axons by WERNER and MOUNTCASTLE (1965) when the stimulus strength was expressed as mechanical deformation of the corpuscle.

4. Logarithmic Functions

It can be seen from the broken curves of Fig. 2 that there is a range of the tanh log function for which the receptor potential is linearly proportional to the stimulus strength. This range covers only about one decade. The range over which the logarithmic relation holds can be extended by several neural mechanisms. One mechanism depends upon the first-order neuron having many branches along which impulses travel from receptor cells. At high discharge rates, these impulses

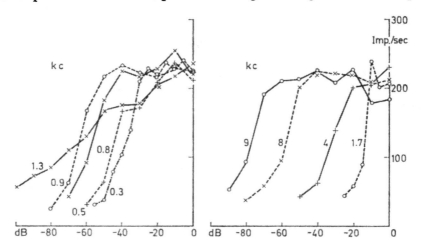

Fig. 6. Response functions of primary auditory nerve fibers. Abscissae are the relative sound intensities in logarithmic units, ordinates are the firing rates in impulses per second. The frequencies of the tone stimuli are marked in *kc* next to the response curve for each. Left: crossed ramp type of fibers. Right: parallel ramp type. From NOMOTO *et al.* (1964)

collide at the branching points, reducing the number of impulses along the main axon (VAN BERGEIJK, 1961; DEUTSCH, 1967; HARMON and LEWIS, 1966). Such a mechanism fits the behavior of the external spiral nerve fibers from the cochlea. (See, for example, the curve labelled "1.3" at the left in Fig. 6.)

A second mechanism leading to an extended range of logarithmic transform requires for receptors having the self-shunting transducer mechanism that: (a) many receptors are excited by the stimulus; (b) these receptors have different sensitivities; and (c) the outputs of these receptors (or of the synapsing neurons)

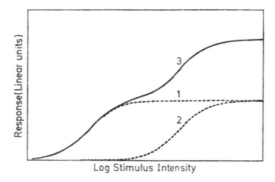

Fig. 7: *Curves 1 and 2:* The stimulus-response functions of two receptors, both having tanh log transfer functions but having different sensitivities. *Curve 3:* The summed response of the two receptors

be summed at a more central portion of the sensory pathway. The effect of summing two such receptors is illustrated in Fig. 7. For a larger number of receptors, the resulting output would approach even more closely a logarithmic relation to stimulus strength. Such a relation was found for the massed response of the cochlear nerve by DESMEDT (1962).

5. Function of a Power Greater than One

All of the cases considered above were functions for which the second derivatives of V and of $(f - f_0)$ with respect to $(S - S_0)$ were negative. It is possible to get positive second derivatives (that is, power functions to a power greater than one) by introducing cooperative mechanisms. For example, several molecules of the chemical stimulant might have to hit the same site on the receptor membrane to activate it. Also, several photoreceptors might each have to be activated by an absorbed photon before their coincident outputs would be sufficient to excite the central visual pathway. Data supporting this example have been obtained by VAN DEN BRINK and BOUMAN (1963) and show that the coincidence requirement rises from 2 to 20 as the eye becomes light adapted.

Another example of such a power law relation is the summated discharge of human gustatory nerve upon stimulation with sucrose, the exponent n being 1.1 (BORG et al., 1967). The sensitivity of the taste receptors to sucrose is low compared to that for other stimulants. A single gustatory nerve fiber innervates more than one taste receptor cell, all within the same taste bud. It may be that partially coincident activity of two of those receptors is required to activate the nerve fiber.

E. A Model of the Neurophysiological Generation of Sensation Magnitude

1. Sensory Magnitude Measurement

There are three classes of sensory scales: (1) discriminability scales which are formed by counting off the number of times that there is a just-noticeable difference (jnd) in the sensation as the stimulus is changed in intensity; (2) category scales which are formed by partitioning a stimulus intensity continuum into equal-appearing intervals of sensation magnitude; and (3) magnitude scales which are formed by estimating the apparent strength of a sensation relative to a standard sensation, usually produced by a standard intensity stimulus. These magnitude scales depend on judging the ratio of magnitudes of sensation (Fig. 8). A discussion of these different classes of scales may be found in articles by STEVENS (1961) and by EKMAN (1961). These three classes of scales do not produce the same relation of sensation to stimulus intensity. This implies that the physiological mechanism involved in the operations that produce these three scales could be different. Examples of these three classes of scales are given in Fig. 8 for the relation of the sensation of vibration to the intensity of the physical stimulus.

For a number of senses it has been found that the estimated magnitude of sensation, ψ, is related to the stimulus strength, S, as

$$\psi = k(S - S_0)^\beta \qquad (9)$$

where k and β are constants and S_0 is the threshold value of S which produces a sensation.

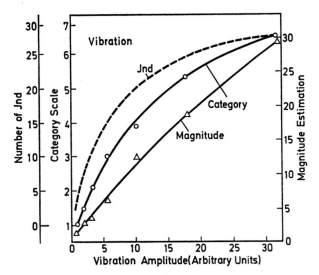

Fig. 8. Three scales of psychological measure of the apparent intensity of a 60 hz vibration applied to finger tip. △: Mean magnitude estimations. ○: Mean category judgements on a scale of 1 to 7. The two end stimuli were presented at the outset to indicate the range. *Dashed curve:* discriminability scale obtained by counting off jnd. From Stevens (1961). Reproduced from *Sensory Communication*, Walter A. Rosenblith (ed.) by permission of the M. I. T. Press, Cambridge, Mass. Copyright 1961, Massachusetts Institute of Technology

2. A Model of the Production of Sensation

A physiological mechanism has been proposed by MacKay (1963) which leads to the relation of Eq. (9) whether or not neural activity at central portions of the sensory pathway is related to stimulus strength as in Eqs. (6), (7), or (8). MacKay's basic idea is that "Perception is regarded as an internal, outwardly

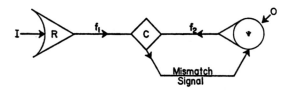

Fig. 9. Model for "matching response" neural mechanism for the perception of stimulus intensity. From MacKay (1963). Copyright 1963, Amer. Assoc. for the Advancement of Science

directed, 'matching' response to stimulation, generated within the organizing system that determines the current 'state of readiness' of the organism. On this view the perceived intensity of the stimulus should reflect, not the frequency of impulses from the receptor organ, but the magnitude of internal 'organizing activity' evoked to match, or in some other sense 'counterbalance', that frequency." A diagram of the system is given in Fig. 9.

R is a given sensory receptor system and its associated central pathways. It responds to S, measured in units of S_0, with a firing rate f_1 according to some function P:

$$f_1 = P(S/S_0) . \tag{10}$$

It is postulated that there is an internal "organizer" or "effort-generator" O which produces a "matching" frequency f_2 according to the relation:

$$f_2 = Q(\psi - \psi_0) \tag{11}$$

where Q is some function, and $(\psi - \psi_0)$ is the internal activity of the organizer which is sufficient to make f_1 match f_2 according to the criteria of the comparator, C.

It is postulated that such an equilibrium is obtained when

$$f_1 = bf_2 + k_3 \tag{12}$$

where b is a factor representing the relative "weight" given by C to the output from O compared to the output from R, and k_3 is the "zero bias" of C.

a) **The Case of a Power Law of Central Sensory Input.** If function P were a power law relation and if function Q were also, then

$$f_1 = k_1 (S - S_0)^{\alpha_1} \tag{13}$$

and

$$f_2 = k_2 (\psi - \psi_0)^{\alpha_2} \tag{14}$$

then for the equilibrium condition of Eq. (12), it follows that

$$\psi - \psi_0 = [(k_1/bk_2)(S - S_0)^{\alpha_1} - (k_3/bk_2)]^{-\alpha_2} . \tag{15}$$

If (k_3/bk_2) is small compared to (k_1/bk_2) $(S - S_0)^{\alpha_1}$, then

$$\psi - \psi_0 = a(S - S_0)^\beta \tag{16}$$

where

$$a = (k_1/bk_2)^{-\alpha_2} \tag{17}$$

and

$$\beta = \alpha_1/\alpha_2 . \tag{18}$$

If $\alpha_1 = \alpha_2$, then

$$\psi - \psi_0 = a(S - S_0) , \tag{19}$$

a linear relation. Such a relation has been observed for the sensation evoked by vibratory stimulation of the finger tip (MOUNTCASTLE, 1967).

If $\alpha_2 = 1$, then at equilibrium

$$\psi - \psi_0 = a(S - S_0)^{\alpha_1} , \tag{20}$$

the same power law exponent as for the relation of f_1 to stimulus strength. This equality has been found for the taste sense by BORG et al. (1967). They measured the massed firing of the gustatory nerve and the estimated magnitude of sensation for various concentrations of chemical applied to the tongue. Both the firing rate and the sensation magnitude varied as the same power of the concentration for each chemical tested. Depending on the chemical, the value of the exponent varied from 0.55 to 1.1.

b) The Case of a Logarithmic Function of Central Sensory Input. If both function P and function Q were logarithmic, then

$$f_1 = k_1 \log(S - S_0) + a_1 \tag{21}$$

and

$$f_2 = k_2 \log(\psi - \psi_0) + a_2. \tag{22}$$

For the equilibrium condition of Eq. (12) it follows that

$$\psi - \psi_0 = a(S - S_0)^\beta \tag{23}$$

where

$$\log a = (a_1 - ba_2 - k_3)/bk_2 \tag{24}$$

and

$$\beta = k_1/bk_2. \tag{25}$$

A sensory system described by such a set of relations is the auditory system. DESMEDT (1962) found the massed discharge of the cat cochlear nerve to have a relation to sound intensity such as that in Eq. (21). STEVENS and GUIRAO (1964) and KEIDEL and SPRENG (1965) found the relation between loudness and acoustic intensity to be of the form of Eq. (23).

F. Application of these Models to the Gustatory System

If the models developed in the previous sections are applied to sample data from sensory systems, they assist in organizing the data and they lead to predictions of relations between psychophysical and neurophysiological data. The first system to be considered is the gustatory.

1. Saltiness

In the taste buds of the tongue there are receptor cells, some of which respond maximally to salt solutions although they will respond to a lesser extent to other substances, such as sucrose. In the rat and hamster, these salt receptors have been shown to have the tanh log function of a self-shunting transducer mechanism [Eq. (3)] between receptor potential and salt concentration (KIMURA and BEIDLER, 1961). The massed discharge of the gustatory nerve was found to fit the same relation (BEIDLER, 1954).

BORG et al. (1967) measured the massed discharge of the gustatory nerve (chorda tympani) of humans. The one set of data they obtained for the response to salt (their Fig. 4) is shown as being fitted poorly by a power law relation with an exponent of 1.38. The five data points are fitted better by the tanh log function, but they are too few for such curve-fitting to be conclusive.

They also measured the relation of magnitude of sensation to salt concentration and found it was fitted by a power law with an exponent of 1.38 for one subject and of 1.05 for the average of 14 other subjects. EKMAN and ÅKESSON (1965) have found the same relation with the exponent varying greatly between individuals, from 1.11 to 1.97. It is quite possible, although not definitively demonstrated, that the exponent is the same for both massed nerve firing and subjective sensation in each individual. BORG et al. (1967) mentioned that the amplitude of the postsynaptic potential evoked in the cerebral cortex varied linearly with the massed firing.

The data reported above indicate the following for the salt taste system. (1) The receptor potential is a tanh log function of salt concentration in rat and hamster. (2) The massed firing of the gustatory nerve is proportional to the receptor potential in rat and hamster. (3) The massed firing of the gustatory nerve in man is a power function of salt concentration, with the exponent greater than one. This suggests that the nerve firing depends on cooperative action of the receptor cell potentials. (4) In man the evoked cerebral cortical output potential is probably linearly related to the massed firing of the gustatory nerve. (5) In man the magnitude of sensation is probably linearly related to the massed firing. (6) This suggests that if f_1 of Eq. (13) is the massed nerve discharge with a power law exponent α_1, then β of Eqs. (16) and (18) must equal α_1, so α_2 of Eq. (14) must equal 1. The cortical evoked potential output could be a measure of $(\psi - \psi_0)$ in Eq. (14).

2. Other Tastes

BORG et al. (1967) measured the massed gustatory discharge and the subjective sensation on the same two humans in response to citric acid or sucrose applied to the tongue. Both types of response were power law relations with the same exponent for a given stimulus in a given individual. The exponents were 0.85 and 1.47 for citric acid and 1.1 and 1.15 for sucrose. The same considerations as discussed above for saltiness appear to apply to these other taste modalities.

G. The Touch Systems

A number of different types of touch (cutaneous mechanoreceptor) receptor cells are known. There also seem to be a number of distinct touch sensations, e.g., pressure, light touch, and prickle. There is no reason to believe that each sensation is served by only one type of receptor. Therefore, experiments in which the mechanical stimulus is applied simultaneously to more than one type of receptor are very difficult to interpret. Even when the stimulus is limited to a single receptor type, all that can be measured is its contribution to the particular sensation being thus evoked; this is not necessarily the same as its contribution to one of the distinct touch modalities evoked by multireceptor stimulation.

1. Receptor Potential

The only touch receptor for which the receptor potential has been measured is the Pacinian corpuscle. As discussed above, its receptor potential is a tanh log function of the deformation of the bare nerve ending within the corpuscle.

2. Touch Systems of Glabrous Skin

MOUNTCASTLE et al. (1966) measured the number of nerve impulses at an axon of the palmar branch of the median nerve of a monkey upon indenting the skin of the ball of the thumb. The number of impulses was linearly proportional to the skin indentation above a threshold value of indentation (see their Fig. 2), the same form of relation as that of Eq. (6).

A similar measurement was made of the firing of a single neuron in the ventro-basal nuclear complex of the thalamus of a monkey upon indenting the glabrous skin of the hand (Mountcastle, 1967). The relation was that of Eq. (7) because the rate of spontaneous firing, f_0, had to be taken into account.

These results can be compared with the magnitude of sensation invoked in humans by indentation of the ball of the finger. Mountcastle (1967) found the magnitude to be linearly proportional to the amplitude of indentation.

It would appear that for this sensory system linear transforms are involved all the way from physical stimulus to sensation. If the thalamic neuronal output corresponds to f_1 of Eq. (13), then it is required by the proposed model of sensation generation that α_2 of the organizer's output [see Eqs. (14), (18), and (19)] must equal 1. Thus, the organizer also has a linear output. Of course, the sensory input to the comparator may occur even more centrally than the thalamus, and a power law transform of exponent α_1 could occur between the thalamic neuronal output and the input reaching the comparator. The observed linear relation of sensation magnitude to stimulus amplitude would still occur if α_2 equalled α_1.

3. The Iggo Corpuscles Touch-sensitive System

The Iggo corpuscle is a dome-like elevation in the hairy skin of the cat and monkey and contains the ending of a myelinated fiber. This fiber fires when the corpuscle is indented. Werner and Mountcastle (1965) measured the firing rate of the fiber when a single Iggo corpuscle was stimulated. They found that the rate adapted rapidly during the first 4 sec and then became almost constant (see their Fig. 14). The steady state firing rate was found to be linearly proportional to the stimulus expressed as force on the corpuscle. A similar relation was found by Darian-Smith et al. (1968) when recording from single slowly adapting mechanoreceptors in the hairy skin of the cat face. They found the same relation when recording from single slowly adapting trigemino-thalamic neurons within the nucleus oralis. It would seem that at least to that stage only linear transforms were involved.

There is no directly comparable psychophysical data, but Uttal's (1959) experiments are suggestive. He stimulated the human ulnar nerve with two brief electric pulses and measured the number of action potentials produced in the nerve and the magnitude of the sensation. He found that the subjective intensity seemed to be linearly proportional to the neural responses.

H. The Auditory System

1. The Receptor Potential

The cochlear microphonic, which is regarded as the receptor potential (Davis, 1965), is linearly proportional to sound pressure and proportional to the square root of sound intensity over a 20,000 to 1 range (Lawrence, 1965).

2. The First Auditory Neurons

The relation between sound intensity and firing rate of the first-order auditory neurons was measured by Nomoto et al. (1964), who found two types of neurons.

The "parallel-ramp type" had the relations illustrated on the right in Fig. 6; at all input sound frequencies, its relation to sound intensity was the same tanh log function, but its input sensitivity differed for each sound frequency. The tanh log function probably arises as the transfer function of the synapse between hair cell and the first-order auditory neuron. The "crossed-ramp type" of neuron had the relations illustrated at the left in Fig. 6; its relation to sound intensity was the same tanh log function except at the sound frequencies which were very effective inputs. The more effective the input, the more the function tended toward a simple logarithmic relation. The possible mechanism of this change was discussed in the section on logarithmic functions.

Also in that section, another mechanism was discussed which would lead the massed discharge, N_1, of the first auditory neurons to be a logarithmic function of sound intensity, as was found by DESMEDT (1962).

J. The Vertebrate Visual System

Neurophysiological measurements have been made at many points along the visual sensory pathway. These data and the extensive psychophysical data provide a broad test of the models being proposed.

1. Receptor Potential

It has already been mentioned that the carp retinal cone shows a hyperpolarizing potential change upon illumination (TOMITA et al., 1967). This change fits a tanh log function (Fig. 3). The retinal rod of the gekko also shows such a hyperpolarizing change (TOYODA et al., 1969).

2. Neurons Connecting to the Photoreceptors

The rods and cones of the vertebrate retina have been shown by electron microscopy to synapse with the bipolar cells and the horizontal cells (e.g., MISOTTEN, 1965; DOWLING and BOYCOTT, 1966). These synapses have the fine structure of chemical transmission types. STELL (1967) found that in the goldfish retina: (1) the inner horizontal cells contacted only cones and no rods; (2) the small bipolar cells contacted only cones; (3) the outer horizontal cells contacted only rods; and (4) the large bipolar cells contacted both rods and cones. Type (1) contacts will be discussed below.

a) Transfer Function of Cone to Horizontal Cell. The L-type S-potential is recorded from a horizontal cell (MACNICHOL and SVAETICHIN, 1958; TOMITA et al., 1959; MITARAI, 1960; SVAETICHIN et al., 1961; WERBLIN and DOWLING, 1969). NAKA and RUSHTON (1966c) found that the L-potential of the tench retina is a tanh log function of light intensity, I. This function is a template curve fitted to all the sets of data points in Fig. 10. This curve has parameter values which correspond to variation of the light-sensitive membranal conductance as the 0.716 power of I. The derivation and method of evaluation of this relationship are given in Appendix 3.

The L-potentials of Fig. 10 had a maximum sensitivity to light of wavelength about 550 nm and are presumed to arise from excitation transmitted from cones having that spectral sensitivity. Other L-potentials in the tench retina were found by NAKA and RUSHTON (1966c) to be tanh log functions of I. The light-sensitive conductance varied as the 0.735 power of I for excitation from cones having maximal sensitivity at 680 nm, and as the 0.975 power of I for excitation from cones having a maximum sensitivity at 620 nm. The carp cone measured by TOMITA et al. (1967) had its maximum sensitivity at 610 nm, and the fitted tanh log I curve (see Fig. 3) corresponded to a variation of the light-sensitive conductance as the 0.774 power of I.

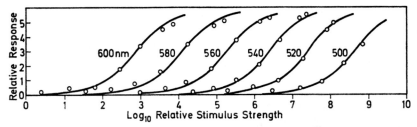

Fig. 10. L-type S-potentials of the tench retina. Ordinate: hyperpolarizing potential change in response to illumination. Abscissa: logarithm of light intensity. Open circles are experimental data. The data for each wavelength are displaced horizontally an arbitrary amount. The response had its maximum sensitivity at about 550 nm. The curve drawn through the points is a fixed tanh-shaped template. Redrawn from NAKA and RUSHTON (1966c)

This data can be interpreted in two different ways, depending upon whether the source of the change in cone potential is considered to originate from a change in the cone's membrane or from a change in the membrane of a cell to which the cone connects in a passive electrical manner. The change in cone potential was found to have a shorter latency than L-type S-potentials for cat retina (BROWN and MURAKAMI, 1968), fish retina (MURAKAMI and KANEKO, 1966), and mudpuppy retina (WERBLIN and DOWLING, 1969). Furthermore, the S-potentials of fish are selectively abolished by ammonia vapor without abolishing the potential changes of the photoreceptors (MURAKAMI and KANEKO, 1966). These findings indicate that the cone potential change originates prior to and independent of the S-potentials. WERBLIN and DOWLING (1969) found that the light-evoked potential changes in the horizontal cells and bipolar cells of the mudpuppy had a longer latency than the changes evoked in photoreceptor potentials. If this is also true for fish, then, since these are the only excitable cells contacting the photoreceptors in fish (STELL, 1967), it must be concluded that the cone potential change originates as a change in the membrane of the cone in response to illumination.

TOYODA et al. (1969) found that upon illumination the potential hyperpolarized and the membrane resistance increased. Then, the fit of tanh log I to the cone potential change could imply that illumination shuts off a process which operates in darkness and which results in a "self-shunting" conductance increase in the cone's membrane. The transfer function from light to cone potential includes the series of processes: (a) light absorption → (b) photochemical reaction → (c) chemical reactions → (d) change of permeability of cone's membrane → (e) change of cone potential.

It is also possible that illumination reduces the active transport of the cone's membrane and that a portion of the membrane has a voltage-dependent conductance. This would lead to an approximation to the self-shunting relation as in Eq. $(8a_6)$.

If it is assumed that the cones of the tench have the same form of potential response as those of the carp, then, within the variability of the data, it is possible that the transfer function of the synapse from cone to horizontal cell is a linear one. This transfer function, on the other hand, includes the series of processes (e) change of cone potential → (f) release of chemical at synapse → (g) change of permeability of postsynaptic membrane → (h) change of potential of postsynaptic cell. A linear transfer function could result from two hyperbolic tangent transfer functions acting in series if one of those tanh functions was being used only over the very beginning of its curve. (See the solid curves of Fig. 2 and note that the response is nearly linear for stimulus intensities up to those giving a response of about half the value of the upper asymptote of response.) If at the synapse the maximum change in neuronal excitation (postsynaptic potential) produced by maximum light intensity corresponded to this limited range, then the transfer function of the synapse would be linear.

A linear transfer function could also result from a synapse with a postsynaptic membrane having no change in total conductance upon excitation, but only a change in active transport. [See Eq. $(8a_1)$].

b) The C-Type S-Potential. The C-type S-potential is a light-evoked potential change consisting of a slow maintained change, which is either depolarizing or hyperpolarizing depending on the stimulating wavelength, and which is detected

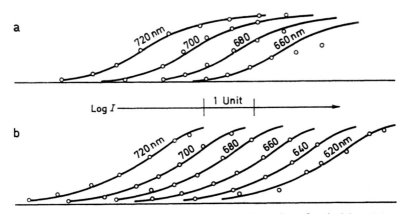

Fig. 11a and b. C-type S-potentials of the tench retina. Data for a depolarizing response component with a maximum sensitivity at 680 nm plotted as in Fig. 10. The curves are a fixed tanh-shaped template. From NAKA and RUSHTON (1966b)

only from small regions in the inner nuclear layer of the vertebrate retina. It seems to originate from single cells in that region, but which type of cells has not been conclusively established, leaving the bipolar, horizontal, amacrine, and Müller cells as possibilities (MacNICHOL and SVAETICHIN, 1958; TOMITA et al., 1959; MITARAI et al., 1961; VILLEGAS, 1961).

NAKA and RUSHTON (1966a, b) found that in the tench retina both the depolarizing and hyperpolarizing components of the C-response fit a tanh log I relation (Fig. 11). Moreover, both components showed spectral-sensitivity curves that matched the absorption curve of the green cones (maximum at about 550 nm) determined microspectrophotometrically. For some C-response sites, the fitted tanh log I curve had parameter values corresponding to variation of the membranal conductance as the 1.56 power of I; in other sites, the variation was as the 0.89 power of I. The depolarizing component with maximal sensitivity at 550 nm had this variation as the 0.67 power of I. The depolarizing component with maximal sensitivity at 680 nm had this variation as the 1.20 power of I.

Thus, the same conclusions can be drawn regarding the transfer function from cone to C-response site as for that from cone to horizontal cell.

3. The Electroretinogram

The electroretinogram consists of at least four rapid and successive waves: R_1, R_2, the a-wave, and the b-wave. The first two, R_1 and R_2, have such short latencies that they are probably generated in the outer or inner segments of the receptors. It has been shown that the amplitude of each is linearly proportional to the number of pigment molecules excited by the stimulus flash (CONE, 1965). The a-wave has recently been identified as a receptor potential (BROWN and WATANABE, 1962a, b). In its isolated form, the a-wave has been labeled "P III" by GRANIT (1933) and the "late receptor potential" by BROWN and WATANABE

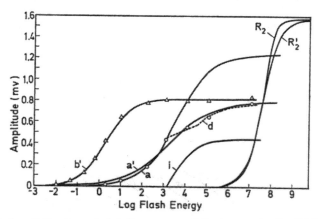

Fig. 12. Amplitudes of R_2, the unisolated a-wave, and the b-wave in the ERG of the dark-adapted albino rat as functions of the logarithm of flash energy, E, R'_2, a', and b' are tanh log E curves fitted to the data for the respective components. a is a tanh log E curve fitted to the left-hand four data points for the a-wave. i is a tanh log E curve which has the value zero for less than a critical value of E and which when subtracted from curve a gives curve d, which closely fits all the data points for the a-wave. Log flash energy $E = 0$ corresponds to one quantum per rod. Based on CONE (1965)

(1962a, b). It is probably generated at the synaptic end of the visual receptors. The b-wave is generated in the inner nuclear layer (bipolar cell layer) of the retina (NOELL, 1954). Data for the amplitudes of R_2, the unisolated a-wave, and the b-wave as a function of light energy, E, are shown in Fig. 12 for the albino rat. It can be seen that the b-wave data are fitted closely by a tanh log curve. The unisolated a-wave data are fitted poorly by a tanh log curve, suggesting that a nonlinear transform is involved between the tanh log transform of the cone action potential and the a-wave. The unisolated a-wave is thought to be the late receptor potential which is cut short by the onset of the b-wave. Since the b-wave has a shorter latency at higher values of E, this cuts down the late receptor potential by an increasing factor up to the high values of E at which the latency of the b-wave becomes constant (CONE, 1964).

The unisolated a-wave data cannot be fitted by a tanh curve multiplied by a constant or by another tanh curve, or by a tanh curve multiplied by a tanh curve having a constant added to or subtracted from it. The data can be fitted, as shown in curve d of Fig. 12, by a tanh curve from which is subtracted a tanh curve which varies with a different power of E and has a threshold E below which its amplitude is zero. This could be interpreted as the a-wave response being a tanh log E relation which is being reduced by another tanh log E process at high energies, possibly the b-wave, as discussed above.

The human electroretinogram has been measured under conditions of moderate light adaptation by BIERSDORF (1958) (Fig. 13). Both the a-wave (Fig. 13a) and the x-wave (Fig. 13c) fit a tanh log E curve which never reaches saturation at the light energies used. The b-wave (Fig. 13b) also fits a tanh log E curve up to the light energy at which there is a dip followed by a secondary rise in the response amplitude. This secondary rise is thought to be primarily the super-

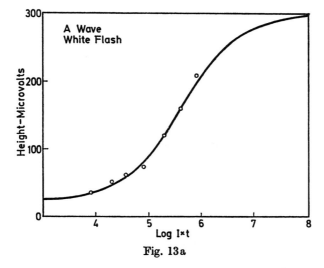

Fig. 13a

Fig. 13a—c. Amplitudes of the human ERG components during moderate light adaptation as a function of the energy of the test flash (the product of the luminance I in ft-L and duration t in msec for a fixed area of illumination. a Tanh log energy curve fitted to data for the a-wave amplitude in response to a white test flash of 25 msec duration. b Data for the b-wave response to a white flash. The broken curve has the tanh log (It) relation and is fitted to the 25 msec flash data. c Data for the x-wave response to a red flash. The broken curve has the tanh log (It) relation and is fitted to the 25 msec flash data. b and c The solid curves are lines through the experimental data, each curve for the flash duration labeled at its top. Adapted from BIERSDORF (1958)

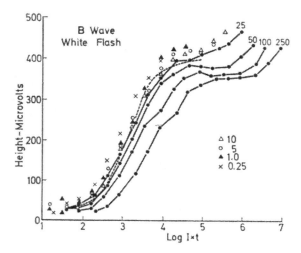

Fig. 13b

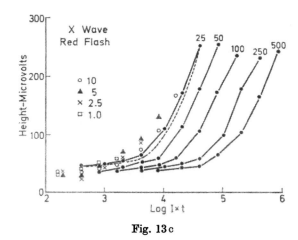

Fig. 13c

position of the x-wave on the b-wave. The x-wave has been related to photopic and especially to red cone activity, whereas the b-wave is considered primarily scotopic in nature. At stimulus light energies high enough to emphasize the photopic activity relative to the scotopic activity, the amplitude of the x-wave would be expected to become significant with respect to the amplitude of the b-wave.

4. Ganglion Cells

As far as is known, the horizontal, bipolar, and possibly some amacrine cells respond only by graded changes of potential. Some amacrine cells respond by a few spikes and graded potentials (WERBLIN and DOWLING, 1969). The only retinal cells known to produce only all-or-none propagated action potentials are the ganglion cells, whose axons form the optic nerve.

Ganglion cells receive synapses from bipolar cells and amacrine cells. The transfer function of these synapses cannot readily be tested because most ganglion cells receive inputs from more than one type of bipolar cell, and this input may be modified by the lateral effects transmitted by horizontal cells and amacrine cells (DOWLING and BOYCOTT, 1965, 1966; DOWLING et al., 1966; DOWLING, 1967; MISSOTTEN, 1965; STELL, 1967).

The bipolar-ganglion cell synapse can be studied in those ganglion cells for which this lateral interaction is apparently absent. Such seem to be the ganglion

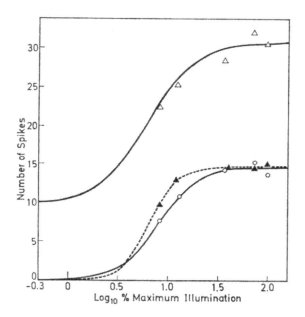

Fig. 14. Responses by ganglion cells of the frog retina to increases of blue illumination as a function of the intensity increase, the cells being of the blue-sensitive on-type and connecting to the diencephalon. ▲, △, O: experimental data. The curves have the relation, tanh log I.
▲, △: Responses to reflected light from Munsell paper 10B, dominant wavelength 482 nm.
O: Responses to light through an interference filter with maximum transmission at 478 nm.
Replotted from MUNTZ (1962)

cells of the frog's retina whose axons go not to the optic tectum but to the diencephalon. These ganglion cells are almost entirely simple on-response cells that respond maximally to light in the blue spectral range. They respond only to an increase of illumination and do not respond to moving objects or contours. Their receptive fields are 10° to 17° in diameter and do not appear to be organized into central and peripheral zones having different properties. The amplitudes of response (number of spikes) of these cells for flashes of various light intensities, I, were measured by MUNTZ (1962) and are plotted in Fig. 14.

The data for each of the three ganglion cells fits a tanh log function. For all three, the parameters of the function correspond to a variation of the stimulus-sensitive conductance as the 0.89 power of I. This is in the range found above for the C-potential sites (of fish). This agreement leads to the inference that the bipolar-ganglion cell synapse, where not modified by lateral interactions, tends to have a linear transform.

5. The Optic Nerve

In general, ganglion cells do not fire at a rate proportional to the intensity of a steady illumination, but fire only when some aspect of the illumination is changed.

The behavior of the entire population of ganglion cells can be studied by measuring the total discharge rate for the whole optic nerve. The cat optic nerve maintains a steady rate of firing in the dark, the "dark discharge". When the retina is stimulated with steady diffuse illumination, the onset of that illumination evokes a strong "on" discharge to well above the dark discharge level. This subsides in about a second or less and then plateaus at a level well below that of the dark discharge. This lower discharge level is maintained as long as the illumination is maintained. Upon the offset of the illumination, there is an "off" discharge which raises the discharge to a level above that of the dark discharge. From that level it decays exponentially over a period of minutes to the dark discharge level (ARDUINI and PINNEO, 1962, 1963a). The interesting point is that for steady diffuse illumination the level of optic nerve discharge *decreases* as the level of light adaptation increases (ARDUINI and PINNEO, 1962, 1963a, b).

6. The Lateral Geniculate Nucleus

Most of the optic nerve fibers of mammals go to the lateral geniculate body (LGN). The discharge rate of these fibers, individually (SUZUKI and ICHIJO, 1967) and as a group (ARDUINI and PINNEO, 1963a; ARDUINI and CAVAGGIONI, 1965), was proportional to that of the whole optic nerve.

The steady discharge rate (tonic activity) of the LGN neurons (on which those optic fibers terminate) was proportional to the steady discharge rate of the optic nerve (SUZUKI and ICHIJO, 1967; ARDUINI and PINNEO, 1963a). In fact, the ratio of LGN input to output was a constant over the tested adapting light-intensity range of 300 to 1 (ARDUINI and CAVAGGIONI, 1965). The discharge rate of the LGN neurons was reduced when the steady discharge of the optic nerve was reduced by light adaptation in the cat (ARDUINI and PINNEO, 1963a; ARDUINI and CAVAGGIONI, 1965) and in the squirrel monkey (BROOKS, 1966), or by blocking the optic nerve by mechanical pressure in the cat (SUZUKI and ICHIJO, 1967).

Light adaptation produced by a steady background illumination was found to reduce the transient increase in discharge rate (phasic activity) evoked by a superimposed light flash at the optic nerve and the LGN neurons of the cat (ARDUINI and PINNEO, 1963b) and at the LGN neurons of the monkey (BROOKS, 1966).

7. The Visual Cortex

The massed tonic discharge of the fibers from the visual cortex to the thalamus in the cat was found by ARDUINI and CAVAGGIONI (1965) to decrease as the discharge from the LGN to the cortex was decreased by light adaptation to steady illumination. However, the cortical-thalamic discharge did not decrease as rapidly as did the tonic LGN discharge when the cat was under light anesthesia, although it decreased at the same rate when the cat was under deep Nembutal anesthesia (Fig. 15).

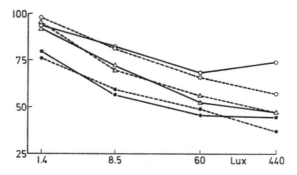

Fig. 15. Effect of intravenous injection of Nembutal on tonic activity in the visual system. Ordinate: activity during illumination as a percentage of the preceding dark adapted activity. Abscissa: intensity of the steady illumination on a logarithmic scale. ● : optic chiasm; △ : geniculo-cortical radiations; ○ : cortico-thalamic tracts. Solid lines: before Nembutal; broken lines: after Nembutal. From ARDUINI and CAVAGGIONI (1965)

8. The Psychophysical Relations of Brightness

The magnitude estimation procedure can be used to determine the relation between brightness, ψ, and luminance (light intensity), I. The relation is of the form shown in Fig. 16 when plotted in log-log coordinates. The function starts

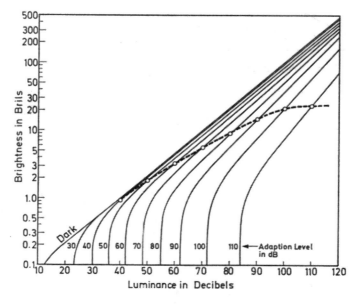

Fig. 16. Relative brightness as a function of luminance on log-log coordinates. For a 1-sec flash superimposed on various levels of adapting illumination. From STEVENS and STEVENS (1963)

out as an abruptly rising increase in log brightness with increase in log luminance and changes over to a straight line with a positive slope at higher luminances. All of these studies determined the brightness of a light flash superimposed on a steady illumination, not the brightness of the steady illumination.

If the luminance is plotted in units of the threshold luminance, I_0, which just evokes a sensation, then the data points lie along a straight line on log-log coordinates. This is shown in Fig. 17a and b. Thus, the relation between brightness and luminance is

$$\psi/\psi_0 = k_4(I - I_0)^\beta , \qquad (26)$$

the same as Eq. (23).

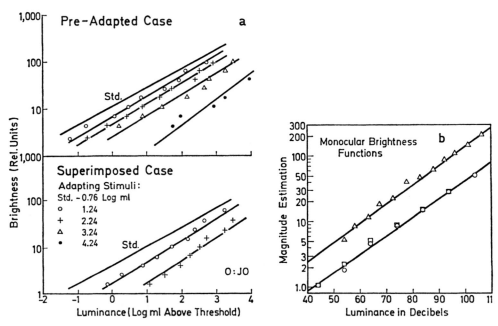

Fig. 17a and b. Log brightness v. log luminance, with luminance plotted as luminance above threshold. *a* For 1° test flashes of 275 msec duration against a 10° background of various adapting luminances presented until just before the test flash (pre-adapted case) or continually (superimposed case). From ONLEY (1961). *b* △, data from ONLEY. ○, Interocular comparisons, □, monocular magnitude estimation of brightnesses; both for a 5.7° field with a dark surround. From STEVENS and STEVENS (1963)

9. Interpretation in Terms of MacKay's Model

The psychophysical relation of Eq. (26) could result according to MacKay's model if the comparator received an input f_1 from the peripheral sensory system, R, and input f_2 from the organizer, O. These inputs are represented by Eqs. (21) and (22), respectively. These are logarithmic transforms, but MACKAY (1963) shows that the results also hold for near-logarithmic transforms. Then, according to Eq. (25), $\beta = k_1/bk_2$, where k_1 is the relative gain of R, k_2 is the relative gain of O, and b is the relative weight given by the comparator to the input from O compared to the input from R.

a) **Effect of Area of Illumination.** If the O side of the comparator remains unchanged, then β can be expected to be proportional to k_1. For example, for the dark adapted observer the psychophysical power law of brightness has a β of 0.50 for a point source of light (STEVENS, 1961), but a β of 0.33 for a 5° field of illumination. This can be interpreted as the action of the additional stimulated receptors around a given illuminated point on the retina, reducing the effectiveness of the coupling from the receptors at that point to the comparator. This is what would be expected as a result of lateral inhibition from the adjacent illuminated receptors.

b) **Fechner's Law.** It can be seen from the diagram of MacKay's model (Fig. 9) that a certain minimum difference, Δf, between f_1 and f_2 is required to cause the comparator C to signal a mismatch. This Δf must correspond to the smallest noticeable difference in stimulus strength, in other words, to the jnd. If Δf remains constant as f_1 and f_2 change, then the relation of stimulus strength, S, to the sum of the jnd is the transfer function of the entire peripheral sensory pathway, R, up to the comparator. If R has a near-logarithmic transfer function, then the change ΔS required to produce the mismatch Δf will be nearly proportional to S (more exactly, to $S - S_0$). As S is increased from a low level and the number of jnd are summed, the total number is related to S by a near-logarithmic curve. This is Fechner's law. Thus, the model gives a physiological meaning to Fechner's law, while also leading to a power law relation between stimulus strength and sensation magnitude.

Jnd values can be determined from measurements of DI versus I, where DI is the increment threshold intensity for behavioral response to a stimulus superimposed on an adapting stimulus of intensity I. Such data for the rhesus monkey and the human (CRAWFORD, 1935) and for the squirrel monkey (BROOKS, 1966) showed a decrease of DI/I with increasing I down to some ratio which remained constant for all higher I's. The summed jnd curve has been constructed as a function of log I from Brooks' behavioral DI versus I data for the squirrel monkey and is plotted as the solid curve of Fig. 18. At high values of I the curve becomes a straight line with a positive slope, as do all such curves for monkeys and humans. The curve represents the transfer function of R, yet it clearly is not the tanh log function of the photoreceptors. Therefore, the peripheral visual pathway to the comparator must include at least one additional nonlinear transform. A nonlinear mechanism which could produce this overall transfer function has been described in the section on logarithmic functions.

The DI used to construct the jnd curve were based on the behavioral responses of monkeys in choosing the one of two illuminated panels on which a 1-sec flash of intensity DI had been superimposed. Therefore, the jnd curve of Fig. 18 is the transfer function of R for phasic (ON/OFF) activity of the visual system.

The massed discharge of the lateral geniculate nucleus neurons of these squirrel monkeys was measured for the tonic activity resulting from the steady illumination and for the superimposed phasic activity resulting from the light flashes (BROOKS, 1966). This data is made comparable with the summed jnd curve by setting the point for the activity for the smallest DI on the jnd curve, since that activity corresponded to a just discriminable difference. The scale of the activity units was then adjusted to make the points for the largest DI fall on the jnd curve. It can be seen in Fig. 18 that the phasic activity values (open triangles) are a good fit to the curve, but the values for the sum of phasic and tonic activities (crosses) and for the tonic activity alone (circles) are a poor fit. This suggests two things; first, that the phasic activity of the lateral geniculate nucleus is proportional to the activity reaching the comparator; and second, that for this particular discrimination task only the phasic activity is used by the comparator, the tonic activity somehow being kept separate from the phasic.

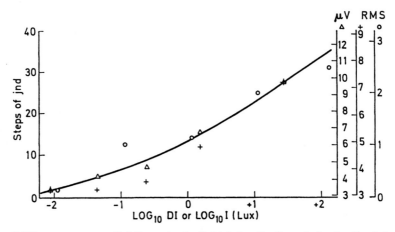

Fig. 18. Solid curve: summed jnd v. steady light intensity from behaviorally determined brightness discrimination function. ○ : Average decrease in rms voltage of tonic massed activity of lateral geniculate cells compared with activity in dark evoked by steady illumination with intensity I. △ : Average increase in rms voltage of phasic discharge to 1 second flash of intensity DI superimposed upon the same steady illumination as used for that DI in the brightness discrimination study. + : Average net change in rms voltage of the sum of the tonic and phasic activities for each DI. Data averaged for three squirrel monkeys; calculated from the data of BROOKS (1966)

This latter deduction is not surprising. DE VALOIS et al. (1962) found that in the lateral geniculate of the macaque monkey a minority of the neurons (non-opponent type) fired continually at a rate that was dependent on the light intensity. The rate decreased (inhibitory) or increased (excitatory) for an increase in light intensity, depending upon the particular neuron. By assuming that luminosity information was coded at these cells in the form of rate of change of firing frequency rather than as absolute frequency, a luminosity curve was constructed from the light intensity required to produce a criterion number of spikes in the averaged population of excitatory non-opponent cells. The resulting curve matched the behaviorally determined curves for these monkeys and for humans (DE VALOIS, 1965). However, the majority of the geniculate cells showed very little change in firing upon change of light intensity.

At the visual striate cortex (area 17) of the cat (HUBEL, 1959; HUBEL and WIESEL, 1959) and of the macaque and squirrel monkeys (HUBEL and WIESEL, 1968) the cells did not respond to steady diffuse illumination, and they gave unclear or inconsistent responses to changes in intensity of diffuse illumination. The cells responded briskly only to specific spatiotemporal patterns of light intensity at their receptive fields. These cortical cells gave no indication of responding to the (reduction of) tonic activity from the lateral geniculate produced by diffuse steady illumination.

c) **Effect of Light Adaptation.** During adaptation to steady illumination the geniculo-cortical phasic discharge in response to a superimposed light flash decreases, and (as can be seen from Fig. 17a) the slope, β, of the psychophysical relation increases. If the geniculo-cortical discharge is considered to be a measure of f_1, then the coupling, k_1, of the peripheral sensory pathway must be considered to decrease with light adaptation. Then, from Eq. (25), the observed increase in β must be attributed to an even greater decrease in bk_2 upon light adaptation. The activity $(\psi - \psi_0)$ of the organizer would have to decrease more slowly with light adaptation than did f_1 in order to keep the signals matched at the comparator. The fact that the activity in the cortico-thalamic tract showed this behavior (ARDUINI and CAVAGGIONI, 1965) suggests that this tract is part of the organizer or of its pathway to the comparator.

Several mechanisms can be postulated by which light adaptation could reduce bk_2, the coupling between the organizer and the comparator. One possibility is that the coupling is facilitated by the tonic activity from the lateral geniculate, since the tonic activity is known to decrease with light adaptation. However, this seems unlikely since the cortical response to direct electric stimulation of the geniculate is *increased* whenever that tonic activity is decreased by blocking the optic nerve (ARDUINI and PINNEO, 1963a; KASAMATSU et al., 1967).

A second possibility is that the tonic activity from the retina that passes along the nonspecific visual pathways *increases* as light adaptation increases. This activity would act to reduce (inhibit) the coupling from the organizer to the comparator. This rather broad suggestion is given some plausibility by the finding that deep barbiturate anesthesia, which blocks the nonspecific pathways, causes the cortico-thalamic discharge to remain linearly proportional to the geniculate activity as light adaptation increases (ARDUINI and CAVAGGIONI, 1965). If the preceding discussions of this section are combined with this possible mechanism, then the organizer pathway should include a structure that is excited (directly or via relays) by the cortico-thalamic tract and is inhibited by the visual non-specific pathways. The comparator should receive activity from both that structure and directly or via the visual cortex from the lateral geniculate nucleus and should itself send activity to the visual cortex.

K. Summary

The relations between stimulus strength, neural activity amplitude, and subjective sensation magnitude have been considered for several sensory systems. Those relations were found to be different, depending on the sensory system. A "self-shunting" transducer mechanism of receptor membrane action was shown to describe the receptors of many systems and to give a tanh log transfer function or a linear transfer function, as determined by the dependence of the membrane conductances and active transport on excitation and voltage. A number of examples were given of mechanisms by which the overall transfer function could be further modified more centrally in the sensory pathway to achieve the observed linear, logarithmic, and power law relations of neural activity amplitude to stimulus strength.

By use of MacKay's neural circuit model for the generation of sensation, these observations of the relations in the overall sensory pathway could be compared with the observed psychophysical relations between sensation magnitude and stimulus strength. The comparisons permitted deductions regarding the neural functions of the sensation generator and the possible location of some of its components in the visual system. In addition, it was shown how this model permitted the use of psychophysical experiments to extend the studies of lateral inhibition and the pathways of light adaptation.

Appendix 1

The relation between strength of applied stimulus and resulting change in the membrane potential of a receptor cell can be derived using text Eq. (2). The receptor cell's membrane is considered to be made up of two portions; one inexcitable, the other excitable and consisting of m unit areas which are capable of being excited in an all-or-none fashion. For a unit area of membrane we denote Σg_γ by G, ΣJ_γ by J, ΣI_γ by I, and $\Sigma_\gamma g_\gamma / G$ by $\overline{E}$. The subscript (1) is used for an unexcited excitable unit area, the subscript (2) for an excited unit area, and the subscript (3) for the entire inexcitable portion of the membrane.

It is assumed that the internal impedance of the receptor cell is negligible compared to the resistance across the membrane. For simplicity it is assumed that the resting potential of the inexcitable membrane is the same as that of the unexcited excitable membrane, so that for zero excitation $I_1 = I_3 = 0$. The voltage transients, such as would be produced by the membrane's capacitance are neglected, so this derivation applies only for the steady state.

From Eq. (2)

$$I = GV + (J - G\overline{E}). \tag{1a}$$

The circuit analog for Eq. (1a) is shown in Fig. 1a.

Now let n be the number of excited unit areas. Then the membrane of the active receptor cell is modeled by the circuit of Fig. 1b. For convenience we have shifted the voltage scale so that the resting potentials $(G_1\overline{E}_1 - J_1)/G_1$ and $(G_3\overline{E}_3 - J_3)/G_3$ are zero.

The output voltage is then given by

$$V_{\text{OUT}} = \frac{n(G_2\overline{E}_2 - J_2)}{(mG_1 + G_3) + n(G_2 - G_1)}. \tag{4a}$$

Since the output voltage at zero excitation, V^0_{OUT}, is zero, V_{OUT} equals the change in measured membrane voltage, $\Delta V_m = (V_{\text{OUT}} - V^0_{\text{OUT}}) = V_{\text{OUT}}$, produced by the stimulus.

Now define

$$b = \frac{m(G_2 - G_1)}{mG_1 + G_3} \tag{5a}$$

$$kb = \frac{m(G_2\overline{E}_2 - J_2)}{mG_1 + G_3} \tag{6a}$$

$$x = n/m. \tag{7a}$$

Then Eq. (4a) may be rewritten as

$$V_{\text{OUT}} = \frac{kbx}{1 + bx}, \tag{8a}$$

which is the result given in the text. The "self-shunting" non-linear relation between V_{OUT} and x will then be observed, for example, if kb and b are independent of voltage.

It is interesting to consider the special case in which G_1 and G_2 are equal. Then b is zero and

$$V_{\text{OUT}} = \frac{m(G_2\overline{E}_2 - J_2)}{mG_1 + G_3} \cdot x. \tag{8a_1}$$

This gives a linear relation unless some term in the coefficient of x is dependent on V or x.

Consider the special case in which G_1 equals G_2, and G_3 is voltage-dependent, so that it changes with x. A first order (in x) approximation for G_3 can then be written as

$$G_3 \cong A + Bx \tag{8a_2}$$

where A and B are constants. This approximation need not remain exact as x approaches 1. Using this approximation one obtains

$$V_{\text{OUT}} \cong \frac{m\,(G_2\,\overline{E}_2 - J_2)\,x}{(m\,G_1 + A) + Bx}\,. \tag{8a_3}$$

Now define the new constants

$$\beta = \frac{B}{m\,G_1 + A} \tag{8a_4}$$

$$\beta K = \frac{m\,(G_2\,\overline{E}_2 - J_2)}{m\,G_1 + A}\,. \tag{8a_5}$$

Then

$$V_{\text{OUT}} \cong \frac{K\beta x}{1 + \beta x}\,, \tag{8a_6}$$

which is the same form as Eq. (8a). This result is especially important for the consideration of receptors whose response is caused by a change in active transport current rather than by a selective change in ionic conductance.

Appendix 2

The theoretical relation between the generator potential and the logarithm of the stimulus strength has been derived for an analog of the self-shunting receptor membrane [see electric analog at bottom of Fig. 1 of NAKA and RUSHTON (1966a)]. They start with the relation

$$V_0 = J/(1 + J) \tag{9a}$$

where V_0 is the ratio of generator potential to the maximal generator potential, and J is the stimulus strength, S, in units such that for $J = 1$, $V_0 = \frac{1}{2}$.

Thus, in the terms used in Appendix 1,

$$V_0 = k\varDelta V_m / k\varDelta V_m(\text{max}) = J/(1 + J)\,. \tag{10a}$$

It can be seen by comparison that Eqs. (8a) and (9a) are the same except for scale factors.

If S (and therefore also J) is plotted on a logarithmic scale, and if the scale units are such that $\ln J = 2\,w$ (so $J = e^{2w}$), then from Eq. (9a)

$$2\,V_0 = 2\frac{J}{J + 1} = \frac{2e^{2w}}{e^{2w} + 1} = \frac{2e^w}{e^w + e^{-w}} = \frac{e^w + e^w}{e^w + e^{-w}} = \frac{e^w + e^{-w} + e^w - e^{-w}}{e^w + e^{-w}}$$
$$= 1 + \frac{e^w - e^{-w}}{e^w + e^{-w}} \tag{11a}$$

and since the last term on the right is the hyperbolic tangent of w,

$$2\,V_0 = 1 + \tanh w \tag{12a}$$

then

$$V_0 = \tfrac{1}{2} + \tfrac{1}{2} \tanh w\,. \tag{13a}$$

If the w axis is shifted an amount w_0, where w_0 is the value of w for $V_0 = \frac{1}{2}$,

$$V_0 = \tfrac{1}{2} + \tfrac{1}{2} \tanh (w - w_0)\,. \tag{14a}$$

This gives a curve which is symmetrical about the point ($V_0 = \frac{1}{2}$, $w = w_0$), and which asymptotes at $V_0 = 0$ and $V_0 = 1$. This is the tanh log template curve of NAKA and RUSHTON (1966a, b, c). For this template curve

$$J = e^{2(w - w_0)}\,. \tag{15a}$$

Appendix 3

To apply the tanh log template curve to a set of data: (1) plot the data with the response amplitude on the ordinate and the logarithm to base 10 of the stimulus strength, S, on the abscissa. Use any scale factors that spread the points sufficiently for easy measurement. (2) Fit a rough curve by eye to the data. (3) Estimate from this curve the maximum response asymptote, the minimum response asymptote, and the point of symmetry of the curve. At least two of these must be obtained in order to proceed.

(4) If the two asymptotes are obtained, set the upper to $V_0 = 1$ and the lower to $V_0 = 0$. If only one asymptote is obtained, set it to $V_0 = 0$ or 1 for lower or upper, and set the point of symmetry to $V_0 = 0.5$. Use the points so determined to lay out a linear ordinate scale of V_0.

(5) Mark the log S scale where the rough curve intersects $V_0 = 0.5$. This gives log S_0 on the log S scale and w_0 on a w abscissa scale.

(6) Mark the log S scale where the rough curve intersects $V_0 = 0.119$. The distance to log S_0 on the log S scale is one unit on the w scale. As a check, this should be the same as the distance along the log S scale from log S_0 to the abscissa value at which the rough curve intersects $V_0 = 0.881$. Use this unit distance to lay out a linear abscissa scale of w.

(7) From the table below, determine for various values of w the corresponding values of V_0 and plot the points so determined on the V_0 and w plane. Connect the points by a smooth curve to get the tanh log I template curve.

Coordinates of tanh log template curve

w	V_0	w	V_0
$w_0 - 2.538$	0.00675	$w_0 + 0.20$	0.60
$w_0 - 2.185$	0.0125	$w_0 + 0.42$	0.70
$w_0 - 1.84$	0.025	$w_0 + 0.69$	0.80
$w_0 - 1.47$	0.05	$w_0 + 1.00$	0.881
$w_0 - 1.10$	0.10	$w_0 + 1.10$	0.90
$w_0 - 1.00$	0.119	$w_0 + 1.47$	0.95
$w_0 - 0.69$	0.20	$w_0 + 1.84$	0.975
$w_0 - 0.42$	0.30	$w_0 + 2.185$	0.9875
$w_0 - 0.20$	0.40	$w_0 + 2.538$	0.99325
w_0	0.50		

To determine the relation between J and stimulus strength S:

(8) Measure c, the ratio between one unit on the $\log_{10} S$ scale and one unit on the w scale.

(9) From Eq. (15a),

$$J = e^{2(w - w_0)} = e^2 \left(\frac{\log S}{c} - \frac{\log S_0}{c} \right) = e^2 \left(\frac{\ln S}{2.3c} - \frac{\ln S_0}{2.3c} \right) = (S/S_0)^{\frac{2}{2.3c}} . \tag{16a}$$

From Eqs. (5a) and (8a),

$$J = bx = m(G_2 - G_1)/(mG_1 + G_3) . \tag{17a}$$

If G_1 is assumed to be small compared to G_3, then Eq. (17a) becomes

$$J \cong mG_2/(mG_1 + G_3) . \tag{18a}$$

From Eqs. (16a) and (18a),

$$G_2 \cong J(mG_1 + G_3)/m \cong \text{constant} \cdot S/S_0^{2/2.3c} , \tag{19a}$$

so G_2 is approximately proportional to the $2/2 \cdot 3c$ power of the stimulus strength, S. This power can be evaluated by step (8) above.

Acknowledgements :

Supported in part by a joint program of the Army Research Office and The National Aeronautics and Space Administration under Contract NASW 1815.

The formalism and derivations used in Appendix 1 were suggested by Dr. KARL KORNACKER.

Some of this material was previously published in *Vision Research* (LIPETZ, 1969) and is reproduced by permission of the editors.

References

ARDUINI, A., CAVAGGIONI, A.: Transmission of tonic activity through lateral geniculate body and visual cortex. Arch. ital. Biol. 103, 652—667 (1965).

— PINNEO, L.: Properties of the retina in response to steady illumination. Arch. ital. Biol. 100, 425—448 (1962).

— — The tonic activity of the lateral geniculate nucleus in dark and light adaptation. Arch. ital. Biol. 101, 493—507 (1963a).

— — The effects of flicker and steady illumination on the activity of the cat visual system. Arch. ital. Biol. 101, 508—529 (1963b).

BEIDLER, L. M.: A theory of taste stimulation. J. gen. Physiol. 38, 133—139 (1954).

— Mechanisms of gustatory and olfactory receptor stimulation. In: Sensory communication. New York: John Wiley & Sons 1961.

BIERSDORF, W. R.: Luminance-duration relationships in the light-adapted electro-retinogram. J. Opt. Soc. Am. 48, 412—417 (1958).

BORG, G., DIAMANT, H., STRÖM, L., ZOTTERMAN, Y.: The relation between neural and perceptual intensity: a comparative study on the neural and psychophysical response to taste stimuli. J. Physiol. (Lond.) 192, 13—20 (1967).

BORSELLINO, A., FUORTES, M. G. F., SMITH, T. G.: Visual responses in Limulus. Cold Spr. Harb. Symp. quant. Biol. 30, 429—443 (1965).

BROOKS, B. A.: Neurophysiological correlates of brightness discrimination in the lateral geniculate nucleus of the squirrel monkey. Exp. Brain Res. 2, 1—17 (1966).

BROWN, K. T., MURAKAMI, M.: Rapid effects of light and dark adaptation upon the receptive field organization of S-potentials and late receptor potentials. Vision Res. 8, 1145—1171 (1968).

— WATANABE, J.: Isolation and identification of a receptor potential from the pure cone fovea of the monkey retina. Nature (Lond.) 193, 959—960 (1962a).

— — Rod receptor potential from the retina of the night monkey. Nature (Lond.) 196, 547—550 (1962b).

CONE, R. A.: The rat electroretinogram. I. Contrasting effects of adaptation on the amplitude and latency of the b-wave. J. gen. Physiol. 47, 1089—1105 (1964).

— The early receptor potential of the vertebrate eye. Cold Spr. Harb. Symp. quant. Biol. 30, 483—491 (1965).

CRAWFORD, M. P.: Brightness discrimination in the rhesus monkey. Genet. Psych. Monogr. 17, No. 2, 71—162 (1935).

DARIAN-SMITH, I., ROWE, M. J., SESSLE, B. J.: "Tactile" stimulus intensity: information transmission by relay neurons in different trigeminal nuclei. Science 160, 791—794 (1968).

DAVIS, H.: A model for transducer action in the cochlea. Cold Spr. Harb. Symp. quant. Biol. 30, 181—190 (1965).

DESMEDT, J. E.: Auditory-evoked potentials from cochlea to cortex as influenced by activation of the efferent olivo-cochlear bundle. J. Acoust. Soc. Am. 34, 1478—1496 (1962).

DEUTSCH, S.: Models of the nervous system. New York: John Wiley & Sons 1967.

DE VALOIS, R., JACOBS, G. H., JONES, A. E.: Effects of increments and decrements of light on neural discharge rate. Science 136, 986—988 (1962).

— Analysis and coding of color vision in the primate visual system. Cold Spr. Harb. Symp. quant. Biol. 30, 567—579 (1965).

DÖVING, K. B.: Studies of the relation between the frog's electro-olfactogram (EOG) and single unit activity in the olfactory bulb. Acta physiol. scand. 60, 150—163 (1964).

DOWLING, J. E.: The site of visual adaptation. Science 155, 273—279 (1967).

— BOYCOTT, B. B.: Neural connections of the retina: fine structure of the inner plexiform layer. Cold Spr. Harb. Symp. quant. Biol. 30, 393—402 (1965).

— — Organization of the primate retina: electron microscopy. Proc. roy. Soc. B 166, 80—111 (1966).

— BROWN, J. E., MAJOR, D.: Synapses of horizontal cells in rabbit and cat retinas. Science 153, 1639—1641 (1966).

EKMAN, G., ÅKESSON, C.: Saltiness, sweetness and preferences; a study of quantitative relations in individual subjects. Scand. J. Psychol. 6, 241—253 (1965).

Fuortes, M. G. F.: Initiation of impulses in the visual cells of Limulus. J. Physiol. (Lond.) **148**, 14—28 (1959).

Granit, R.: The components of the retinal action potential in mammals and their relation to the discharge in the optic nerve. J. Physiol. (Lond.) **77**, 207—240 (1933).

Harmon, L. D., Lewis, E. R.: Neural modeling. Physiol. Rev. **46**, 513—591 (1966).

Hubel, D. H.: Single unit activity in striate cortex of unrestrained cats. J. Physiol. (Lond.) **147**, 226—238 (1959).

— Wiesel, T. N.: Receptive fields of single neurones in the cat's striate cortex. J. Physiol. (Lond.) **148**, 574—591 (1959).

— — Receptive fields and functional architecture of monkey striate cortex. J. Physiol. (Lond.) **195**, 215—243 (1968).

Kasamatsu, T., Kiyono, S., Iwama, K.: Electrical activities of the visual cortex in chronically blinded cats. Tohoku J. exp. Med. **93**, 139—152 (1967).

Katz, B.: Depolarization of sensory terminals and the initiation of impulses in the muscle spindle. J. Physiol. (Lond.) **111**, 261—282 (1950).

— Miledi, R.: Input-output relation of a single synapse. Nature (Lond.) **212**, 1242—1245 (1966).

Keidel, W. D., Spreng, M.: Neurophysiological evidence for the Stevens power function in man. J. Acoust. Soc. Am. **38**, 191—195 (1965).

Kimura, K., Beidler, L. M.: Microelectrode study of taste receptors of rat and hamster. J. cell. comp. Physiol. **58**, 131—139 (1961).

Kornacker, K.: Physical principles of active transport and electrical excitability. In: Biological membranes. R. M. Dowben, editor Boston: Little-Brown 1969.

Lawrence, M.: Dynamic range of the cochlear transducer. Cold Spr. Harb. Symp. quant. Biol. **30**, 159—167 (1965).

Lipetz, L. E.: The transfer functions of sensory intensity in the nervous system. Vision Res. **9**, 1205—1234 (1969).

Loewenstein, W. R.: The generation of electrical activity in a nerve ending. Ann. N. Y. Acad. Sci. **81**, 367—387 (1959).

— Excitation and inactivation in a receptor membrane. Ann. N. Y. Acad. Sci. **94**, 510—534 (1961).

MacKay, D. M.: Psychophysics of perceived intensity: a theoretical basis for Fechner's and Steven's laws. Science **139**, 1213—1216 (1963).

MacNichol, E. F., Jr.: Visual receptors as biological transducers. Molecular structure and functional activity of nerve cells. Amer. Inst. Biol. Sci., Publ. No. 1, Washington 1956.

— Svaetichin, G.: Electric responses from the isolated retinas of fishes. Amer. J. Opthal. **46**, 26—40 (1958).

Missotten, L.: The synapses in the human retina. In: The structure of the eye. II. Stuttgart: Schattauer-Verlag 1965.

Mitarai, G.: Determination of the ultramicroelectrode tip position in the retina in relation to S potential. J. gen. Physiol. **43**, 95—99 (1960).

— Svaetichin, G., Vallecalle, E., Fatehchand, R., Villegas, J., Laufer, M.: Glia-neuron interactions and adaptational mechanisms of the retina. In: Neurophysiologie und Psychophysik des Visuellen Systems. Berlin-Göttingen-Heidelberg: Springer 1961.

Mountcastle, V. B.: The problem of sensing and the neural coding of sensory events. In: The neurosciences. New York: Rockefeller Univ. Press 1967.

— Talbot, W. H., Kornhuber, H. H.: The neural transformation of mechanical stimuli delivered to the monkey's hand. In: Ciba Found. Symp.: Touch, heat and pain. Boston: Little-Brown 1966.

Muntz, W. R. A.: Microelectrode recordings from the diencephalon of the frog (*Rana pipiens*) and a blue-sensitive system. J. Neurophysiol. **25**, 699—711 (1962).

Murakami, M., Kaneko, A.: Differentiation of P III subcomponents in cold-blooded vertebrate retinas. Vision Res. **6**, 627—636 (1966).

Naka, K. I., Rushton, W. A. H.: S-potentials from colour units in the retina of fish (Cyprinidae). J. Physiol. (Lond.) **185**, 536—555 (1966a).

— — An attempt to analyze colour reception by electrophysiology. J. Physiol. (Lond.) **185**, 556—586 (1966b).

NAKA, K. I., RUSHTON, W. A. H.: S-potentials from luminosity units in the retina of fish (Cyprinidae). J. Physiol. (Lond.) 185, 587—599 (1966c).

NOELL, W. K.: The origin of the ERG. Amer. J. Ophthal. 38, 78—90 (1954).

NOMOTO, M., SUGA, N., KATSUKI, Y.: Discharge pattern and inhibition of primary auditory nerve fibers in the monkey. J. Neurophysiol. 27, 768—787 (1964).

ONLEY, J. W.: Light adaptation and the brightness of brief foveal stimuli. J. Opt. Soc. Am. 51, 667—673 (1961).

PURPLE, R. L.: The integration of excitatory and inhibitory influences in the eye of Limulus. Ph. D. thesis. New York: Rockefeller Institute 1964.

— DODGE, F. A.: Interaction of excitation and inhibition in the eccentric cell in the eye of Limulus. Cold Spr. Harb. Symp. quant. Biol. 30, 529—537 (1965).

RUSHTON, W. A. H.: A theoretical treatment of Fuortes's observations upon eccentric cell activity in Limulus. J. Physiol. (Lond.) 148, 29—38 (1959).

SMITH, T. G., STELL, W. K., BROWN, J. E.: Conductance changes associated with receptor potentials in Limulus photoreceptors. Science 162, 454—456 (1968).

— — — FREEMAN, J. A., MURRAY, G. C.: A role for the sodium pump in photoreception in Limulus. Science 162, 456—458 (1968).

STELL, W. K.: The structure and relationships of horizontal cells and photoreceptor-bipolar synaptic complexes in goldfish retina. Amer. J. Anat. 121, 401—424 (1967).

STEVENS, J. C., GUIRAO, M.: Individual loudness functions. J. Acoust. Soc. Am. 36, 2210—2213 (1964).

— STEVENS, S. S.: Brightness function: effects of adaptation. J. Opt. Soc. Am. 53, 375—385 (1963).

STEVENS, S. S.: The psychophysics of sensory function. In: Sensory communication. Cambridge, Mass.: M. I. T. Press 1961.

STIEVE, H.: Interpretation of the generator potential in terms of ionic processes. Cold Spr. Harb. Symp. quant. Biol. 30, 451—456 (1965).

SUZUKI, H., ICHIJO, M.: Tonic inhibition in cat lateral geniculate nucleus maintained by retinal spontaneous discharge. Jap. J. Physiol. 17, 519—612 (1967).

SVAETICHIN, G., LAUFER, M., MITARAI, G., FATEHCHAND, R., VALLECALLE, E., VILLEGAS, J.: Glial control of neuronal networks and receptors. In: Neurophysiologie und Psychophysik des visuellen Systems. Berlin-Göttingen-Heidelberg: Springer 1961.

TOMITA, T., KANEKO, A., MURAKAMI, M., PAUTLER, E. L.: Spectral response curves of single cones in the carp. Vision Res. 7, 519—531 (1967).

— MURAKAMI, M., SATO, Y., HASHIMOTO, Y.: Further study on the origin of the so-called cone action potential (S-potential): Its histological determination. Jap. J. Physiol. 9, 63—69 (1959).

TOYODA, J., NOSAKI, H., TOMITA, T.: Light-induced resistance changes in single photoreceptors of Necturus and Gekko. Vision Res. 9, 453—463 (1969).

UTTAL, W. R.: A comparison of neural and psychophysical responses in the somesthetic system. J. comp. physiol. Psychol. 52, 485—490 (1959).

VAN BERGEIJK, W. A.: Studies with artificial neurons: II. Analog of the external spiral innervation of the cochlea. Kybernetik 1, 102—107 (1961).

VAN DEN BRINK, G., BOUMAN, M. A.: Quantum coincidence requirements during dark-adaptation. Vision Res. 3, 479—481 (1963).

VILLEGAS, J.: Studies on the retinal ganglion cells. In: Neurophysiologie und Psychophysik des visuellen Systems. Berlin-Göttingen-Heidelberg: Springer 1961.

WERBLIN, F. S., DOWLING, J. E.: Organization of the retina of the mudpuppy, Necturus maculosus: II, Intracellular Recording. J. Neurophysiol. 32, 339—355 (1969).

WERNER, G., MOUNTCASTLE, V. B.: Neural activity in mechanoreceptive cutaneous afferents: stimulus-response relations, Weber functions, and information transmission. J. Neurophysiol. 28, 359—397 (1965).

Chapter 7

Sensory Power Functions and Neural Events

By
S. S. Stevens, Cambridge, Mass. (USA)

With 4 Figures

Contents

How the sensory systems distinguish light from sound has traditionally seemed a more urgent question than how the brain can tell a loud sound from a faint one, or a bright light from a dim one. Of course, the light and sound can be omitted in favor of electrical stimulation. Then, depending on which of the sensory systems receives the current, the sensation is one of seeing or hearing. Experiments of that kind have been thought to dispose of the quality problem, for it has been made to seem that sensory quality depends on which nerves are actuated and where in the brain they lead.

But what about magnitude? A whisper, a shout, a crash of thunder — all three stimuli fire messages through the same sense organ and up the same tract of nerves. How do we perceive the vast intensity differences? In the scientific sense, that question became a question only about half a century ago. Prior to the discovery of the all-or-none law, it was supposed that whatever the nerves did, they could presumably do a bit more of the same whenever the stimulus became stronger. But once the code was broken and it was disclosed that each nerve fiber carries a train of impulses, all the impulses very much alike, and that the train spaces itself out to a limit of a few hundred impulses per second, then it became plain that the mediation of sensory intensity demands its place among the major neurological puzzles.

Fechner's Conjecture

For many years following the advent of the all-or-none principle, it was thought that the demands on the all-or-none transmission system would fall within a tractable range, because the magnitude we perceive was supposed to grow only as the logarithm of the stimulus intensity. That principle would provide for a slow nonlinear growth, causing thereby an enormous compression of the energies that strike the receptors. The idea that the "psychophysical law", as it came to be called, could represent a compression as severe as a logarithmic function derives from the century-old notion bequeathed by the physicist G. T. FECHNER (1860), who proposed that a *relative* increment in the external energy corresponds to a *constant* increment in the apparent intensity. That idea, FECHNER tells us, occurred to him as he lay abed on the morning of October 22, 1850. He later perceived its relation to WEBER's principle, which says that, in order to produce a series of just noticeable differences, we must make a fixed *relative* increase in the stimulus. FECHNER put those two conceptions together, added the requisite machinery of experimental methods, and thereby founded the science of psychophysics, progenitor of experimental psychology.

So unlikely an idea as FECHNER's did not fail to stir challenge, of course, but it was the engineering problems of acoustics that first generated a sufficient scientific effort to threaten the logarithmic law. In the 1930's several laboratories produced functions showing that the growth of apparent loudness does not follow the logarithm of the stimulus. Loudness, it was found, does not increase in proportion to decibels, as FECHNER's law would have it. But laws have a way of refusing to expire merely because they have been found wrong. In order for a punctured law to deflate and vanish, there must be something to fill its place.

The Power Law

The replacement process began in 1953 when, in pursuit of a curious but minor phenomenon, three different scaling procedures were applied to stimuli for both vision and hearing. The procedures were bisection, fractionation, and magnitude estimation (STEVENS, 1953). The similarities in the quantitative results obtained with light and sound appeared striking enough to demand a closer look to see what it was that nature was trying to disclose. The consensus of the measurements supported the proposition that the sensation ψ for both brightness and loudness grows in proportion to stimulus intensity ϕ raised to a power. This may be written

$$\psi = \varkappa \, \phi^\beta$$

where β is the exponent and $\varkappa$ is a constant. A convenient feature of the power law is that in log-log coordinates it describes a straight line whose slope is the exponent. Thus

$$\log \psi = \beta \log \phi + \log \varkappa \, .$$

The results from both sense modalities, vision and hearing, appeared to have the same exponent of about 1/3, provided the stimulus was measured in terms of energy flow. (In terms of sound pressure, the loudness exponent is 2/3.)

The next task was to test the generality of the power principle. Does it hold only for the two main senses, or is it a universal principle? In those days it seemed hardly likely that a single quantitative rule would be found to govern every sense modality, but the exploration of more than three dozen sensory and perceptual continua has failed thus far to uncover a firm exception. In general, each sense

15*

modality has its own exponent, but the value of an exponent may depend on many parameters, such as adaptation, contrast (inhibition), area, duration, and so forth.

Cross-Modality Comparisons

The relations among the exponents of the sensory power functions can be established by a direct comparison from one modality to another. For example, you can place your finger on a vibrator and a pair of earphones on your ears and turn a knob to adjust the loudness until the sound and the vibration seem to be of equal strength. When that kind of cross-modality matching is carried out over a wide range of stimulus intensities, there results a matching function whose exponent is predicted by the ratio between the exponent for loudness and the exponent for vibration (STEVENS, 1959).

Dozens of such cross-modality tests have confirmed the capacity of the power function exponents to define an interconnected net of transitive relations (J. C. STEVENS and MARKS, 1965; STEVENS, 1966a, 1967, 1969). Examples of the matching of four different continua to visual brightness are shown in Fig. 1.

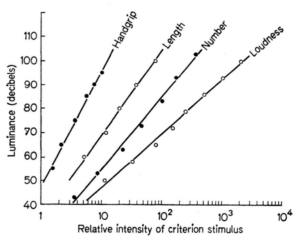

Fig. 1. Equal-sensation functions determined by cross-modality matches between brightness and four other continua. Values on each of the four continua were adjusted to match the brightness of a circular luminous target viewed in a dark room. Each point is the geometric mean of matches made by 10 or more subjects in separate experiments. Subjects squeezed a precision hand dynamometer to match the apparent force to the apparent brightness. Other subjects adjusted the length of a line of light projected on a wall to match the apparent brightness of a target. Numbers were matched to apparent brightness by the method known as magnitude estimation. For the loudness-brightness matches, the subjects adjusted the level of a band of noise to match the apparent brightness of a luminous target. Luminance is measured in decibels re 10^{-10} lambert, which is close to the absolute threshold

In summary, then, the functional operating characteristics of the whole sensory domain can be expressed in one simple and pervasive invariance: equal stimulus ratios produce equal sensory ratios. Measured quantities of neural action ought somehow to manifest a similar invariance if we are to understand the basis of the power law.

Most of the laws of physics, it is interesting to note, also define power functions. The exponents in physical laws are usually integers or simple ratios of integers, whereas the measured values of sensory exponents seldom exhibit any transparent simplicity. Yet we must remain alert to the possibility of simplicity.

Table 1. *Measured exponents and their possible fractional values for power functions relating subjective magnitude to stimulus magnitude*

Continuum	Measured exponent	Possible fraction	Stimulus condition
Loudness	0.67	2/3	3000-Hz tone
Brightness	0.33	1/3	5° Target in dark
Brightness	0.5	1/2	Very brief flash
Smell	0.6	2/3	Heptane
Taste	1.3	3/2	Sucrose
Taste	1.4	3/2	Salt
Temperature	1.0	1	Cold on arm
Temperature	1.5	3/2	Warmth on arm
Vibration	0.95	1	60 Hz on finger
Vibration	0.6	2/3	250 Hz on finger
Duration	1.1	1	White noise stimuli
Finger span	1.3	3/2	Thickness of blocks
Pressure on palm	1.1	1	Static force on skin
Heaviness	1.45	3/2	Lifted weights
Force of handgrip	1.7	5/3	Hand dynamometer
Vocal effort	1.1	1	Vocal sound pressure
Electric shock	3.5	3	Current through fingers
Tactual roughness	1.5	3/2	Feeling emery cloths
Tactual hardness	0.8	4/5	Squeezing rubber
Visual length	1.0	1	Projected line
Visual area	0.7	2/3	Projected square

Table 1 presents a sample of typical measured values together with the values that might be tentatively suggested under the hypothesis that ideal conditions and greater accuracy would produce power functions with simple exponents.

In those sense modalities that must cope with enormous ranges of energy — sometimes exceeding 10^{12} — there is an obvious need for low exponents in order to provide a compressor action. But the rule does not necessarily hold in reverse, as is evidenced, for example, by the sense of smell. There the exponent is less than 1.0 despite the fact that the effective range of stimulus concentrations is comparatively limited. Nevertheless, the low exponents in vision and hearing appear to be nature's way of providing a sufficient nonlinearity to match the input from the outside world to the needs of the central nervous system. By means of a non-linear interface at some point in the system, a billionfold change in light energy, or in sound energy, becomes a thousandfold change in apparent magnitude. The direct central processing of intensity ranges that exceed a billionfold would seem to lie beyond the capacity of the brain. Changes of a few thousandfold appear to lie within bounds.

In some of the sense modalities, no compressor action occurs, and the corresponding exponents in Table 1 have the value 1.0. Other modalities exhibit an expander action, with exponents greater than 1.0. What, one may wonder, would

Fechner have thought of that kind of acceleration in functions where only deceleration was supposed to occur?

Physiological Measures

Although the sensory power law provides no automatic solution to the underlying physiological puzzle, it lays useful guidelines for research on the problem; we now know more explicitly what we are looking for, namely, processes that are consonant, not with logarithmic functions, but with power functions. That is not to say, however, that many physiological functions do not approximate a logarithmic form, for many such instances have been reported. In this context it should be noted that the logarithmic function becomes the limiting case of a power function when the exponent decreases toward zero. Consequently it may prove difficult to distinguish a log function from a power function when the exponent is small.

The sensory receptors have often seemed to act logarithmically when they have been studied by means of electrical recording techniques. Piéron (1952, p. 381) summarized a widespread view, "The intensity of the excitation at the level of the receptor mechanism is proportional to the logarithm of stimulus intensity". That view has led to the question whether a logarithmic stage would be possible in a sensory system even though the overall characteristic followed a power function. As an answer to that question, an instructive model was formulated by MacKay (1963) to show that a logarithmic transducer does not preclude a power-law response. If the output of the transducer goes to a comparator and is there balanced against a centrally generated signal that also approximates a logarithmic function, then the resultant psychophysical relation may be a power function. The basic principle is that two logarithmic functions operating back to back, so to speak, would generate a power function. According to that model, sensation magnitude is regarded as determined by a centrally generated matching response, so that perceived intensity is not so much the inward flow from the sense organ as it is the strength of the internal activity evoked to match, or in some sense to counterbalance, the inward flow. MacKay observed that his hypothesis "requires nothing but approximately logarithmic or linear transformations to take place at receptors or at any point within the nervous system, yet leads directly ... to a prediction of Stevens' power law ...". MacKay further pointed out that "no firm inference to the transfer function of the sensory channel can be drawn from demonstrations that the curve of perceived intensity is best fitted by a power function. If the present model is valid in its simplest form", he added, "it is a mistake to suppose that *any* stage or chain of stages performs a power-law transformation; but in any case it is clear that no stage or chain *need* do so in order to produce Stevens' results".

The advantage served by the foregoing model, and some of the other models addressed to the same problem (e. g. Marimont, 1962), is to alert the experimenter to pitfalls in the hypothesis that a well-placed electrode must necessarily record a function that conforms to the psychophysical power law. There exist many other possible outcomes, as experience has already demonstrated. Nevertheless, the effort to understand the internal dynamics of the sensory systems requires an exploration of the issue: can the sensory systems generate power-law transformations at the physiological level?

Early Studies

An early attempt was made by STEVENS and DAVIS (1936) to relate the cochlear microphonics and the action potentials of the auditory nerve to the scale of subjective loudness. The electrical measurements happened at that time to be noisy, and the loudness function was an early version whose lower end was excessively steep. Partly because of those two deficiencies, the cochlear microphonic appeared to resemble the loudness function closely enough to suggest that, "as a first approximation, the form of the loudness function is imposed by the behavior of the cochlear mechanism". That particular conclusion may be essentially correct, but three decades of research have improved our understanding to the point where it now seems less likely that the loudness function, with its exponent 2/3, can derive in any direct fashion from the cochlear microphonic with its exponent 1.0. In any case, that early excursion into "explanation" has provided the present author with a cautionary directive against premature attempts to anchor psychophysical functions in neurelectric effects.

When the signal leaves the cochlea to ascend the auditory nerve it finds itself subject to the all-or-none principle of nerve transmission, the principle that transformed sensory intensity from a nonproblem into an issue of critical difficulty. A single first-order neuron in the cat responds preferentially to a narrow range of frequencies, and its rate of firing increases with increasing sound pressure over a range of some 20 to 50 db (KIANG, 1965, p. 82). As a steady discharge, the firing rate does not exceed 200 per second. And with added stimulus intensity, the firing rate may decrease from its maximum. Whether human neurons behave differently is not known, but through some such limited pulse-transmitting system there must pass, presumably, all the rich and varied information that we hear.

The loudness of a sound has been thought to correlate with the total number of impulses, but that hypothesis remains an open issue. To be sure, STEVENS and DAVIS (1936) found a rather close correlation between the early version of the loudness function and the response in the auditory nerve. The nerve response, recorded by small concentric (coaxial) electrodes inserted in the nerve, probably represented a sampling of the impulses rather than a full inventory of the active fibers. As has always seemed to be the case, the typical nerve response reached a maximum at what was only a moderate stimulus level, and at higher levels the response declined. A more searching examination of the same basic question was made three decades later by BOUDREAU (1965), this time with stainless steel microelectrodes introduced into the superior olivary complex. There the picture remained much the same as in the auditory nerve. BOUDREAU collected interesting evidence that the growth in the electrical responses is a power function and that the average exponent is similar to that of the loudness function (as listed in Table 1). Nevertheless, the typical growth curve exhibited the usual short range and the ubiquitous turnover at the top. In one exceptional cat, the power function held nicely for 60 db before it turned over.

In the complex domain of the nervous system, it has often been hard for experimenters to see the thing they were not looking for. To illustrate, consider a classic study of the eye of the horseshoe crab *Limulus* — a study that has provided data to support two opposing conceptions. By carefully teasing out a single fiber from the optic nerve, Hartline and Graham (1932) were able to record the all-or-none action potentials produced by the fiber in response to a wide range of light intensity, in some instances as great as 60 db. At the onset of the stimulus there was a burst of rapid activity, which later subsided to a steady value. It was found that the frequency of the impulses in the initial discharge grows according to one function, whereas the frequency of impulses in the steady discharge grows according to a quite different function, as shown in Fig. 2.

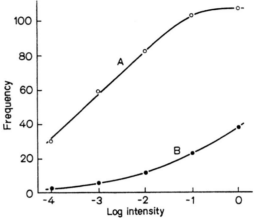

Fig. 2. The relation between frequency of impulses (number per second) and logarithm of the intensity of stimulating light. Intensity in arbitrary units (1 unit = 630,000 meter candles). Curve *A*, frequency of the initial maximal discharge. Curve *B*, frequency of discharge 3.5 seconds after onset of illumination (from Hartline and Graham, 1932). — In these semi-logarithmic coordinates, the straight part of curve *A* describes a logarithmic function. Curve *B* describes a power function with an exponent of approximately 0.25

Different authors have cited one or the other of the curves in Fig. 2 as support for a point of view. Granit (1955) reproduced the figure and noted that the form of the upper curve conforms to the principle that "spike frequency is proportional to the logarithm of the stimulus intensity". From that and other evidence, Granit said, "It seems reasonable to conclude that Fechner did arrive at a sound generalization and its basis is laid down in the receptor mechanism". In a similar vein, Piéron (1951) cited the upper curve in Fig. 2 among pieces of evidence favoring the logarithmic law.

On the other hand, in developing a theory of the brightness response, Diamond (1960) made use of the lower curve in Fig. 2. He showed that a power function provides an excellent fit to the steady response of the optic fiber, and that the frequency of the nerve impulses increases as the 0.25 power of the light intensity. Then by assuming "that psychophysical brightness is directly proportional to the frequency of nerve discharge in the optic nerve", Diamond deduced a power law for apparent brightness — a correct conclusion, perhaps, but from rather tenuous evidence.

It is noteworthy that several years earlier a somewhat similar search for a plausible operating characteristic of the human eye had been made by Hunt

(1953). Having assumed that "we cannot measure sensation magnitudes with any accuracy", HUNT shifted the argument to the physiological level and proposed that a *relative* change in the light intensity must correspond to a *relative* change in the number of nerve impulses. As a consequence of that form of mathematical assumption, the number of impulses would grow as a power function of the light intensity. In order to estimate the value of the exponent, HUNT examined published data relating to cat, frog, and Limulus, and concluded that a reasonable consensus existed to support an exponent of about 1/3.

The Taste Nerve

The sense of taste has provided a unique testing ground for the hypothesis that subjective magnitude is mediated by the total activity in a nerve, or at least that part of the activity that can be measured as a summated electrical response. By a quirk of nature, the gustatory nerve fibers from the anterior part of the tongue pass through the cavity of the middle ear. During certain types of middle ear surgery, this taste nerve, the chorda tympani, may be exposed in a way that permits a direct electrical recording of the response to substances applied to the tongue. The summated neural responses may then be compared with the quantitative estimates of taste intensity made by the same patients for the same substances. A series of such experiments has been performed with the taste substances sucrose, sodium chloride, and citric acid. Both the subjective estimates and the neural responses could be described by power functions. The authors (BORG *et al.*, 1967a, b) summed up their work by saying, "Quite aside from the question whether the function describing the relation between the strength of a

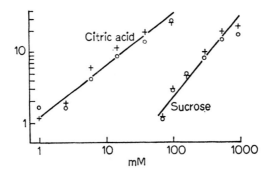

Fig. 3. Mean values of neural response (o o) and of subjective response (+ +) from two patients, plotted against molarity of citric acid and sucrose solution, in log-log coordinates (from BORG *et al.*, 1967a, b)

sapid solution and the summated electrical response satisfies a Stevens power function or a Fechnerian log function, it is apparent that there is a fundamental congruity between neural activity and perceptual intensity". Averaged data from two patients for citric acid and sucrose are shown in Fig. 3.

A demonstration as direct and dramatic as that shown in Fig. 3 may or may not prove reproducible. In other sense modalities anatomy has not favored us with sensory nerves that are so easily accessible to electrodes.

Averaged Cortical Potentials

The development of computer techniques for signal averaging has launched a burgeoning effort to study the cortical potentials that are evoked by brief repetitive stimuli and are recorded from electrodes applied directly to the scalp of a normal conscious subject. The averaging procedure is in effect a cancellation process that suppresses those brain potentials which are more-or-less random. With the noise thus suppressed, the potentials created by a repeated sensory stimulus become distinct and measurable. It was natural that the development of averaging should lead promptly to a study of the relation between stimulus intensity and the resulting evoked potential. Among the early triumphs was the demonstration that the evoked potential from an auditory stimulus could be detected when the stimulus was only a few decibels above the minimum psychophysical threshold (GEISLER et al., 1958).

Although some of the earliest records of evoked potentials appeared to conform to a power law, there were large individual differences among subjects. Furthermore, since the evoked potential is itself a summation of potentials that may have separate origins, it is not obvious what aspect, if any, of the evoked potential should correlate with perceived magnitude.

By measuring the amplitude of one of the slower components (delayed by 130 to 170 msec), KEIDEL and SPRENG (1965) showed that the growth in response to stimulation by a tone was a power function of the stimulus. The same relation held for stimulation by electric current and by vibration, but the three exponents differed. All three exponents were smaller than the corresponding psychophysical values, but it is interesting to note that the relative values of the three exponents were approximately the same as those obtained in psychophysical experiments. A further study of vibration produced families of power functions relating averaged evoked potentials to stimulus amplitude (EHRENBERGER et al., 1966). Five different frequencies of vibration were used. The exponent was smallest at 200 Hz (0.52) and largest at 50 Hz (0.62). In that respect, the relation to frequency was not unlike the relation found in psychophysical experiments (STEVENS, 1968), but the absolute values of the exponents for the evoked potentials were smaller than those obtained in the psychophysical experiments.

Like other evoked potentials, the cortical response to a flash of light is a complex wave that changes form and amplitude when the light intensity is altered. When the potential was measured at an appropriate latency (190—300 msec), the response was found to grow as a power function of light intensity, with an exponent of 0.21 (LOEWENICH and FINKENZELLER, 1967). The power function was shown to hold over a wide stimulus range, 48 db, which is a range of about 65,000 to 1. Approximately the same exponent was obtained with red, blue, green, and white light. Here again, the exponent for the visual evoked potential was smaller than the psychophysical value, but it was consistent in *relative* size with the exponents determined by evoked potentials for sound, vibration, and electric shock obtained by KEIDEL and his collaborators.

It appears, then, that there are at least four modalities in which some particular aspect of the evoked potentials has been shown to grow according to a power function and in which the four exponents exhibit the same *relative* values as those obtained in psychophysical experiments.

Numerous experiments have been carried out by DAVIS and his collaborators to determine the nature and properties of the vertex or V potential — an evoked potential that seems to be generated rather diffusely in the cerebral cortex and is best recorded by an active electrode on the top of the head, with a reference electrode placed near the ear. When the stimuli consist of repeated bursts of tones, the V potential grows slowly with sound pressure, so slowly and variably, in fact, that its path cannot be readily distinguished from a logarithmic growth curve. Power functions fitted to the data for five subjects gave exponents that ranged from 0.10 to 0.18 (DAVIS et al., 1968). Power functions with such low exponents can be distinguished from logarithmic functions only if the variability is extremely well behaved.

In a summary of their work on the maturation of the auditory V potential, DAVIS and ONISHI (1969) distinguished between the physiological maturation of the V potential and the maturation of auditory sensation. They went on to say, "It would be doubly interesting if the V potential did parallel some psychological phenomenon, but apparently it does not. For a time it seemed that the voltage of the V potential might parallel the loudness of the auditory sensation, but the slower growth of the V potential with intensity, its more rapid summation in time, its phasic character and long recovery process, and its great variability are all quite different from the properties of loudness, or of any other known psychological process. The V potential begins about 60 msec after the onset of an auditory stimulus and appears widely over the frontal and parietal lobes. The response is probably a series of overlapping electrical waves that have not yet been effectively separated".

The failure of the cortical V potentials to exhibit growth functions having the same exponents that appear to govern perceived sensory magnitude does not, of course, rule out other interesting comparisons. As noted above, the *relative* values of the exponents for some aspects of the cortical responses may accord with the *relative* values obtained in psychophysical experiments. Another interesting question has been formulated by DAVIS et al. (1968), who asked if stimuli that appear subjectively equal produce similar V potentials. Despite a considerable variability, it appeared that when sounds of different spectra were equated for loudness, they gave rise to approximately equal voltages at the cortex.

The same question can be extended to cross-modality comparisons. If stimuli in different modalities have been equated for apparent magnitude, do they produce the same cortical potential? The answer given by DAVIS et al. (1968) is as follows: "In a set of cross-modality comparisons of V potentials evoked by sounds, flashes of light, vibration, or electric shock ... we found that stimuli that were adjusted to equal subjective magnitude evoked similar V potentials".

Visual Latency

Another aspect of the evoked potential that changes with the intensity of a light flash is latency (TEPAS and ARMINGTON, 1962; VAUGHAN and HULL, 1965). Latency may also be measured in a behavioral experiment, as was done, for example, by LIANG and PIÉRON (1942—43), who made use of the Pulfrich effect to determine how light intensity affects the delay in the visual response. More recently MANSFIELD (1970) employed a conventional reaction time procedure to measure visual latency under a variety of conditions. There seems to be a remarkable agreement between the behavioral measures of latency and those determined by means of evoked potentials. Except for a small and irreducible latent period, the delay in the visual reaction varies with light intensity. The variable portion of the latency decreases as a power function of intensity with an exponent approximately equal to 1/3, which is the same exponent that governs the growth of subjective brightness with intensity. Thus it appears that, to a first approximation, visual latency, measured either behaviorally or by cortical potentials, is inversely proportional to subjective brightness.

Although Piéron, the most erudite psychophysicist of them all, seems never to have accepted the power function as the correct form of the psychophysical law, it is noteworthy that he once entertained a particular hypothesis that would lead to a power law. If, he said, "a strict parallelism were admitted between reaction speed and sensation intensity, then the law governing the variation of the speeds should be taken to govern also that of the perceived intensities". He then asked, "What, therefore, is this law, derived as a reciprocal function, from that which links the reaction latencies to the stimulus intensities"? The answer he gave was that the variable portion of the latency "is inversely proportional to the stimulus intensity raised to a certain power . . ." (Piéron, 1952, p. 352). That close brush with the sensory power law was dismissed by Piéron with the observation that latency and sensation are two different effects, closely related, to be sure, but showing no "direct quantitative dependence one on the other".

Cutaneous Senses

Tactile pulses on the finger produce evoked cortical potentials that have been averaged and compared with psychophysical functions determined with the same subjects (Franzén and Offenloch, 1969). The stimuli were produced by a moving-coil vibrator activated by a brief electrical pulse of 1.5 msec. The subjects made magnitude estimations of the apparent intensity of the pulses, and, on the same day, the potentials evoked by series of pulses were recorded and averaged. It was found that the growth of the overall peak-to-peak amplitude of the evoked potentials as a function of stimulus amplitude could be described by power functions with exponents of approximately 0.5. The magnitude estimations gave exponents of approximately 0.6. Since the method of magnitude estimation under-estimates the size of the exponent, owing to the everpresent regression effect (Stevens and Greenbaum, 1966), it appears that the cortical potential grows with a lower exponent than the experienced intensity.

A similar conclusion was reached by Beck and Rosner (1968) who applied pulses of electric current directly to the skin. The subjects made magnitude estimations of the apparent intensity of the shocks, and the cortical potentials from eight of the subjects were recorded and averaged. Although the exponents for the psychophysical data "were somewhat larger than those for evoked potentials", the authors noted that, of all the formulas they applied to the data, "the power function $[\psi = \varkappa(\phi - \phi_0)^\alpha]$ emerges as the most parsimonious description for results on scaling and on evoked potentials". Beck and Rosner also called attention to the fact that the difference in the values of the exponents suggests that "the electrophysiological and psychophysical responses to intensity of stimulation are nonlinearly related".

That outcome has thus far proved to be the most general finding with averaged evoked potentials: in those rather numerous instances in which the growth of a cortical response has seemed to follow a power function, the value of the exponent has fallen systematically below the corresponding value of the exponent obtained in psychophysical studies. When two power functions differ in exponent, they are nonlinearly related.

Let us consider one more instance. Electrical pulses applied to the tongue produce taste sensations, and at the same time the potentials evoked at the cortex may be recorded and averaged. By measuring appropriate features of the resulting evoked potentials, PLATTIG (1967) found that the amplitude grew as a power function with an exponent of 0.8. Once again, the exponent at the cortex is lower than the psychophysical exponents for either electric current or taste, as they are listed in Table 1.

Neural Responses to Light

Electrical recording directly from nerves and cells can be most easily accomplished with animals, but the implantation of electrodes for the stimulation and recording of neural activity in the human brain is a developing art. An example of the possibilities for sensory investigations with implanted electrodes is presented by PINNEO and HEATH (1966), who recorded from stainless steel depth electrodes located near the left optic tract. The patient made judgments of the apparent brightness of a flickering field. The judgments were consistent with the well-known enhancement effect that occurs when the frequency of the flicker falls in the vicinity of 10 per second.

Corresponding to the brightness enhancement, there was a change in the average amount of electrical activity measured by means of a recording voltmeter. As the frequency of the flicker changed from 1.5 to 50 flashes per second, the recorded voltage passed through a maximum corresponding approximately to a frequency of 10 flashes per second. Above about 35 flashes per second, the voltage was comparable to that produced by a steady light of the same average intensity — a direct neural verification of the Talbot-Plateau law.

The response to steady illumination is often a decrease in the ongoing activity of the visual system. When the activity is recorded with gross electrodes in suitable parts of the visual system, the overall tonic activity decreases with increasing levels of illumination (ARDUINI and PINNEO, 1962; BROOKS, 1964, 1966). A remarkable feature of the decreased neural activity appears to be its quantitative relation to the light intensity: the relation is a power function with an exponent comparable to that observed in psychophysical experiments. ARDUINI and PINNEO are led thereby to the novel conjecture that the level of the tonic activity within the visual system may provide the mechnism for the perception of brightness.

In the single cell, however, a different picture emerges. Single cells in the lateral geniculate of the monkey respond to steady light with a steady rate of discharge. Depending on the type of cell involved, an increment in the light intensity may cause an increase or a decrease in the firing rate (DE VALOIS et al., 1962). The changes were said by the authors to be roughly proportional to the logarithm of the changes in the light intensity, but a replotting of the data in the published recordings shows that the rate of discharge can be described quite well by a power function with an exponent of about 0.4. DE VALOIS (1965, p. 160) has also noted that the data fit a power function.

Such functions, made over relatively short ranges of intensity and response, are afflicted with much variability and consequent uncertainty. It is not surprising, therefore, that tests for the goodness of fit of power and logarithmic functions led to indeterminate results when applied to the data relating light intensity to neuronal discharge in the optic tract and in the lateral geniculate body. The power function was not notably superior to the logarithmic function (CREUTZFELDT et al., 1966).

It has been suggested that the uncertainty of the power function for short ranges and low exponents makes the question of the form of the function a topic suited only for academic argument (KUHNT, 1967). It is indeed apparent that the issue cannot be firmly settled by the present evidence based on potentials recorded in the visual system, but the question still remains: how does the visual system manage to mediate a sensation of brightness that grows as the cube root of the stimulus intensity? Perhaps the ability of some kinds of recorded potentials to define a power function, despite much noise and uncertainty, provides at least a first step toward understanding.

In any case, we must consider the problem raised by CREUTZFELDT and SAK-MANN (1969), who said that "log and power functions with an exponent near 0.3 (0.2—0.4) have an almost identical course over several log units". If "several" is interpreted as three, the relative forms of the two functions, log and power, may

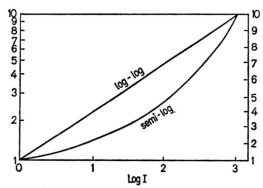

Fig. 4. Showing the forms taken by a power function, exponent 1/3, in two sets of coordinates, log-log and semi-log

be depicted by the lines in Fig. 4. For an assumed exponent of 0.33, the straight line in Fig. 4 shows a power function covering a 10-to-1 change on the ordinate. The curve shows the same function plotted against a linear ordinate (right-hand scale). In most psychophysical experiments, it is extremely easy to decide which of the two ordinate scales produces the straighter line, but in many physiological experiments, the decision is less certain, and the uncertainty increases when either the range or the exponent decreases.

Nevertheless, in several studies of the visual system, the variability has seemed to permit the determination of power functions despite rather low exponents. Experiments with squirrels, Sciurus notatus, produced potentials at the visual cortex that increased as the 0.16 power of the light intensity (VATTER, 1966). The 11 stimulus values covered a range of 2.6 log units, and the response measures defined rather closely a straight line in log-log coordinates.

When the exponents are larger, the power function may appear well determined for even shorter stimulus ranges. HUGHES and MAFFEI (1966) imposed a slow sinusoidal variation on a light stimulus and recorded the impulse responses from retinal ganglion cells of the cat. As the light intensity varied up and down, the number of spikes rose and fell. Over a range of variation of about one log unit, the rate of response followed a power law with an exponent of about 0.7.

A novel approach to the problem of the operating characteristic of the visual system was undertaken by EASTER (1968) who recorded the spike activity produced by single ganglion cells in the goldfish retina in response to brief flashes of red light carefully localized on either one or two retinal areas. The problem was to compare the responses to single-spot and double-spot stimuli. What intensity falling on a single spot is needed to produce a response as large as that produced by light falling on two spots ? The two small spots, it should be said, were equally sensitive points within the same receptor field, in fact, within the same "critical area" as would be defined by Ricco's law. The experimental paradigm employed here is analogous to the procedure of equating binaural to monaural loudness in order to determine the loudness function, given the assumption that the loudnesses in the two ears add. If it is assumed that the stimulation of two equally excited spots in the retina produces twice the value of E (an intervening, hypothetical excitation variable), then it becomes possible to determine how E varies with light intensity. EASTER showed that the excitation function, thus defined, is a power function of intensity with an exponent of about 0.5. It is interesting to note that when the value of the exponent for a point source, or a brief flash, is determined in psychophysical experiments, the exponents turn out also to be about 0.5 (STEVENS, 1966b).

The temptation is great to conclude from the coincidence of exponents that a powerful method for the analysis of the operating characteristic of the visual transducer has at last been formulated by EASTER's splendid experiments, and that the site of the psychophysical power law has been pushed into the retina. Caution must prevail, however, for the abundant richness of current physiological findings do no more, at the present stage of knowledge, than signal directions for future excursions.

Somatic Responses

ROSNER and GOFF (1967, p. 181) made the observation that "MOUNTCASTLE and his co-workers apparently have provided the first major experimental insight into physiological mechanisms underlying the psychophysical power law". That view finds its justification in numerous experiments in which the relation between stimulus magnitude and neural response has been traced in such a way that lawful transformations are shown to govern kinesthetic movement. For example, the relation between joint position and the activity of third-order neural elements can be usefully described by a power function. "What is of additional importance", said MOUNTCASTLE et al. (1962, p. 207) "is that this transformation has occurred so early as the thalamocortical stage of the chain of sequential neural transformations leading to the behavioral response. What it may mean in terms of neural mechanism" is that those neural transformations subsequent to the cortical input stage *may* occur along linear coordinates, so far as the value 'intensity' is concerned".

The same suggestion of linearity in the central transformations recurs in other contexts, for example, in the studies of the neural response to mechanical pressure on the glabrous skin of the monkey's hand (MOUNTCASTLE et al., 1966). The number of impulses in a myelinated axon of the monkey's median nerve was found to grow as a power function of stimulus magnitude, measured as amount of indentation of the skin. The exponent was close to 1.0, which is close to the exponent of 1.1 obtained when human subjects judged the apparent intensity of pressures applied to the palm of the hand (J. C. STEVENS and MACK, 1959).

Interestingly, pressure on human hairy skin does not show the nearly linear response exhibited by the glabrous skin. On the hairy skin of the forearm, the exponent drops to about half the value observed on the glabrous skin of the middle finger. A similar change in the exponent was evident in the neural responses in the monkey. Glabrous skin, as noted above, gave a linear response to degree of indentation whereas hairy skin gave a response characterized by an exponent that averaged about 0.5 (HARRINGTON and MERZENICH, 1970).

How far beyond the receptor the power relation in the neural response can be traced has been said by DARIAN-SMITH (1969) to extend to the level of the medial lemniscal neurons and to the postcentral gyrus.

In a quantitative study of a different stimulus dimension, TAPPER (1965) measured the input-output relations for rate of movement when carefully controlled stimuli were applied to the hairy skin of the cat. The stimulus was produced by an electromechanical driver that could be varied in rate of displacement over a wide range. The frequency of the recorded action potentials was found to increase as a power function of the rate of displacement, with an exponent that ranged from approximately 0.3 to 0.6, depending mainly on the tactile unit stimulated.

Movement applied to visual systems may also produce power-function responses under certain circumstances. Thus STRASCHILL and TAGHAVY (1967) reported that, in some 10 different units of the cat's Tectum opticum, the discharge rate varied with the angular velocity of the stimulus. The exponent averaged about 0.67.

Of course, the occurrence of such power functions may or may not relate to psychophysical functions. That question stands wide open. What the neural power functions demonstrate is a capability: sensory systems are capable of power-function transformations. The precise role of the recorded transformations remains to be determined.

One final observation. At the Ciba Symposium (1966) there was a general discussion on the topic "Linearity of Transmission Along the Perceptual Pathway". In that discussion, and elsewhere at the Symposium, Sir JOHN ECCLES turned forceful attention to the question whether the sense organ could account adequately for the nonlinearity in the coupling between stimulus and sensation, leaving the central nervous system with the task of performing only linear transformations. He observed that "there is no great impediment to the idea that ... the transfer functions across the synaptic mechanism are approximately linear". To which Professor MOUNTCASTLE added, "The interesting point for me here is the great importance that we must now place upon the transducer process itself, at the periphery" (p. 23). Therein lies a critical problem. Is it at the interface between man and his world, at his peripheral sense organs, that the receptor process undergoes the transform that permits the eye and the ear to couple the organism to dynamic ranges of stimuli that exceed billions to one ?

References

ARDUINI, A., PINNEO, L. R.: Properties of the retina in response to steady illumination. Arch. ital. Biol. **100**, 425—448 (1962).

BECK, C., ROSNER, B. S.: Magnitude scales and somatic evoked potentials to percutaneous electrical stimulation. Physiol. Behav. **3**, 947—953 (1968).

BORG, G., DIAMANT, H., STRÖM, L., ZOTTERMAN, Y.: Neural and perceptual intensity. J. Physiol. (London) **191**, 118—119 (1967a).

— The relation between neural and perceptual intensity: a comparative study on the neural and psychophysical response to taste stimuli. J. Physiol. (London) **192**, 13—20 (1967b).

BOUDREAU, J. C.: Stimulus correlates of wave activity in the superior-olivary complex of the cat. J. Acoust. Soc. Amer. **37**, 779—785 (1965).

BROOKS, B. A.: Neural correlates of brightness discrimination in the squirrel monkey *(Saimiri sciureus)*. Ph. D. thesis. Florida State University (1964).

— Neurophysiological correlates of brightness discrimination in the lateral geniculate nucleus of the squirrel monkey. Exp. Brain Res. **2**, 1—17 (1966).

Ciba Foundation Symposium, Touch, Heat and Pain. A. V. S. DE REUCK and J. KNIGHT, eds. London: Churchill 1966.

CREUTZFELDT, O., FUSTER, J. M., HERZ, A., STRASCHILL, M.: Some problems of information transmission in the visual system. In: Brain and conscious experience. J. C. ECCLES, ed. pp. 138—164. New York: Springer 1966.

— SAKMANN, B.: Neurophysiology of vision. Ann. Rev. Physiol. **31**, 499—544 (1969).

DARIAN-SMITH, I.: Somatic sensation. Ann. Rev. Physiol. **31**, 417—450 (1969).

DAVIS, H., BOWERS, C., HIRSH, S. K.: Relations of the human vertex potential to acoustic input: loudness and masking. J. Acoust. Soc. Amer. **43**, 431—438 (1968).

— ONISHI, S.: Maturation of auditory evoked potentials. Int. Audiol. **8**, 24—33 (1969).

DE VALOIS, R. L.: Behavioral and electrophysiological studies of primate vision. In: Contributions to sensory physiology, Vol. 1. W. D. NEFF, ed. pp. 137—178. New York: Academic Press 1965.

— JACOBS, G. H., JONES, A. E.: Effects of increments and decrements of light on neural discharge rate. Science **136**, 986—988 (1962).

DIAMOND, A. L.: A theory of depression and enhancement in the brightness response. Psychol. Rev. **67**, 168—199 (1960).

EASTER, S. S., JR.: Excitation in the goldfish retina: evidence for a non-linear intensity code. J. Physiol. (London) **195**, 253—271 (1968).

EHRENBERGER, K., FINKENZELLER, P., KEIDEL, W. D., PLATTIG, K. H.: Elektrophysiologische Korrelation der Stevensschen Potenzfunktion und objektive Schwellenmessung am Vibrationssinn des Menschen. Pflügers Arch. ges. Physiol. **290**, 114—123 (1966).

FECHNER, G. T.: Elemente der Psychophysik, 1860 (Vol. I: Available in English translation as Elements of Psychophysics. New York: Holt, Rinehart and Winston, 1966).

FRANZÉN, O., OFFENLOCH, K.: Evoked response correlates of psychophysical magnitude estimates for tactile stimulation in man. Exp. Brain Res., 1969, **8**: 1—18.

GEISLER, C. D., FRISHKOPF, L. S., ROSENBLITH, W. A.: Extracranial responses to acoustic clicks in man. Science **128**, 1210—1211 (1958).

GRANIT, R.: Receptors and sensory perception. New Haven: Yale University Press, 1955.

HARRINGTON, T., MERZENICH, M. M.: Neural coding in the sense of touch. Exp. Brain Res. **10**, 251—254 (1970).

HARTLINE, H. K., GRAHAM, C. H.: Nerve impulses from single receptors in the eye. J. cell. comp. Physiol. **1**, 277—295 (1932).

HUGHES, G. W., MAFFEI, L.: Retinal ganglion cell response to sinusoidal light stimulation. J. Neurophysiol. **29**, 333—352 (1966).

HUNT, R. W. G.: Characteristic curves of the human eye. J. Photo. Sci. **1**, 149—158 (1953).

KEIDEL, W. D., SPRENG, M.: Neurophysiological evidence for the Stevens power function in man. J. Acoust. Soc. Amer. **38**, 191—195 (1965).

KIANG, N. Y.-S.: Discharge patterns of single fibers in the cat's auditory nerve. Cambridge, Mass.: The M. I. T. Press 1965.

Kuhnt, U.: Visuelle Reaktionspotentiale an Menschen und Katzen in Abhängigkeit von der Intensität. Pflügers Arch. ges. Physiol. **298**, 82—104 (1967).

Liang, T., Piéron, H.: Recherches sur la latence de la sensation lumineuse par la methode de l'effet chronostéréoscopique. Ann. psychol. **43—44**, 1—53 (1942—43).

v. Loewenich, V., Finkenzeller, P.: Reizstärkeabhängigkeit und Stevenssche Potenzfunktion beim optisch evozierten Potential des Menschen. Pflügers Arch. ges. Physiol. **293**, 256—271 (1967).

MacKay, D. M.: Psychophysics of perceived intensity: a theoretical basis for Fechner's and Stevens' laws. Science **139**, 1213—1216 (1963).

Mansfield, R. J. W.: Intensity relations in vision: analysis and synthesis of a nonlinear sensory system. Ph. D. thesis. Harvard University (1970).

Marimont, R. B.: Model for visual response to contrast. J. Opt. Soc. Amer. **52**, 800—806 (1962).

Mountcastle, V. B., Poggio, G. F., Werner, G.: The neural transformation of the sensory stimulus at the cortical input level of the somatic afferent system. In: Information processing in the nervous system. R. W. Gerard and J. W. Duyff, eds. pp. 196—217. Amsterdam: Excerpta Medica 1962.

— Talbot, W. H., Kornhuber, H. H.: The neural transformation of the mechanical stimuli delivered to the monkey's hand. In: Ciba Foundation Symposium, Touch, heat and pain. pp. 325—345. London: Churchill 1966.

Piéron, H.: Les problèmes fondamentaux de la psychophysique dans la science actuelle. Actualités scientifiques et industrielles, No. 1136. Paris: Hermann 1951.

— The sensations, their functions, processes, and mechanisms. New Haven: Yale University Press 1952.

Pinneo, L. R., Heath, R. G.: Electrophysiology of the human visual system in perception of flicker, fusion, and brightness. Bull. Tulane Univ. Med. Fac. **25**, 255—266 (1966).

Plattig, K. H.: Subjektive Schwellen- und Intensitätsabhängigkeitsmessungen am elektrischen Geschmack. Pflügers Arch. ges. Physiol. **294**, 76 (1967).

Rosner, B. S., Goff, W. R.: Electrical responses of the nervous system and subjective scales of intensity. In: Contributions to sensory physiology, Vol. 2. W. D. Neff, ed. pp. 169—221. New York: Academic Press 1967.

Stevens, J. C., Mack, J. D.: Scales of apparent force. J. exp. Psychol. **58**, 405—413 (1959).

— Marks, L. E.: Cross-modality matching of brightness and loudness. Proc. nat. Acad. Sci. (Wash.) **54**, 407—411 (1965).

Stevens, S. S.: On the brightness of lights and the loudness of sounds. Science **118**, 576 (1953).

— Cross-modality validation of subjective scales for loudness, vibration, and electric shock. J. exp. Psychol. **57**, 201—209 (1959).

— Matching functions between loudness and ten other continua. Percep. Psychophysics **1**, 5—8 (1966a).

— Duration, luminance, and the brightness exponent. Percep. Psychophysics **1**, 96—100 (1966b).

— Intensity functions in sensory systems. Int. J. Neurol. **6**, 202—209 (1967).

— Tactile vibration: change of exponent with frequency. Percep. Psychophysics **3**, 223—228 (1968).

— On predicting exponents for cross-modality matches. Percep. Psychophysics **6**, 251—256 (1969).

— Davis, H.: Psychophysiological acoustics: pitch and loudness. J. Acoust. Soc. Amer. **8**, 1—13 (1936).

— Greenbaum, H.: Regression effect in psychophysical judgment. Percep. Psychophysics **1**, 439—446 (1966).

Straschill, M., Taghavy, A.: Neuronale Reaktionen im Tectum opticum der Katze auf bewegte und stationäre Lichtreize. Exp. Brain Res. **3**, 353—367 (1967).

Tapper, D. N.: Stimulus-response relationships in the cutaneous slowly-adapting mechanoreceptor in hairy skin of the cat. Exp. Neurol. **13**, 364—385 (1965).

Tepas, D. I., Armington, J. C.: Properties of evoked visual potentials. Vision Res. **2**, 449—461 (1962).

Vatter, O.: Das photisch evozierte Rindenpotential und die optische Sensitivität des Platanenhörnchens (Sciurus notatus). Vision Res. **6**, 61—81 (1966).

Vaughan, H. C., Hull, R. C.: Functional relation between stimulus intensity and photically evoked cerebral responses in man. Nature (Lond.) **206**, 720—722 (1965).

Chapter 8

Generation of Responses in Receptor

By

M.G.F. Fuortes, Bethesda, Md. (USA)

With 17 Figures

Contents

Receptors are specialized nerve cells and share with them a number of important properties. Therefore, a brief description of the general features of nerve cells may be useful as an introduction to the study of receptors. A diagram of a nerve cell and its connections is shown in Fig. 1. The cell includes a dendritic tree, the

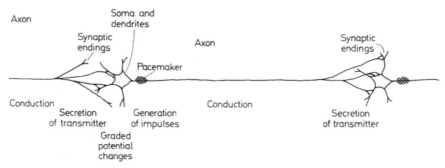

Fig. 1. Diagram of nerve cells. The synaptic endings of nerve cells are supposed to be secretory elements which release a "transmitter" substance when their membrane is depolarized. The transmitter evokes a change of permeability of the postsynaptic membrane which may produce depolarization or hyperpolarization of the postsynaptic cell. Impulses are usually generated at the initial portion of the axon (cross-hatched area)

soma, the axon and terminal branches. In vertebrates, synaptic contacts occur mostly at the soma and dendrites; in invertebrates they may be located at the axon. In both vertebrates and invertebrates, dendrites may be absent. The axon can be over 1 m long, but in many cells it is so short (about 1 mm) that it cannot be easily distinguished from the dendritic branches (GOLGI, 1883).

1. Properties of Nerve Cells

It was once thought that membrane properties were essentially uniform throughout the extension of a nerve cell, and the nerve impulse was considered to be the only important sign of activity of a neuron (GASSER, 1939). More recently it was recognized that different parts of the neuronal membrane have essentially different properties (see GRUNDFEST, 1957) and that the all-or-none nerve impulse is only one of the several types of response of a nerve cell. It is now generally accepted that the synaptic endings release a transmitter substance when their membrane potential is decreased (FATT and KATZ, 1951; ECCLES, 1952 DEL CASTILLO and KATZ, 1954). The depolarization is frequently brought about by the arrival of nerve impulses at the terminals, but it is important to realize that nerve impulses are not essential and that transmitter release can occur as a consequence of other types of depolarization (DEL CASTILLO and KATZ, 1954; LILEY, 1956; KATZ and MILEDI, 1967).

The transmitter substance acts as a stimulus for the subsynaptic regions of the membrane, and produces a change of their permeability to ions. Excitatory transmitters increase permeability to Na^+ and other ions and thereby decrease membrane potential (FATT and KATZ, 1951; COOMBS, et al., 1955b; TAKEUCHI and TAKEUCHI, 1960); inhibitory transmitters increase permeability to ions such as K^+ or Cl^- and thereby stabilize or increase membrane potential (FATT, 1960; COOMBS, et al., 1955a). Production of an electrical change as a consequence of a chemical stimulus can be regarded as a transducer action, and in this sense the properties of the subsynaptic membrane of a central nervous system neuron are analogous to those of a receptor membrane.

The subsynaptic membrane differs from the membrane of the axon because it does not have the properties of regeneration and inactivation (HODGKIN, 1951; HODGKIN and HUXLEY, 1952) which lead to the nerve impulse. Therefore, it produces graded and sustained depolarizations rather than the explosive and transient potential changes characteristic of impulse activity (see GRUNDFEST, 1957, 1959). The potential change evoked by synaptic activity spreads to a special region where impulses are generated following depolarization. This region is frequently located in the initial part of the axon (ECCLES, 1957; FUORTES, et al., 1957). The properties of this region — which may be called the *pacemaker* of the neuron — are essentially the same as those of the axon. Accommodation to long-lasting depolarizations, however, may be slower at the pacemaker, permitting development of sustained trains of impulses following prolonged depolarization (ADRIAN, 1928). Impulses produced in this manner are conducted without decrement to the terminals of the axon where they evoke transmitter release as dexcribed above.

When the axon is very short, the synaptic or generator potential can spread effectively to the terminals and cause transmission without the intermediary of nerve impulses (Bush and Roberts, 1968). Although direct evidence is still lacking, it is quite possible that many short-axon neurons function without ever producing nerve impulses.

2. Definiton of Receptors

According to this description, transducer actions are not a specific feature of receptors but are common to all neurons. The characteristic feature of receptors is that they are normally activated by specific changes of their surroundings such as temperature, light, pressure, etc., rather than by synaptic activity as is the case for other neurons. Receptors can, therefore, be defined as nerve cells which are normally activated by stimuli other than synaptic activity.

3. Morphology of Receptor Cells

The morphology of receptors varies extensively with relation to the function subserved and in different animal species. In some cases (see Fig. 2), the specific receptive membrane is connected to the synaptic ending by long processes. In these cells (typically represented by the receptors innervating skin, muscles, joints

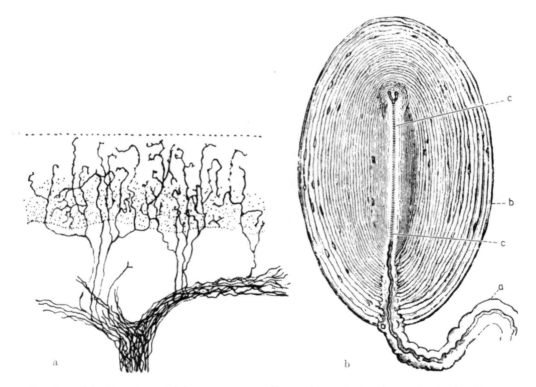

Fig. 2a and b. Receptors with long processes. a Free endings of a touch receptor in the skin of cat's paw (Ramon y Cajal, 1909, 1911). b Pacinian corpuscle (pressure receptor) in the mesentery of cat. The receptive terminal c is surrounded by a capsule b formed by concentric laminae, and continues in a long afferent fiber a (Sobotta, 1928). In both cases, the cell body is situated in the dorsal root ganglia

and viscera of vertebrates), transmission of information from the receptive membrane to the synaptic ending clearly requires nerve impulses since the distance separating the two membranes is too long for effective transmission by decremental conduction.

Other receptor cells (such as the visual, vestibular, auditory or gustatory receptors illustrated in Fig. 3) are short (less than 1 mm) and do not possess either dendritic aborizations or a well-recognizable axon. In these cells, as in other nerve

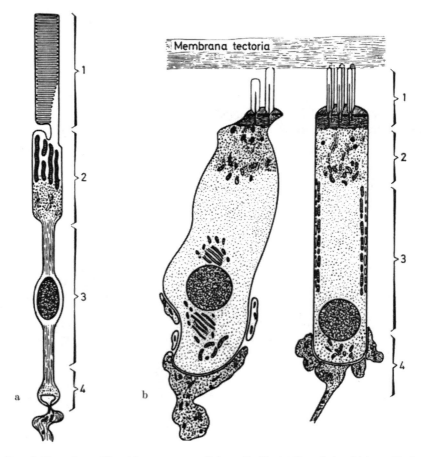

Fig. 3a—d. Receptors without long processes. Schematic illustration of visual (a), vestibular (b), auditory (c) and gustatory (d) receptors. These cells include (1) a specialized receptive area; (2) a region rich in mitochondria and other cytoplamatic organelles; (3) an area with clearer cytoplasm, containing the nucleus; and (4) a synaptic area. The receptive region (1) differs characteristically for different functions (from Bairati, 1961)

cells with short axons, transmission may occur without nerve impulses (see Grundfest, 1961, 1965). In a discussion on auditory receptors, Davis (1965, p. 185) summarizes this view as follows:"In the hair cells of the cochlea there are no spikes or all-or-none nerve impulses. Several investigators, from Békésy to

TASAKI onward, have placed microelectrodes in hair cells and not one of them has ever reported an axon-like spike. The hair cell has no axon, and we can reasonably assume that its receptor potential acts directly on the synaptic junctions at the base of the hair cell to cause liberation of chemical mediator in the same way that the nerve impulse acts on the presynaptic terminals of a neuron."

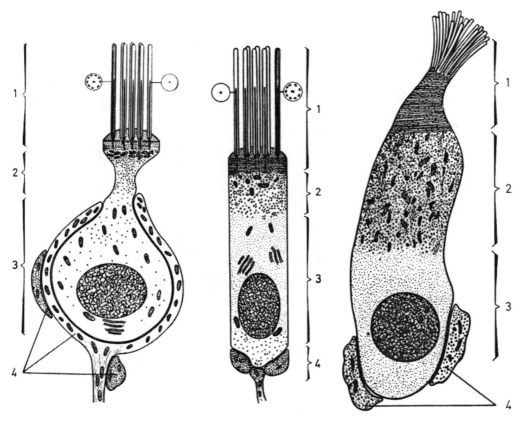

Fig. 3 c   Fig. 3 d

The receptive membrane itself is structually similar to other cell membranes; it consists of a sheet about 70 Å thick in which three layers can be resolved by electron microscopy (Fig. 4).

The arrangement of the receptive membrane differs for different functions subserved. In some mechanical receptors it takes the form of free endings which are rather similar to the dendritic arborization of central neurones (Fig. 2). In other cases, specific forms of differentiation are found. For instance, the receptive membrane of visual cells presents elaborate infoldings which give rise to discs, sheets or microvilli, such as those illustrated in Fig. 4. Tight membrane contacts (called "tight junctions") are quite frequent in these structures (LASANSKY, 1967). This configuration increases membrane surface and may have the purpose of supporting a large number of molecules of visual pigment.

In auditory, vestibular and lateral-line receptors, the membrane is associated with cilia which presumably have the function of transmitting vibrations (Fig. 3).

In other receptor cells, the membrane is surrounded by accessory structures which form the "corpuscles" of classical histology. A typical example is the

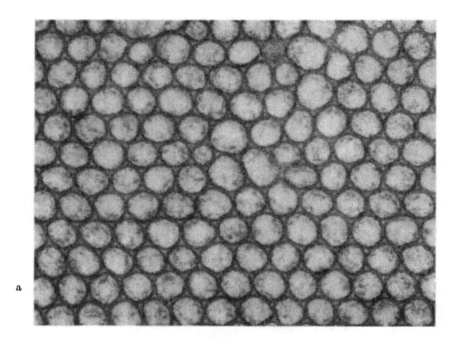

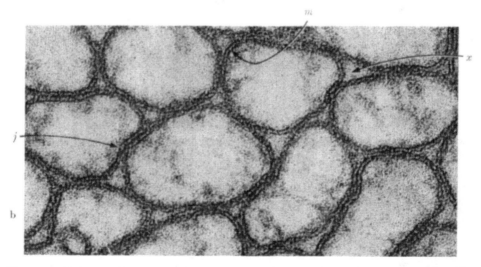

Fig. 4a and b. Microvilli in visual receptors of *Limulus*. Linear magnification is 75,000 in a and 250,000 in b. Cross section of microvilli shows hexagonal symmetry, membrane contacts or tight junctions (*j*), and small extracellular channels (*x*). The membrane (*m*) has the same appearance as other cell membranes (Lasansky, unpublished)

Pacinian corpuscle (Fig. 2) which encapsulates the sensitive nerve endings and renders them specifically sensitive to changes of pressure (see Fig. 12).

4. Nerve Impulses Generated by Receptors

Pioneering investigations on the signals generated by receptor cells have been performed by ADRIAN, (1928, 1931, 1947) who recorded impulse discharges from single fibers originating from different types of receptors (Fig. 5a). Using this

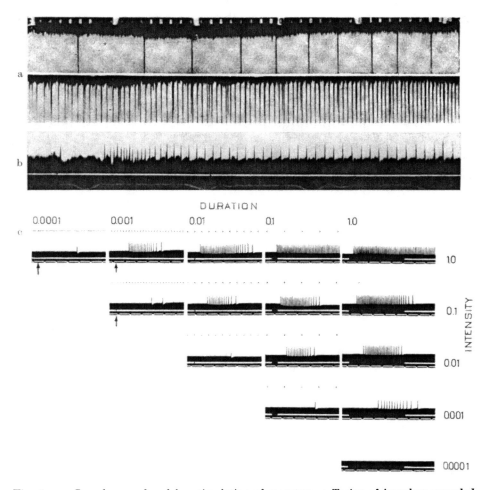

Fig. 5a—c. Impulses produced by stimulation of receptors. a Trains of impulses recorded from dorsal root fibers of cat. The responses were evoked by pressure on the paw: weak pressure produced the low-frequency discharge shown in the upper record; with stronger pressure (lower record) frequency reached almost 100 imp/sec and a second nerve ending started to fire impulses at lower frequency. Time marks: 0.25 sec (ADRIAN, 1928, p. 35). b Impulses recorded from fibers originating in stretch receptors of frogs. Time marks: 0.2 sec (MATTHEWS, 1931). c Discharges recorded from a small filament of the optic nerve of *Limulus*. The responses were evoked by flashes of light of different intensities and duration. The periods of illumination are signalled by breaks in the white line under each record. Time marks: 0.2 sec (HARTLINE, 1934). The photographs in this figure are among the first published records of unitary activity of receptors

method, Adrian reached the conclusion that "the impulses in sensory fibers are of the same type as those ... in motor fibers ... and there is the same all-or-nothing relation between the stimulus and the impulse ... A preparation containing only one end organ can be made from one of the small cutaneous muscles in the frog and it is found that the impulses produced by stimulating this end organ recur at regular intervals ... The frequency varies with the strength of the stimulus and the time over which it has been in action ... The production of the rhythmic discharge may be explained by regarding the end organ as a structure having the same properties of the nerve fiber, but differing in its rate of recovery and in its rate of adaptation to the stimulus." (Adrian, 1928, pp. 70—71.) In discussing the information that the sensory message can convey to the central nervous system, Adrian concludes that "the arrival of a discharge in a particular sensory nerve fiber will mean of course that a particular receptor has been excited and this will

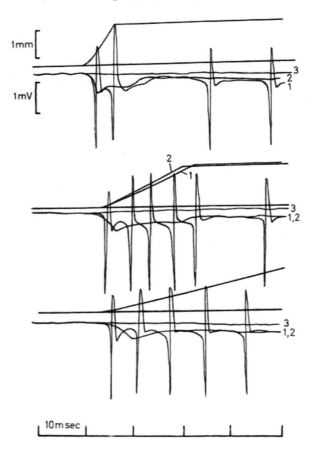

Fig. 6. Generator potential of the stretch receptor of frog. Potentials were recorded by means of external electrodes located near the origin of the sensory fiber. Each set of tracings shows the time course of the stimulus and three superimposed records. Record *(1)* taken from the normal preparation shows sustained depolarization (generator potential) and repetitive spikes; after application of 0.35% procaine *(2)* the spikes are abolished but the generator potential persists; the generator potential disappears after crushing the nerve *(3)* (Katz, 1950)

imply a particular form of a stimulus and a particular locality ... The accuracy of localization will naturally depend on the number of receptors of the same type in the area which contains the stimulated region — or rather it will depend on the number of nerve fibers which lead from these receptors. If there is much overlapping in the area supplied by the terminal branches of different fibers, it might be possible to localize the stimulus more exactly by comparing the relative intensities of discharge in the different fibers." (ADRIAN, 1928, pp. 91—92). ADRIAN'S conclusions were confirmed by later research including the elegant studies by MATTHEWS (1931) on stretch receptors of frogs and HARTLINE (1934) on visual receptors of *Limulus* (Fig. 5 b and c).

5. Generator Potentials

It was presumed that the sensory stimulus evokes depolarization of the peripheral terminals, but the experimental demonstration of occurrence of this potential change came only several years later when KATZ (1950), recording from the sensory fiber in the immediate vicinity of the stretch receptor of frog, showed that the impulse discharge was generated by a graded and sustained depolarization of the sensory terminals (Fig. 6). This depolarization became known as the "generator potential" (a term introduced by BERNHARD and GRANIT, 1946) or "receptor potential" (DAVIS, 1961).

Systematic studies on generator potentials were performed soon after KATZ'S (1950) early work, either using extracellular electrodes positioned in the immediate vicinity of the sensory terminals or with the aid of intracellular micropipettes. The first method usually requires delicate microdissection; the second is simpler and for many purposes more advantageous, but it can be applied only to structures large enough to be penetrated by micropipettes without irreparable damage. Intracellular techniques are, therfore, unsuitable for the study of the potentials arising in fine terminals such as those of the stretch receptors of vertebrates, but they can be conveniently applied to receptors where the transducer membrane is immediately adjacent to the cell soma, since the soma can be penetrated by microelectrodes without apparent injury. The first intracellular studies of electrical responses of receptors were performed by HARTLINE et al., (1952) on visual cells of *Limulus* and by EYZAGUIRRE and KUFFLER (1955a, b) on the stretch receptor of lobster and crayfish.

In agreement with the results obtained by KATZ (1950), it was found that sensory stimuli produce depolarization of the cell membrane and that nerve impulses arise if the depolarization attains sufficient amplitude (Fig. 7).

6. Hyperpolarizing Generator Potentials

Similar depolarizing generator potentials were recorded later from a variety of receptor cells. In some receptors, however, different results have been reported. BORTOFF (1964), TOMITA (1965), KANEKO and HASHIMOTO (1967), WERBLIN and DOWLING (1969) and BAYLOR and FUORTES (1970) have presented evidence showing that the cones in the retina of vertebrates are hyperpolarized by illumination (Fig. 8). Cones apparently have a low (30 to 40 mV) membrane potential and the internal negativity is increased by illumination. Since the response of the cones must be com-

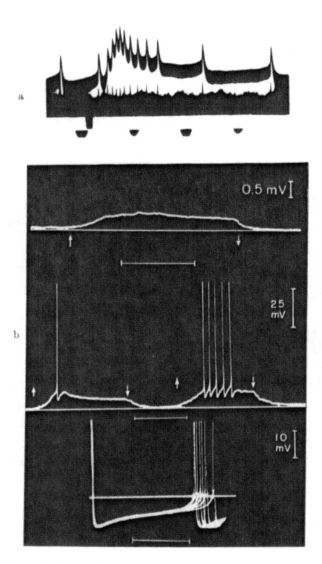

Fig. 7a and b. Intracellular records of generator potentials. a Potentials recorded from a visual cell of *Limulus* by an intracellular micropipette (upper trace) and from optic nerve fibers by means of external electrodes (lower trace). Activity was evoked by a flash of light, indicated by the signal under the records. Upward deflection of the top trace indicates depolarization of the cell. Spikes are attenuated due to low frequency response of the recording system. Time marks: 0.2 sec. This is probably the first published intracellular record of a generator potential (HARTLINE, *et al.* (1952). b Intracellular records of responses of the stretch receptor of the crayfish. Start and end of stimulus are indicated by arrows. A weak stimulus (top) evokes a sustained generator potential but no spikes; stronger stimuli (middle) produce larger generator potentials and nerve impulses. Time lines: 1 sec. In the superimposed records at bottom repetitive firing was produced by prolonged constant stretch and the sweep of the cathode ray oscilloscope was started in coincidence with one of the spikes. It is seen that intervals between successive spikes are approximately constant and that in these conditions all spikes originate at the same membrane potential indicated by the horizontal line. Time line: 0.1 sec. (from EYZAGUIRRE and KUFFLER, 1955)

municated to other cells of the retina, these findings suggest that in the cones transmission is associated with hyperpolarization. As GRUNDFEST (1958, 1961) pointed out several years ago, it is possible that some synaptic endings release a transmitter substance when they are hyperpolarized.

Alternatively, one might suppose that a signal is continuously released in darkness and is interrupted or decreased during illumination. Recent experiments

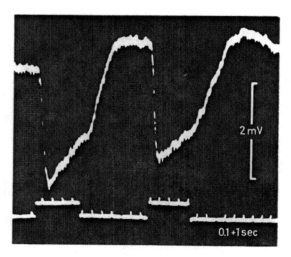

Fig. 8. Hyperpolarizing generator potentials in the cones of fish. Potentials were led off by a micropipette inserted in the inner segment of a cone (histological controls performed by KANEKO and HASHIMOTO, 1967). Illumination indicated by the upward deflection of the lower beam. The light was localized to a small area of the retina for the first record and covered a larger area for the second (TOMITA, 1965)

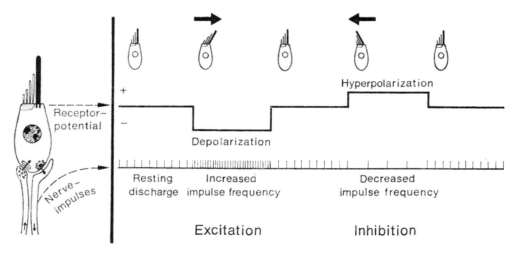

Fig. 9. Diagram of potentials presumed to occur in hair cells of the lateral line. It is supposed that displacement of the cilia in one direction produces depolarization of the hair cell and increase of firing frequency of the second-order neuron. Displacement in the opposite direction is presumed to produce hyperpolarization and decrease of frequency (FLOCK, 1965)

have shown that the hyperpolarizing responses of cones are accompanied by decrease of membrane conductance (BORTOFF and NORTON, 1967; TOYODA et al., 1969; BAYLOR and FUORTES, 1970). Hyperpolarizing responses to light also occur in visual cells of the scallop (TOYODA and SHAPLEY, 1967; GORMAN and McREYNOLDS, 1969) but in this case the generator potential is associated with an increase of membrane conductance (TOYODA and SHAPLEY, 1967; McREYNOLDS and GORMAN, 1970a and b).

It has also been suggested (FLOCK and WERSALL, 1962; FLOCK, 1965) that the hair cells of the lateral line of fish are depolarized by bending their cilia in one direction and hyperpolarized by bending in the opposite direction. As a conse-

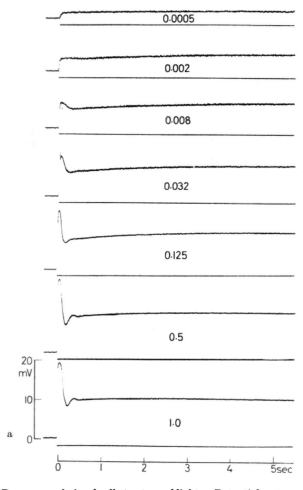

Fig. 10a and b. Responses of visual cells to steps of light. a Potentials were recorded by means of an intracellular micropipette inserted in a photoreceptor in the eye of the bee. The stimulus was a step of light starting at time 0. Light intensity in arbitrary units indicated by the figures near each record. With the dimmer lights, the response rises monotonically to an approximately constant level; with brighter lights, the transient phase includes a peak and damped oscillations (from BAUMANN, 1968). b A similar experiment performed on a visual cell of *Limulus*. In this case, spikes are superposed on the generator potential. 8.4 seconds elapse between the two sections of each record (from FUORTES and POGGIO, 1963)

quence, firing of the sensory cells making contact with these receptors is increased in one case and decreased in the other (Fig. 9). One may suppose either that an excitatory signal is continuously released by the hair cell when the cilia are in their resting position and that this signal is increased or decreased by bending in opposite directions, or that no signal is produced by the hair cell at rest and two different signals (for instance, two different transmitters) are produced by the two directions of bending. Interesting as these suggestions may be, it should be noted that they are highly speculative at this time.

7. Depolarizing Generator Potentials

In the majority of receptor cells investigated so far, the membrane presents normal resting potential in the absence of stimulation and becomes depolarized following application of stimuli. Some features of these depolarizing generator potentials are common to many receptors and will be briefly described in this section,

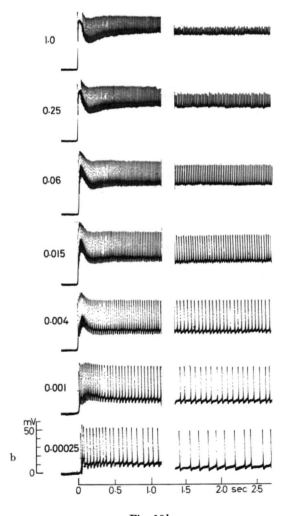

Fig. 10b

taking as an illustration responses produced by stimuli in the form of a step. Fig. 10 shows responses recorded from visual cells following illumination at different intensities. With dim lights the response to steps grows gradually (transient) and remains constant thereafter (steady state); with brighter lights the response grows rapidly at the beginning and later decays to an approximately steady level. Oscillations may be apparent before steady level is reached. These general characteristics of responses have been observed in a variety of "slowly adapting" receptors (see ADRIAN, 1928), as one may see in Fig. 11. Other receptors ("rapidly adapting") give instead only a transient response to a sustained stimulus; responses of this type are characteristic of touch and pressure receptors (ADRIAN, 1928, 1931). In the Pacinian corpuscle, rapid adaptation is due in large part to the properties

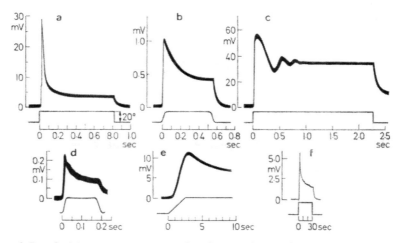

Fig. 11 a—f. Depolarizing generator potentials. a Potentials recorded from a mechanoreceptor associated with the hair of the bee. Bending of the hair is recorded in the lower trace. The record in the upper trace was led off by a fine pipette located near the receptor ending (retraced from THURM, 1965). b Record from the stretch receptor of the frog. A single spindle was dissected from the muscle and potentials were led off from the afferent fiber in the immediate vicinity of the receptor. Stretch of the spindle is signalled by the lower trace (retraced from OTTOSON and SHEPHERD, 1965). c Response of a photoreceptor in the eye of the Dragonfly. Intracellular record. A step of bright light (lower trace) produced a transient peak decaying with oscillations to a constant level (FUORTES, unpublished). d Response of a decapsulated Pacinian receptor (see Fig. 12). Record taken from the afferent nerve in the vicinity of the receptor ending. Pressure (lower trace) was applied by means of a piezoelectric crystal. e Response of the stretch receptor of the crayfish. Intracellular record of the generator potential evoked by stretch, after application of tetrodotoxin in concentration 5×10^{-6}. The drug abolishes nerve impulses but does not affect the generator potential (retraced from LOEWEN-STEIN et al., 1963). f Potentials recorded by a pipette in contact with the olfactory epithelium of the turtle. A solution of amylacetate was applied to the olfactory mucosa for 30 sec as indicated by the lower trace (retraced from TUCKER and SHIBUYA, 1965)

of the capsule surrounding the receptive membrane. During prolonged compression the deeper laminae of this capsule are deformed only transiently (HUBBARD, 1958) and the generator potentials decays rapidly (GRAY and SATO, 1953; LOEWEN-STEIN, 1959). If the capsule is removed, however, the generator potential is prolonged as shown in Fig. 12 (MENDELSON and LOEWENSTEIN, 1964).

8. Input-output Analysis

Fig. 10, 11, and 12 illustrate responses to a particular type of stimulus — a step function. Simply by examining these records we may reach some limited conclusions on the properties of the receptors considered. A more complete understanding of these properties could, however, be obtained if one could describe the *general* relations between stimulus and response rather than the particular response to a particular stimulus.

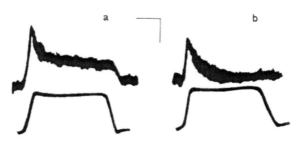

Fig. 12. Responses of Pacinian corpuscle. The normal corpuscle gives only a transient response to a prolonged stimulus (pressure). When the capsule is removed, however, a sustained response is obtained as shown in a. Record b was taken after assembling an artificial capsule around the receptor ending. Sustained pressure on this capsule evoked only a transient response resembling the response of the normal receptor. Calibration: 5 msec and 50 μV. (from LOEWENSTEIN and MENDELSON, 1965)

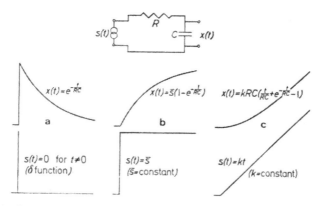

Fig. 13. A simple electric circuit as an example of input-output systems. A voltage generator produces the input function $s(t)$. The response of the system is the voltage $x(t)$ at the output. The general relation between input and output is given by Eq. (2). From this relation, it is possible to calculate the output produced by any input: the plots show responses to inputs in the form of a unit impulse (DIRAC's, 1947, δ-function), of a step and of a ramp

In order to reach this generalization it is useful to apply to the study of receptors some of the notions of "input-output analysis" (see KAPLAN, 1962, Chapters 3 and 5) and of model-building (see BERMAN, 1963). As an introduction to this type of study, we may consider the simple electrical circuit of Fig. 13, in which $s(t)$ is the voltage of the input or stimulus and $x(t)$ is the output or response. The properties of the components are known, and it is also known (from KIRCHHOFF's

Law) that the current through the resistance R is equal to the current through the condenser C:

$$\frac{s-x}{R} = C\frac{dx}{dt}. \tag{1}$$

Rearranging terms we obtain a general relation between input and output in the form:

$$RC\frac{dx}{dt} + x = s. \tag{2}$$

From this relation we can calculate the output produced by any input. Solutions for inputs in the form of an impulse ($s = 0$ when $t \neq 0$ and $\int_{-\infty}^{\infty} s\, dt = 1$), of a step ($s = $ constant) or of a ramp ($s = kt$) are illustrated in Fig. 13. With more complex systems, the input-output relations become correspondingly more complicated: a general expression applicable to a large variety of physical systems[1] is given by the equation:

$$A_0\frac{d^n x}{dt^n} + A_1\frac{d^{n-i} x}{dt^{n-i}} + \ldots + A_{n-1}\frac{dx}{dt} + A_n x = B_0\frac{d^m s}{dt^m} +$$

$$+ B_1\frac{d^{m-1} s}{dt^{m-i}} + \ldots + B_{m-1}\frac{ds}{dt} + B_m s \tag{3}$$

where $s(t)$ is the input, $x(t)$ is the output, and the coefficients A_i (are constants Eq. (2) is a special case of this general equation.

In stable systems and for inputs of the type shown in Fig. 13 the derivatives of x vanish or become constant for large values of t and the output $x(t)$ becomes linearly related to the input $s(t)$ when t is large. The output will then consist of a *transient* phase, during which it deviates from the input and a *steady-state* phase during which it is a linear function of the input. As Kaplan (1962) notes, the terms including the derivatives can be regarded as obstacles which prevent the output from following the input or, in other words, as expression of the transformations introduced by the system. Thus, from the properties of the coefficients, we may be able to infer some important properties of the system considered. For example, the network of Fig. 13 is described by Eq. 3 with all coefficients equal to zero except A_{n-1}, A_n and B_m. This system gives a sustained output to a step input and thus it may be adequate for reproducing the responses of the "slowly-adapting" receptors. The network obtained by interchanging the resistance and the condenser in Fig. 13 is described by Eq. 3 with all coefficients equal to zero except A_{n-1}, A_n and B_{m-1}. Its output to a step is unsustained and thus it may simulate "rapidly-adapting" responses.

It is clear that if we know Eq. (3) we can, at least in principle, calculate the output evoked by any given input, even if the nature and structure of the interposed system are unknown. For instance, we may not know if the process responsible for the observed input-output relations is the diffusion of particles, the spread of an electric current or some different event, and yet we may find a mathematical expression suitable for describing and predicting all our experimental results.

---

[1] The following discussion applies only to linear systems, defined as follows: if the inputs $s_1(t)$, $s_2(t)$ ... give the corresponding outputs $x_1(t)$, $x_2(t)$ ... and if the input $s_1(t) \pm s_2(t)$ ... gives the output $x_1(t) \pm x_2(t)$ ... then the system is linear.

9. Model Building

We may conclude from the above discussion that knowledge of the general input-output relations of a system permits one to find the output to any input and may give valuable insight on some important properties of the system under investigation. The set of equations which describe the general input-output relations of a system is called a "model" and the process by which these equations are found is called "model building". Models are often presented in the form of an electrical network, of a system of "compartments" or as an arrangement of other components possessing defined properties. These components, however, do not necessarily imply a physical analogy with the actual system, and the essential characteristic of a model is the mathematical relation between input and output. For example, a system of compartments may be used to represent the spread of an electric current (see RALL, 1964), an electrical network may be employed as a model of chemical reactions, and so on.

Model-building is essentially a mathematical problem which can be defined as follows: given response curves $r_i(t)$ to corresponding inputs $s_i(t)$, devise a set of equations capable of reproducing the given relations. Formulation of a model is based upon the observed input-output relations as well as on other known properties of the system under investigation but must resort in addition to intuitive considerations. A model may, therefore, be regarded as a hypothesis which is proposed as an aid for the interpretation of the experimental data.

As an illustration of model-building, we refer again to the example of Fig. 13, but now we consider as known only response curves given by plots such as those shown in this figure and the corresponding inputs. The equations accompanying the plots, the general relations (1) and (2) and the structure of the system are all unknown (the unknown system is often represented by a "black box"). Our problem is to construct a system which will give the observed response curves.

The first step (called "curve fitting") is to find mathematical expressions suitable for describing the observed curves. In the example of Fig. 13a, a semilogarithmic plot of the response curve would be sufficient for determining that the response is fitted by the expression: $x = e^{-t/\tau}$. In general, however, curve fitting is a more laborious problem and no formal techniques are available for solving it.

The mathematical equation describing a response gives information on the complexity and general structure of the model required for reproducing the experimental data. Based on this and other information, a model is postulated; for instance, the relation describing the response curve of Fig. 13a would immediately suggest the discharge of a condenser or some similar process. Once a model is devised, it must be tested against all available data and, if discrepancies are observed, it must be modified until it simulates the experimental observations with the desired degree of accuracy.

If this stage can be reached, the model will reproduce the kinetics of the processes occurring in the actual system under investigation. It should not be expected, however, to give direct information on the nature of these processes, since as already mentioned, the same kinetic relations may apply to a variety of events.

By studying the actual system with reference to a successful model, however, one may gain sufficient insight to plan experiments directed at finding out what physical entities or processes in the actual system correspond to the parameters

17*

of the model. If this can be done, a satifactory physical description of the system is obtained.

10. Steaty-State Relations between Stimulus and Response. Non-linearities.

Whether or not models are used as a framework for analysis, it is usually convenient to investigate first the linear operating range of the system. This is the range over which the properties of the system (or the parameters of the model) remain constant; in these conditions the amplitude of the response remains at all times proportional to the strength of the stimulus. The network of Fig. 13, for example, is a linear system for inputs which do not evoke breakdown of the condenser or of the resistance. It is well known, however, that the normal operations of many systems are often characterized by nonlinear properties. In sensory physiology it has long been known that many sensations are approximately proportional not to the intensity of the stimulus but to the logarithm (Fechner, 1860) or to some power (Stevens, 1961) of the intensity. In some cases these

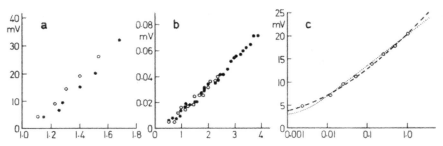

Fig. 14 a—c. Relation between generator potential and stimulus. a Intracellular measurements from stretch receptors of crayfish. Abscissa: relative length of stretch receptor; ordinate: amplitude of generator potential in steady state. The two symbols refer to two different units (from Terzuolo and Washizu, 1962). b Records taken from the nerve fiber attached to a Pacinian corpuscle. Abscissa: deflection of the piezoelectric crystal used to apply pressure; ordinate: peak height of generator potential. Data taken before (○) and after (●) application of tetrodotoxin as in Fig. 11 e (from Loewenstein, et al., 1963). c Intracellular measurements from a visual cell of *Limulus*. Amplitude of generator potential was measured in steady state. Abscissa gives light intensity in relative units (Note logarithmic scale). The points can be roughly fitted by the logarithmic function proposed by Rushton in 1960 (dotted line) or by the function of Eq. (4) with $n = 10$ (dashed line) (modified from Fuortes, 1958)

nonlinearities may arise from the transformations occurring in the central nervous system. In other instances they are already observed in the relation between stimulus and generator potential of a receptor cell.

Analysis of nonlinear systems is usually difficult. For this reason it is often preferable to consider at first only steady-state relations, postponing the more laborious analysis of transients until some features of the nonlinearities are understood.

Steady-state relations between stimulus strength and amplitude of generator potentials are illustrated in Fig. 14 for three different receptors. It will be observed that some receptors (like the Pacinian corpuscles) normally respond to a small range of stimuli. In these cases the relation between stimulus and generator potential may be approximately linear over the whole normal range of operation.

In other receptors (such as auditory or visual receptors), the normal operating range may extend over several decades and the relation between stimulus and generator potential is non-linear over most of the range.

In a number of photoreceptors, the nonlinearities of the response become evident only when the response exceeds a certain amplitude. Fig. 15 b shows responses recorded from visual cells of *Limulus* following a flash delivered at time 0. Superposed on these responses are theoretical curves representing the output curves of the model shown in the figure (FUORTES and HODGKIN, 1964). Both the experimental and the theoretical responses were obtained by doubling stimulus intensity for each successive curve, as indicated by the figures near each pair of tracings. It is seen that the theoretical responses are at all times proportional to the intensity of the stimulus, as required by linearity. The experimental responses follow roughly the theoretical curve up to a height of about 8 mV, but above this amplitude they deviate sharply, growing less than they would in a linear system. These nonlinearities can be reproduced by the model if one adds to it a feedback loop by which the value of the resistances R is decreased as a function of the voltage at the last stage (or at one of the last stages — MARIMONT, 1966). In this way, as the output voltage grows, the leaks are increased and the gain[2] is decreased in each stage.

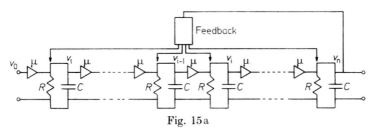

Fig. 15a

Fig. 15a—c. Simulation of visual responses. a Electrical analogue of visual cells of *Limulus*. The circuit consists of n (approximately 10) identical stages and a feedback loop. Each stage includes a resistance R, a condenser C, and an element μ which generates a current proportional to the voltage at its input. When the feedback loop is disconnected, the system is linear and the equation characterizing the network is: $\mu v_{i-1} = C \dfrac{dv_i}{dt} + \dfrac{v_i}{R}$. The feedback loop changes the value of R in accordance with the relation: $\dfrac{1}{R} = \dfrac{1}{R_0}\left(1 + \dfrac{v_n}{w}\right)$ where R_0 and w are constants. With this addition, the system becomes non-linear and simulates the responses of visual cells over a wide range of light intensities. b Tracings of visual responses (———) and output of linear model (∘∘∘∘) for different intensities of the stimulus. Visual responses were produced by brief flashes and were recorded by means of an intracellular electrode. The input of the linear model was an instantaneous impulse. Stimulus strength in arbitrary units, indicated by the figures near each pair of records. The experimental responses roughly follow the responses of the linear model up to amplitudes of 8 mV. c Output of the model with or without feedback. Responses to impulses are illustrated at left and responses to steps at right. Without feedback (∘∘∘∘) the responses grow in proportion with the strength of the input (given by the figures near the tracings). When the feedback loop is connected, the output changes as shown by the continuous line (———). Responses to steps have several features in common with the experimental responses illustrated in Figs. 10 and 11. Time (abscissa), voltage (ordinate) and stimulus strength in arbitrary units

---

[2] Gain of a stage is defined as the ratio between output and input voltage for that stage:

$$g = \frac{v_i}{v_{i-1}}$$

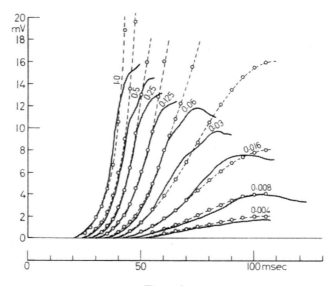

Fig. 15b

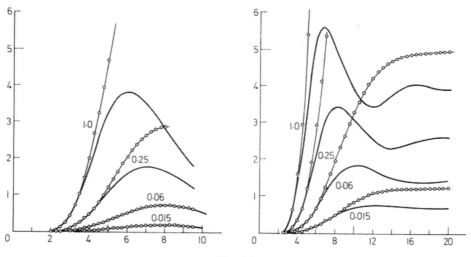

Fig. 15c

For low output voltages, the gain-control loop is almost ineffective and the non-linearities of the responses remain inappreciable; as the control voltage increases, however, its effect on the gain grows very rapidly and the deviations from linearity become more and more evident. The responses of this nonlinear model to impulses or to steps resemble the experimental responses of visual cells.

The responses of other types of receptors present some similarities with the responses of visual cells, but it is not known if their time course can be reproduced by a model similar to that proposed for photoreceptors.

It can be seen, however, that a model of this type can reproduce the steady state relations between stimulus and response which have been observed in a variety of receptors. In the model this relation is:

$$s = \alpha x \,(1 + \alpha x)^n \qquad (4)$$

where s is the intensity of the stimulus, x is the amplitude of the response, α is a constant reflecting the effect of the output upon the gain, and n is the number of stages. It is seen that if αx is much smaller than unity, the equation reduces to $s = \alpha x$ and the response will be simply proportional to the stimulus. If instead αx is much larger than unity, the equation becomes $s = (\alpha x)^{n+1}$ describing a power relation between stimulus and response. If n is considerably larger than unity (for instance, if $n = 10$), a slightly different type of gain control incorporated in the model of Fig. 13 will give a logarithmic relation between stimulus and response (see FUORTES and HODGKIN, 1964).

11. Relations between Generator Potentials and Nerve Impulses

When the receptor cell is long, generator potentials are inadequate for transmission and nerve impulses are needed. Since impulses have fixed size and shape, the amplitude of a generator potential cannot be reflected by the features of individual impulses and it is signalled instead by the intervals between impulses (ADRIAN, 1928). Thus, the intensity of the stimulus $s(t)$ controls the amplitude of the generator potential $v(t)$, and this controls in turn the frequency of impulses $f(t)$.

The relation between generator potential amplitude and frequency of impulses has been studied almost exclusively in steady state. In these conditions, an approximately linear relation has been observed:

$$v = Af + B \qquad (5)$$

where A and B are constants.

Both in visual cells of *Limulus* (MACNICHOL, 1956; FUORTES, 1959) and in the stretch receptor of the crayfish (TERZUOLO and WASHIZU, 1962) changes of 1 mV

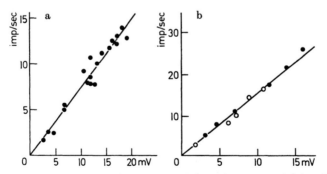

Fig. 16a and b. Relation between generator potential and frequency of firing. Measurements taken from visual cells of *Limulus* (a) and from stretch receptors of crayfish (b) by means of intracellular electrodes. In both cases the data are approximately fitted by a straight line. (a: from MACNICHOL, 1956; b: from TERZUOLO and WASHIZU, 1962)

in the amplitude of the generator potential are accompanied by changes of 0.75 to 2.5 impulses/sec in the steady state discharge of impulses (Fig. 16).

In order to understand this relation, it is useful to remember that generator potentials and nerve impulses originate in different parts of the cell. This situation can be conveniently represented by the diagram of Fig. 17, in which SD represents the site of origin of the generator potentials and P represents the pacemaker region. An electrode in the soma will record generator potentials without serious distortions, but the spikes generated at the pacemaker will be attenuated as shown in Fig. 17a due to the resistance separating the pacemaker from the site of recording. If the resistance of the somatic membrane decreases, the amplitude of the recorded spikes will also decrease as shown in the same figure. Similarly, generator potentials originating near the soma will be attenuated when they reach the pacemaker, and this attenuation will be a function of the resistance of the membrane at the pacemaker region.

During impulse activity, the resistance between inside and outside of the cell is sharply reduced at the pacemaker but remains appreciable at the soma. Consequently, both generator potentials and impulses may be large at the soma; at

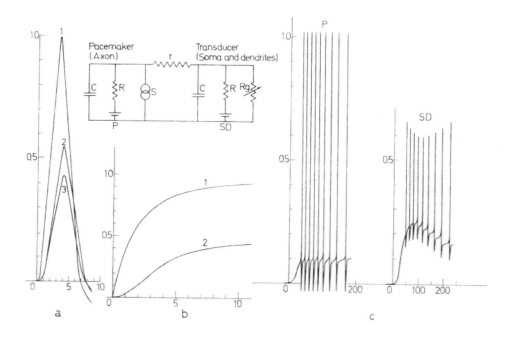

Fig. 17a—c. Calculated potential changes at soma and axon. The electrical network represents the soma-dendritic region (SD) and the initial part of the axon (P) of a receptor cell. It is assumed that generator potentials originate in SD and are associated with decrease of the resistance Rg. Nerve impulses are supposed to originate in a region of the axon (P) connected to the soma by a resistance r. The resistance of the spike generator S becomes negligible during impulse activity. a A spike (curve 1) originating at the pacemaker region produces the voltage of curve 2 at the soma if the resistance $\bar{R} = \dfrac{R\,Rg}{R + Rg}$ is high ($\bar{R} = 2r$) and the smaller voltage of curve 3 if the value of $\bar{R}$ is low ($\bar{R} = r$). b Voltage drops produced at SD (curve 1) and at P (curve 2) by a step of current applied at the soma. c Generator and spike potentials at pacemaker region (P) and at soma (SD). The spikes are large and they all start at the same membrane potential at the pacemaker. At the soma, the spikes are attenuated and are superposed on the generator potential. Near the peak of the slow potential the spikes are smaller because membrane resistance of the soma is lower

the pacemaker instead, the spikes will dominate and the generator potential will produce only a small additional potential drop.

As a consequence of these interactions, the potential changes which occur during firing at the soma and at the pacemacer area will have the general features illustrated in Fig. 17 c. At the soma, the spikes will be superposed on the generator potential as though they were initiated at different membrane potentials. At the pacemaker, they arise instead all at the same level of depolarization; different amplitudes of the generator potentials will produce, however, different intensities of the depolarizing current through the pacemaker and consequently different frequencies of firing.

12. Impulses Evoked by Extrinsic Currents

Both in the photoreceptor of *Limulus* and in the stretch receptor of the crayfish, sustained trains of impulses can be evoked in the absence of sensory stimulation by depolarizing currents applied through a microelectrode in the soma. Currents delivered in this way will evoke essentially the same phenomena described with reference to generator potentials; the potential drop evoked by the current at the soma will spread with decrement to the pacemaker region and the spikes will be larger at the pacemaker than at the soma. If membrane resistance is different during the generator potential or during application of currents, the decrements will also be different in the two conditions. For instance, it has been observed in the eccentric cell of *Limulus* that the resistance between inside and outside is less during the generator potential than it is when the cell is depolarized by external currents (FUORTES, 1959). Consequently, the potential drop recorded at the soma when a spike is generated at the pacemaker is smaller during illumination than it is when spikes are generated by external currents (see Fig. 17 a).

The steady-state relation between frequency of firing and current intensity is approximately linear in both the stretch receptor and the photoreceptor. Changes of 10^{-9} A in the current applied through a microelectrode in the soma produce changes of 5 to 15 impulses/sec in different cells. It is not clear how this simple relation between firing frequency and depolarizing current comes about. Apparently some balance is reached in steady state between the excitatory effects of depolarizing currents and the inactivations (HODGKIN and HUXLEY, 1952) resulting from firing of impulses (refractoriness) and from persistence of depolarizing current (accommodation). These interactions are difficult to analyze and have not been worked out in detail up to now. The mechanisms underlying repetitive firing have been discussed, however, by ADRIAN (1928), FESSARD (1936), MATTHEWS (1937), ARVANITAKI (1938), HODGKIN (1948), ECCLES (1953, p. 176), FITZHUGH (1961), FUORTES and MANTEGAZZINI (1963), and others.

It is interesting to note that an approximately linear relation between frequency and current intensity, such as found in receptors, has been observed also in peripheral axons of *Carcinus* (HODGKIN, 1948), although in these nerves the frequency of firing is controlled by the slow development of a subthreshold response rather than by accommodation and refractoriness, as seems to be the case in receptors (FUORTES and MANTEGAZZINI, 1963).

13. Relation between Stimulus Intensity and Firing Frequency

Since frequency of firing is linearly related to the amplitude of the generator potential, the relation between frequency and stimulus intensity should have the same form as that between generator potential and stimulus. Recent experiments on slowly-adapting skin receptors (Mountcastle, 1965; Werner and Mountcastle, 1965); show that the steady-state relation between frequency and intensity of the stimulus agrees well with the expression:

$$s = Af^m . \tag{6}$$

The value of the exponent m was found to be between 1 and 2 in most of the skin receptors studied. Choosing an appropriate value for α, the steady state equation of the model mentioned above [Eq. (4)] fits the experimental data about as accurately as Eq. (6). It is possible, therefore, that the results on skin receptors can be interpreted by the same model proposed for photoreceptors, but further studies will be required in order to support this assumption.

References

Adrian, E. D.: The basis of sensation. London: Christophers 1928.
— The mechanism of nervous action. Philadelphia: Univ. Penn. Press 1931.
— The physical background of perception. Oxford: Clarendon 1947.
Arvanitaki, A.: Les variations graduees de la polarization des systemes excitables. Paris: Hermann & Cie. 1938.
Bairati, A.: Anatomia Umana. Torino: Minerva Medica 1961.
Baumann, F.: Slow and spike potentials recorded from retinula cells in the honeybee drone in response to light. J. gen. Physiol. 52, 855—875 (1968).
Baylor, D. A., Fuortes, M. G. F.: Electrical responses of single cones in the retina of the turtle. J. Physiol. (Lond.) 207, 77—92 (1970).
Berman, M.: The formulation and testing of models. Ann. N. Y. Acad. Sci. 108, 182—194 (1963).
Bernhard, C. G., Granit, R.: Nerve as model temperature end organ. J. gen. Physiol. 29, 257—265 (1946).
Bortoff, A.: Localization of slow potential responses in the Necturus retina. Vision Res. 4, 627—635 (1964).
— Norton, A. L.: An electrical model of the vertebrate photoreceptor cell. Vision Res. 7, 253—263 (1967).
Castillo, J. del, Katz, B.: Quantal components of the end-plate potential. J. Physiol. (Lond.) 124, 560—573 (1954).
Coombs, J. S., Eccles, J. C., Fatt, P.: The specific ionic conductances and the ionic movements across the motoneuronal membrane that produce the inhibitory post-synaptic potential. J. Physiol. (Lond.) 130, 326—373 (1955a).
— — — Excitatory synaptic action in motoneurones. J. Physiol. (Lond.) 130, 374—395 (1955b).
Davis, H.: Some principles of sensory receptor action. Physiol. Rev. 41, 391—416 (1961).
— A model for transducer action in the cochlea. Cold Spr. Harb. Symp. quant. Biol. 30, 181—189 (1965).
Dirac, P. A. M: The principles of quantum mechanics. London: Oxford Univ. Press 1947.
Eccles, J. C.: The electrophysiological properties of the motoneurone. Cold Spr. Harb. Symp. quant. Biol. 17, 175—183 (1952).
— The Neurophysiological Basis of Mind: The principles of Neurophysiology. Oxford: Clarendon Press 1953.
— The physiology of nerve cells. Baltimore: Johns Hopkins Univ. Press 1957.
Eyzaguirre, C., Kuffler, S. W.: Process of excitation in the dendrites and in the soma of single isolated sensory nerve cells of the lobster and crayfish. J. gen. Physiol. 39, 87—119 (1955).

FATT, P.: Alterations produced in the post-junctional cell by the inhibitory transmitter. In: Inhibition in the nervous system and γ-aminobutyric acid. Ed. by ROBERTS, E. New York: Pergamon Press 1960.
— KATZ, B.: An analysis of the end-plate potential recorded with an intra-cellular electrode. J. Physiol. (Lond.) 115, 320—370 (1951).
FECHNER, G. T.: Elemente der Psychophysik. Leipzig: J. C. Barth 1860.
FESSARD, A.: Proprietes rythmique de la matiere vivante. Paris: Herman & Cie. 1936.
FITZHUGH, R.: Impulses and physiological states in theoretical models of nerve membrane. Biophys. J. 1, 445—466 (1961).
FLOCK, A.: Transducing mechanisms in the lateral line canal organ receptors. Cold Spr. Harb. Symp. quant. Biol. 30, 133—144 (1965).
— WERSALL, J.: A study of the orientation of the sensory hairs of the receptor cells in the lateral line organs of fish with special reference to the function of the receptors. J. Cell Biol. 15, 19—27 (1962).
FUORTES, M. G. F.: Electric activity of cells in the eye of Limulus. Amer. J. Ophthal. 46, 210—223 (1958).
— Initiation of impulses in visual cells of Limulus. J. Physiol. (Lond.) 148, 14—28 (1959).
— FRANK, K., BECKER, M. C.: Steps in the Production of Motoneuron Spikes. J. gen. Physiol. 40, 735—752 (1957).
— HODGKIN, A. L.: Changes in time scale and sensitivity in the Ommatidia of Limulus. J. Physiol. (Lond.) 172, 239—263 (1964).
— MANTEGAZZINI, F.: Interpretation of the repetitive firing of nerve cells. J. gen. Physiol. 45, 1163—1179 (1962).
— POGGIO, G. F.: Transient responses to sudden illumination in cells of the eye of Limulus. J. gen. Physiol. 46, 435—452 (1963).
GASSER, H. S.: Axons as samples of nervous tissue. J. Neurophysiol. 2, 361—369 (1939).
GOLGI, C.: Recherches sur l'histologie des centres nerveux. Arch. ital. Biol. 3, 285—317 (1883).
GORMAN, A. L. F., McREYNOLDS, J. S.: Hyperpolarizing and depolarizing receptor potentials in the scallop eye. Science 165, 309—310 (1969).
GRAY, J. A. B., SATO, M.: Properties of the receptor potential in Pacinian corpuscles. J. Physiol. (Lond.) 122, 610—636 (1953).
GRUNDFEST, H.: Electrical inexcitability of synapses and some consequences in the central nervous system. Physiol. Rev. 37, 337—361 (1957).
— An electrophysiological basis for cone vision in fish. Arch. ital. Biol. 96, 135—144 (1958).
— Synaptic and ephaptic transmission. In: Handbook of physiology. Neurophysiology. Ed. by FIELD, J. Washington: Amer. Physiol. Soc. 1959.
— Excitation by hyperpolarizing potentials. A general theory of receptor activities. In: Nervous inhibition. Ed. by FLOREY, E. London: Pergamon Press 1961.
— Electrophysiology and pharmacology of different components of bioelectric transducers. Cold Spr. Harb. Symp. quant. Biol. 30, 1—13 (1965).
HARTLINE, H. K.: Intensity and duration in the excitation of single photoreceptors. J. cell. comp. Physiol. 5, 229—274 (1934).
— WAGNER, H. G., MacNICHOL, E. F.: The peripheral origin of nervous activity in the visual system. Cold Spr. Harb. Symp. quant. Biol. 17, 125—141 (1952).
HODGKIN, A. L.: The local electric changes associated with repetitive action in a non-medullated axon. J. Physiol. (Lond.) 107, 165—181 (1948).
— The ionic basis of electrical activity in nerve and muscle. Biol. Rev. 26, 339—409 (1951).
— HUXLEY, A. F.: A quantitative description of membrane current and its application to conduction and excitation in nerve. J. Physiol. (Lond.) 117, 500—544 (1952).
HUBBARD, S. J.: A study of rapid mechanical events in a mechano-receptor. J. Physiol. (Lond.) 141, 198—218 (1958).
KANEKO, A., HASHIMOTO, H.: Recording site of the single cone response determined by an electrode marking technique. Vision Res. 7, 847—851 (1967).
KAPLAN, W.: Ordinary differential equations. Reading: Addison and Wesley 1962.
KATZ, B.: Depolarizarion of sensory terminals and the initiation of impulses in the muscle spindle. J. Physiol. (Lond.) 111, 261—282 (1950).

Katz, B., Miledi, R.: A study of synaptic transmission in the absence of nerve impulses. J. Physiol. (Lond.) **192**, 407—436 (1967).

Lasansky, A.: Cell junctions in ommatidia of Limulus. J. Cell Biol. **33**, 365—383 (1967).

Liley, A. W.: The quantal components of the mammalian end-plate potential. J. Physiol. (Lond.) **133**, 571—587 (1956).

Loewenstein, W. R.: The generation of electric activity in a nerve ending. Ann. N. Y. Acad. Sci. **81**, 367—387 (1959).

— Facets of a transducer process. Cold Spr. Harb. Symp. quant. Biol. **30**, 29—43 (1965).

— Mendelson, M.: Components of receptor adaptation in a pacinian corpuscle. J. Physiol. (Lond.) **177**, 377—397 (1965).

— Terzuolo, C. A., Washizu, Y.: Separation of transducer and impulse generating processes in sensory receptors. Science **142**, 1180—1181 (1963).

MacNichol, E. F.: Visual receptors as biological transducers. In: Molecular structure and functional activity of nerve cells. Eds. Grenell, R. G., Mullins, L. J. Washington: Amer. Inst. Biol. Sci. 1956.

Marimont, R.: Numerical studies of the Fuortes-Hodgkin Limulus model. J. Physiol. (Lond.) **179**, 489—497 (1965).

Matthews, B. H. C.: The response of single end organ. J. Physiol. (Lond.) **71**, 64—110 (1931).

— Do the rhythmic discharges of sense organs and of motor neurones originate in the same way? Proc. roy. Soc. (London) B. **123**, 416—418 (1937).

McReynolds, J. S., Gorman, A. L. F.: Photoreceptor potentials of opposite polarity in the eye of the scallop, *Pecten irradians*. J. gen. Physiol. 1970a (in press).

— — Membrane conductances and spectral sensitivities of *Pecten* photoreceptors. J. gen. Physiol. 1970b (in press).

Mendelson, M., Loewenstein, W. R.: Mechanisms of receptor adaptation. Science **144**, 554—555 (1964).

Mountcastle, V. B.: The neural replication of sensory events in the somatic afferent system. In: Semaine d'étude sur cerveau et expérience consciente. Vatican City: Pontificia Academia Scientiarum 1965.

Ottoson, D., Shepherd, G. M.: Receptor potentials and impulse generation in the isolated spindle during controlled extension. Cold Spr. Harb. Symp. quant. Biol. **30**, 105—113 (1965).

Rall, W.: Theoretical significance of dendritic trees for neuronal input-output relations. In: Neural theory and modeling. Ed. by Reiss, R. F. Stanford: Stanford Univ. Press 1964.

Ramon y Cajal, S.: Histologie du systeme nerveux de l'homme et des vertébrés. Paris: Malone 1909—1911.

Rushton, W. A. H.: The intensity factor in vision. In: Light and Life. Eds.: McElroy, W. D., Glass, B. Baltimore: Johns Hopkins Univ. Press 1961.

Sobotta, J.: Handbuch der Mikroskopischen Anatomie des Menschen. Hrg. Müllendorf, W. von. Berlin: Springer 1928.

Terzuolo, C. A., Washizu, Y.: Relation between stimulus strength, generator potential and impulse frequency in stretch receptor of crustacea. J. Neurophysiol. **25**, 56—66 (1962).

Thurm, U.: An insect mechanoreceptor. II. Receptor potentials. Cold Spr. Harb. Symp. quant. Biol. **30**, 83—94 (1965).

Tomita, T.: Electrophysiological study of the mechanisms subserving color coding in the fish retina. Cold Spr. Harb. Sympl quant. Biol. **30**, 559—566 (1965).

Toyoda, J., Nosaki, H., Tomita, T.: Light-induced resistance changes in photoreceptors of *Necturus* and *Gekko*. Vision Res. **9**, 453—463 (1969).

— Shapley, R. M.: The intracellularly recorded response in the scallop eye. Biol. Bull. **133**, 490 (1967).

Tucker, D., Shibuya, T.: A physiologic and pharmacologic study of olfactory receptors. Cold Spr. Harb. Symp. quant. Biol. **30**, 207—215 (1965).

Werblin, F. S., Dowling, J. E.: Organization of the retina of the mud puppy, *Necturus maculosus*. II. Intracellular recording. J. Neurophysiol. **32**, 339—355 (1969).

Werner, G., Mountcastle, V.: Neural activity in mechanoreceptive cutaneous afferents: stimulus-response relations, Weber-functions and information transmission. J. Neurophysiol. **28**, 359—397 (1965).

Chapter 9

Mechano-electric Transduction in the Pacinian Corpuscle. Initiation of Sensory Impulses in Mechanoreceptors

By

Werner R. Loewenstein, New York (USA)

With 16 Figures

Contents

Mechanical stimuli have so widespread an effect on living things that there is hardly an organism which is not endowed with some mechanism to sense such stimuli. Even primitive single-cell organisms are mechano-receptive, and several kinds of non-nervous cells of higher organisms show this property at least to some degree. As a further refinement, specialized and highly mechanosensitive structures have evolved in higher organisms. The essential element in all of these structures is a dendrite, alone or associated with satellite cells.

This article deals mainly with a structure of this kind in which the satellite cells are arranged concentrically around the end portion of a dendrite, forming a lamellated fluid-filled system, the Pacinian corpuscle (Figs. 1, 2). The whole structure is unusually large (1 mm long; 0.5 mm diameter); and because of this it has offered the rare opportunity for studying biological transducer mechanisms in some detail.

1. Sites and Modes of Transduction

The transducer mechanisms, i.e., the processes that convert mechanical into electrical energy (*generator current*), appear to be contained in the dendrite ending; most of the satellite structure can be removed or destroyed without loss of transducer capability (Fig. 1) (Loewenstein and Rathkamp, 1958). The transducer processes are very sensitive to assymmetric (with respect to the dendrite terminal) pressure fields, but quite insensitive to axially symmetric fields (Fig. 3). Thus only stresses that tend to strain the dendrite in the direction parallel to its surface appear to generate current (LOEWENSTEIN, 1965). The strain sensitivity (computed) is of the order of 10^{-5} cm or better.

The conversion of strain into current occurs presumably in the surface membrane of the dendrite ending. This relatively elastic membrane bounds a rather

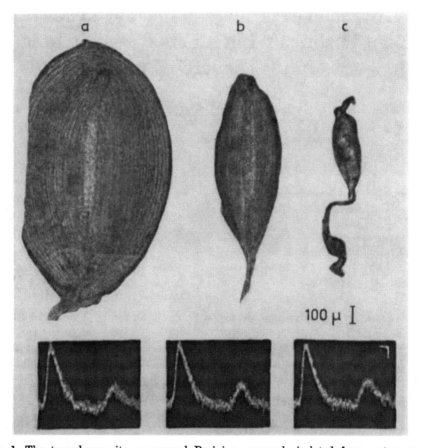

Fig. 1. The transducer site. *a*, normal Pacinian corpuscle isolated from cat mesentery (phase contrast); *b* and *c*, after elimination of capsular structure. In *c*, only the core of the corpuscle remained intact. Below are the corresponding electrical responses to two pulses of compression. The pulses of compression are delivered by a piezoelectric crystal and the responses are led off the dendrite near its point of emergence from the corpuscle. The first response is a generator potential plus a nerve impulse; the second is a generator potential alone. Calibration: 1 msec, 25 μV (from LOEWENSTEIN and RATHKAMP, 1958)

incompressible cylinder of axoplasm of elliptical cross section (Fig. 2). Because of its elliptical shape, small compressions of the ending along its minor elliptical axis cause the membrane to distend whereas compression along its major elliptical axis has the opposite mechanical effect. The former mode of compression produces a predominantly inward current through the membrane and the latter, a predominantly outward current (ILYINSKY, 1964). The inward current constitutes the generator current that triggers the nerve impulse. This current is readily led off the outside of the dendrite, and much of what we know about the transducer process derives from it.

The evidence in several receptor structures is consistent with the idea that the membrane operates like a variable conductance transducer. Like the remainder

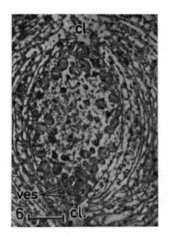

Fig. 2a

Fig. 2. a, Electron micrograph of the transducer region. The unmyelinated dendrite terminal of a Pacinian corpuscle is seen in cross section together with some core lamellae, Calibration: 1 μ (from Pease and Quilliam, 1957). b, Schematic view of a Pacinian corpuscle. L, outer lamellae; C, core lamellae; the unmyelinated dendrite is at the center. The arrows b and a

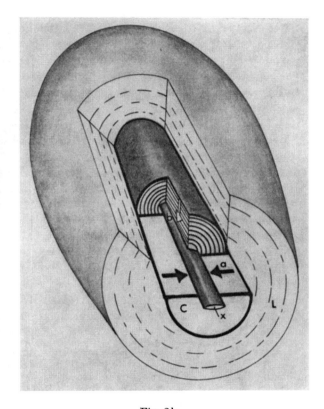

Fig. 2b

illustrate situations of compression in the plane of the minor and major elliptical axes of the dendrite respectively

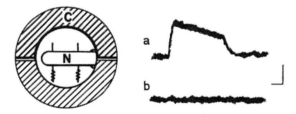

Fig. 3. Strain vs. stress sensitivity of transducer. The decapsulated dendrite of a Pacinian corpuscle is compressed in a chamber made of two piezoelectric half-cylinders (C). a, Generator potential in response to asymmetric pressure (pulse) produced by one half-cylinder; and b, to symmetric pressure distribution by both half-cylinders. Calibration: 50 mV, 10 msec
(from Loewenstein, 1965)

of the surface membrane of the nerve cell, the transducer membrane is presumably electrically charged and of low conductance at rest. Distension of the membrane causes its conductance to rise, the net rate of charge transfer through the membrane constituting the generator current. A simple possibility is that charge is transferred by ions moving along their electrochemical gradients across the transducer membrane. The sodium ion is a good candidate here but other ions are probably involved also (FATT, 1950; cf. FATT and KATZ, 1951; CASTILLO and KATZ, 1954; ECCLES, 1957; TAKEUCHI and TAKEUCHI, 1960; DIAMOND et al., 1958; OTTOSON, 1964). As to how the conductance change comes about, the answer is as yet uncertain, as it is to the related question concerning other excitable membranes. The conductance change very likely involves a change in molecular configuration. In the case of a mechanoreceptor, one may suspect the simplest transducer mode, namely that the configurational change is itself the membrane strain. It is suggestive here that mechanoreceptive membranes are not sensitive to pure stress (LOEWENSTEIN, 1965), and that current is produced with extremely small latencies (GRAY and SATO, 1953; OTTOSON and SHEPHERD, 1965; THURM, 1964). A simple possibility is that distension reduces the lateral attractive forces between constituent molecules of the membrane and thereby lowers the resistance to ion flow across the membrane (ISHIKO and LOEWENSTEIN, 1961). An example of such a mechanosensitive diffusion system is provided by certain fatty acid monolayers in which the diffusion resistance decreases with decreasing surface pressure (ROSSANO and LaMER, 1956). Such a homogeneous transducer system model is probably the simplest conceivable model; conductance here is simply determined by the molecular packing in the membrane. A more structured model is the one suggested by KATZ (1950) in which membrane distension stretches out pre-existing diffusion pores. For other plausible mechanisms, the reader is referred to the chapters in this volume by KATCHALSKY and OPLATKA and by TEORELL.

It is possible that a change in membrane capacitance contributes to the generator current. The capacitance may be expected to increase by membrane distension (thinning) and the resulting charge redistribution to cause a transient fall in membrane potential. However, with the small distensions involved in a sensitive mechanoreceptor, such contibutions are likely to be very small, if at all significant (KATZ, 1950).

2. Properties of the Transducer Process

Transduction appears to be highly focalized. Unlike the conductance changes associated with the nerve impulse which propagate along the membrane, those associated with mechano-electric transduction seem to be confined to the distorted membrane regions. This is brought out conveniently by explorations of the electric field around a distorted membrane spot, which show that the current sink (the site of the conductance change) is located at and restricted to this spot (LOEWENSTEIN, 1961; Fig. 4a). Thus, in contrast to the nerve impulse processes, there appears to be no electrical or chemical linkage in the transducer processes along the membrane; different spots on the membrane must be activated independently of each other. An immediate functional consequence of this is spatial summation of the output currents of such spots; the total (instantaneous) generator current increases with the area of membrane activated (LOEWENSTEIN, 1959; Fig. 5).

A plausible model of the transducer membrane in which these properties are incorporated is a network of the kind shown in Fig. 6. Each unit (U) in the model represents an area of the receptor membrane incorporating a fraction of the membrane conductance and capacitance. Transduction is represented by a definite rise in conductance in each unit area, as effected by shunt r_2. The units are not coupled

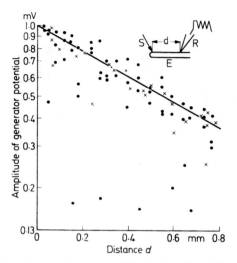

Fig. 4. Focal transduction. A membrane spot (S) of about 20-μ diameter of a decapsulated. unmyelinated dendrite terminal of a Pacinian corpuscle is stimulated with a series of equal pulses of compression; the resulting generator potentials (circles, logarithmic scale) are led off the membrane surface at varying distances from the stimulated spot with a microelectrode (R). A current is then applied to the surface at this spot, and the resulting passively dissipating (electrotonic) potentials (crosses) are led off as before (from LOEWENSTEIN, 1961)

in this operation. However, because of the large number of units, the generator current increases rather smoothly with the increasing area of membrane excited. The peak generator potential (V) recorded between two external points of the nerve fiber is given approximately by

$$V = \frac{Ebx}{1 + bx} \qquad (1)$$

where x is the fraction of area excited; E, the resting potential; and $1 + b$ the ratio between E and the final value to which the membrane potential falls in the maximally excited state (LOEWENSTEIN, 1961a). This equation accounts satisfactorily for the experimental area-generator potential curve (Fig. 5, bottom).

Many mechanoreceptor systems are so structured that a change in strength in the physiological stimulus must result in a change in active area of transducer membrane. In the Pacinian corpuscle, for instance, the area of activated transducer membrane will evidently increase as the degree of compression of its lamellated structure increases. The concomitant increase in intensity of generator current (ALVAREZ-BUYLLA and DE ARELLANO, 1953; GRAY and SATO, 1953; Fig. 5, bottom) is thus probably determined by an increase in the active area of transducer membrane. Since the frequency of the nerve impulses from a variety of receptors is in turn determined by the intensity of the generator current (e.g., KATZ, 1950; EYZAGUIRRE and KUFFLER, 1955; FUORTES, 1959; THURM, 1965), this would imply that the frequency coding in receptors depends in the last analysis on the area of active transducer membrane.

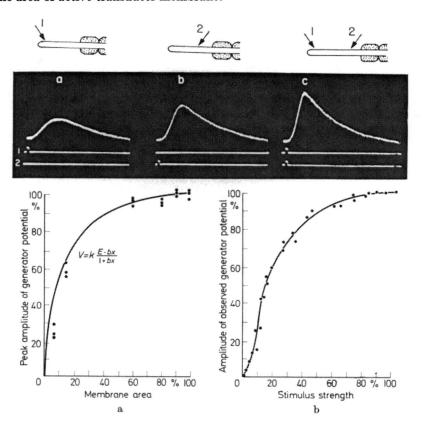

Fig. 5. *Top*, spatial summation on the dendrite membrane (decapsulated Pacinian corpuscle). Stylus 1 (about 30-μ diameter) and 2 (about 20-μ diameter) driven by independent crystals compress two membrane spots on the unmyelinated ending about 400 μ apart. Generator potentials (upper beam) in response to mechanical pulse applied in *a* to spot 1, in *b* to spot 2, and in *c* to both spots simultaneously. Generator potentials are led off the myelinated portion of the dendrite. Lower beam signals compression pulses. Calibration: 1 msec, 50 μV. *Bottom, a*, The generator potential-excited membrane area relation (decapsulated corpuscle). The area of excited transducer membrane is varied by applying equal mechanical pulses with styli of different diameters. The amplitude of the maximal (saturated) generator potential is 100%. The solid line is the theroretical generator potential-area relation as given by Eq. (1). *b*, A generator potential—amplitude of compression (stimulus strength) relation (intact Pacinian corpuscle) (from LOEWENSTEIN, 1961a)

Two other important determinants of the generator current are the membrane potential and temperature. Eq. (1) specifies the generator current-membrane potential relationship to be one of direct proportionality. This is supported by the findings that applications of polarizing currents to the dendrite terminal cause proportional increments in generator current (LOEWENSTEIN and ISHIKO, 1960);

by direct measurement of membrane potentials, a similar relationship is shown in *Limulus* photoreceptor (FUORTES, 1959) and crustacean stretch receptor (TERZUOLO and WASHIZU, 1962). Temperature affects both the intensity and the rate of rise of the generator current, presumably chiefly through an action on the conductance change during transduction. A change of $10°$ in the range 15 to $35°$ C

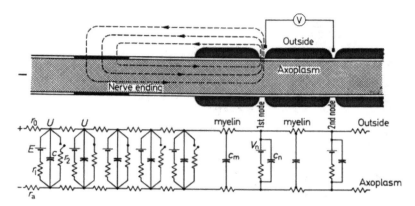

Fig. 6. Model of a transducer membrane. Description in text (from LOEWENSTEIN, 1961a)

more than doubles the generator current in the Pacinian corpuscle (Fig. 7; INMAN and PERUZZI, 1961; ISHIKO and LOEWENSTEIN, 1961). This contrasts with the slight or even negative effects on the action current (TASAKI and FUJITA, 1948; HODGKIN and KATZ, 1949; AUTRUM and SCHNEIDER, 1950).[1]

A further interesting feature of the transducer process is brought out by treatment with tetrodotoxin, a specific blocker of sodium-conductance increase in various excitable membranes (NARAHASHI et al., 1964; GRUNDFEST, 1966). Tetrodotoxin abolishes impulse activity in the Pacinian corpuscle and crayfish stretch receptor (LOEWENSTEIN et al., 1963), lobster stretch receptor (ALBUQUERQUE and GRAMPP, 1968), and frog muscle spindle (ALBUQUERQUE et al., 1969), but it leaves the generator potential little if at all affected[2] (Fig. 8). Part of the generator current is known to depend on extracellular sodium in at least three of these receptors

---

[1] It may at first sight seem surprising that an essentially physical process (see p. 270) should have such a high temperature coefficient. The coefficient would be high indeed in the case of diffusion through bulk solution, where the Einstein equation approximately holds. But the coefficient is not unusually high in diffusion through thin membranes. In monolayers, because of lateral association of the molecules and solvation of polar groups, diffusion may have very high activation energies. An instructive example is diffusion of water through a fatty acid monolayer, with an activation energy of 14,500 cal/mole (ARCHER and LaMER, 1955), which is of the same order as that calculated from the temperature dependence of the rate of rise of the generator current (ISHIKO and LOEWENSTEIN, 1961).

[2] The only apparent exception is that reported by NISHI and SATO (1966) who observed a 30 to 40% decrease in a graded response of the Pacinian corpuscle. However, this response was obtained by strong mechanical stimulation during the refractory period of the propagated impulse. It is therefore likely that the response contained a substantial (abortive) impulse component in addition to the generator potential. In fact, strong stimuli falling during the latter part of the relative refractory period of the impulse will elicit abortive impulses amounting to a substantial fraction of the total response (LOEWENSTEIN and ALTAMIRANO, 1958).

18*

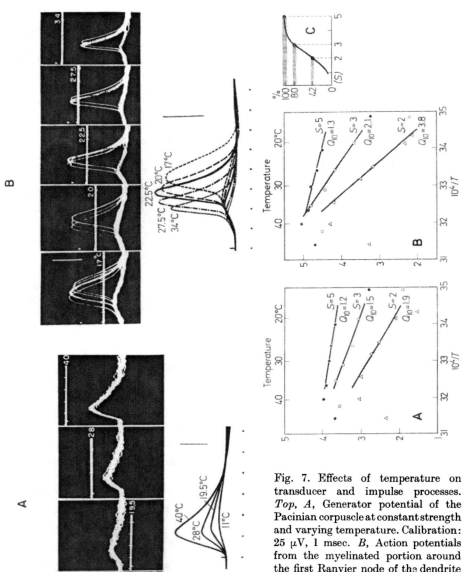

Fig. 7. Effects of temperature on transducer and impulse processes. *Top, A,* Generator potential of the Pacinian corpuscle at constant strength and varying temperature. Calibration: 25 μV, 1 msec. *B,* Action potentials from the myelinated portion around the first Ranvier node of the dendrite terminal at varying temperature. The strength of the compression is adjusted at each temperature to be at threshold for firing of action potentials. Calibration: 100 μV, 1 msec. *Bottom,* Temperature coefficients at three compression amplitudes (*S*) in ratios of 2:3:5 (*C* shows the corresponding *S* vs. generator potential curve). Abscissae *A* and *B,* reciprocal of absolute temperature. Ordinate *A,* log$_n$ of amplitude of generator potential; ordinate *B,* log$_n$ of rate of rise of generator potential. Each point is the mean of 30 to 50 cases. Solid lines are least-square curves. The slopes give the activation energies; the corresponding temperature coefficient is indicated on each curve (from ISHIKO and LOEWENSTEIN, 1961)

(DIAMOND *et al.*, 1958; TERZUOLO and WASHIZU, 1962; OTTOSON, 1964). Unlike the sodium-dependent part of the action current, however, this part seems insensitive or inaccessible to tetrodotoxin.

All these properties set the processes producing the generator current apart from those producing the nerve impulse, and provide a basis for supposing that

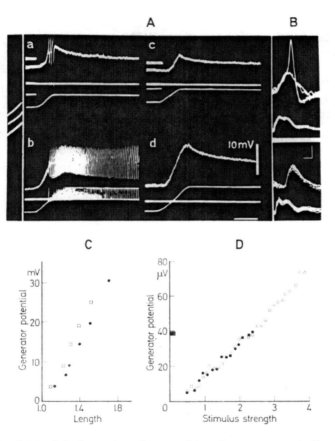

Fig. 8. Effects of tetrodotoxin on transducer and impulse processes. *A*, Crayfish stretch receptor neuron. Response to stretch before (*a, b*) and after (*c, d*) application of tetrodotoxin, 5×10^{-6}. Upper beam, intracellular record from cell soma; middle beam, record from axon, about 2 mm away from soma; lower beam, strain gauge record of mechanical displacement (stretch) to receptor organ. Calibration: 2 sec, 10 mV. *C*, the generator potential stimulus strength relation in two slowly adapting stretch receptors after treatment with tetrodotoxin *B*, Pacinian corpuscle. Response to a threshold mechanical stimulus before (upper photograph) and after (lower photograph) application of tetrodotoxin, 5×10^{-4}. Upper beam in each photograph is the electrical response recorded externally from the axon near its junction with the non-myelinated portion of the ending; lower beam is the photoelectric record of stimulus. Calibration: 1 msec; 15 µV. *D*, the generator potential-stimulus strength relation (low strength) before (•) and after (○) tetrodotoxin (from LOEWENSTEIN *et al.*, 1963)

the two arise in different membrane components. The question which then presents itself is how are the components distributed on the membrane of the dendrite — are they interspersed on a given membrane region or are they segregated ? In dealing with this question, one is severely handicapped by the poor spatial resolution of the electrical methods and by the limited information on membrane structure provided by the present electron-microscopic techniques. The most obvious anatomical differentiation at the dendrite ending is the lack of a myelinated Schwann cell cover on the terminal half-millimeter (PEASE and QUILLIAM, 1957). This correlates with a difference in mechanosensitivity: the unmyelinated terminal is highly sensitive; the myelinated part, including the nodes of Ranvier, is not (LOEWENSTEIN; 1961b). Mechanotransducer components are clearly present on the unmyelinated terminal; if such components are at all present on the myelinated dendrite, they must be sparse or inaccessible to mechanical stimulation. As to the distribution of the impulse component, the evidence is not so clear. In the case of the crustacean stretch receptor (EYZAGUIRRE and KUFFLER, 1955; EDWARDS and OTTOSON, 1958), as well as in many other nerve cells (see ECCLES, 1957; GRUNDFEST, 1957; FURSHPAN and FURUKAWA, 1962), the evidence suggests a coarse separation of the two components; the impulse is initiated at the axon origin, far away from the receptor site (EDWARDS and OTTOSON, 1958). In the case of the Pacinian corpuscle, the early evidence also suggested a coarse separation. The results of two kinds of experiment — in which impulse activity was selectively blocked by polarizing currents (DIAMOND et al., 1956) or by pressure (LOEWENSTEIN and RATHKAMP, 1958), and in which the block seemed attributable to the first Ranvier node — pointed to the node as the site of impulse initiation. Neither experiment, however, entirely excluded the possibility of spread of the blocking effects beyond the node. More recently, impulse activity was led off the outside of a partly decapsulated corpuscle (HUNT and TAKEUCHI, 1962). This interesting finding would have been quite conclusive in showing the presence of impulse components in the membrane of the unmyelinated terminal if there were certainty that the sinks of electrical activity were on the terminal, i.e., if the principal current path between dendrite and electrodes were radial in respect to the dendrite. However, the current flow may have been primarily axial and the current sinks far away. The difficulty here is that one is dealing not with a bare dendrite, but with one enclosed in a laminated sheath (the residue of the capsule, 50 to 100 μ in diameter) with a cross-sectional area 150 to 600 times that of the dendrite itself. The sheath is made of many insulating lamellae, each lamella pair bounding a coaxial fluid space that appears to open onto the dendrite at the distal and proximal ends. The array of the lamellar system appears to be such that the farther out the interlamellar spaces are in the radial direction, the closer to the myelinated portion are the openings on the proximal side. In a typical sheath of 100-μ diameter, the outermost spaces open onto the myelinated region around the first Ranvier node; these may thus carry current to a more distally located electrode. The situation is not essentially different when microelectrodes are inserted into the sheath (OZEKI and SATO, 1964). A weighty argument for the idea that impulse activity originates at the unmyelinated terminal is HUNT and TAKEUCHI's (1962) result that currents between an electrode on the tip and one on a proximal part of the sheath flow in opposite directions during dromic and antidromic impulse activity. But even this could conceivably result from axial projection if the proximal electrode were to collect current from the myelinated dendrite via the outer interlamellar spaces, and the tip electrode to collect current from a more distal region of the dendrite. (This may also possibly explain why the apparent impulse velocities estimated under such conditions are an order of magnitude greater than those obtained with microelectrodes in presumably deeper sheath locations; OZEKI and SATO, 1964.) A result which does not simply fit the assumption of radial current flow and the idea of impulse origin on the unmyelinated terminal is that exposure of the sheath to Na-free media abolishes impulse activity rapidly, whereas the effects (depression) on the generator current take many times longer to reach steady state (SATO and OZEKI, 1963). This would be readily explained by differences in diffusion time if the impulse originated on the myelinated portion, which is nearer to the outside axially and has less lamellar insulation than the unmyelinated terminal; this was the explanation originally advanced by DIAMOND et al. (1958) when these marked differences were first observed in perfused intact corpuscles. The problem is similar with regard to the action of procaine which acts much faster on the impulse than on the generator current.

To sum up, there are a mechanoelectric transducer process and an impulse-producing process with distinctly different properties in the dendrite of the Pacinian corpuscle. The two processes are so coupled that activation of the transducer process leads to activation of the impulse process, but not the reverse. The link between the two processes appears to be the output current of the transducer process. The overall sequence is

$$\underset{\text{external stimulus}}{\overset{1}{}} \rightarrow \underset{\text{membrane strain}}{\overset{2}{}} \rightarrow \underset{\text{transducer process}}{\overset{3}{}} \rightarrow \underset{\text{generator current}}{\overset{4}{}} \rightarrow$$

$$\underset{\text{impulse process.}}{\overset{5}{}}$$

The first four steps can be bypassed and impulses elicited by currents from an external source. The transducer and impulse processes may therefore be represented as arising in parallel membrane components of the dendrite ending. The spatial distribution of these components on the dendrite ending remains to be clarified.

3. Information Losses, Adaptation

a) Stimulus→Strain. The amount of information transferred in the sequence external stimulus→impulse is rather limited in the Pacinian corpuscle. Much information is lost already in the step external stimulus→membrane strain, in the transmission of mechanical forces from the surface of the corpuscle to its core where the dendrite is located. The corpuscle is so structured that only sufficiently fast compressions produce significant strain at the dendrite level, and this is the chief cause of adaptation in this sense organ. The capsule operates like a multi-layered squeeze bulb that transmits viscous forces well, but not elastic ones. A mechanical equivalent is a system of springs and dashpots as shown in Fig. 9 (LOEWENSTEIN and SKALAK, 1966). One class of springs (M) represents the elastic components of the lamellae. The underlying notion is that when a closed elastic sheet filled with an incompressible fluid is compressed along a certain diameter, the cross section perpendicular to this diameter will increase, and the resulting tension in the sheet will tend to restore the initial shape. This restoring force is one of the two principal elastic forces operating in the capsule. Another class of springs (S)

represents the much more compliant interlamellar connections. During compression the lamellar surfaces function like a series of dashpots which, in displacing fluid through narrow constraints (the interlamellar spaces), generate viscous pressure. During a static compression, part of the pressure applied to the outer lamella

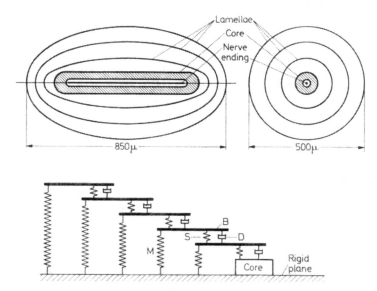

Fig. 9. Mechanical model of the corpuscle. The model incorporates the main structural elements of the capsule: rigid bars B represent the positions of the lamellae; springs M, their compliance; springs S, the compliance of the weak lamellar interconnections; and dashpots D, the resistance of the interlamellar fluid (viscosity of water) (from LOEWENSTEIN and SKALAK, 1966)

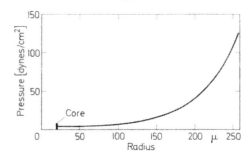

Fig. 10. Computed pressures in the corpuscle under static compression. Pressures at various levels inside the capsule for a static external compression of 20 μ. Center of the dendrite is at radius 0 and outermost lamella at radius 250 (from LOEWENSTEIN and SKALAK, 1966)

is carried by this lamella, and part is passed on to the next lamella by the interconnection, and so on. The interconnections, however, are so compliant in relation to the lamellae themselves that almost the entire load is carried by the outer lamellae. Analysis predicts that the pressure finally reaching the dendrite at the core of the corpuscle is attenuated two orders of magnitude (Fig. 10; LOEWENSTEIN and SKALAK, 1966), and high-speed photography shows that lamella displacement attenuates markedly from the outermost lamella to the more internal ones, at least within the peripheral zone (HUBBARD, 1958). The corpuscle is thus an excellent filter for elastic forces and therefore cuts out slow stimulus components.

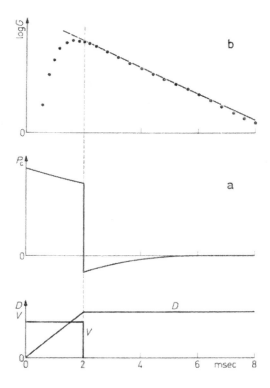

Fig. 11. Pressure input and current output at the transducer in dynamic compression. *a*, Time course of pressure (computed) at the core for a compression of 20 μ applied uniformly in 2 msec and then sustained. (*D*, displacement and *V*, velocity at outermost lamella.) *b*, Generator potential (experimental) recorded externally in response to a compression of this kind. Vertical dotted line gives the duration of the generator current on the abscissa, i.e., the duration of the transducer action, on the assumption that the exponential part of the falling phase of the generator potential corresponds to the passive capacitative and resistive components of the dendrite membrane (from LOEWENSTEIN and SKALAK, 1966)

For fast stimulus components, force transmission is good. The fluid velocities then set up significant viscous pressures which attenuate little from periphery to core. Fig. 11 illustrates this for an example of stimulation in the physiological range, a sustained compression of 20 μ reaching peak at uniform velocity in 2 msec. The result of such a sustained stimulus is a brief pulse of pressure at the core. Only during the dynamic phase of compression, while there are fluid velocities in the system, is there significant pressure (non-uniform), and hence strain, at the dendrite. During the static phase there is simply no significant pressure to be transduced at the dendrite; and, indeed, the generator response to such a stimulus is a current pulse closely matching the pressure pulse in its duration (Fig. 11). The duration of the current pulse here is not limited by processes in the transducer membrane; the same kind of stimulus acting on a decapsulated corpuscle produces a relatively long-lasting current (Fig. 12). Thus, the rate-limiting factor in receptor adaptation, as manifested at the level of the generator current, is clearly the filter action of the capsule.

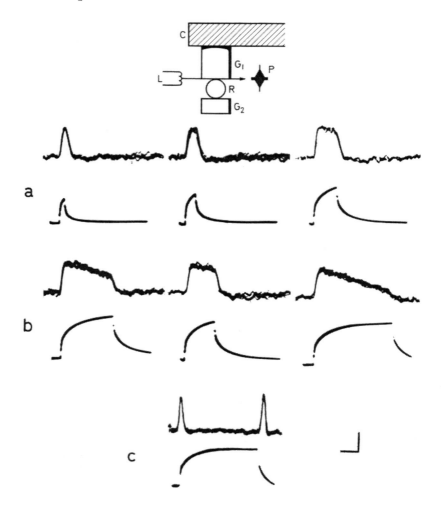

Fig. 12. Generator response after elimination of filter action of the capsule. A decapsulated Pacinian corpuscle is compressed between a fixed (G_2) and a movable plate (G_1) driven by a piezoelectric crystal (C) and the resulting displacements monitored by a photoelectric system consisting of a miniature light source (L) and a phototransistor (P). Upper oscilloscope beam in each row gives the generator potential, and lower beam gives the photoelectric record of the displacement. c, Generator responses to a compression pulse of an intact corpuscle; a and b, after removal of the capsule. Calibration for all records: 10 msec, 50 μV upper beam (from LOEWENSTEIN and MENDELSON, 1965)

It will be interesting to see to what extent adaptation of other mechanoreceptors is determined by mechanical factors. That such factors may be at play (though not necessarily rate-limiting) is indicated directly or indirectly by results obtained on crustacean stretch receptors (EYZAGUIRRE and KUFFLER, 1955; WENDLER and BURKHARDT, 1961), frog skin touch receptors (LOEWENSTEIN, 1956; CATTON, 1966), and frog and cat muscle spindles (KATZ, 1950; EDWARDS, 1955; LIPPOLD et al., 1960; OTTOSON and SHEPHERD, 1968).

b) Generator Current → Impulse. A duration of the order of 2 msec is the maximal possible duration of the generator current in response to a single com-

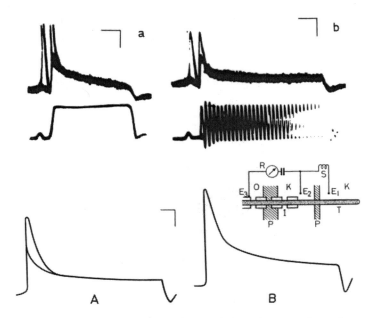

Fig. 13. Adaptation of impulse production. *Top, a*, Adaptation to prolonged generator potentials produced by stimulation of a decapsulated corpuscle with a long mechanical test stimulus. *b*, Adaptation to prolonged generator potentials produced by repetitive stimulation of an intact corpuscle. The test stimulus in *a* and *b* is preceded by an auxiliary stimulus of short duration critically at threshold for production of action potentials. The auxiliary pulse serves to raise the threshold of the test responses, which start during the relative refractory period of the action potential, so as to display the test generator response. The test generator response is just threshold for impulse firing at its beginning, but, thereafter, for at least 30 msec in *a* and through the entire period of stimulation in *b*, the generator response is of an amplitude greater than the minimum required for triggering of impulses at the beginning of a single stimulus. This minimum is displayed by the auxiliary response. Calibration: 5 msec, 50 μV. *Bottom*, Adaptation of impulse production to a constant current. Tracings of impulse responses to square pulses of current (1.5×10^{-8} A) passed outward through the membrane regions of the dendrite in pool E. A, Just threshold current; B, About $2 \times$ threshold. Lower trace in A is the resistive and capacitative coupled potential spread. Calibration: 5 msec, 50 μV (from LOEWENSTEIN and MENDELSON, 1965)

pression in the intact corpuscle in the absence of oscillations. The corresponding impulse response would thus be limited to very few impulses for this reason alone. In fact, the response to a single compression typically consists of one or two impulses. But aside from this cause and apparently quite independently of it, the number of impulses is limited by another mechanism operating at the level of impulse production. There are still only few impulses in response to long generator currents produced by single compression in decapsulated corpuscles (Loewenstein and Mendelson, 1965) or by repetitive compression at high frequency in intact corpuscles (Loewenstein, 1958), and only few impulses in response to a prolonged current from an external source, bypassing steps 1 to 4 in the transducer sequence (Mendelson and Loewenstein, 1964; Fig. 13).

Thus a second filter for slow electrochemical events operating at the level of step 5 limits the information flow in the receptor. The mechanisms underlying this filter action are not known. They are probably akin to those underlying 'accommodation' in various kinds of nerve fibers. Among several possibilities is a mechanism of slow sodium-inactivation[3], such as seen in certain crustacean nerve fibers (Narahashi, 1964).

The presence of two low-cut filters in the transducer sequence may at first seem surprising. But the arrangement may not be entirely redundant if the second filter has a lower frequency cutoff than the first. Because of the capacitance of the dendrite membrane, sustained depolarizations are built up in response to repetitive stimuli of high frequency that pass through the first filter (see Fig. 13

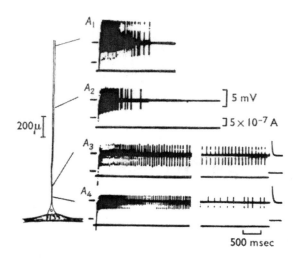

Fig. 14. Local membrane differentiation in adaptation. Impulse discharges in response to a constant current from various parts of the axon of the slowly adapting crayfish stretch receptor. The locations of the lead-off points are indicated on the left diagram. Note that only the axon portion close to the cell body gives a sustained response (from Nakajima and Onodera, 1969)

top b). The second filter would then prevent impulse production in response to such stimuli (which may be of particular physiological importance in the filtering out of the effects of stimulus after-oscillations). As a result of this arrangement of two filters in series, information transfer is restricted to single and rapid stimuli, which makes this an extraordinary phasic sensory system.

---

[3] The time constants of such an inactivation process would have to be slower than those in sodium-inactivation considered in Hodgkin and Huxley's (1952) theory; fast sodium-inactivation does not alone give rise to adaption (Fitzhugh, 1961; Stein, 1967).

A filter action at step 5 is probably quite common in sensory systems; in receptors more slowly adapting than the Pacinian corpuscle, the rate of adaptation may be determined by this action alone (e.g., HARTLINE et al., 1952; MACNICHOL, 1956; WENDLER and BURKHARDT, 1961; FUORTES and MANTEGAZZINI, 1961; FUORTES and POGGIO, 1962). Particularly revealing here is the recent work by NAKAJIMA and ONODERA (1969) on the crayfish stretch receptor neurons. In the RM_2 receptor, this kind of filter action is clearly the rate-limiting factor in adaptation. The filter action (with respect to constant current) appears to be present all along the afferent axon of the receptor, as it is also in the case of the Pacinian corpuscle (GRAY and MATTHEWS, 1951). Interestingly, it is also present along the axon of the much more slowly adapting RM_1 receptor, except for the initial axon portion where the impulse normally originates. The membrane of this portion appears to be different from that of the rest of the axon; it responds with a sustained impulse discharge to constant currents (Fig. 14). It will be interesting to see if other very slowly adapting receptors, such as the muscle spindle, share this differentiation.

4. Off-Response

It is a widespread property of phasic receptors to give generator currents both at the application (on-phase) and at the removal (off-phase) of the external stimulus. Part of the underlying mechanisms are now understood for the case of the Pacinian corpuscle. Energy stored in the elastic elements of the corpuscle during

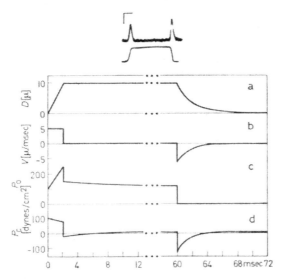

Fig. 15. Off-stimulus. Time course of computed velocities and pressures at the core of a Pacinian corpuscle under a compression of 20 μ applied uniformly in 2 msec and released suddenly during steady state conditions. D, Displacement of outer lamella (the stimulus); V, velocity of outer lamella; P_o, pressure on outer lamella; P_c, pressure on core. Top inset, Experimental on- and off-generator responses of a Pacinian corpuscle to compression and release similar to D. Upper oscilloscope beam, generator potential (calibration: 50 μV); lower beam, photoelectric record of displacement. Time calibration: 10 msec (from SKALAK and LOEWENSTEIN, 1966)

the on-phase is set free during the off-phase, and viscous pressure is produced anew. The overall pressure distribution is then rotated by 90° about the dendrite axis in this axially symmetric system, and the local velocity vectors are reversed; in other words, whereas the on-pulse was maximal in the plane defined by the axis and the point of stimulus application, a positive pressure pulse occurs once again during the off-phase, but this time in the plane through the axis and perpendicular to that during the on-phase. Analysis predicts that at a rate of release determined

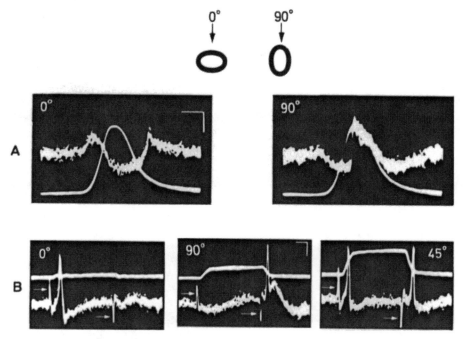

Fig. 16. Responses to compression and decompression at various angles of rotation of the dendrite of Pacinian corpuscle. One oscilloscope trace gives the photoelectric record of the displacement of the stimulating stylus; compression upwards. The second trace gives the electrical responses of the dendrite, local in A (calibration: 4 msec, 5 μV) and propagated in B (calibration: 2 msec, 30 μV). The value on the records gives the angle of rotation as defined in the inset where the arrow indicates the assumed direction of compression (after Ilyinsky, 1966a, b)

by the passive mechanical constants of the capsule (unrestrained by external forces), the amplitude of the pressure pulse is of a similar order to that during the on-phase (Loewenstein and Skalak, 1966; Fig. 15). Thence, if the off-pressure pulse is in the direction of the minor elliptical axis of the dendrite, membrane distension and consequently generator currents ensue (see p. 270). This provides a simple explanation for the on- and off-responses of the receptor when the elliptical axis is obliquely oriented with respect to the direction of compression – probably the most common case in experimentally isolated corpuscles. However, on- and off-responses appear to result also with parallel or perpendicular orientations. Indeed, it is a common observation that sufficiently strong stimuli elicit on- and off-responses at any orientation. Ilyinsky (1966) has therefore advanced the hypothesis that the pressure field inside the capsule is sufficiently undamped so that the first oscillation in opposite phase to the initial pressure pulse gives rise to a substantial generator current. Fig. 16 illustrates experimental results that are in good qualitative agreement with the hypothesis.

References

ALBUQUERQUE, E. X., GRAMPP, W.: Effects of tetrodotoxin on the slowly adapting stretch receptor neurons of lobster. J. Physiol. (Lond.) 195, 141—156 (1968).
— CHUNG, S. H., OTTOSON, D.: The action of tetrodofoxinon the frog's isolated muscle spindle. Acta physiol. scand. 75, 301—312 (1969).
ALVAREZ BUYLIA, R., ARELLANO, I. R. DE: Local responses in Pacinian corpuscles. Amer. J. Physiol. 172, 237—250 (1953).
— REMOLINA, J.: The initiation of action potentials at Pacinian corpuscles. Acta Physiol. Lat.-Amer. 9, 178—187 (1959).
ARCHER, R. J., LAMER, V. K.: The rate of evaporation of water through fatty acid membranes. J. Phys. Chem. 59, 200—208 (1955).
AUTRUM, H.-J., SCHNEIDER, D.: Der Kälteblock der einzelnen markhaltigen Nervenfaser. Naturwissenschaften 37, 21—22 (1950).
BERNHARD, C. G., GRANIT, R., SKOGLUND, C. R.: The breakdown of accomodation. Nerve as model sense organ. J. Neurophysiol. 5, 55—68 (1942).
BULLOCK, T. H.: Initiation of nerve impulses in receptor and central neurons. Rev. mod. Phys. 31, 504—514 (1958).
BURKHARDT, D., GEWECKE, M.: Mechanoreception in arthropoda: The chain from stimulus to behavioral pattern. Cold Spr. Harb. Symp. quant. Biol. 30, 601—614 (1965).
CATTON, W. T.: A comparison of the responses of frog skin receptors to mechanical and electrical stimulation. J. Physiol. (Lond.) 187, 23—33 (1966).
Ciba Foundation Symposium: Touch, Heat and Pain. Eds.: REUCK, A. S. DE, KNIGHT, J. Boston: Little, Brown and Co. 1966.
DEL CASTILLO, J., KATZ, B.: The membrane change produced by the neuromuscular transmitter. J. Physiol. (Lond.) 125, 546—565 (1954).
— — Biophysical aspects of neuromuscular transmission. Progr. Biophys. 6, 121—170 (1956).
DIAMOND, J., GRAY, J. A. B., INMAN, D. R.: The relation between receptor potential and the concentration of sodium ions. J. Physiol. (Lond.) 142, 382—394 (1958).
— — SATO, M.: The site of initiation of impulses in Pacinian corpuscles. J. Physiol. (Lond.) 133, 54—67 (1956).
ECCLES, J. C.: The physiology of nerve cells. Baltimore: The Johns Hopkins Press 1957.
EDWARDS, C.: Changes in the discharge from a muscle spindle produced by electrotonus in the sensory nerve. J. Physiol. (Lond.) 127, 636—640 (1955).
— OTTOSON, D.: The site of impulse initiation in a nerve cell of a crustacean stretch receptor. J. Physiol. (Lond.) 143, 138—148 (1958).
EYZAGUIRRE, C., KUFFLER, S.: Processes of excitation in the dendrites and in the soma of single isolated sensory nerve cells of the lobster and crayfish. J. gen. Physiol. 39, 87—119 (1955).
FATT, P.: The electromotive action of acetylcholine at the motor end-plate. J. Physiol. (Lond.) 111, 408—422 (1950).
— KATZ, B.: An analysis of the end-plate potential recorded with an intracellular electrode. J. Physiol. (Lond.) 115, 320—370 (1951).
FITZHUGH, R.: Impulses and physiological states in theoretical models of nerve membrane. Biophys. J. 1, 445—466 (1961).
FLOCK, A.: Transducing mechanisms in the lateral line canal organ receptors. Cold Spr. Harb. Symp. quant. Biol. 30, 133—144 (1965).
— WERSALL, J.: A study of the orientation of the sensory hairs of the receptor cells in the lateral line organs of fish with special reference to the function of receptors. J. Cell Biol. 15, 19—27 (1962).

FUORTES, M. G. F.: Initiation of impulses in visual cells of Limulus. J. Physiol. (Lond.) **148** 14—28 (1959).
— HODGKIN, A. L.: Changes in time scale and sensitivity in the ommatidia of Limulus. J. Physiol. (Lond.) **172**, 239—263 (1964).
— MANTEGAZZINI, F.: Interpretation of the repetitive firing nerve cells. J. gen. Physiol. **45**, 1163—1179 (1962).
— POGGIO, G. F.: Transient responses to sudden illumination in the cells of the eye of Limulus. J. gen. Physiol. **46**, 435—452 (1962).
FURSHPAN, E. J., FURUKAWA, T.: Intracellular and extracellular responses of the several regions of the Mauthner cell of the goldfish. J. Neurophysiol. **25**, 732—771 (1962).
GRANIT, R.: Receptors and sensory perception. New Haven: Yale University Press 1955.
GRAY, J. A. B.: Mechanical into electrical energy in certain mechanoreceptors. Progr. Biophys. biophys. Chem. **9**, 285—324 (1959). London: Pergamon Press 1959.
— DIAMOND, J.: Pharmacological properties of sensory receptors. Brit. med. Bull. **13**, 185—188 (1957).
— MATTHEWS, P. B. C.: A comparison of the adaptation of the Pacinian corpuscle with the accomodation of its own axon. J. Physiol. (Lond.) **114**, 454—464 (1951).
— RITCHIE, J. M.: Effects of stretch on single myelinated nerve fibres. J. Physiol. (Lond.) **124**, 84—99 (1954).
— SATO, M.: Properties of the receptor potential in Pacinian corpuscles. J. Physiol. (Lond.) **122**, 610—636 (1953).
GRUNDFEST, H.: Electrical inexcitability of synapses and some consequences in the central nervous system. Physiol. Rev. **37**, 337—361 (1957).
— Heterogeneity of excitable membrane: electrophysiological and pharmacological evidence and some consequences. Ann. N. Y. Acad. Sci. **137**, 901—949 (1966).
— Tetrodotoxin: action on graded responses. Science **156**, 1771 (1967).
HARTLINE, H. K., COULTER, N. A., JR., WAGNER, H. G.: Effects of electric currrent on responses of single photoreceptor units in the eye of Limulus. Fed. Proc. **11**, 65—66 (1952).
— WAGNER, H. G., MacNICHOL, E. F.: The peripheral origin of nervous activity in the visual system. Cold Spr. Harb. Symp. quant. Biol. **17**, 125—141 (1952).
HODGKIN, A. L.: The conduction of the nervous impulse. Springfield, Ill.: Charles C. Thomas, Publ. 1964.
— The local electric changes associated with repetitive action in a non-medullated axon. J. Physiol. (Lond.) **107**, 165—181 (1948).
— HUXLEY, A. F.: A quantitative description of membrane current and its application to conduction and excitation in nerve. J. Physiol. (Lond.) **117**, 500—544 (1952).
— KATZ, B.: The effect of temperature on the electrical activity of the giant axon of the squid. J. Physiol. (Lond.) **109**, 240—249 (1949).
HUBBARD, S. J.: A study of rapid mechanical events in a mechanoreceptor. J. Physiol. (Lond.) **141**, 198—218 (1958).
HUNT, C. C., TAKEUCHI, A.: Responses of the nerve terminal of the Pacinian corpuscle. J. Physiol. (Lond.) **160**, 1—21 (1962).
ILYINSKY, O. B.: Process of excitation and inhibition in single mechano-receptors (Pacinian corpuscles). Nature (Lond.) **208**, 351—353 (1964).
— Nekotorie aspektyi deyatelnosti retseptornyikh apparatov. Sechenov Physiol. J. U.S.S.R. **52**, 360—369 (1966a).
— Electrophisiologia mechanoreceptornich elementov. In: Pervitchnie processi v receptornich elementach organovtchuvst. Eds.: BRONSTEIN, A. A., ILYINSKY, O. B., SAMSONOVA, V. G. Moscow-Leningrad: Otdelnie Ottisk Isdatelstvo "Nauka" 1966b.
INMAN, D. R., PERUZZI, P.: The effects of temperature of the responses of the Pacinian corpuscle. J. Physiol. (Lond.) **155**, 280—298 (1961).
ISHIKO, N., LOEWENSTEIN, W. R.: Effects of temperature on the generator and action potentials of a sense organ. J. gen. Physiol. **45**, 105—124 (1961).
JULIAN, F., GOLDMAN, D.: The effects of mechanical stimulation on some electrical properties of axons. J. gen. Physiol. **46**, 297—313 (1962).
KATZ, B.: Depolarization of sensory terminals and the initiation of impulses in the muscle spindle. J. Physiol. (Lond.) **111**, 261—282 (1950).

KATZ, B.: Nerve, muscle, and synapse. New York: McGraw-Hill Book Co., Inc. 1966.
— MILEDI, R.: Propagation of electric activity in motor-nerve terminals. Proc. Roy. Soc. B 161, 453—482 (1965a).
— — The measurement of synaptic delay and the time course of acetylcholine release at the neuromuscular junction. Proc. Roy. Soc. B 161, 483—495 (1965b).
LIPPOLD, O. C. J., NICHOLLS, J. G., REDFEARN, J. W. T.: Electrical and mechanical factors in the adaptation of a mammalian muscle spindle. J. Physiol. (Lond.) 153, 209—217 (1960).
LOEWENSTEIN, W. R.: Excitation and changes in adaptation by stretch of mechanoreceptors. J. Physiol. (Lond.) 133, 588—602 (1956a).
— Modulation of cutaneous mechanoreceptors by sympathetic stimulation. J. Physiol. (Lond.) 132, 40—60 (1956b).
— Generator processes of repetitive activity in a Pacinian corpuscle. J. gen. Physiol. 41, 825—845 (1958).
— The generation of electric activity in a nerve ending. Ann. N. Y. Acad. Sci. 81, 367—387 (1959).
— Biological transducers. Sci. Am. 203, 98—107 (1960).
— Excitation and inactivation in a receptor membrane. Ann. N. Y. Acad. Sci. 94, 510—534 (1961a).
— On the specificity of a sensory receptor. J. Neurophysiol. 24, 150—158 (1961b).
— Biological transducers. In: The encyclopedia of electrochemistry. New York: Reinhold Publ. Corp. 1964.
— Facets of a transducer process. Cold Spr. Harb. Symp. quant. Biol. 30, 29—43 (1965).
— Rate sensitivity in a biological transducer. Ann. N. Y. Acad. Sci. 156, 892—900 (1969).
— ALTAMIRANO-ORREGO, R.: The refractory state of the generator and propagated potentials in a Pacinian corpuscle. J. gen. Physiol. 41, 805—824 (1958).
— MENDELSON, M.: Components of receptor adaptation in a Pacinian corpuscle. J. Physiol. (Lond.) 177, 377—397 (1965).
— RATHKAMP, R.: The sites for mechano-electric conversion in a Pacinian corpuscle. J. gen. Physiol. 41, 1245—1265 (1958).
— SKALAK, R.: Mechanical transmission in a Pacinian corpuscle. J. Physiol. (Lond.) 182, 346—378 (1966).
— TERZUOLO, C. A., WASHIZU, Y.: Separation of transducer and impulse-generating processes in sensory receptors. Science 142, 1180—1181 (1963).
MACNICHOL, E. F., JR.: Visual receptors as biological transducers. In: Molecular structure and functional activity of nerve cells. Eds.: GRENELL, R. G., MULLINS, L. J. Washington: Amer. Inst. Biol. Sci. Publ. 1956.
MENDELSON, M., LOEWENSTEIN, W. R.: Mechanisms of receptor adaptation. Science 144, 554—555 (1964).
NAKAJIMA, S., ONODERA, K.: Membrane properties of the stretch receptor neurones of crayfish with particular reference to mechanisms of sensory adaptation. J. Physiol. (Lond.) 200, 161—185 (1969a).
— — Adaptation of the generator potential in the crayfish stretch receptors under constant length and constant tension. J. Physiol. (Lond.) 200, 187—204 (1969b).
NARAHASHI, T.: Restoration of action potential by anodal polarization in lobster giant axons. J. cell. comp. Physiol. 64, 73—96 (1964).
— MOORE, J. W., SCOTT, W.: Tetrodotoxin blockage of sodium conductance increase in lobster giant axon. J. gen. Physiol. 47, 965—974 (1964).
NISHI, K., SATO, M.: Blocking of the impulse and depression of the receptor potential by tetrodotoxin in nonmyelinated nerve terminals in Pacinian corpuscles. J. Physiol. (Lond.) 184, 376—386 (1966).
OTTOSON, D.: The effect of sodium deficiency on the response of the isolated muscle spindle. J. Physiol. (Lond.) 171, 109—118 (1964).
— SHEPHERD, G. M.: Receptor potentials and impulse generation in the isolated spindle during controlled extension. Cold Spr. Harb. Symp. quant. Biol. 30, 105—114 (1965).
— — Changes of length within the frog muscle spindle during stretch as shown by stroboscopic photomicroscopy. Nature (Lond.) 220, 912—914 (1968).

Ozeki, M., Sato, M.: Initiation of impulses at the non-myelinated nerve terminal in Pacinian corpuscles. J. Physiol. (Lond.) **170**, 167—185 (1964).

Pacini, F.: Nuovi organi scoperti nel corpo umano. Pistoja: Tipografia Cino 1840.

Pease, D. C., Quilliam, T. A.: Electron microscopy of the Pacinian corpuscle. J. biophys. biochem. Cytol. **3**, 331—357 (1957).

Perkel, D. H., Bullock, T. H.: Neural coding. Neurosci. Res. Progr. **6**, 221—348 (1969).

Rosano, H. L., LaMer, V. K.: The rate of evaporation of water through monolayers of esters, acids and alcohols. J. phys. Chem. **60**, 348—353 (1956).

Sato, M., Ozeki, M.: Response of the non-myelinated terminal in Pacinian corpuscles to mechanical and antidromic stimulation and the effect of procaine, choline and cooling. Jap. J. Physiol. **13**, 564—582 (1963).

Schumacher, S.: Beiträge zur Kenntnis des Baues und der Funktion der Lamella Körperchen. Arch. mikr. Anat. **77**, 157—193 (1911).

Stein, R. B.: The frequency of nerve action potentials generated by applied currents. Proc. Roy. Soc. B. **167**, 64—86 (1967).

Stoney, S. D., Machne, X.: Mechanisms of accommodation in different types of frog neurons. J. gen. Physiol. **53**, 248—262 (1969).

Takeuchi, A., Takeuchi, N.: On the permeability of end-plate membrane during the action of transmitter. J. Physiol. (Lond.) **154**, 52—67 (1960).

Tasaki, J., Fujita, M.: Action currents of single nerve fibers as modified by temperature changes. J. Neurophysiol. **11**, 311—315 (1948).

Teorell, T.: Some biophysical considerations on presso-receptors. Arch. Int. Pharmacodyn. Therapie **140**, 563—576 (1962).

— Electrokinetic considerations of mechano-electrical transduction. Ann. N. Y. Acad. Sci. **137**, 950—966 (1966).

Terzuolo, C. A., Washizu, Y.: Relation between stimulus strength, generator potential and impulse frequency in stretch receptor of crustacea. J. Neurophysiol. **25**, 56—66 (1962).

Thurm, U.: Das Rezeptorenpotential einzelner mechanorezeptorischer Zellen von Bienen. Z. vergl. Physiol. **48**, 131—156 (1964).

— An insect mechanoreceptor. Part II: Receptor potentials. Cold Spr. Harb. Symp. quant. Biol. **28**, 83—94 (1965).

Vallbo, A. B.: Accommodation of single myelinated nerve fibres from Xenopus laevis related to type of end organ. Acta physiol. Scand. **61**, 413—428 (1964).

Wendler, L., Burkhardt, D.: Zeitlich abklingende Vorgänge in der Wirkungskette zwischen Reiz und Erregung (Versuche an abdominalen Streckrezeptoren dekapoder Krebse). Z. Naturforsch. **16b**, 464—469 (1961).

Wiersma, C. A. G., Furshpan, E., Florey, E.: Physiological and pharmacological observation on muscle receptor organs of the crayfish, Cambarus clarkii Girard. J. exp. Biol. **30**, 136—150 (1953).

Wolbarsht, M. L., Hanson, F. E.: Electrical activity in the chemoreceptors of the blowfly. III: Dendritic action potentials. J. gen. Physiol. **48**, 673—683 (1965).

Chapter 10

A Biophysical Analysis of Mechano-electrical Transduction

By

Torsten Teorell, Uppsala (Sweden)

With 29 Figures

Contents

Introduction

This treatise is an attempt to approach some basic problems of biological excitability in terms of nonlinear oscillation theory applied to membrane systems where there is a coupling between electrical and mechanical forces. The basic concept assumes that the excitable membrane contains *fixed charges*, i.e., ionic membranes of ion-exchange character. Such membranes are known to exhibit *electrosmosis*, i.e., transport of water owing to the combined effect of an electrical potential gradient and a hydrostatic pressure gradient. This mutual interdependence is the basis of the essential coupling between the mechanical and electrical events in the pressoreceptor analogs to be treated in this paper. In essence, our analysis is a combination of classical electrokinetics and nonlinear mechanics. It is believed that this gives a better overall description of the salient membrane phenomena. It deals with the total effects of the participating ions rather than with specific actions of particular ions such as sodium and potassium (as in the dominating Hodgkin-Huxley concepts). It should be stressed that the Hodgkin-Huxley mathematical description as well as the one presented here are formal representations and not necessarily mechanistic descriptions of events on the molecular membrane level. The main difference lies in the basic assumptions. The Hodgkin-Huxley hypothesis describes the membrane conductance in terms of specific voltage-time-dependent differences in permeabilities of the potassium and sodium ions. The electrokinetic theory employs a fixed-charge character of the membrane and water transport concomitant with the ion transport. An important feature is also the consideration of the *viscoelastic properties* of the cell membrane; in fact, this leads to the creation of the "frequency modulation", the signal code of the biological transducers.

Papers dealing with the development of the presented concept have previously been presented (TEORELL, 1957, 1959a, b, 1962a). These dealt with the description of a physical-chemical model, the "membrane oscillator", which exhibits oscillatory phenomena and which can be stimulated by both electrical and mechanical means (pressure). A series of papers dealt with the conversion of the physical model into the "electrohydraulic excitability analog" (TEORELL, 1959c, 1960). Two recent papers deal specifically with mechano-electrical transduction problems and should be consulted for details pertaining to this presentation (TEORELL, 1962b, 1966).

I. The Sequence of Events in Mechano-electrical Transduction

The following scheme is based on opinions expressed by a great number of previous authors in the field, but it is modified in some respects.

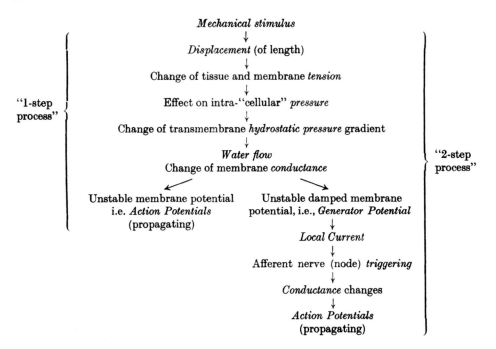

The scheme above is only concerned with the biophysical transduction actions of the mechanoreceptors. Possible interference from chemical events, perhaps one source of adaptation or fatigue, will be tacitly neglected. In accordance with a previous paper (TEORELL, 1966), we will consider both a "1-step" process, where the excitable element is directly susceptible to mechanical stimuli and is able to produce conducted action potentials (JULIAN and GOLDMAN, 1962), and a "2-step" process involving a primary process, i.e., the formation of a generator potential[1] (current) and the subsequent stimulation of the nerve element. These two alternatives are schematically sketched in Fig. 1. A salient feature of this scheme

---

[1] This paper adheres to the discrimination suggested by WOLBARSHT (1960). He reserved the term "generator potential" for those potentials existing intracellularly at the site of impulse initiation, whereas the term "receptor potentials" apply to those nonpropagated potentials recorded near a receptor. Hence, the receptor potential acts as a local current source to cause depolarization of some part of the neuronal membrane, which can, when sufficiently depolarized, initiate propagated impulses. The site of origin for the impulses may be quite different and remote from the site of origin for the receptor potential. In this paper, "generator potential" (GP) will refer to the actual sensor potential (generator potential proper); that of the afferent node potential is denoted "receptor potential proper" (RP). When the node is used for the recording of the GP it has usually been "anesthetized" or otherwise treated so that *spike* initiations are abolished. See also the classical work by GRANIT (1955) which should be consulted for all problems in receptor physiology.

is that the fullblown action potential (which can propagate) and the local graded generator potential are both regarded as membrane oscillation systems belonging to the same type of oscillator family; the only difference is the degree of damping, which is larger in the case of generator potentials.

This close relationship between the origin of the generator potential and the action potential, different as it might appear on a recording, makes it possible first to analyze the "1-step process" and then to use the results for a following extension to the "2-step mode." It may well be envisaged that both processes might be utilized in biology. The 1-step can be a more "primitive" mechanism; the 2-step,

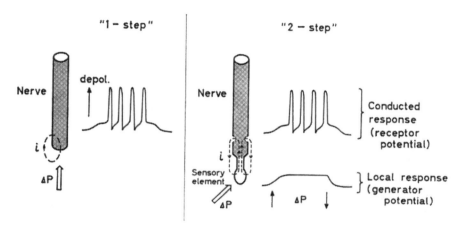

Fig. 1. Scheme illustrating two ways of transforming the pressure stimulus to a conducted electrical response. *Left side*: A direct-coupling, "1-step process". Here the stimulus acts at the same nerve element as is initiating the conducted potential responses. *Right side*: A spatial or functional separation of the receptor element and the conductile element ("2-step process"). Here the generator potential is elicited in a specific receptor from which a generator current triggers the repetitive response in the conductile element proper. ΔP = pressure stimulus.

with morphologically specialized sensor elements, may be used when certain requirements such as "filtering" are fulfilled (as in the Pacinian corpuscles; (LOEWENSTEIN and SKALAK, 1966; LOEWENSTEIN, 1966a,b). The novelty in the picture is, of course, not in the pictorial division of two processes but rather in the proposed intimate relation in the mechanism of the two types of potentials.

Before an account of the description of the pressure-sensitive membrane oscillator and its theory is presented, it may be appropriate to discuss first some current opinions about the purely mechanical properties of cell and tissue structures, i.e., the visco-elastic properties.

II. The Mechanical Properties of Mechanoreceptors

1. The Relation between Mechanical Displacement (Stretch) and Tension

In the experimental procedures used to study stretch-sensitive isolated receptors or compound tissues, one has most commonly employed some form of mechanical device which has caused a measurable *length* displacement. Most workers, however, agree that it is the ensuing change in *tension* rather than the actual length displacement which acts as a primary stimulus (see the review in OTTOSON's article in this volume). Those who have measured both displacement and tension simultaneously find that length and tension curves run only very approximately a similar course. There are, in fact, quite important differences which are shown by the results of many. For example, TERZUOLO and WASHIZU (1962) spoke of an "overshoot of tension" at a ramp stretch. KRNJEVICS and VAN GELDER (1961) stated that "tension did not increase linearly but more and more steeply and it reached a peak and then more or less slowly declined" (see also HUBBARD, 1958). They showed also hysteresis phenomena at a cycling of the length changes. BROWN and STEIN (1966) and BROWN (1967) also described a similar relation, which had already been anticipated by MATTHEWS (1931, 1962). Most workers ascribe these differences to interference of "viscoelastic" properties of

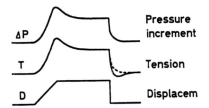

Fig. 2. Scheme of the relation between length displacement, tension and assumed trans-membrane pressure change

the tissue elements. To find examples of more well-defined models, one might refer the thorough papers of HOUK (1966); MILHORN (1966); and CATTON and PETOE (1967). Fig. 2 is a scheme redrawn from a recent paper by RACK and WESTBURY (1966). The scheme represents only a gross picture of the relationship. The tension response depends both on the amplitude of the displacement and, above all, on the *rate* of length change. One deals with a typical "derivative" sensitive process. A negative going derivative is equally important, as the length-tension curve might form a hysteresis loop (ROBERTS, 1963).

On the nature of the mechano-electrical transduction. Exactly how the tension changes in a tissue or sensor element elicit the electrical response is not known at the present time. There exist many hypotheses as to "opening up" of pores, changes of ion permeability, etc. (see the review articles by GRAY, 1959; LOEWEN-STEIN, 1966a,b; GOLDMAN, 1966; and articles in this volume). The intention of this paper is to suggest a new hypothetical model for mechano-electric transduction. From the point of view of this "electrokinetical" receptor analog, the interrelation between "cell" membrane tension and the transmembrane pressure difference comes into focus. In fact, the transmembrane pressure force appears here as the final trigger in the long chain of transformations from the external mechanical stimulus to the local electrical events (see tabulation on p. 293).

2. The Forces Acting on a Sensory Unit

A scheme of this is epicted in Fig. 3; a more detailed picture of the "anatomy" is given in Fig. 9. The sensory unit is here envisaged as a small bag suspended in extracellular fluid. The bag membrane is regarded as sufficiently porous to admit both ions *and* water. The membrane also has a transmembrane electrical potential which will be elaborated upon in more detail later (see chapter on the "Membrane Oscillator", p. 297). *Electroosmosis*, driving liquid inwardly, will pump up the bag and inflate it, hence distending the wall structures which are regarded to have certain viscoelastic properties. In a steady state, the electroosmotic pumping force will be balanced by the counterpressure exerted via the wall tension. The exact

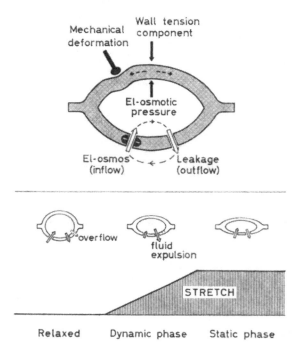

Fig. 3. Scheme of the forces acting on a pressure-sensitive sensory unit. The "turgor" pressure inside is created mainly by an electroosmotic inflow of fluid and is counterbalanced by the wall tension forces. An external deformation force, or stretch, disturbs the magnitude and sign of the various types of bulk flow, thereby creating conductance and potential changes as indicated in Fig. 8. *Bottom*: Scheme of the deformation of the sensory unit subject to a ramp-shaped stretch stimulus. Note that the "overflow" channel, which "locks" oscillations, becomes inoperative during stretch. The diagrams are purely schematic

relation between the external stretch force and the transmembrane pressure difference is certainly complicated, particularly for transient states which occur at a dynamic change of the length displacement. Furthermore, it should be observed that the bag is regarded as leaky (Fig. 9). Some "pores" (charged) convey inward or outward movements of fluid; other pores (noncharged) convey outward movements of fluid, i.e., "leakage". Physically, the problem is related to that of a leaky elastic ballon, simultaneously subject to filling and externally applied deformation. It is to be expected that the character of the viscoelasticity may be of great importance for the outcome of the relations between length-tension-internal pressure. Since at the present time one entirely lacks information pertaining to cell membrane levels, we will make a tacit approximation in this article that *the change of wall tension will be equivalent to change of transmembrane pressure.*

3. The Relation between the Sensor Volume and the Transmembrane Pressure Difference

is a very important one in the present context. Obviously, this is again dependent on the mechanical viscoelastic properties of the "cell" envelope (the "bag membrane"). Offhand, quite different types of relations can be anticipated as sketched in Fig. 4. All three types are known from macroscopical biological tissues

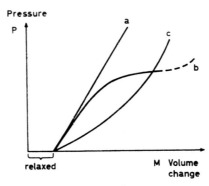

Fig. 4. Various relations between "cell" volume and transmembrane pressure. Line *a* gives a very limited range of frequency modulation. Line *b* has been assumed in the present analysis because it yields a marked frequency modulation. All types may have physiological counterparts. Hysteresis phenomena arising from time-dependent viscoelastic properties have been neglected

(see, for instance, REMINGTON's book). In the mechanoreceptor analog of this paper, we have elected type *b* of the diagram, because this is the one which induces the proper *frequency modulation*, i.e., increase of firing frequency with pressure. In fact, the assumption of a nonlinear relationship between sensor volume and pressure is the backbone of the whole presentation. The linear relation, type *a*, yields an almost constant frequency of firing. Further discussion of these problems will be taken up later (p. 299).

III. Description of a Pressure-sensitive Membrane Oscillator

The following is a recapitulation of a series of previous papers dealing with a physical model, which has a built-in coupling between external pressure and electrical events (TEORELL, 1957—1962). A special modification of this model called the *membrane oscillator* has recently been described (TEORELL, 1966).

1. A Scheme of the Physical Model

is shown in Fig. 5. It contains two compartments, one "inside" (left) and one "outside" (right). The two compartments are separated by a membrane which is porous enough to permit passage of both water and ions. Some of the membrane pores contain fixed charges, in this case negative; other membrane pores have no charge ("leakage"). The two compartments are filled with an electrolyte solution (e.g., NaCl), so that the inside has a higher concentration than the outside. This requirement is not obligatory, because with some types of fixed-charge membranes the model works equally well with no concentration differences (TEORELL, 1961a). The membrane is polarized by an externally applied constant electrical current. Provided that the current density exceeds a certain threshold value, this device will exhibit oscillatory movements both of water and of ions

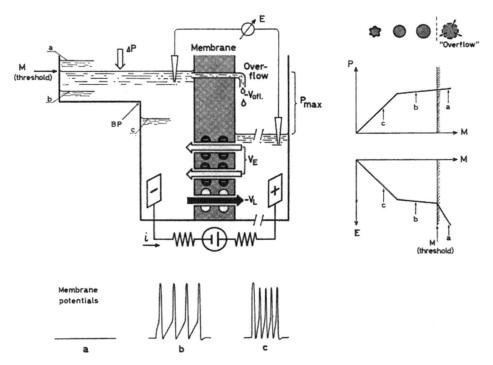

Fig. 5. Scheme of the physical membrane oscillator. — The left hand chamber represents the "receptor unit", the right-hand side the "extracellular medium". For a further description, see the text. *Bottom*: The membrane potentials obtained at different free-liquid levels indicated a, b and c. Note the frequency modulation. The *inset* (right) illustrates the nonlinear pressure-volume relationship of the model. The cell symbols above denote schematically the shape changes of the "cell". The second inset shows the relation between the cell volume change, M, and the membrane potential ,E, at steady state condition (voltage clamp). a, b and c refer to corresponding positions in the main figure

across the membrane, which lead to periodic conductance changes within the membrane. As a consequence, the transmembrane potential will also exhibit a rhythmical "response" of a type similar to nerve action potentials.

A *detailed theory* of this "membrane oscillator", later modified as an "electro-hydraulic excitability analog", can be found in previous publications (TEORELL, 1959b,c, 1962). The theory is based on seven fundamental equations derived from the laws of electrodiffusion and electrokinetics[2]. The salient feature of the

---

[2] See Appendix, p. 334.

membrane is the presence of fixed charges, by which movements of water are induced, i.e., *electroosmosis*. The electroosmotic process creates a difference in the hydraulic pressure between the two compartments as indicated in the figure by P_{max}. In fact, it is the *coupling* between the electrical potential gradient, the pressure gradient, and the chemical potential gradient which leads to the *metastability* of the system, in the sense that it can be triggered to produce transient or periodic oscillations when subject to various types of stimuli (electrical, pressure, or chemical). The conditions for metastability and rhythmicity are governed by various parameters, such as the "hydraulic permeability" and the "electroosmotic permeability" (proportional to fixed charge density), the membrane thickness and other factors (see p. 304).

In order to incorporate the required nonlinear relationship between pressure (P) and sensor volume (M), the prerequisite for *frequency modulation* (see p. 297), the model of Fig. 5 has been given two important features. First, the left-hand compartment has an enlargement on the top: secondly, there is an "extra" leakage channel, the "overflow", which becomes effective only when the solution volume (M) becomes sufficiently large. The spatial geometry of this "inside" chamber leads to the break in the *pressure-volume* relationship (indicated in the right part of the figure). In fact, such a nonlinear pressure-volume relationship is similar to those found in biological tissues (see p. 297). Furthermore, the overflow may simulate the "overstretching" of cell membranes, which may occur when the cells are in a swollen very leaky state; for instance, red blood corpuscles in hypotonic solutions exhibit this overstretching when they are on the verge of hemolysis. The hypothetical simulation of the left-hand chamber by actual "cell" elements is indicated in the figure by the cell symbols, see top right. It should be noted that the model operates with some limitation, both of the maximal hydraulic pressure gradient, and of the "cell" volume.

The *functional relationships* between the various factors which operate in the model are schematically demonstrated in Fig. 6a. It is based on the basic equation systems referred to above, and is at the same time a provisional *operational program* for analog computation (TEORELL, 1959b,c, 1966). It is evident from the figure that one deals with a tightly coupled system between the transmembrane potential, E, the hydraulic pressure gradient, P, the transmembrane resistance, R, and the net water transport velocity, W. The volume changes of the system, M, are obtained as the $\int W$ by means of integration. The figure is identical with our previous schemes of the electrohydraulic nerve analog with the exception of an added overflow loop, marked in the figure by a diode symbol and a device for generating the non-linear P-M relation.

Fig. 6b depicts a three-dimensional representation of the coupling between the potential, E, the pressure, P, and the current I (for various "clamped" P values). One should note the *negative* slope conductance values (dI/dE) at the higher pressures (*i. e.*, a "dynatron" or tunnel diode characteristic).

The results to be presented have been obtained by an *analog computer*, which simulates the physical model. Experiments with a real physical model (see model set up by TEORELL, 1959a) have shown that the theoretical equations quite well represent the actual behavior of the physical model.

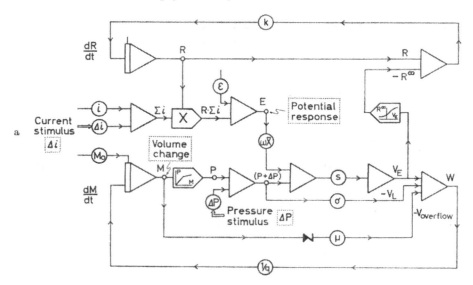

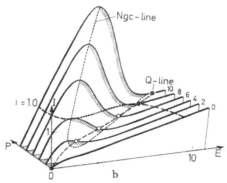

Fig. 6a and b. a) Operational scheme of the modified membrane oscillator, i.e. the artificial pressoreceptor analog. The scheme is based on the seven mathematical equations which govern the system (see Appendix). The scheme is also a *program for the analog computation* procedure (although necessary inverters, etc. have been omitted). The scheme also comprises two function generators for the P-M and R^∞-V_E relation and a multiplier, X. The main *symbols* are: $E =$ membrane potential; $R =$ instantaneous, integral membrane resistance; $R^\infty =$ steady state membrane resistance; $P =$ transmembrane pressure difference; $V_E =$ electroosmotic bulk flow velocity; $V_L =$ leakage channel flow velocity; $V_{overflow} =$ overflow leakage velocity, operative only at $M > M_{threshold}$; $W =$ net transmembrane flow velocity; and $M =$ volume change from the initial volume, M_0. The *coefficients* are: $s =$ hydraulic permeability constant of fixed charge route, $\sigma =$ hydraulic permeability constant for main leakage route; $\mu =$ hydraulic permeability constant for overflow route. $w\overline{X}$ denotes the sign and magnitude of the fixed charge density controlling the electroosmosis; $\varepsilon =$ intramembrane "diffusion" potential; $k =$ an integration constant dependent on membrane thickness, concentration and ion mobilities (k controls the rate of resistance change). $1/q =$ rate constant controlling the volume change, $i =$ a constant polarization current. $\Delta i =$ current stimulus; $\Delta P =$ pressure stimulus, both fed from suitable waveform generators. After the coefficients have been set at selected values, the external stimuli (ΔP or Δi) were applied and the resulting effects read off, for example, as a "potential response" at E or at any other desired "state variable" available as a read-out. For further information, see text and Appendix

b) A three-dimensional representation of the interrelations between the membrane potential, E, the transmembrane pressure, P, and the polarization current, I. These graphs are generated from the basic equations and represent *current-voltage characteristics* at various clamped P-values. The shadowed contours embodied by the "*Ngc-line*" denote the phase-space where "negative conductance" is operative. The "*Q-line*" represents the steady state current-voltage characteristic. The filled points are stable quiescent points, the unfilled are unstable; i.e., oscillations of E (and P) take place around these. See also Appendix part 3 and Fig. 29, for a section perpendicular to the E-axis, representing an E-P phase plane. This phase-space plot is based on the somewhat simpler conditions used by Teorell (1959b)

2. Some Illustrations of the Features of a Membrane Oscillator as a Mechanoreceptor Analog

Since a mathematical description in terms of the theoretical equations of this non-linear oscillation system would be too cumbersome, we will resort to a qualitative and graphical description of some conspicuous features for the purpose of the present paper. First, it should be stated that the quiescent or *resting state* (the Q-point) is a damped although highly metastable condition, with a maximal volume filling as indicated by the free liquid level (a) in Fig. 5 (insets). This quiescent state is conditioned by the effect of the extra leakage channel operating as the "overflow" in a model. An external tension increase corresponds to the application of a pressure load ΔP on top of the liquid level of the left-hand chamber[3]. If the pressure addition is sufficient, the free liquid level (a) will be pressed down to the level indicated by (b) where the overflow becomes inoperative. Hence, the main stabilizing damping factor (i.e., the overflow leakage) will be removed, and the system goes into a regenerative, sustained and *slow* oscillation. A further increase of external pressure will expel more water out of the cell by overriding the inwardly directed electroosmotic flow. A new mean equilibium is reached at the level (c). The oscillation frequency will now be *faster* as the pressure excursions will take place more rapidly, because one now operates on the steeper part of the pressure-volume relationship. In fact, the theory anticipates that the slopes of the P-M relation determine the oscillation frequency, other conditions remaining unaltered. The membrane-potential patterns for the three conditions (a, b, c) are depicted in the lower part of the diagram.

Briefly stated, the electrical potential spikes are accompanied by volume changes of the sensory unit cell, owing to a cyclical outward and inward flow of liquid across the excitable membrane. The *"frequency span"*, i.e., the ratio between the minimum and maximal frequency, is mainly determined by the slopes of the extreme parts of the P-M relation. A continuous variation of frequency is due to overlapping of the oscillations of the different volume domains.

Phase plots. (a) Effects of Pressure Stimuli. To further illustrate the behavior of the system, a *phase plot* is presented of the *membrane potential* vs. the *volume change* (Fig. 7, *Top*). Here case (a) represents the subthreshold "graded" depolarization at a small pressure stimulus, ΔP. Note that the "path" (dotted) now moves slightly on the left side of the "closing" value of the overflow, hence a small graded oscillation. The large "limit cycle" and the full-sized potentials belong to case (b) (solid line), just above threshold. A point in the phase plane has to travel over a quite wide range in the E-M domain before it returns to the starting point. This requires a long time; hence, the time for a full cycle, the period time (inversely related to the frequency), will be long. Compare the potential response in the time domain to the right of the phase plot. On the other hand, case (c) (dashed line) at a high pressure stimulus gives an appreciably smaller "limit cycle" with a smaller period of time, hence a higher frequency (compare the oscillations in the time domain adjacent to the phase diagram). It should also be noted that in this case the mean "depolarization" amplitudes are slightly less in (c) than in (b). This is also evident in Fig. 10 among the "application pictures". The reason for this lies in the nature of the fundamental equations governing this particular oscillator. Case (d), finally at pressure excess, yields a damped "firing" because of adaptation (see p. 311).

(b) *Effects of Electrical Stimuli.* For the sake of the forthcoming discussion of the application of the present concepts to the 2-step process (i.e., one where the afferent, conductile neural

---

[3] This can be physically realized by adding a nonaqueous liquid, which does not enter or "wet" the overflow channel.

element is stimulated by *electrical* stimuli arising from the neighboring generator potential source), a phase plot is presented in Fig. 7, *Bottom*. Again there exists a threshold stimulus which induces a "local response" (not depicted). Just above threshold, a wide low-frequency oscillation occurs (case *a*). At still stronger current stimuli, (case *b*), the limit cycle is smaller and of higher frequency. The most noticeable difference between current-

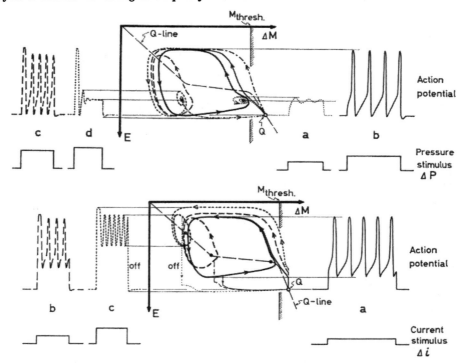

Fig. 7. *Top*. A phase plot demonstrating how the action potential trajectories are generated in the E-M plane. In the quiescent state, a stable condition prevails at Q. Four different step stimuli of *pressure*, ΔP, were applied. The resulting trajectories showing the concurrent changes of potential and colume are indicated as arrow-marked "paths". The potential excursions in the time domain are indicated to the right and the left of the phase plane. The partition denoted M_{thres} indicates approximately the separation of a stable zone, right, and an unstable one, left. To the right of this value, the "overflow" is operative. *Bottom*. A phase plot similar to *Top*. Here, however, three different *current* stimuli (*a*, *b*, *c*) have been applied instead of pressure stimuli. Note the different locations of the trajectories and the limit cycles. In this figure, the off-responses are also indicated

induced and pressure-induced stimuli is that in the former case the "undershoot" level will move in the depolarizing direction with increased stimuli, whereas a direct pressure effect leads to an undershoot level remaining at a fairly constant potential, regardless of the stimulus intensity [4].

An interesting case appears when the resting potential E has fallen to the extent that the Q-point becomes located on the low (left) side of M threshold. The system then exhibits slow, somewhat irregular spontaneous activity (not depicted). It reacts, however, with increased regular frequency for further depolarization or pressure stimuli. This may simulate many cases reported in the biological literature on spontaneous activity of pressure sensors (see discussion of Fig. 14).

---

[4] This fact was stressed in a previous paper (TEORELL, 1966), where diagrams were also given for the concomitant excursions of the potential E, the total pressure differential P, and the volume changes M (see Figs. 6, 9, 10).

3. Some Remarks on the Membrane Events During Firing Activity

As may be understood from a study of the theoretical physicochemical background of the membrane oscillator, the main reason for the oscillations is deformation of concentration profiles within the membrane, particularly within the fixed charge channels (TEORELL, 1959b). These channels are sites of a concentration gradient induced either by *external* concentration differences of one or several permeable ions or by a constant current flow within the membrane even at equal external concentrations (TEORELL, 1961a). These concentration gradients are distorted by the simultaneous operation of electrodiffusion and fluid movement (created by

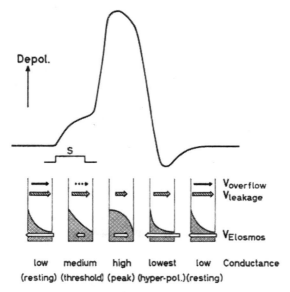

Fig. 8. Scheme of the membrane events during an action potential according to the electro-hydraulic analog. After a stimulus, S, a series of events takes place within the membrane as indicated in the figure. The electroosmotic component of bulk flow is particularly affected, causing distortion of the *concentration profiles* (stippled) which results in conductance changes, the main cause of the potential excursion. Note that the leak flows remain always unidirectional

electroosmosis minus hydrostatic pressure). Depending on the deformation, these profiles may be concave toward one side or they may be convex. The "integral resistance" of the membrane will therefore become different under different conditions. This is schematically illustrated in Fig. 8 which shows the *concentration profiles* during various phases of an action potential cycle. The stippled areas indicate the distribution of the salt content, and from this it can be inferred that the *conductance* varies as indicated[5]. The direction and magnitude of the various contributions to the net fluid flow are also schematically indicated.

Since it is assumed that the system has a constant current source "driving" the oscillations, it is understood that the *potential excursions* are mainly an ohmic

---

[5] In our treatment the conductance contributions of the leakage channels have been neglected since it is conceivable that these channels may admit only water (i.e., "salt filtration" or "hyperosmosis"). Also, the allowance for leakage conductance does not seriously change the general behavior.

voltage drop, (current × resistance). Further discussions on the "driving current" are deferred to a later section (p. 336). Here we will also dwell on the importance of other potential sources as the diffusion potential, ε, (which, in fact, is embodied in the basic theoretical equations as well as in the analog scheme).

IV. The Membrane Oscillator Converted into a Biological Excitability Analog

In the physical membrane oscillator model discussed above, the transmembrane pressure difference has been represented by the difference in the heights between free levels of liquid inside and outside the membrane. The excess pressure inside was maintained by an inwardly directed electroosmotic pressure (being proportional to the fixed charge density, $w\overline{X}$, the transmembrane potential, E, and the hydraulic permeability coefficient, s). When making a mental conversion of this model into a closed biological cell analog, one may envisage that the "free" hydrostatic pressure differential will be substituted for by the internal pressure created by the *viscoelastic tension of the cell wall*, which has already been alluded to (Fig. 3). This wall tension conveys the counterpressure against the electroosmotic pressure as required by the electrokinetic equation for the *net flow velocity*, W, across the membrane:

$$W = -sw\overline{X}E - (s + \sigma + \sigma_{\text{overflow}}) \cdot P .$$

Here we have $E = iR + \varepsilon$, where R is the integral membrane resistance, i the current, and ε a "diffusion potential". The coefficient s denotes the hydraulic permeability in the charged pores, and the σ and σ_{overflow} denote the leakage hydraulic coefficients in the uncharged channels, respectively the overflow channels and $w\overline{X}$ denotes the sign (w) and the fixed charged density $\overline{X}$. Only in the quiescent state does $W = 0$. During oscillation W shifts value and sign. The viscoelastic properties of the cell wall have, as mentioned above, such properties that the pressure differential P is a nonlinear function of the cell volume M (Fig. 5).

Fig. 9 visualizes purely schematically the membrane oscillator as an electrohydraulic excitability analog, sensitive for both electrical and mechanical stimuli. In passing, it should be noted that there exist other ways of triggering this type of oscillator (see Teorell, 1962a). One especially interesting way is by *chemical operations*, which may change the value of $w\overline{X}$, i.e., the fixed charge density. Such a change is equivalent to a depolarization-hyperpolarization, which can be realized by an examination of the electrokinetic formula given above (the term $sw\overline{X}E$). In fact, alteration of almost any of the state variables or parameters in the operational scheme of Fig. 6a may act as a stimulus trigger. For the sake of maintaining reasonable simplicity, we will only study pressure or current perturbations in this context.

Some remarks on the possible validity of the electrohydraulic analog for biological systems. It should be strongly emphasized that the membrane oscillator and its proposed equivalent, the electrohydraulic analog, are presented just as *formal* representations of a purely biophysical system. The theory condenses basically to a nonlinear, second-order differential equation. The "state variables" R, E, P, etc., as well as the various descriptive physical parameters, s, σ etc., can be substituted by noncommitting symbols like x, y, etc., deprived of any particular physical or chemical meaning, or they might be identified in terms of quite different processes than those elected here. Nevertheless, it is believed—imaginative and unproven

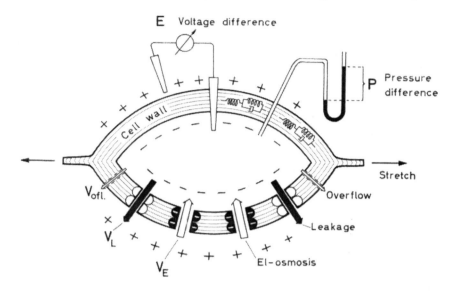

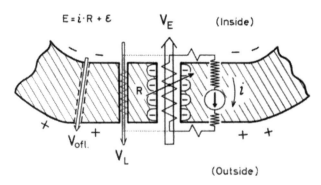

Fig. 9. Scheme of the membrane oscillator converted into an electrohydraulic excitability analog. The electroosmotic inflow in the charged channels induces an internal pressure excess, which is balanced by the viscoelastic elements of the wall. An applied stretch upsets the various bulk flows leading to changes of membrane conductance, etc., as pictured in Fig. 8. *Bottom*: An "equivalent scheme" of the proposed model. Note the built-in *current source* generating the polarization current, corresponding to a "fuel cell". The main mechanism is centered in the time and voltage varying resistance of the charged channels

from the point of view of biology as the model might appear—that it may have a few points of merits. It is based on some characteristics which could be present in natural excitable structures; it equations have a reasonably clear physicochemical meaning; and, above all, as an excitation system it shows the properties of frequency modulation and pressure sensitivity. These features have hitherto not been too well covered by the well-known Hodgkin-Huxley hypothesis. It should be observed that this is also, in essence, an oscillation theory although it is based on different assumptions. In fact, it is probable that a great number of oscillation theories with different sets of assumption may lead to alternative physical and chemical interpretations of biological excitability.

In pursuing the critical evaluation, a few *crucial points* should be discussed briefly. They deal with the convertibility of the assumptions valid for the real physical model into biological counterparts.

(1). The idea of the existence of *fixed charges* in cell membranes, long since advocated by the present author, seems to be gaining wider acceptance (TEORELL, 1953, 1958a,b, 1961b, 1962a). HODGKIN and CHANDLER (1965) have recently incorporated such ideas into their formulations. The existence of the electroosmosis in animal membranes has not as yet been unequivocally proven by direct methods. [STALLWORTHY and FENSOM (1966), however, claim that they have demonstrated electroosmosis in axons of freshly killed squids.]

(2). If one grants the presence of electroosmosis, then the question arises whether animal cells or membranes have the required *pressure conditions*. Experiments on perfused axons, as discussed in the review by CALDWELL (1968), have not yet furnished any positive information. Although the present excitability analog for simplicity has been depicted as a "cell" or "bag", it can also be conceived of as being a small vesicle or a spatium *within* an excitable membrane structure. It can also be represented by a *"swellable"* gel element (see TEORELL, 1962a; Fig. 18). In fact, it has been shown that the gel type of a charged membrane may maintain an *intramembrane* pressure gradient of an appreciable size when subject to the constant current flow, even with zero pressure difference between the "inside" and "outside" bounding the membrane (TEORELL, 1961a). It seems not improbable that unstable electrokinetic processes may exist inside those structures.

(3). In the physical model, the *concentration gradients* of the conducting salt ions were maintained by differences in inside and outside concentration of the order of 10:1. However, experiments with different kinds of ion-exchange membranes have demonstrated that, even with *equal* total concentrations bounding the membrane, steep gradients may be built up *within* the membrane when this is subject to a constant current flow (TEORELL, 1961a, and *unpublished results*).

(4). Finally, the most conspicuous problem is concerned with the assumption of the more-or-less constant *"resting" current* flow in the excitable membrane. Although *local* current during excitation is a necessary ingredient in all excitation theories, the dominating view is that the net membrane current is zero during the quiescent conditions. This poses the problem of the "driving" *energy* for the oscillations. All self-sustained oscillatory phenomena require some form of energy for their maintenance. The contemporary view is that this energy is furnished by the running down of electrochemical energy stored in the form of a different salt distribution inside and outside. The maintenance of this difference is ascribed to the interference of metabolic processes driven by "active processes" termed "sodium pumps", etc. The actual nature of the coupling between metabolic (chemical) energy and electrical energy (in biology) remains unknown. In modern technology there exist "fuel cells" which can directly convert chemical into electrical energy. Perhaps it is conceivable that localized biological structures might act as such fuel cells and introduce electrical currents on the submicroscopical, almost molecular levels, thus giving rise to the "polarization" of the membranes, i.e., the necessary requirement for metastable or oscillatory states. Such considerations, admittedly vague, have led to the insertion of a "current generator" symbol in the equivalent scheme of the membrane, as depicted in Fig. 9. This current source represents a fuel cell, i.e., an ion current pump. These kind of speculations are related to recently emerging ideas on rhythmical chemical reactions (see for instance, CHANCE, 1965; HIGGINS, 1965; RAPOPORT, 1970).

(5). *The specific ion selectivity*, particularly of potassium and sodium, so often employed in modern concepts, has not been given any particular role in the present electrokinetical model. Certainly, different behavior owing to individual mobility, hydration, etc. of ions of different natures may complicate the picture. It should be observed, however, that the system presented here operates on the basis of the *overall* conductance of all permeable ions. Recent experiments with ion mixtures on the physical membrane oscillator seem to indicate the possibility of a marked separation of different alkali ions due to the complicated interplay of forces in this electrodiffusion convection system (*unpublished material*). Perhaps apparently selective ion distribution patterns may be a *symptom* rather than the *cause* of the electrical membrane events arising from the joint interactions between chemical properties and physical driving forces.

V. Analog Computer Simulation of the Mechanoreceptor Properties

The following examples of applications of the electrokinetical (electrohydraulic) mechanoreceptor analog have been processed by conventional analog machine computation.

The *programming* was done according to the *operational scheme* of Fig. 6a. The numerical values for the various parameters are given in the Appendix. The voltage and time scaling of the problem were arbitrary, since no attempt was made to match a real biological condition. When the results are evaluated, it should be remembered that the assumptions employed have been highly simplified. Parameters such as hydraulic permeability coefficients, fixed charge density, etc. have also been considered to remain constant during any particular run. The pressure-volume relation, although nonlinear, consists essentially of two straight-line segments, and, in particular, no hysteresis has been taken in account. Hysteresis is likely to occur because of time-dependent and rate-dependent viscoelastic properties. A purely technical deficiency depends on the fact that the function generator used for the $R = f(V_E)$ backbone curve consists of 20 straight-line segments rather than being a perfectly smooth somewhat S-shaped curve (see Appendix). This introduces a certain stepwise "noise", which can be seen on the records of comparatively small voltage variations as those of the generator potentials (Fig. 18a).

The general aim of the computer procedure is to show the possible potentialities of *describing some major mechanoreceptor phenomena in terms of nonlinear oscillation theory*, and not to reproduce all the singular patterns obtained by the experimentalist on biological objects.

20*

The presentation is given in the following sequence:

A. *Repetitive all-or-non discharges (mechanical stimuli).*

B. *The generator potential (sensory element).*

C. *The "receptor" potential (afferent nerve element).*

D. *The afferent nerve response (with umimpaired firing ability).*

E. *Miscellaneous phenomena (summation, voltage clamp).*

A. Repetitive All-or-none Discharges (Mechanical Stimuli)

1. Simulation of Baroreceptor Activities as an Example of Pressure Stimuli (1-step Process)

A case demonstrating the effects of pressure stimuli according to the present system has been previously published (TEORELL, 1966). It showed a straight-line relation between stimulus intensity vs. firing frequency over a wide range. The frequency span, i.e., the relation between maximal and minimal frequency, was of the order of 1:5 to 8.

In Fig. 10, a similar experiment is recorded, but with slightly changed parameters causing a lower pressure threshold (the membrane potential was set just above the level where spontaneous discharge occurs (see frame *f*). This system was characterized by the absence of "adaptation"; i.e., a sustained rhythmical firing took place as long as the stimulus was on, regardless of the level of pressure stimulation. There was, however, a maximum firing rate attained at high stimuli levels (see Fig. 11 a). This firing pattern may simulate the behavior of the blood pressure-sensing baroceptors, which show no "cease of firing" even at high blood pressure levels although they do show a "postexcitatory inhibition" for dynamic stimuli.(In Fig. 10, frame *e*, note a special form of inhibition, called single pulse "annihilation", this particular phenomenon will be discussed on p. 315).

2. Simulation of Slow- and Fast-adapting Pressoreceptors. Adaptation and the "Overstretch" Phenomenon

It has long been shown that certain types of receptors can show a fairly constant firing, with regard to both amplitude and frequency as long as the stretch is maintained reasonably constant (*"nonadapting* or *slowly adapting"*). On the other hand, for similar conditions, other types may respond with only a short-duration burst of firing, which rapidly declines in amplitude and vanishes (*"rapidly adapting"*). Many workers have been inclined to derive the adaptation from viscoelastic changes or to physical factors changing the tension arising from viscoelastic creep, i.e., time-dependent slackening. Concentration changes and permeability changes have also been discussed. No doubt, all these factors may be working. However, from the point of view of pure oscillation theory (the approach taken in this paper), it can be anticipated that long- and short-term oscillations may also depend on the *damping parameters* of the basic oscillator. To illustrate this point, a case similar to the nonadapting one discussed in the previous section is shown in Fig. 12. The main difference is that now an approximate 10% increase of the leakage factor,

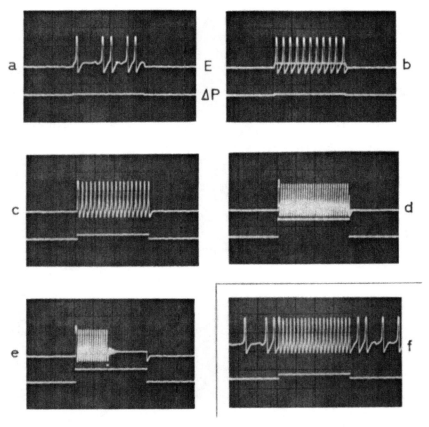

Fig. 10. The effect of increasing pressure stimuli, ΔP, on the frequency of the membrane potential discharges, E. Frame a, threshold response; b-d, increased stimulus strengths. In frames b and d there is also a superposed trace showing the approximate firing level, i.e., a highly damped response ("q-damping", see p. 322). Frame e, a superposed brief pressure pulse causes an inhibition ("annihilation"). Frame f, identical condition with a-e, except the membrane potential was depolarized to a level where spontaneous activity appeared. Note the regularization during the pressure stimulus. The *scales*: E:2, ΔP:2, time: 50 relative units per screen scale division. The σ-damping was "medium", (see Fig. 14); thus, this is a "*nonadapting*" case, simulating, for instance, baroreceptor activity

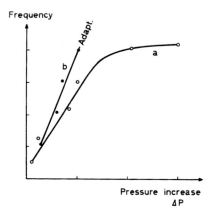

Fig. 11. Relation between the stimulus strength (pressure increase) and the spike frequency. The data for line a were obtained from the *nonadapting* case of Fig. 10. Note the linear stimulus strength-frequency relationship and the "ceiling effect" on the frequency at high stimulus strengths. No threshold is discernable in this case because the membrane potential was set very near the spontaneous activity level. Line b, represents the same relations for the *adapting* case (data obtained from Fig. 12, frames a—d). Note the absence of a "ceiling"

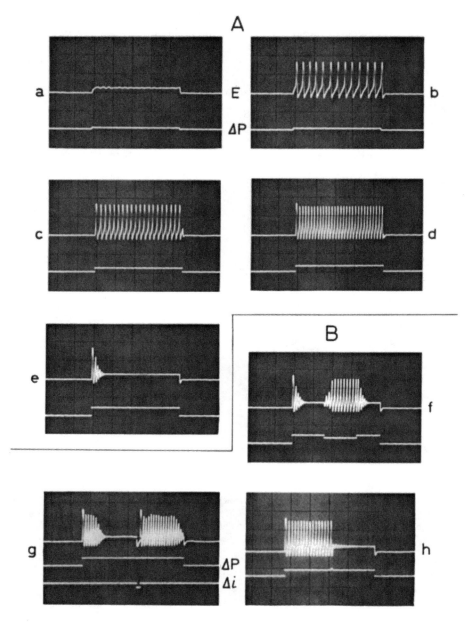

Fig. 12. Relation between pressure stimulus strength and frequency in an *adapting* case. Frames *a — e* represent the same conditions as in the nonadapting case, Fig. 10, but for higher damping (corresponding to "σ high"; see Fig. 14). Note the rapid cessation of firing in *e*. The relation between frequency and stimulus strength is shown in Fig. 11, line *b*. Section B has the same basic conditions as A. In frame *f* is demonstrated a "postinhibitory excitation" by a small relaxation of the pressure stimulus. Frame *g* demonstrates an *anodal current* pulse restoration of firing, which again shows an adaptation. Frame *h* is another example of a single *ΔP* pulse annihilation. *Scale units* are as in Fig. 10

σ, has been introduced. Furthermore, although less important, there is the condition that this case has a very low threshold; i.e., it is set to a membrane potential level hyperpolarized just enough to prevent a spontaneous firing (see Fig. 10, frame f, which is a case depolarized just above the "critical" potential level).

Again one notices an increase of firing frequency with increased pressure. Above a certain high critical pressure, however, only a brief burst of more-or-less rapidly fading out oscillations results (frame e), and at further "stretch" only a single "make" response would appear. Fig. 11, graph b, shows a straight-line frequency-stimulus relation (from the records of Fig. 12). In comparison with graph a relating to the run of Fig. 10, one notices that the range with constant maximal firing is now absent. Instead, the *adaptation*, or overstretch inhibition, takes place primarily when the level of the most rapid firing is reached. The type of "adaptation" shown is fully reversible, because a small relaxation of the stimulus restores firing (Fig. 12 B, frame f).

Only a very slight decrease of the damping factor will give oscillation periods with less time decrement, very much of the slow-adaptation type. Small perturbations as system "noise" may also make the transitions between slow and rapid adaptations irregular and graded.

Some Relations to Biological Observations. The demonstrated feature that an excessive pressure, equivalent to an excessive stretch, can extinguish discharges may have some bearing on biological observations called *"overstretch"*, originally described by WIERSMA et al. 1953), and EYZAGUIRRE and KUFFLER (1955a). It is also interesting to note that Ottoson in experiments with hypertonic solutions found a firing rate, that was high but of a very short duration. He suggested that the sensory bulbs of the frog muscle stretch receptors become shrunken (OTTOSON, 1965, 1969). This may be in accord with our model concepts. TERZUOLO and WASHIZU (1962), working on crayfish stretch receptors, found that firing block due to overstretch could be reversed by pulses of anodal current. This may be related to the above demonstration of the reversibility of adaptation to a pressure release. In fact, the application of an anodal pulse in the silent period showed a reinitiation of firing as seen in Fig. 12 B, frame g. The simulation of *nonadaptaing* and *adapting receptors* will be commented in Section 4 below.

3. On the Relation between the Resting Membrane Potential and Threshold Intensity of Pressure Stimuli

In the system presented, it is to be expected that there will be a spontaneous spontaneous discharge activity even in the absence of any pressure stimulus (no tension, relaxed state) when the membrane potential is low (see Fig. 10, frame f). When the membrane is repolarized to a quiescent state, the higher the resting potential level is set, the more pressure is required to bring about a *threshold* discharge (Fig. 13). This corresponds to a "nonadapting" case; i.e., the damping is moderate. From these series, it could also be inferred that the firing level and the undershoot level will be shifted towards hyperpolarized values with increasing resting potentials. Noticeable also is the marked after-hyperpolarization, or underswing, which is present at low membrane potential values (Fig. 13, frames a and b), and which is absent at higher resting potentials (frames c and d). Grossly speaking, the *threshold for pressure stimuli increases approximately linearly with an increase of the membrane potential* (referred as to the critical value, i.e., the membrane potential which barely elicits spontaneous discharges).

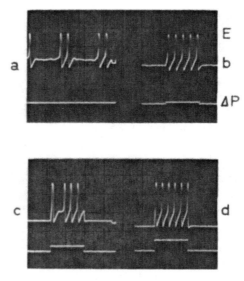

Fig. 13. The dependence of threshold intensity of pressure stimuli on the resting membrane potential. Frame *a*, the resting membrane potential was depolarized just to the level for spontaneous activity. Frames *b*, *c* and *d* represent increasing resting potentials, case *d* being the most hyperpolarized one. The ΔP heights indicate the threshold just for all-or-none firing. Note the progressive increase of this threshold at increasing resting potentials of *E*. Observe also the different relative locations of the resting, firing and undershoot levels. Damping corresponds to "σ-low" (see Fig. 14). *Scales* units are as in Fig. 10

4. On the Relation between Thresholds for Pressure and Current Stimuli and Dependence on the Damping Factors

A closer examination of Fig. 13 reveals that good straight-line relationships also exist between the polarizing resting current *i* and the pressure level required to elicit firing. This is true regardless of the value of the damping factors (here σ). The diagram in Fig. 14, interrelates the pertaining variables. In particular, the diagram is useful because it demonstrates comprehensively the conditions for the various response patterns discussed above, e.g., the overstretch inhibition and the related types of adaptation. The diagram is also helpful for understanding how pressure and current may interfere with one another with regard to the resulting discharge pattern.

As an example, let us consider the following case. Assume that the sensor element is in a relaxed state and quiescent at point *Q* in the diagram; i.e., the membrane potential, is higher than the critical level for spontaneous discharge (point *S* in the diagram). Now a moderate constant pressure increase will bring about a just-threshold low frequent firing (point *a*). Application of a further pressure stimulus will correspond to the position marked *b* in the diagram. The "receptor" will then fire with rather high frequency, without adaptation. An excessive pressure increase may move the point to position *c*, and then only a few discharges will ensue; i.e., there occurs a fast adaptation. The same effect will be observed if, instead, point b is moved to d, by now a moderate depolarization (here caused by a decrease of the membrane current i). In fact, even hyperpolarization, from position b to e may cause the same effect.

One can now also understand the case of "anodal pulse" restoration [according to TERZUOLO and WASHIZU (1962), simulated in Fig. 12, frame g]. In the diagram (Fig. 14), this case is located at point d during the onset of the pressure step. The brief anodal current pulse brought it "south" into the oscillation zone (somewhere between points b and e), and it was then returned to d for a new state of adaptation.

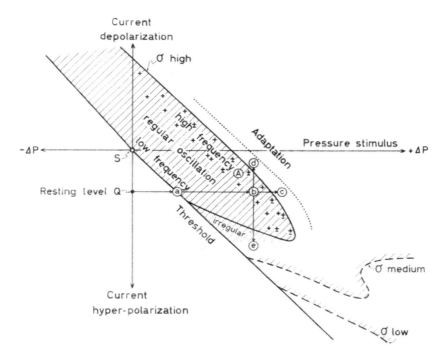

Fig. 14. A comprehensive diagram showing the relation between the types of potential discharge in relation to pressure stimulus and membrane current polarization. This aims primarily at showing the influence exerted by one oscillation "damping factor", σ, on the type of discharge, which arises from manipulation with pressure and/or current stimuli, particularly at increased membrane leakage (i.e., "σ high"). At "medium" and "low" damping, the oscillatory area is widened to the right of the "threshold" line, and adaptation becomes increasingly more difficult. The dotted line, belonging to the "σ-high" case, marks approximately the area of a transition from more gradual to an abrupt cessation of the firing. S indicates the state where spontaneous activity takes place and is equivalent with frame a of Fig. 13. Q is represented by frame b. The graphs are processed by analog computation

In the diagram, there is also an area shadowed with ($\pm$) signs. This indicates a region where it is possible to "annihilate" a regular sustained train of discharges with a brief "positive" or ("negative") pressure pulse, that should be positioned in the refractory period and of a selected amplitude and duration (this case corresponds to location A in Fig. 14 and the oscillogram Fig. 12B, frame h). This particular area is the site of "quiescent zones" (see p. 336) which are also of great importance when dealing with continuously changing pressure-current stimuli as ramps or triangular or any "soft" wave form. The background for all exemplified events is the particular course on the limit that bounds the *oscillation zone* in the diagram.

The examples refer to moderate damping ($\sigma = 0.5$); with less damping ($\sigma = < 0.5$) somewhat similar effects can be deducted although depolarization does not cause any inhibition. In such cases, hyperpolarization is always an inhibition factor (see the "outfolded" boundary limit for "σ medium" and "σ low" in Fig. 14). One may venture to characterize the more damped case as capable of *slow/rapid adaptation*, and the less damped as a *nonadapting* system. This classification, however, is very approximate because the oscillatory states also depend on the level of current polarization, i.e., on the prevailing level of the membrane potential. This may be relevant to the observations on differences in firing levels of slow and fast receptors by Eyzaguirre and Kuffler (1955a).

The *inhibitory effect of current stimulation* is further illustrated in Fig. 15. Here a constant step pressure, which normally results in a constant firing over a stimu-

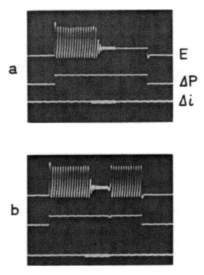

Fig. 15. A demonstration of the inhibitory effect of current stimulation on a pressure-induced discharge. Regular response elicited by the step pressure ΔP is blocked by a small high frequent current train, Δi (frame a). Frame b demonstrates how the block can be relieved by a small pressure perturbation. *Scales* units are as in Fig. 10

lation period, was applied. However, in the midst of such a period, an extra stimulus was applied, now as a *current*. This current was a small-amplitude triangular wave with about four times the spike discharge frequency (see the indication on the bottom trace). The effect was an immediate cessation of the discharge, which in this case, however, outlasts the external current stimulus. That the system is very labile even in the "postinhibitory" state is evident from frame b, of Fig. 15, which illustrates how a small brief pressure relaxation was injected, which restored the normal discharge. This case may perhaps be relevant to the findings of Eyzaguirre and Kuffler (1955b); they demonstrated that a stimulation of the inhibitory *efferent* nerve blocked the firing from the stretch receptors. These authors and many others generally conceive of the role of efferent nerve inhibition as affecting the generator potential of the sensor element. In biological objects, the generator potential appears as a highly damped, graded and small potential.

In our examples up to now, the "sensor" potential has been considered fully oscilla-tory. In a special section, we will consider the transformation of oscillatory sensor potentials into small graded generator potentials (see "Generator Potential" p. 317). Also in those cases, there may arise interaction from an applied "efferent" current stimuli.

The problems of inhibition phenomena and annihilation will be discussed again with reference to postexcitatory inhibition and "silent intervals" (page 328). The "applications" examples above were intended only to show that there exist *theoretically* complicated interactions between pressure excitation, current exci-tation or inhibition, and the degree of oscillatory damping of the metastable presso-electric transducer analog, which has been analyzed.

5. The Dynamic Response Pattern of Direct Pressure Stimulation

It is a well-established fact in receptor physiology that the *rate of change* of a stimulus is of decisive importance. In previous sections were examined the features of long-lasting *static* pressure stimuli of varying intensity, applied step-wise. Now the same system will be subject to *dynamic*, i.e., time-varying, stimuli. The first case is a slowly rising stimulus followed by a static (i.e., ramp stimuli) and then by a linearly rising and falling stimulus (i.e., triangular stimuli). The resulting oscillations are shown in Fig. 16. The amplitude of the static stimulus was kept just below the "overstretch" threshold in frames *a—e*. In frames *f* and *g* the pressure stimulus peak was just above this threshold (which corresponds very closely to point *c* in the diagram of Fig. 14). Some features of the results deserve comments:

(1). The firing level, as well as the undershoot level, remains constant regardless of the discharge frequency. However, the spike amplitude diminishes clearly with increasing frequency. (2) The overstretch inhibition (frame *f*) prevails during the decline of the stimulus, but can be "reversed" by a sharp small pressure pulse as shown in frame *g*.

Again the diagram in Fig. 14 is helpful. The *postexcitatory inhibition* in this case is due to the fact that the oscillations, once "adapted", do not regain sufficient "momentum" enabling a rapid passage back to the oscillation zone (from point *c* towards *a*), unless they get boosted by an extra pulse (see the existence of "quiescent zones", also discussed in the Appendix, part 3). Note also that the membrane potential remains constant during the greater part of the "silent period". (3) An especially interesting phenomena can be observed in frame *c*, where a brief pressure increment, strategically placed, can *annihilate* the static train of spikes The brief duration of the pulse precludes an explanation in terms of overstretch; compare the effect of the long pulse injection in frame *d*. A similar annihilation was shown in Fig. 12 B, frame *h*, and both have the same mechanism. (4) Case *h* is in-cluded for comparison with case *e*. Now the stimulus was a cathodal triangular *current* instead of pressure. In this case, with electrical instead of pressure stimulation, one should notice the depolarization of both the firing level and the underswing level, accompanied by an increasing discharge frequency. In fact, this difference between the behavior of the firing, respectively the underswing level, was suggested in a previous publication (TEORELL, 1966) as *a means of discrimi-nating between the* 1-*step process* (*pressure* stimuli) *and the* 2-*step process* (*current*

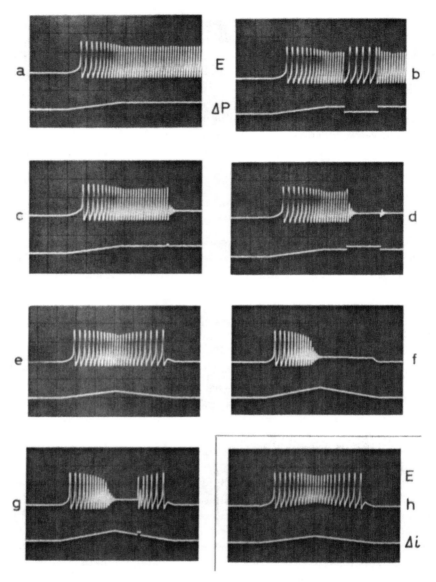

Fig. 16. Response patterns from dynamic direct pressure stimuli. The conditions correspond to Fig. 12, i.e., an adapting case. For detailed comments, see text. Note, however, the difference of the E response pattern between frame e and h, the latter being stimulated by a *current*, Δi, instead of pressure. *Scale* units are as in Fig. 10

stimuli). Further studies of current stimuli will be commented later on in this article (p. 327).

With regard to a *comparison with biological objects* great caution must be exerted. The simulation results described above have hitherto referred to direct stimulation with pressure, i.e., a 1-step process. At the present time, hardly any physiological case is known where a genuine full-sized oscillatory spike discharge occurs in the sensory element proper. Exceptions may be the baroceptors in the carotis sinus and possibly the crustacean muscle stretch receptors. The majority of the biological reports seem to be concerned with a 2-step process, where the oscillatory discharge is triggered at the adjacent afferent conductile nerve by a generator potential or generator current, which is graded, i.e., a degenerate oscillation in terms of the present analysis. On the other hand, the technical procedures used for the recording of the activity of mechanoreceptors may have precluded the analyses of the *actual* properties of the intact sensory element without any external disturbances. In spite of this critical attitude, it may be tempting to make a few references to some reports by experimentalists bearing on some of the more specific model results commented above.

In the crayfish (slow) stretch receptor, EDWARDS and HAGIWARA(1959) observed: "The firing level and the undershoot level were fairly constant, independent of the degree of stretch", compare the statement (1) above.

Regarding the decreasing spike amplitude at high firing rates, it is of interest to compare Fig. 16, frame *e*, with the "famous" picture of EYZAGUIRRE and KUFFLER (1955a, their Fig. 5), they are rather similar.

SATO (1961) reported that Pacinian corpuscles often responded to a short single mechanical pulse with *repetitive* discharge, which lasted many times longer than the duration of the single pulse. A somewhat similar case of "single-pulse initiation" can be seen in Fig. 16, frame *g*, (in fact, both positive or negative pressure pulses can be triggering). For an extensive discussion of repetitive receptor potentials in response to short mechanical stimulus, one may refer to SATO's publication, which also cites earlier observations in this respect. The extensive report of TERZUOLO and WASHIZU (1961) on "overstretch" block and "pulse" triggering of repetitive activity is also of great interest in the present context. Other reports on overstretch were alluded to earlier (p. 311).

Biological counterparts to the brief pulse "annihilation" (see Fig. 10*e*, 12*h*, 16*c*) seem not to have been reported, at least not on mechanoreceptors. It would perhaps be worthwhile to look for this somewhat paradoxical phenomenon in neurophysiological work.

LIPPOLD et al. (1960) found that a steady discharge due to stretch, when subject to current flow, increased in frequency during depolarization, and hyperpolarization stopped the discharge (see their Fig. 3). This interesting observation may be understood also in the framwork of the diagram, Fig. 14.

B. The Generator Potential (Sensory Element)

1. Introduction

In the vast literature on generator potentials in different transducers, in particular those of stretch receptors, it is generally agreed that the generator potential (GP) is of a graded type and quite moderate in size as compared with the afferent spikes (see reviews by GRAY, 1959; FLOREY, 1961; LOEWENSTEIN, 1959, 1961; EYZAGUIRRE, 1961; MATHEWS, 1961; GRUNDFEST, 1966; and several contributions in the present volume, e.g., that of OTTOSON). The amplitude value ranges between fractions of millivolts to about 20 mV (LOEWENSTEIN, 1963), i.e., less than about one-fifth of the afferent spike amplitude. It should be remembered, however, that most of the recordings have been made either at some distance from the sensory ending proper, as in the soma, or at the first Ranvier afferent node after blocking this with local anesthetics, tetrodotoxin or other means. The recorded potentials have also been subject to a spatial decrement owing to an electrotonic spread in the sense organ. EYZAGUIRRE and KUFFLER (1955a) estimated a reduction of the order of 20 to 80 %. Accordingly, it seems that there exist no really reliable figures of the actual generator potential. The authors mentioned have also measured the *membrane potential* at the sensor unit proper to be perhaps 80 mV or even higher, i.e., of the *same order* as that of the adjacent afferent node membrane. This information gives an important starting

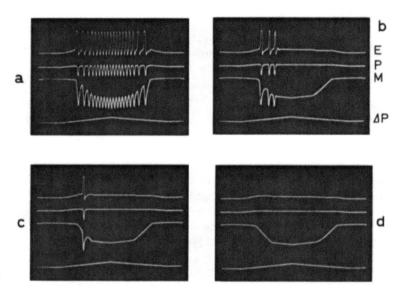

Fig. 17. The effects of increased damping on the responsiveness of the artificial pressoreceptor analog. The same triangular pressure stimulus, ΔP, is applied in all four records. The leakage factor σ, is increased from a normal value (frame a) by 30, 40 and 42% in the subsequent records. Note the absence of spikes in frame d. Included here are typical patterns of the variations of the *transmembrane pressure difference* P and of the *volume changes* M accompanying the membrane potential excursions, E (reproduced from TEORELL, 1966)

point for analysis of the possible relationships between all-or-none responding elements (as discussed in the previous sections) and the generator potential. Our analysis will attempt to show that the *generator potentials could be characterized as highly damped manifestations of the same type as a "repetitive" oscillator.* Such an aspect was introduced in a previous publication (TEORELL, 1966).

Fig. 17 is reproduced from this earlier paper. Focusing attention only on the E records (the *membrane potential*) in the frame series $a-d$, it is evident that the firing pattern changes markedly from a continuous one to one with only a few spikes on the rising part of the stimulus; finally a single spike remains, which eventually disappears, and only a small elevation of the membrane potential remains. These effects are caused only by a stepwise increase of the *damping factor* σ for each picture; the system otherwise remained the same as in the "repetitive case". It was subject to the same triangular pressure stimulus, ΔP. The last frame, d, has a close resemblance to many generator potentials described in the biological literature, both as to the shape and the magnitude, relative to the full firing spikes.

In this figure are also recorded the corresponding changes in *total pressure differential*, P, and the *change of volume*, M. As these two variables have not been included in most of the figures here, it is of interest to notice that the total trans-membrane pressure difference changes but little, with the exception of during spike discharges. In contrast, there is a marked volume change also in case d, i.e., during the "generator potential" formation.

2. Note on the Physical Nature of Damping Factors

The analysis of this paper is performed in terms of nonlinear oscillation phenomena in a well-specified electrokinetical system. There are many parameters, which by a change of their numerical values can exert a profound influence on the type of oscillations which follow. In fact, the conditions for full sustained oscillations depend on an adequate balancing of all the parameters to optimized values. In contrast, damped or overdamped oscillations are the common ones (for fuller information, see TEORELL, 1957–1962, and the Appendix). The most important *external* variable is the *polarizing current density* (i). Of those parameters which characterize the membrane properties, factors such as the *hydraulic permeability coefficient* (s), the *fixed-charge density* ($\overline{X}$), and the *leaky pore coefficients* (σ) are all possible candidates for "physiological" variations. But also of importance are parameters such as k and q which mainly depend on the "geometry" of the system. Both are related to *rate* constants and determine primarily the oscillation frequency. The q, in particular, may operate as a damping factor, which arises from the relation between available pore area and the pressure restraint exerted at a given cell volume. Accordingly, it is understood that there are many choices for electing a damping parameter suitable for a pilot study as the present one. In our previous papers, the leakage factor σ has been preferred because it operates in parallel with the "overflow" leakage, a condition assumed to exist in the resting case (see description of the membrane oscillator on p. 297). A change of σ simply means an effect on the number of leaky pores per unit area membrane, which may be a function of its physicochemical architecture. In this context, it should be noted that LOEWENSTEIN (1961) has discussed the importance of the active membrane area for the "charge transfer" in the receptor membrane.

It should be emphasized strongly that we have made a somewhat arbitrary choice in selecting the σ-factor as the main damping regulator. The results obtained by employing the σ-factor seem, however, to be justified by a good agreement with the characteristics given in the biological literature on GP in the *sensor elements proper*. On the other hand, the continued analysis will indicate that the q-factor is a better choice when discussing the "secondary" GP, which appear in the first part of the afferent nerve element, those which many authors designate as "receptor potentials". In summary, it is ventured that both *the generator potentials proper* and *the receptor potentials are damped manifestations of the same type of* oscillator, although the dominating damping factor may be of different types (σ-damping and q-damping respectively)[6].

In the following Sections 3—5, we will produce "artificial" GP's and study their relations to stimulus strength, mode of mechanical stimuli (static and dynamic), and the effect of "polarizing" currents. First the σ-factor should be employed for damping, then the geometry factor (q); the results will be compared with relevant biological observations by other authors.

3. The Generator Potential at Different Velocities of Stretching and During Current Polarization (σ-damping)

In order to reconstruct the generator potential in a pressure-sensitive element, the same model system has been used as discussed earlier in this paper but with the introduction of an appropriate damping (σ, the main leakage coefficient was doubled)[7].

First, the effect of triangular pressure impulses was examined, i.e., the dynamic response of the receptor potential with *different rates* of "stretching". The results with a *constant peak stimulus* level are seen in Fig. 18a. With the fastest rate of stimulus, the potential has a rapidly rising initial phase; at lower rates of pressure stimulus, the onset of the response is more gradual, the potential rises more slowly, and the peak amplitude of the potential is low.

In Fig. 18b, a "hump"-shaped stimulation of *varying strength* at constant duration has been used. As can be observed from the ensuing potentials, there is a rough proportionality between the stimulus strength and the GP-amplitude. The time characteristic shows, on the whole, a more rapid rise and somewhat delayed decay, which interestingly enough already starts during dynamic phase. In the figures, it is clearly seen that the rate of rise of the GP is more or less linearly related to the height and rate of stimulus. No quantitative evaluation of this will

---

[6] Theoretically, the q-factor should vary in some proportion together with the σ-factors, i.e. the sum of all pore areas (expressed in s, $σ$ and $σ_{overflow}$). Some information on the interdependence of the parameters is given by the condition for sustained oscillations (for the "linearized" case, see Teorell, 1957, 1959c):

$$i = [1 + (s + σ)/kq]/rl .$$

Here i is the current density; s and $σ$, "hydraulic" conductances also including the pore areas; k and q, the "delay" and the "geometry" constant; $l = - sw\overline{X}$ (where $w\overline{X}$ is the sign (ω) and density $\overline{X}$ of the fixed charges); and r is the slope of the membrane resistance "backbone" curve at the "singular point" of the oscillation (also see Appendix).

[7] Another item was the employment of the somewhat higher quiescent (resting) potential required to compensate for the cell volume loss caused by the increased leakage; i.e., the quiescent point was, as usual, brought on the higher side of the M_{thres}, (see "phase plot" in Fig. 19). This may be compared with the previous plots in Fig. 7 where a different inclination of the Q-line can be observed. The Q-line indicates the locus of the possible quiescent points obtained by a voltage clamp procedure; its shape is closely relevant to the pressure-volume relation assumed in Fig. 5 and refers, strictly speaking, to the current-stimulated system (see Fig. 7). It serves here merely as an "orientation mark", and it should be observed that the quiescent point (Q) of the system lies on this E-M characteristic. Note also that the Q-point is located on the high side of the M_{thres}-value, which means that the overflow condition is an operation (see Fig. 5). The gap on the shadowed M_{thres}-line indicates roughly the position of the *threshold region* of this system.

be given, since two stimulation factors have been varied concurrently, namely the rate of rise and the static stimulus amplitudes. However, Fig. 18a lends itself to a correlation between the peak pressures and the GP-amplitudes during the dynamic phase and the admittedly short static period. Such a diagram is found in frame c of Fig. 18. From this, it can be concluded that the dynamic peaks vary linearly with the stimulus, whereas the static part of the response is smaller and shows a definite ceiling at higher stimuli levels.

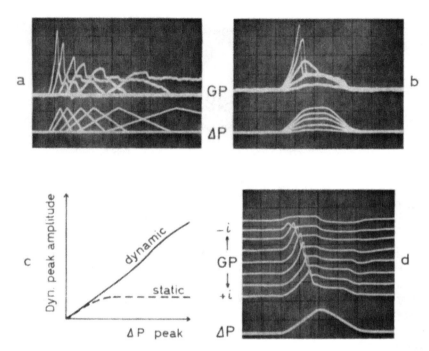

Fig. 18. Simulation of generator potentials. Frame a, the effect of triangular pressure stimuli, ΔP, at different rates of rise but constant peak amplitude. Frame b, responses to an "experimental" pressure stimulus of varying maximum intensity. *Scales*: GP:, 0.5, P: 2, time: 20 (a), 10 (b) units per scale division. Frame c, diagram of GP peak-stimulus peak relation calculated from frame b. Frame d: The influence of *current polarization* on the generator potentials. The same pressure stimulus was applied at different membrane potential levels, induced by various degrees of current polarization, the top curve being the most depolarized state. Equal current steps between each subsequent trace. *Scale*: same as frame b. The main difference between this series and the previous all-or-none spike records was the introduction of a 100% increase in the damping factor, σ ("sigma" damping)

Regarding *biological comparisons*, several similarities may be found in the literature. First, the general *shape* resembles that reported by LOEWENSTEIN (1959) for the Pacinian corpuscle; the GP-peak and stimulus intensity are roughly linear over a range. Furthermore, LOEWENSTEIN and ISHIKO stated that the GP-potential amplitude, as well as its rate of rise, was a function of a polarizing current, increasing with hyperpolarization. Also the simulated GP-peaks increased with hyperpolarization as seen in Fig. 18, frame d. However, this conclusion applies only to the dynamic peaks (with not excessively high rates of rise, then a tendency to all-or-none spikes may appear in the very depolarized conditions, thus reversing our conclusions). Comparison with OTTOSON's (1970) careful observations may not be appropriate at this stage because he recorded mostly from anesthetized nodes.

Another interesting detail has been observed by LIPPOLD et al. (1960) who frequently found that the receptor potential after stretch showed an "off-effect", which consisted of a brief hyperpolarization. This was previously found in the

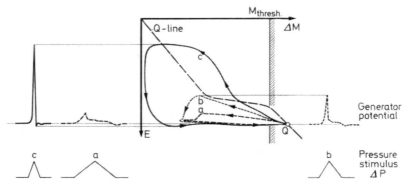

Fig. 19. A phase plot demonstrating the GP-trajectories in the *E-M* plane. The diagram is similar to that of Fig. 7, *Top*. Three pressure stimuli of equal amplitude but varying durations were applied. The ensuing GP are shown in the time domain to the right and the left of the phase plane. Corresponding oscillograms are found in Fig. 18, frame *a*

frog by KATZ (1950). In the simulated GP of Fig. 18, there is also evidence of brief underswings, particularly at "normal" or slightly depolarized membrane potentials (frame *d*).

In order to make our analysis more comprehensible, a *phase plot of the potential trajectories* has been made in the *E-M* plane (see previous phase plots in Fig. 7). The results are depicted in Fig. 19. The paths marked *a* and *b* yield GP which correspond to some of the patterns in the oscillogram of Fig. 18a. It should be noted that the GP amplitudes here are of a considerably smaller amplitude than "action potentials" (see Fig. 21 *b*). The path marked *c* yields, however, a quite large all-or-none spike, this is induced by a very rapid rising pressure impact (see *c* to the extreme left). A detail which can be understood from the diagram is the small after-hyperpolarization which occurs during the regress of the GP. This may correspond to the "off-effect" mentioned above (LIPPOLD et al., 1960).

Already at this stage, we may conclude that the employment of "σ-damping" may simulate many of the features of real generator potentials. The fact that under excessive circumstances it theoretically may give rise to an all-or-none potential is perhaps a minor objection (see case *c* of Fig. 19). A further evaluation of the possible merits of the σ-damping will be postponed until the characteristics of the other choosen form, the *q*-damping, have been analyzed.

4. The Properties of Simulated GP with "q-damping" and its Relation to the Afferent Node Response

It should be remembered that the significance of *q* is related to a *rate* constant of volume changes, induced by the net fluid velocity across the sensory membrane (see p. 319). In the fundamental equations employed, q has a more composite physical meaning than the simple leakage constant σ. It is dependent on all the hydraulic permeability coefficients, as well as on the geometry relation, i.e., the ratio between totally available pore area and the volume of the sensor elements, or, still more correctly, the pressure restraint controlling this volume (the restraint being exerted by the viscoelastic properties of the wall elements of the sensory unit). This complicated situation may make it conceivable that effects of external agents or procedures which affect the permeability properties or the viscoelastic properties, or both, can be simulated by an appropriate change of the q-value. For the pilot study of this paper, we have now entered the q with a 10 times lower value in the analog computation procedures, this being the only change as the σ factor was restored to normal value. With the physical meaning attached to the q, it can be anticipated that the computation results would model that of an afferent node in anesthetized conditions, rather than those of a sensory ending proper. The arguments for this view will be somewhat justified by the results to be presented. However, it should also be inferred that an unequivocal assignment of a different type of damping factor to the generator process and the afferent process is not possible.

The stimulation with different modes of pressure stimuli reveals the following properties which be read off from Fig. 20. The change of rate of rise of stimuli with constant peak amplitudes influences the rate of rise of the GP in a fairly parallel fashion (frame a). However, in sharp contrast to the previously discussed σ-case, the dynamic GP-peak amplitudes have, on the whole, an almost constant level.

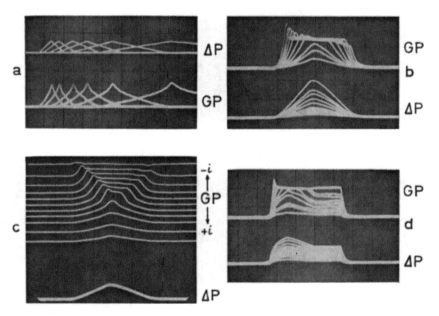

Fig. 20. Simulation of generator potentials by a different mode of damping, i.e., q-damping. This set of records should be compared with those of Fig. 18, where σ-damping was used. Frames a, b and c here correspond to records a, b and d of Fig. 18. Frame d employs pressure stimuli assumed to be similar in shape to actual *tension* changes resulting from a ramp-shaped external displacement. *Scales*: identical to Fig. 18 except for doubled P-sensitivity in d

The *static* GP level varies for small peak stimuli in proportion to these; however, for greater stimuli a marked "ceiling" effect occurs (frame *b*).

The effect of various degrees of current *polarization* is also different from the σ-case, as demonstrated in frame *c*. The average amplitude of the GP, for a constant pressure stimulus, is small at highly depolarized and at highly hyperpolarized states and has a maximum in between.

Although it is tempting to compare the GP patterns of Fig. 20 with many similar ones in the biological literature, we will not pursue the study of pressure effects any further, mainly because of the failure ot the *q*-factor to reproduce the

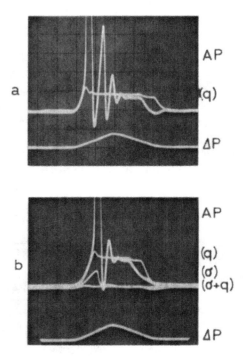

Fig. 21. The effect of various types of damping on the response to identical pressure stimuli. Frame *a*: full-sized dynamic response of an action potential *AP* and a superposed trace where *q*-damping was employed. Frame *b*: a similar case, although slightly more hyperpolarized resting potential. Superposed traces of *q*-damping, σ-damping and a sum of both illustrating the relative sizes of the various potential excursions. *Scales*: For potentials: 0.5, *ΔP*: 2, time: 10 units per scale division. Note that the firing level of the AP is approximately indicated by the *q*-responses, which also resemble a "receptor potential" as recorded in an "anesthetized" afferent node

dynamic GP properties as they are known from sensory endings. This attitude is not weakened by the fact that pressure stimuli (of a type which might run parallel with the actual *tension* changes) during maintained stretch show some conspicuous GP responses (Fig. 20, frame *d*).

However, it might still be of interest to compare the simulated GP with a few *biological cases*. Nakajima (1964), working on crayfish stretch receptors treated with tetrodotoxin, found GP-patterns which closely resemble those in frames *b* and *d* of Fig. 20.

Although the simulated response patterns may also appear similar to those on frog muscle spindles discussed by OTTOSON *et al.* (1965), it should be remembered that these authors measured the "receptor potential", (see OTTOSON, 1970, his Fig. 14). The receptor potential is recorded as the effect of a local current flow from the sensory ending to the first Ranvier node. Such cases will be analyzed in Chapter C below.

In summary, the main differences between the σ- and the q-damping cases, when subject to direct *pressure* stimuli, lie in the dynamic performance and in some differences of shape and size. This statement is born out by Fig. 21 in which are superposed oscillograms of the full-sized action potential and three of its degenerates, namely the q-damping, the σ-damping, and a combination of both. It is also interesting that the top part of the q-excursions coincides with the firing level of the AP. After having examined some consequences of q-damping for the pressure stimuli, the interest will now be shifted over to the response of a q-damped system to *current* stimuli, i. e., the simulation of the afferent node potential under "anesthetized" conditions.

C. The "Receptor" Potential (Afferent Nerve Element)

Simulation of the Influence of Current Stimuli on the Afferent Node, "Receptor" Potential. This study was performed under exactly the same conditions as those in the previous section; the only difference is that here an *electric current* perturbation

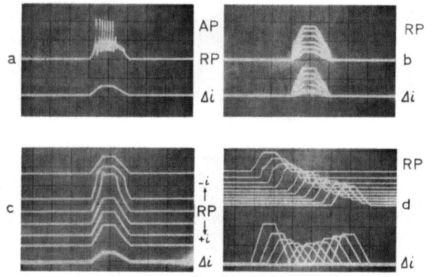

Fig. 22. Simulation of receptor potentials. The *q-damped* system was subject to various *current* stimuli, Δi. Frame *a*: superposed traces of an undamped action potential, *AP*, and a damped response, *RP*. Frame *b*: the influence of varying change of current stimuli. Frame *c*: the *RP*-response to a constant current stimulus, Δi, at various membrane resting potentials. Depolarization upwards, hyperpolarization downwards. Frame *d*: the intensity of "generator current", Δi, required to elicit a constant peak response of a "receptor potential", *RP* (top records), at various degrees of current polarization. From this record, data were obtained for the construction of Fig. 23. *Scales*: Frames *a* and *b*; *AP* and *RP* 2, Δi 2, time 50; frame *c*; *RP* 1, Δi 2, time 50; and frame *d*; *RP* 2, Δi 1, time 50 units per scale division

ΔI, was injected into the analog computer system instead of a pressure perturbation. The damping conditions were still a 10 fold decrease of the q, the σ remaining "normal". In Fig. 22a few characteristic features are demonstrated.

In frame a, it can be seen that the effect of the q-damping is an abolition of the all-or-none firing, but again the resulting *receptor potential follows the firing level* of the undamped spikes.

In frame b, the response to a current stimulation of varying intensity is seen. These stimuli might simulate the effect of varying generator currents arising from the neighboring GP-source. Again it is seen that the potential rather closely follows the shape of the current stimulus, and the degree of polarization (i.e., the peak of the "receptor potential") is rather linearly related to the current stimulus strength.

In frame c, the amplitude of the node receptor potential is demonstrated as a function of the level of the resting membrane potential, i.e., with current polarization. It is evident that the node receptor potential has a maximum at intermediate membrane potentials and falls off both when the membrane is depolarized and

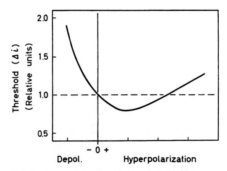

Fig. 23. The relative threshold of current stimuli for eliciting a constant amplitude of receptor potentials at various degrees of current polarization. The data were obtained from Fig. 22 d

hyperpolarized. This finding points to a peculiar relation between the threshold value for current stimuli required to produce a certain "critical depolarization" as a function of the current depolarization. Such a relation is depicted in frame d and summarized in Fig. 23. This shows that there is a threshold minimum, and that both hyperpolarization and depolarization in particular increase the threshold values. This picture is interesting in connection with another finding of Loewenstein and Ishiko (1960); they applied a current flow across the first Ranvier node of a Pacinian corpuscle subject to a presumably constant generator potential arising from a constant mechanical stimulus. Their Fig. 10 inspired the drawing of a similar *computed* relation, Fig. 23 here.

In conclusion, the *q-damped* pressoreceptor analog seems to be a reasonably good model for the *receptor* potential as measured at the anesthetized afferent nerve, whereas the *σ-damping* seems to be a better candidate in the simulations of actual GP at the sensory membrane proper.

Some further considerations on this topic will be mentioned later under the heading "summation" phenomena (p. 330).

D. The Afferent Nerve Response
(with Unimpaired Firing Ability)

In the previous section was studied the simulation of the receptor potentials of the afferent node nerve element in nonphysiological conditions, i.e., under "anesthesia". This particular manifestation of the node activity was denoted "receptor potential". In this section, the attention will now be devoted to the repetitive discharge properties of the unimpaired afferent and conductile nerve element. The technique used is identical with that of the previous presentation. The damping factors σ and q were restored to "normal" values which permitted all- or-none repetitive activity, similar to those of nerve axons. Current stimuli, Δi, were used. Since a full study of the simulated axon activity was outside the scope of this paper, only a few representative simulations will be presented, those which bear direct relationships to experience gained by experimentalists in the mechanoreceptor field. The literature here is enormous, for more information, one may refer to the review articles mentioned earlier in this paper.

1. The Firing Frequency of the Afferent Node Element as a Function of the Generator Current Strength

A typical example of the firing frequency as a function of the intensity of the generator current is shown in Fig. 24. Here, a stepwise increase of the current,

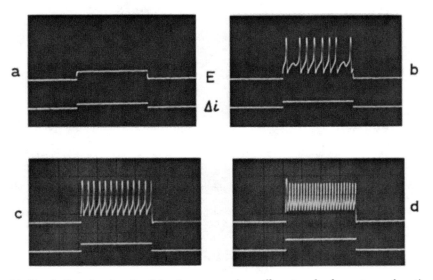

Fig. 24. Simulation showing the firing frequency of an afferent node element as a function of the generator current strength. E, the membrane potential; Δi, current stimuli. The damping conditions were "normal". Note the increasing frequency above a certain stimulus threshold as well as the gradual depolarization of the average position for the E. Scale units are as in Fig. 10

Δi, was employed. In frame a, the stimulus is subthreshold; in frame b, irregular responses just above threshold; and then in frames c and d regular increased frequency is seen. It can be shown that the current intensity-frequency relation is straight linear over a wide range but has a definite ceiling, just as demonstrated with pressure stimuli in Fig. 11a. In this particular series, no adaptation was achieved, but it could be easily demonstrated at a proper choice of parameters.[8]

An obvious feature with direct *current* stimulation is that the average *firing level, as well as the undershoot level shift towards depolarized values.* This feature was already stressed as a possible means of discrimination between a 1-step and a 2-step process (p. 315 and Fig. 16, frame $a-g$, vs. frame h). It should also be noted that the *amplitude decreased with increasing depolarization,* a feature which was brought out by the phase plot Fig. 7, *Bottum.*

2. Some Dynamic Properties of the Firing Activity During Dynamic and Static Current Stimuli

Previously, when the properties of firing activity following pressure stimuli in the 1-step process were analyzed, a distinction was made between dynamic and static response patterns (Section A, p. 315). Phenomena denoted "postexcitatory

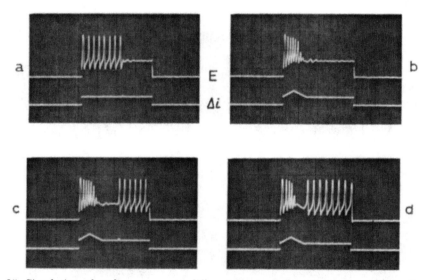

Fig. 25. Simulation of node response to different types of dynamic-static current stimuli. Frame *a*: a short superposed test pulse annihilates the response. A similar inhibition appears in *frame b* as a "postexcitatory inhibition". This is released by a small current pulse in frame *c*. Frame *d* demonstrates the appearance of a "firing gap" or "silent interval". *Scales*: see Fig. 10

inhibition", "pulse initiation" and "annihilation" were mentioned. These phenomena occur also during direct current stimuli (now thought of as applied to the first part of the afferent intact nerve element).

In the series of records in Fig. 25, frame *a* demonstrates an annihilation of firing by a small brief test pulse, injected into the refractory period, just before firing level was attained.

---

[8] The effect of the σ-damping for nonadaptation/adaptation was clearly brought out in connection with the comprehensive excitation scheme of Fig. 14. In this context, it should be added that the inherent "diffusion" membrane potential, ε, is also of decisive importance. Comments on the mechanism of this fall outside the scope of this paper.

A case of postexcitatory inhibition is exemplified in frame *b*. Here a dynamic stimulus component was added at the beginning of the step stimulus. It is clearly seen that the rising part of the stimulus caused a depolarization and a tendency to frequency increase. When the dynamic part of the stimulus resided, the firing stopped, although the potential showed a small "noise" indicating a labile condition. That the condition in this particular case of postexcitatory inhibition is indeed very labile is demonstrated in frame *c* where a small extra current pulse was again injected during the silent phase, and it caused a return to full firing for the rest of the static period, similar in frequency to that of the corresponding current stimulus of frame *a*.

The prolonged inhibitions here can be very transient and delicate. A slightly lower average current stimulus produces a much more common pattern, that of frame *d*. This is characterized by a *dynamic* phase with high-frequency firing of diminishing amplitude and a static low-frequency pattern. The conspicuous feature is the appearance of the transient inhibition, a "silent interval". Even this condition is more or less labile, as will be commented on below.

The mechanism of the initiation and inhibition properties here exemplified can be understood in accordance with the principle outlined for the corresponding phenomena in the 1-step process (Section A, p. 308—351). Again, the existence of "quiescent zones" plays a decisive role as mentioned in the discussions of the Fig. 14. However, it should be clearly understood that it may be difficult, when seeing a firing pattern like that of frame *b*, to decide whether this is a "real" adaptation (a comparatively stable state) or an annihilation phenomenon owing to the entrance of the path into a quiescent zone (a more labile state). At this point, the reader may be referred to some definitions of the concept "quiescent zone" given in the *Appendix*, Part 3.

3. Comments on the "Silent Interval" or "Firing Gap"

This particular type of postexcitatory inhibition has recently been thoroughly investigated by OTTOSON and his co-workers, who also cite earlier literature on this topic (OTTOSON and SHEPHERD, 1965b; OTTOSON, 1970, this volume)[9].

These authors investigated the influence of many factors such as rate of rise, degree of stretch, etc. By and large, it was observed that the firing "pause" was transient and, as to its appearance, very much dependent on the stimulus conditions. Since these observations have a close resemblance to those discussed in the section above, an attempt has been made here to simulate OTTOSON's results. The simulations are reproduced in Fig. 26. In this figure, an applied current (marked GP) was of the shape to be expected from a ramp-shaped mechanical displacement (denoted D in the figure). The relations between mechanical displacement of length, the resulting change in tension changes and the ensuing GP have been discussed earlier in this paper as well as in OTTOSON's (1970) contribution; Fig. 26 here is intended to be matched with OTTOSON's Fig. 12. The difference between frames *a*, *b*, and *c* is a small progressive rise in the height of the displacement. The result is quite conspicuous. In frames *a* and *b*, the dynamic and static phases are

---

[9] In this context, the author wishes to express his gratitude to DR. D. OTTOSON for a precommunication of his paper.

clearly separated by a pause or gap, and there is also a markedly different firing frequency of the two phases. In frame c, with the strongest stretch, an "occlusion of firing gap" is produced. Other simulation experiments, not published here, disclose several other features in common with the careful studies of Ottoson et al. Of course, it is not intended to offer the comparison as an "explanation" of the behavior of the frog muscle spindles, but the resemblance shown may be of some interest.

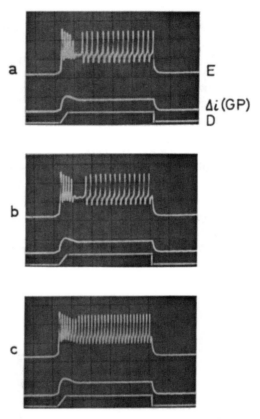

Fig. 26. The appearance and occlusion of the firing gap at an afferent node following a ramp-shaped displacement at a sensor element. The length *displacement D* has been converted into a corresponding *generator current, Δi*, which causes the dynamic and static patterns of E. There is a small gradual increase in the average stimulus intensity from frame a to c. *Scale:* see Fig. 10. This is a simulation of Ottoson's (1970) Fig. 12 contained in this volume (muscle spindles)

E. Miscellaneous Observations ("Summation", Voltage Clamp)

1. On Summation Phenomena

In receptor physiology, it is a well known fact that stimulations which occur concurrently can be added and show up in the generator or receptor potentials as summation effects. Again, it is intended to give only a few comments as they arise from corresponding simulation studies on the presented mechanoreceptor analog system. Only a few citations from the literature will be made, those which bear direct relation to our results.

As before, a distinction has been made between the primarily pressure-sensitive *sensory endings* (which elicit GP) and the *afferent nerve element* which either produces a receptor potential response or an all-or-none firing (depending on the experimental conditions). First, summation phenomena of the GP for pressure stimuli will be considered. Fig. 27, frames a and b, simulates the effect of stretch with superimposed small test stretches. The effect of a single test stretch is indicated at the very left in frames a, b and d. The summation effect of the superposition on the main stretch of the test stretches at different times is clearly seen in these frames. The effect of the added stretches show similar features; the maximal summation effect appears, as expected, around the peak of the main stimulus. The difference between frames a and b is that the former employs σ-damping, the latter q-damping.

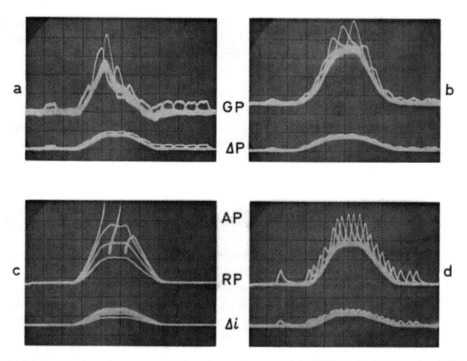

Fig. 27. Summation of generator (*GP*) and receptor potentials (*RP*). A short test pulse, indicated on the left, is superposed at various times on the main stimulus. Records are superposed pictures. Frames a and b employes *pressure* stimuli, *ΔP*, the former with σ-damping, the latter with q-damping. *Scale*: *GP* (a) 0.1, (b) 0.2; *ΔP* 1; time 10 units per scale division. Note in both cases the summation effects and especially the "amplification" around the crest of the *GP*. Frames c and d simulate *afferent node* conditions. Four *current* stimuli of varying strength were applied. The top record in c, with normal damping, showed two full sized action potentials, *AP* (off-screen). The second, third and fourth stimuli employed q-damping, resulting in "receptor potentials", *RP*, of varying heights. The third trace was subject to a summation run in frame d; again a marked summation and amplification effect are noticed. On the crest, the summed amplitudes exceed the firing level of the undamped *AP*. *Scales*: *AP* and *RP*, 0.5; *Δi* 1; time 20 units per scale division

Frames c and d give some information regarding summation phenomena of the *node receptor potential* for current stimuli. In frame c is first seen the unimpaired action potential (AP); q-damping is then applied, resulting in a receptor potential (RP), roughly reaching the firing level of AP. Corresponding RP for less currents are also indicated. The middle RP was subject to *current* test pulses in frame d. It is again seen that this causes summation to a level which here exceeds the all-or-none firing level.

No further comments on this summation phenomena will be made. In the large experimental literature, corresponding summation phenomena have repeatedly been demonstrated. However, in this context a reference will again be made to one of the results of Ottoson and his colleagues. They studied the responsiveness of the sensory ending of frog muscle spindles during stretches with superposed constant stretches at various times, during dynamic and static extension (see Fig. 35 of Ottoson, 1970, this volume). The resemblance between this figure and our Fig. 27 is good. Loewenstein *et al.* (1963) published related findings.

In conclusion, it may be inferred that the presented results with summation may lend some further support to the view advocated in this paper that generator and receptor potentials are but highly *damped* manifestations of the actual nerve potentials.

2. Some Experiments with Voltage Clamp Applied to the Mechanoreceptor Model

As the "voltage clamp" technique has been used so often in the electrophysiological literature as a specific tool for characterizing excitable membranes, it can be of interest to analyze the response of the proposed mechanoreceptor analog in terms of the voltage clamp. The procedure employed was in principle the same as before (Teorell, 1960). The analog computation results are presented in Fig. 28. Different step depolarizations of the membrane potential E were applied, and the corresponding changes of the current were recorded (the upper and lower part of frames).

The *shape* of the current responses show striking similarities with numerous examples in the physiological literature. As usual here an early transient state and a late state are distinguished. Fig. 28c is a current-voltage characteristic obtained from Fig. 28a and b.

A very interesting finding pertains to frame b, which is processed with exactly the same conditions as the "normal" case a except for the introduction of an excessive damping of the q-type. It is interesting to note that a damping causes the "inward current" to disappear almost completely. The late current remains uneffected.

It is tempting to compare the simulated voltage clamp results with the corresponding ones in the literature, in particular with those dealing with axons treated with anesthetics or tetrodotoxin (Taylor, 1959; Nakamura *et al.*, 1965; Moore, 1965; Hille, 1966; and many others). All these authors found a more-or-less complete abolishing of the early "dip" in the transient current, which is usually interpreted as an "inactivation" of the sodium inward current. Without arguing against this explanation, it is interesting to note that the present model does not assume any ion selectivity; it operates with the *overall conductance* (see p. 307).

It should also be observed that the basic concepts assume a resting current in the quiescent state. Other discrepancies could also be pointed out. Nevertheless, it should be emphasized that many *formal* properties of typical voltage clamp behavior are embodied in the suggested mechanoreceptor analog.

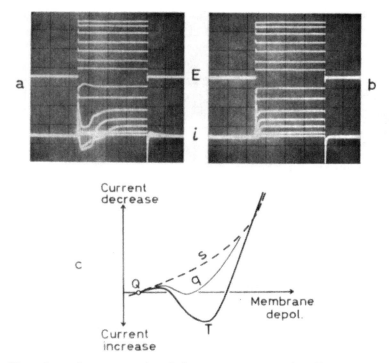

Fig. 28. The voltage clamp properties of the pressoreceptor analog. Frame *a*: rectangular voltage steps were applied, depolarization upward (*E*). The ensuing changes of membrane current were recorded (*i*), normal damping. Frame *b*; the same as *a*, but *q*-damping was employed. *Scale*: (*E*) 0.5; (*i*) 1; time 20 units per scale division. Frame *a* is thought of as representing the *unimpaired* afferent *node* response, whereas frame *b* may represent an "*anesthetized*" node. *Bottom*: *A conventional current-voltage representation* obtained from frames *a* and *b*. *Q* marks the resting or holding potential, *T* is the early transient current (the "dip"), and *S* is the late steady state surrent, equal in both cases. Graph "*q*" is the early transient current of frame *b*

Concluding Comments

This article has dealt with an artificial mechanoreceptor analog which can be described in physical and physicochemical terms, based on relatively simple and fundamental equations. In the most comprehensive form, the whole system can be embodied in a secondary-order differential equation with nonlinear coefficients. Time variation of the parameters can also be expected, owing to particular viscoelastic properties in the receptor organs. The analysis in these biophysical terms has been carried out with the aim of comparing the results with actual biological findings. The behavior of the system has been investigated by the use of analog computation technique. A formal replication of many salient features described in receptor physiology were successfully achieved and explained in terms of the proposed system. In order to help in visualization of the abstract equations, the variables and parameters have been assigned specific meanings such as "pores", "water flow", "ion transport", etc. It should be strongly emphasized that this concretization should be applied to the biological problems with great caution. Nevertheless, it is hoped that the results may be of some value because they stress the great utility of applying oscillation theory to excitability phenomena.

An excitation problem has here been reduced to a type of second-order differential equations. It is quite possible that its variables may be reconstituted in quite different physical and chemical terms than those employed here, depending on the ingenuity of the model maker. Nature has certainly many more means of producing stable and unstable phenomena as revealed by receptor and nerve excitability.

Acknowledgements

This work has been supported by grants from the Swedish Medical Research Council, the Air Force Office of Scientific Research under Grant AF EOAR 66-5 through the European Office of Aerospace Research (OAR), United States Air Force, and the National Institutes of Health grant NB 03712—05, which are gratefully acknowledged. My thanks are also due to Mrs. Ebon Arnelund and Mr. K. Lundegard for valuable assistance.

Appendix

1. The Basic Mathematical Equations governing the membrane oscillator are here slightly modified from earlier publications (Teorell, 1957—1966).

$$V_E = lE - sP\,, \tag{1}$$

$$V_L = -\sigma P\,(-\mu P)\,, \tag{2}$$

$$W = V_E + V_L\,(+\,V_{\text{overflow}})\,, \tag{3}$$

$$E = iR\,(+\,\varepsilon)\,, \tag{4}$$

$$dM/dt = (1/q)\cdot W\,, \tag{5}$$

$$M = f(P)\,, \tag{6}$$

$$dR/dt = k(R^\infty - R)\,, \tag{7}$$

$$R^\infty = f(V_E)\,. \tag{8}$$

The *variables* are: V_E = electroosmotic bulk flow velocity: V_L = leakage flow velocity: W = the net flow velocity: E = transmembrane electrical potential; P = transmembrane total pressure difference; R = instantaneous (integral) membrane resistance; R^∞ = steady state (integral) membrane resistance; M = volume change: i = electric current density; ε = diffusion potential.

The *coefficients* are: s = hydraulic permeability for the electroosmotic channels; σ = hydraulic permeability for the leakage channels; μ = "overflow" leakage coefficient. w = sign (± 1) of the fixed charge of the density $\overline{X}$; l = electroosmotic permeability $[= -\,sw\,\overline{X}]$; and q = geometry (intregation) constant, depending on the P-M relationship (volume change $M = 1/q\cdot\int W\cdot dt$; k = the "time delay" constant (dependent on various membrane factors). The nonlinear function

$R^\infty = f(V_E)$ is the "backbone curve" derived in the paper by TEORELL (1959b). The P-M relation (two linear segments) used in the analog computation (10 V equipment) had the following coordinates: $P = 0/M = 0$; $P = 4/M = 4$; $P = 4.5/M = 8$ (maximum).

Numerical values for the various coefficients were: $k = 4.0$; $s = 1.0$; $\omega X = 1.5$; $\sigma = 0.4$ to 1.0; $q = 2$ (or 0.2); $\varepsilon = +0.5$ to $+0.7$; $\mu = 3.0$.

2. The Notes on Oscillation Theories. The behavior of the membrane oscillator (and the electrohydraulic excitability analog) can be generally described mathematically by a set of equations of the type:

$$dy/dt = ax + by + h, \tag{9}$$

$$dx/dt = cx + ey. \tag{10}$$

The variables y and x can denote any pairwise combination of the "variables of state" present in the system. These are the membrane potential E, the membrane pressure difference P, the electro-endosmotic velocity V, the volume change M, and the membrane electric resistance R. For the purpose of this paper, it appears appropriate to select the potential E and the volume change M as variables and denote them by y and x, respectively. The variables, E-P-V-R-M, were interrelated by the set of simple fundamental equations given above. An important concept was the so-called "characteristic" or "backbone" curve belonging to the actual membrane oscillator. This was a nonlinear S-shaped relation between the membrane electric resistance R and the electro-endosmotic flow velocity V, and corresponds to a "dynatron characteristic", a concept used in electrical oscillation theory. The backbone curve indicated those states where the system theoretically could be at equilibrium (stable or unstable).

Another possible version of the differential equation system above is a single *second-order* differential equation with nonlinear coefficients of the type

$$a_0 \frac{d^2x}{dt^2} + a_1 \frac{dx}{dt} + a_2 x = f(t) \tag{11}$$

In fact, these equations represent in general form a variety of types of oscillators well known within the realm of nonlinear mechanics. Different shapes of oscillations can be conceived e.g., more or less sinusoidal curves, which can appear as "damped", "undamped" or "explosive", when x or y is plotted against the time t. If one instead plots the two variables x and y against one another in a "phase plane", the damped oscillations appear as spirals terminating at a *focus*; undamped, sustained oscillations yield a closed figure called *limit cycle*, whereas a *node* may have such "paths", or trajectories, that irradiate from the node as more or less straight lines and as lines which "turn around". Several other forms are also possible. The nature of the oscillation is determined by the coefficients a, b, c, and e. A form of classification is possible in terms of the so-called *discriminant*, the *determinant* and a *damping term*.

Regardless of the type of trajectories which appear, they are "controlled" by a *singular point*. From the point of view of the physiological application, this will be renamed to *quiescent point*, Q. This Q can be stable in the sense that it is constant if the paths approach it with time, or unstable if the path leaves it with time.

3. On the Nature of the "Quiescent Zone". The type of oscillations discussed in this paper are of the *self-sustained* type. A suitable driving mechanism must be provided constantly to replace the energy dissipated by the damping conditions. A typical example from mechanics is the pendulum clock which may have various types of driving mechanisms. In the membrane oscillator, the energy is supplied by a *current source*, as discussed earlier. In the complicated theory of

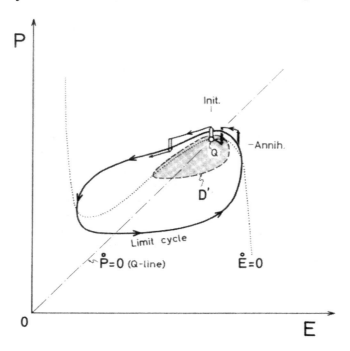

Fig. 29. Diagram showing the location of a "quiescent zone" in the E-P phase plane. At resting membrane potential, the quiescent point Q is located within the *quiescent zone* (stippled). When a small rectangular pressure stimulus (open arrows) is applied, the phase point is brought outside the Q-zone and starts to move on the *limit cycle*, hence eliciting an oscillatory variation of the membrane potential E (and of P). If a new pulse (the thick black arrows). is applied shortly after the maximum hyperpolarization (i.e., the "refractory period"), it is possible to bring the phase point to within the quiescent zone, where it comes to rest at the Q-point. This phenomenon is called "annihilation". The same effect may be brought about by an electrical current stimulus but this requires a three-dimensional diagram to be visualized. $\dot{E}$ and $\dot{P}$ signify, respectively, dE/dt and dP/dt. This phase plane coincides approximately with the horizontal plane in Fig. 6b, marked $i = 1$, and may be called a "current clamp" plane. Both this figure and Fig. 6b refer to a somewhat simpler system, but the inferences for the present analyses remain the same

nonlinear, self-sustained oscillations, one distinguishes between *soft* and *hard* oscillations. The difference lies in the behavior of the limit cycles. Examples of limit cycles have here been given in the E-M phase plane (see Figs. 7 and 19). For the purpose of this discussion, Fig. 29 depicts instead the E-P phase portrait. This may illustrate the situation of a *hard* oscillations. The singular point Q is quiescent because it is located with in an area (stippled), where all trajectories tend to spiral inwardly to Q (a stable focus or spiral point). In order to get an oscillation to take place on the outer stable limit cycle, one must give the point an initial *impulse of disturbance*, such that it can be brought outside of the boundary of the quiescent zone (marked *D*), then it spirals outwardly and goes out "in orbit" on the stable limit cycle as a regular sustained oscillation. On the other hand, it is also possible by a proper perturbation to "take down" the point again to within the quiescent zone and thus "annihilate" the rhythmicity. These two cases are illustrated in the diagram by open and filled arrows, which demonstrate, respectively, "single-pulse initiation" and "brief-pulse annihilation" (both caused here by pressure pulses, but current pulses can have similar effects). It should be observed that these pulses should be placed "strategically", which is evident from the figure. In fact, this "vulnerable" region near the Q-point coincides with the "refractory periods", using physiological terminology. The existence, location and the size of the quiescent zone is dependent on all parameters in the presented case, in particular on i and σ.

In summary, the quiescent zone arises from the fact that hard oscillations can have *two* limit cycles, on large and stable, and one small and unstable (*D*), unstable because all trajectories tend to travel away from it. It is easy to analogize with the pendulum clock; in order to start this, a sort of small push has to be applied. The reverse situation can also be envisaged.

It may be suggested that the important concept of "quiescent zones" should be more commonly introduced in excitation physiology, because it can provide a better formal understanding of "refractoriness", "threshold" and "inhibition" phenomena.

References

BROWN, M.: Some effects of receptor muscle contraction on the responses of slowly adapting stretch receptors of the crayfish. J. Exp. Biol. 46, 445—458 (1967).

— STEIN, R.: Quantitative studies on the slowly adapting stretch receptor of the crayfish. Kybernetik 3, 175—185 (1966).

CALDWELL, P.: Factors governing movement and distribution of inorganic ions in nerve and muscle. Physiol. Rev. 48, 1—64 (1968).

CATTON, W., PETOE, N.: A visco-elastic theory of mechanoreceptor adaptation. J. Physiol. (Lond.) 187, 35—49 (1966).

EDWARDS, C., HAGIWARA, S.: Potassium ions and the inhibitory process in the crayfish stretch receptor. J. gen. Physiol. 43, 315—321 (1959).

EYZAGUIRRE, C.: Excitatory and inhibitory processes in crustacean sensory nerve cells. In: Nervous Inhibition. Proc. Int. Symp. London: Pergamon Press 1961.

— KUFFLER, S.: Processes of excitation in the dendrites and in the soma of single isolated sensory nerve cells of the lobster and crayfish. J. gen. Physiol. 39, 87—119 (1955a).

— — Further study of soma, dendrite and axon excitation in single neurons. J. gen. Physiol. 39, 121—153 (1955b).

FLOREY, E.: Nervous inhibition: Proc. 2nd Friday Harbor Symp. New York: Pergamon Press 1961.

GOLDMAN, D.: The transducer action of mechano-receptor membranes. Cold Spr. Harb. Symp. quant. Biol. 30, 59—68 (1966).

GRANIT, R.: Receptors and sensory perception. London: Yale Univ. Press 1955.

GRAY, J.: Mechanical into electrical energy in certain mechanoreceptors. Progr. Biophys. 9, 286—324 (1959).

GRUNDFEST, H.: Heterogeneity of excitable membranes: Electrophysiological and pharmacological evidence and some consequences. Ann. N. Y. Acad. Sci. 137, 901—949 (1966).

HIGGINS, J.: In: Control of energy metabolism. CHANCE, B. et al., eds. New York: Academic Press 1965.

Hille, B.: Common mode of action of three agents that decrease the transient change in sodium permeability in nerves, Nature (Lond.) **120**, 1220—1222 (1966).

Hodgkin, A.: The conduction of the nervous impulse. The Sherrington Lecture VII. Liverpool: University Press 1967.

— Chandler, W.: Effects of changes in ionic strength on inactivation and threshold in perfused nerve fibers of Loligo. J. gen. Physiol. **48**, 27—30 (1965).

Houk, J.: A model of adaptation in amphibian spindle receptors. J. Theoret. Biol. **12**, 196—215 (1966).

Hubbard, S.: A study of rapid mechanical events in a mechanoreceptor. J. Physiol. (Lond.) **141**, 198—218 (1958).

Ishiko, N., Loewenstein, W.: Electrical output of a receptor membrane. Science **130**, 1405—1414 (1959).

Julian, F., Goldman, D.: The effects of mechanical stimulation on some electrical properties of axons. J. gen. Physiol. **46**, 297—313 (1962).

Katz, B.: Depolarization of sensory terminals and the initiation of impulses in the muscle spindle. J. Physiol. (Lond.) **111**, 261—262 (1950).

Krnjevic, K., Gelder, N. van: Tension changes in crayfish stretch receptors. J. Physiol. (Lond.) **159**, 310—325 (1961).

Lippold, O., Nicholls, J., Redfearn, J.: Electrical and mechanical factors in the adaptation of a mammalian muscle spindle. J. Physiol. (Lond.) **153**, 209—217 (1960).

Loewenstein, W.: The generation of electric activity in a nerve ending. Ann. N. Y. Acad. Sci. **81**, 367—387 (1959).

— Excitation and inactivation in a receptor membrane. Ann. N. Y. Acad. Sci. **94**, 510—534 (1961).

— Permeability of membrane junctions. Ann. N. Y. Acad. Sci. **137**, 441—472 (1966a).

— Facets of a transducer process. Cold Spr. Harb. Symp. quant. Biol. **30**, 29—43 (1966b).

— Ishiko, N.: Effects of polarization of the receptor membrane and of the first Ranvier node in a sense organ. J. gen. Physiol. **43**, 981—998 (1960).

— Skalak, R.: Mechanical transmission in a Pacinian corpuscle. An analysis and a theory. J. Physiol. (Lond.) **182**, 346—300 (1966).

— Terzuolo, C., Washizu, Y.: Separation of transducer and impulse-generating processes in sensory receptors. Science **142**, 1180—1181 (1963).

Matthews, B.: The response of a single end organ. J. Physiol. (Lond.) **71**, 64—110 (1931).

— The differentiation of two types of fusimotor fibre by their effects on the dynamic response of muscle spindle primary endings. Quart. J. exp. Physiol. **47**. 324—333 (1962).

Matthews, P.: Muscle spindles and their motor controls. Physiol. Rev. **44**, 219—288 (1964).

Milhorn, H.: The application of control theory to physiological systems. London: W. Saunders Co. 1966.

Moore, J.: Voltage clamp studies on internally perfused axons. J. gen. Physiol. **48**, 11—17 (1965).

Nakajima, S.: Adaptation in stretch receptor neurons of crayfish. Science **146**, 1168—1170 (1964).

— Takahashi, K.: Post-tetanic hyperpolarization and electrogenic Na-pump in stretch receptor neurone of crayfish. J. Physiol. **187**, 105—127 (1966).

Nakamura, Y., Nakajima, S., Grundfest, H.: The action of tetrodotoxin on electrogenic components of squid giant axons. J. gen. Physiol. **48**, 985—996 (1965).

Ottoson, D.: The effect of osmotic pressure changes on the isolated muscle spindle. Acta physiol. scand. **64**, 93—105 (1965a).

— Shepherd, G. M.: Receptor potentials and impulse generation in the isolated spindle during controlled extension. Cold Spr. Harb. Symp. quant. Biol. **30**, 105—114 (1965b).

— Handbook of Sensory Physiol. Vol. I (1971), p. 442—499.

— — Transducer properties and integrative mechanisms in the frog's muscle spindle. Handbook of Sensory Physiol. Vol. I, pp. 442—499. Berlin-Heidelberg-New York: Springer 1971.

Rack, P., Westbury, D.: The effects of suxamethonium and acetylcholine on the behavior of cat muscle spindles during dynamic stretching, and during fusimotor stimulation. J. Physiol. (Lond.) **186**, 698—713 (1966).

Rapoport, S.: The Na-K exchange pump. Relation of metabolism to electrical properties of the cell. Biophys. J., March 1970.

REMINGTON, J.: Tissue elasticity. Washington: Amer. Physiol. Soc. 1957.

ROBERTS, T.: Rhythmic excitation of a stretch reflex. Revealing (a) hysteresis and (b) a difference between the response to pulling and to stretching. Quart. J. exp. Physiol. **68**, 328—345 (1963).

SATO, M.: Response of Pacinian corpuscles to sinusoidal vibration. J. Physiol. (Lond.) **159**, 391—409 (1961).

SHEPHERD, G., OTTOSON, D.: Responses of the isolated muscle spindle to different rates of stretching. Cold Spr. Harb. Symp. quant. Biol. **30**, 95—103 (1965).

STALLWORTHY, W., FENSOM, D.: Electroosmosis in axons of freshly killed squid. J. Physiol. Pharmacol. **44**, 866—870 (1966).

TASAKI, I.: Nerve excitation: A macromolecular approach. Springfield: Charles Thomas Publ. 1967.

TAYLOR, R.: Effect of procaine on electrical properties of squid axon membrane. Amer. J. Physiol. **196**, 1071—1078 (1959).

TEORELL, T.: Transport processes and electrical phenomena in ionic membranes. Progr. Biophys. Biophys. Chem. **3**, 305—369 (1953).

— On oscillatory transport of fluid across membranes. Acta Soc. Med. upsalien. **62**, 60—66 (1957).

— Transport processes in membranes in relation to the nerve mechanism. Exp. Cell Res. Suppl. **5**, 83—100 (1958a).

— Rectification in a plant cell (Nitella) in relation to electro-osmosis. Z. Phys. Chem. **15**, 385—398 (1958b).

— Electrokinetic membrane processes in relation to properties of excitable tissues. I. Experiments on oscillatory transport phenomena in artificial membranes. J. gen. Physiol. **42**, 831—945 (1959a).

— Electrokinetical membrane processes in relation to properties of excitable tissues. II. Some theoretical considerations. J. gen. Physiol. **42**, 847—863 (1959b).

— Biophysical aspects on mechanical stimulation of excitable tissues. Acta Soc. Med. upsalien. **64**, 341—352 (1959c).

— Application of the voltage clamp to the electrohydraulic nerve analog. Acta Soc. Med. upsalien. **65**, 231—248 (1960).

— Oscillatory electrophoresis in ion exchange membranes. Arkiv Kemi **18**, 401—408 (1961a).

— An analysis of the current-voltage relationship in excitable Nitella cells. Acta physiol. scand. **53**, 1—6 (1961b).

— Excitability phenomena in artificial membranes. Biophys. J. **2**, No. 2, Part 2 (Suppl.), 27—52 (1962a).

— Some biophysical considerations on presso-receptors. Arch. int. Pharmacodyn. **140**, 563—576 (1962b).

— Electrokinetic considerations of mechano-electrical transduction. Ann. N. Y. Acad. Sci. **137**, 950—966 (1966).

TERZUOLO, C., WASHIZU, Y.: Relation between stimulus strength, generator potential and impulse frequency in stretch receptor of Crustacea. J. Neurophysiol. **25**, 56—66 (1962).

WIERSMA, C., FURSHPAN, A., FLOREY, E.: Physiological and pharmacological observations on muscle receptor organs of the crayfish, Cambarus clarkii Girard. J. exp. Biol. **30**, 136 (1953).

WOLBARSHT, M.: Electrical characteristics of insect mechanoreceptors. Naval Med. Res. Inst. Res. Rep. 18, 89—102 (1960).

Chapter 11

Responses of Nerve Fibers to Mechanical Forces

By

David E. Goldman, Philadelphia Pa. (USA)

Living cells respond in one way or another to a wide variety of stimuli. Some cells are highly specialized with respect to both their sensitivity to stimuli and the way in which they respond. Nerve axons respond primarily to certain electrical and chemical stimuli and produce an electrochemical response. However, the specialization is not absolute. Many other kinds of stimuli can, under appropriate circumstances, produce responses in axons although the conditions required are not necessarily those usually found in vivo. In particular, the application of mechanical forces can produce electrical changes; such phenomena are interesting, not only because they may cast light on the structure and behavior of the axon but also because they may be important to the understanding of mechanoreception in general and may have relevance to problems of mechanical trauma.

Mechanical forces are omnipresent, and nerves are usually well protected by surrounding tissues which tend to disperse applied forces. In addition, the sensitivity of nerves is such that they are able to tolerate moderate forces without interference with their proper function. Subjective interest in the action of mechanical forces may indeed begin with the observation that a blow to the ulnar notch produces a sharp pain. On reflection, and with some knowledge of anatomy, this pain may be attributed to the direct action of the force on a nerve trunk. However, modern analytical approaches involve other factors and require more precise characterization. For example, mechanical forces which impinge on an axon must be described by both their temporal and their spatial characteristics. In the time domain, forces may be considered as static, impulsive, periodic or random. In the spatial domain, the region of application, degree of localization and type are important. Fundamentally, mechanical stresses are compressive or shearing. Composite strains include bending, stretching, twisting, etc. Thus, in experimental work in this area, the mechanical forces can, in principle, be characterized in a rather precise way although it is usually very difficult to carry out clean experiments. Studies have been made over a period of years in varying degrees of detail on a variety of living forms including worms, arthropods, mollusks, amphibia and mammals. Observations have usually included microscopic, chemical or electrophysiological procedures. Some studies have dealt with progressive changes such as those of injury and recovery.

There are a number of steps to be considered between the applied force and the biological or other response of the axon. The applied force produces stresses on the axon itself. These stresses in turn produce strains. The stresses and/or strains produce electrochemical or other changes which may then lead to an electrical response or to damage. Since not even a single axon can be considered a completely homogenous isotropic system, stresses of one type may produce strains of several types. Whereas this, of course, is a general property of mechanical structures, it does mean that one must take particular care in following the route through which the applied force generates a response. This will appear in more detail as the specific types of experimental work which have been carried out are discussed.

No systematic studies have been made of the many kinds of mechanical force and the many kinds of nerve and ways of application of force which are possible. The following have been selected as representative of the different types of work which have been done.

The first to be considered are "static forces"; i.e., forces whose rate of application is sufficiently slow so that no significant transient effects are elicited, and in which the observations are made during a steady state. A static force of interest is uniform compression of a nerve in a suitable fluid medium (GRUNDFEST, 1936). As the pressure is increased from atmospheric level to about 10,000 psi, there are progressive changes in excitability, propagation velocity and spike height; these changes are reversible provided that the pressure is not kept too high for too long. The nerve passes through a phase of increased excitability, increased action potential amplitude and increased conduction velocity, but as the pressure continues to rise these are reduced below normal.

Another force of interest is internal pressure applied to a perfused axon. This has been accomplished in squid fibers by BAKER, HODGKIN, and SHAW (1962); they showed that whereas an internal pressure of a few centimeters of water is required to maintain an axon distended and in normal conducting condition, the axon tends to become leaky when the pressure increases to about 10 or more centimeters of water; if the pressure is not soon released, the axon becomes inexcitable.

Another study involves longitudinal stretch of nerves and nerve fibers. GRAY and RITCHIE (1954) applied steady stretch to isolated frog myelinated fibers; they found little, if any, change in conduction or excitability properties until the limits of stretch had been approached, very roughly a 50% length increase, at which time there was a deterioration in performance leading to nerve block and followed shortly thereafter by a breaking of the axon. It is known that the nodal regions stretch much less than the myelinated internodes (SCHNEIDER, 1952). An intermediate stage existed in which, if the stretch were not maintained for too long, the changes in conduction velocity, action potential and the block were reversible. On the other hand, GOLDMAN (1964) has shown that single fibers of *Lumbricus* can be stretched up to at least four times the initial length of the axon with relatively little change in conduction or excitation properties.

Localized compression of long duration is well known to be a cause of conduction block; presumably a significant fractional compression of part of the nerve is required although no measurements appear to have been made. However, no excitation seems to be produced by this or any of the other types of static force referred to above.

An entirely different situation arises if we deal now with impulsive forces. These are forces whose rates of application and/or decay are sufficiently rapid to excite transient responses and whose duration is usually extremely short. Rapid small stretches (about 5% in 1 msec) of single myelinated frog fibers were made by GRAY and RITCHIE (1954) who found that no responses were produced and that there were no observable changes in conduction properties.

The effect of rapid blows in the form of a localized compression of the nerve has been known for a very long time. Such rapid compression can produce action potentials. Experimental interest in this goes back as far as the latter part of the 18th century. More recently, studies have been made by ROSENBLUETH et al. (1953) on cat nerves and by SCHMITZ and WIEBE (1938) and YAMADA and SAKADA (1961) on amphibian nerves. BLAIR (1936) applied short-duration air jets to the whole sciatic nerve of the frog. It has more recently been possible to carry out experiments on isolated fibers of the lobster (JULIAN and GOLDMAN, 1962). Studies of vertebrate nerves have shown that a short rapid mechanical stimulus can produce action potentials, that at subthreshold levels it can facilitate electrical stimuli, and that its effects may thus be described briefly by saying that a mechanical stimulus is analogous to a catelectrotonus. The experiments with the isolated giant axon of the lobster showed that subthreshold mechanical stimuli produce a very lengthy depolarization (up to several seconds), that a superthreshold depolarization persists far beyond the action potential, and that this depolarization results primarily from distortion of the axon membrane. The depolarization is not abolished by a short electrical hyperpolarization or by an action potential. A decrease in membrane resistance accompanies the depolarization process. If the depolarization is minimized by passing an appropriate current, the decrease in resistance still occurs although it may be reduced slightly.

Periodic forces constitute the other main type of mechanical stimulus which has been applied to nerve. In order to obtain sufficiently intense forces, high-frequency acoustic sources are required and these must be applied through direct fluid coupling so that transfer losses will be minimized. There are two regimes which may be involved. If the intensity is great enough, cavitation is produced. This partially decouples the driver from the target but replaces the acoustic wave by repeated bursts of explosions and implosions of small gas-or vapor-containing cavities. These bursts may be very destructive to tissues. However, cavitation is avoided by the use of degassed viscous fluids; the effects seen may be presumed due to the action of the sound waves. In the presence of standing waves, it is possible to separate the effects of the pressure changes from those of the alternating motion. This has been shown in several types of cells (GOLDMAN and LEPESCHKIN, 1952) but has not been tried with axons. If travelling waves are used, the target tissue is exposed to both. The most definitive recent study was carried out by LELE (1963) who used frequencies of 600 kHz, 900 kHz, and 2.7 MHz at intensities up to 40w/cm^2 with varying pulse intervals for intermittent stimulation. Observations were made on giant fibers of *Lumbricus* as well as on peripheral nerves of several mammals. The significant conclusion drawn was that all effects noted could be duplicated by suitable heating; i.e., no changes were observed that were directly attributable to mechanical forces.

The knowledge of mechanical effects on nerve axons is clearly too incomplete to permit any detailed discussion. The basic elements seem to be these: (1) High pressures produce volume changes, and affect chemical reaction rates; they may also produce modifications in macromolecular structures, but considerable time is required for significant changes to occur. (2) Distortion of an axon membrane produces depolarization. The relevant strain seems, however, to be an increase in circumference. Rapid local compression produces circumferential stretching when the axoplasm has no time to flow a significant distance from the compressed area. Longitudinal stretching of the axon (not necessarily the same as stretching the membrane) has very little effect below some characteristic tolerance limit which, however, varies widely with the type of axon used. (3) High-frequency sound absorption generates heat which itself affects the system. Obviously, sufficiently intense forces destroy the nerve although the system may show little apparent behavioral change very nearly up to the failure limit.

The difference between the effects of radial stretching and localized lateral compression on one hand and longitudinal stretching on the other suggests that the membrane is strongly anisotropic, although there is no way as yet to decide to what extent this is a property of the axolemma or of its supporting structures. The effects of local compression of an axon also mimic the responses of certain mechanoreceptors to similarly applied forces (see GRAY, 1959). Pacinian corpuscles in which the nerve ending has been exposed are a particularly relevant example (LOEWENSTEIN, 1965; LOEWENSTEIN and RATHKAMP, 1958). In the lobster axon, a rapid compression appears to produce an increase of ion permeabilities, although it is not known in what way this occurs. The resulting depolarization acts like, and may be the same as, a generator potential (see GOLDMAN, 1965) hence, it is suggested that the axon may be an elementary model for mechanoreception.

References

BAKER, P. F., HODGKIN, A. L., SHAW, T. I.: Replacement of the axoplasm of giant nerve fibres with artificial solutions. J. Physiol. **164**, 330—354 (1962).

BLAIR, H. A.: The time intensity curve and latent addition in the mechanical stimulation of nerve. Amer. J. Physiol. **114**, 586—593 (1936).

GOLDMAN, D. E.: The transducer action of mechanoreceptor membranes. Cold Spr. Harb. Symp. quant. Biol. **30**, 59—68 (1965).

— LEPESCHKIN, W. W.: Injury to living cells in standing sound waves. J. Cell. Comp. Physiol. **40**, 255—268 (1952).

GOLDMAN, L.: The effects of stretch on spike parameters of single nerve fibres; some implications for the theory of impulse propagation. J. Physiol. **175**, 425—444 (1964).

GRAY, J. A. B.: In: Progress in biophysics and biophysical chemistry, Vol. 9. London: Pergamon Press 1959.

— RITCHIE, J. M.: Effects of stretch on single myelinated nerve fibers. J. Physiol. **124**, 84—99 (1954).

GRUNDFEST, H.: Effects of hydrostatic pressures upon the excitability, the recovery, and the potential sequence of frog nerve. Cold Spr. Harb. Symp. quant. Biol. **4**, 179—187 (1936).

JULIAN, F. J., GOLDMAN, D. E.: The effects of mechanical stimulation on some electrical properties of axons. J. gen. Physiol. **46**, 297—313 (1962).

LELE, P. P.: Effects of focussed ultrasonic radiation on peripheral nerve, with observations on local heating. Expl. Neurol. **8**, 47—83 (1963).

Loewenstein, W. R.: Facets of a transducer process. Cold Spr. Harb. Symp. quant. Biol. **30**, 29—43 (1965).
— Rathkamp, R.: Localization of generator structures of electrical activity in a Pacinian corpuscle. Science **127**, 341 (1958).
Rosenblueth, A., Alvarez-Buylla, R., Ramos, J. G.: The responses of axons to mechanical stimuli. Acta Physiol. Latino Amer. **3**, 204—215 (1955).
Schmitz, W., Wiebe, W.: Zur Frage von mechanischer Nervenreizung. Pflügers Arch. **240**, 289—299 (1938).
Schneider, D.: Die Dehnbarkeit der markhaltigen Nervenfaser des Frosches in Abhängigkeit von Funktion und Struktur. Z. Naturforsch. 7b, 38—48 (1952).
Yamada, M., Sakada, S.: Effects of mechanical stimulation on the nerve fiber. Jap. J. Physiol. **11**, 378—384 (1961).

Chapter 12

The Early Receptor Potential

By

Richard A. Cone, Baltimore, Md., and
William L. Pak, Lafayette, Ind. (USA)

With 12 Figures

Contents

Introduction

Of all receptor molecules, visual pigments are the best characterized, yet the the problem of how a pigment molecule initiates an excitatory response in a photoreceptor is still unsolved. It is, however, well established that a photoreceptor can be excited by a single photon (Hecht et al., 1942), and also that the photon is absorbed by visual pigment molecules contained in dense, highly ordered membrane structures such as the rhabdomeres of invertebrate receptors and the outer segments of vertebrate receptors. Within these membrane structures, the pigment molecules are themselves highly ordered (Schmidt, 1938), and since visual pigments are lipoproteins and are major components of the membrane structures, it is likely that they are incorporated structurally in the membrane. (For a discussion of these characteristics, see Wald et al., 1963.) The central problem in the study of visual receptors, therefore, is to determine how a single visual pigment molecule, on absorbing a photon, can initiate an excitatory response. In an elegant series of experiments on squid photoreceptors, Hagins and coworkers (1965) have provided the clearest example of the role the molecule can play in the excitatory mechanism. In these receptors, the visual pigment initiates a transient change in the permeability of the cell membrane which permits sufficient ionic current to flow through the membrane to excite the receptor (Hagins, 1965). It appears likely that in most other photoreceptors the role of the pigment molecule is also to alter membrane permeability. It is clear, therefore, that future research on photoreceptors will require experimental methods which can detect rapid transitions in the pigment molecule after it has absorbed a photon, and also reveal how this activity produces a change in membrane permeability.

After a photon is absorbed by a squid photoreceptor, several milliseconds elapse before ionic membrane current begins to flow. Such a "dead time" appears to be a characteristic of all photoreceptors, and, until recently, little was known about the molecular events which occur during this crucial interval. But in 1964, Brown and Murakami discovered a new electrical response in the eye of the Cynomolgus monkey which has no detectable latency (Brown and Murakami, 1964a). They named this fast photovoltage the early receptor potential (ERP) to distinguish it from the "late" receptor potential which in vertebrate visual receptors presumably arises from a change in membrane permeability. It was immediately apparent that the ERP might help reveal the role played by the pigment molecule in visual excitation. But a significant characteristic of the ERP is that it can only be elicited by unusually intense light flashes, and the possibility that it might be completely unrelated to the excitatory mechanism could not at first be excluded. The primary evidence reported by Brown and Murakami to support their conclusion that this fast photovoltage is generated in the visual receptors may be summarized as follows: 1) The maximum response amplitude occurs when the electrode tip is in the receptor layer of the retina, and at the same depth at which the late receptor potential is maximal. 2) The response has no detectable latency, and is highly resistant to anoxia. 3) The response amplitude can be reduced to below the experimental noise level by repeated stimulus flashes, suggesting that it declines as the visual pigment is bleached by these flashes. 4) The amplitude is larger in the fovea than in the peripheral retina, which indicates that the response originates in the retina.

Fig. 1 shows the ERP in the cone-dominated eye of the Cynomolgus monkey recorded with an intraretinal microelectrode as first reported by Brown and Murakami. The ERP develops concurrently with the flash and then persists for several milliseconds, its final decay phase being masked by the onset of the late receptor potential. The ERP can also be observed in the electroretinogram using a corneal electrode (Cone, 1964), and Fig. 2 shows the ERP as it appears in the electroretinogram of the albino rat. In this figure, all distortions in the waveform of the ERP were eliminated by excising the eye to abolish the late receptor potential, and by using a light flash of a sufficiently short duration (0.7 μsec). The ERP is biphasic (Cone, 1964; Brown and Murakami, 1964b); i.e., there is an initial corneal positive phase, R 1, followed by a slower corneal negative phase, R 2. The rising phase of R 1 is shown to the right in Fig. 2 by several superimposed oscilloscope traces at a high sweep speed. It can be seen from these traces that there is no indication of a latent period preceding R 1. If a latent period exists, it must be shorter than 0.5 μsec (Cone, 1967). Other results obtained from the electroretinogram of the albino rat have also given strong support to BROWN and MURAKAMI's conclusion that the ERP is a receptor potential (CONE, 1964): 1. The ERP has the same action spectrum as the later responses of the electroretinogram, all of which

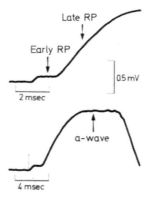

Fig. 1. Time course of the ERP (early RP) and the late receptor potential (late RP) at two different sweep speeds, as first observed with an intraretinal microelectrode in the Cynomolgus monkey. The stimulus is indicated by the vertical line above the lower record, and a stimulus artifact appears as a slight break of the baseline in both records. Reference electrode in vitreous humor, and positive responses recorded upward (from BROWN and MURAKAMI, 1964)

match the absorption spectrum of rhodopsin, the visual pigment in the rods. 2. The ERP can be recorded from the isolated retina. 3. The amplitude of the ERP is linearly proportional to the number of pigment molecules activated by the flash, and it attains a maximum value when the flash energy is just sufficient to saturate the pigment. From these findings, as well as others to be discussed below, it has now become clear that the ERP must depend directly on the activity of the visual

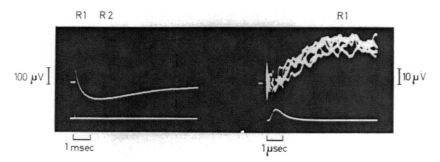

Fig. 2. Time course of the isolated ERP in the electroretinogram of the albino rat at two different sweep speeds. Excised eye at 26° C. Flash duration shown by photodiode traces. Corneal positive responses recorded upward (CONE, 1966)

pigment molecule, and will, therefore, make possible a new and important approach for investigating the pigment molecule and its role in the excitatory mechanism.

The ERP has been observed in every vertebrate eye so far examined: monkey (BROWN and MURAKAMI, 1964a), rat, cat, goldfish, and frog (CONE, 1964), squirrel (Pak and Ebrey, 1966), guinea pig and rabbit (Arden et al., 1966b), and human (Yonemura et al., 1966; 1967; Galloway, 1967; Dawson, 1968). Furthermore, similar fast photovoltages have also been observed in invertebrate photoreceptors: Limulus (Smith and Brown, 1966; Brown et al., 1967), squid (Hagins and McGaughy, 1967), and octopus (Murakami and Pak, 1967). The title "receptor FPV" will be used in this article for all fast photovoltages from visual receptors, and ERP will be used to specify receptor FPV from vertebrate eyes. This terminology will help distinguish receptor FPV from numerous fast photovoltages which have now been observed in a variety of both plant and animal tissues (Brown, 1965; Arden et al., 1966c; Ebrey, 1967; Becker and Cone, 1966; Smith and Brown, 1966). The latter responses will be collectively termed "tissue FPV". In this article we will review all work to date on both receptor and tissue FPV and we will discuss possible generating mechanisms for these responses. Our emphasis, however, will be on the receptor FPV since these are better understood and will be of greater value for investigating visual excitation.

1. Effects of High Flash Energy

The ERP was discovered as a result of employing a gas discharge lamp as a stimulus source. A flash lamp can produce photon fluxes many orders of magnitude greater than those available from conventional arc and filament sources. For example, a millisecond flash from a tungsten filament lamp rarely activates more than 0.1% of the visual pigment in the eye, and this is below the usual threshold of the ERP. But a typical photographic strobe light with a millisecond duration can saturate the visual pigment, every pigment molecule absorbing one or more photons during flash. Fig. 3 illustrates the effect on the electroretinogram produced

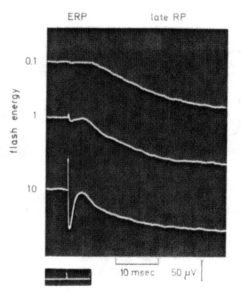

Fig. 3. ERP and late receptor potential as a function of flash energy in the electroretinogram of the frog. Flash duration shown by photodiode trace. Excised eye at 24° C (Cone, 1966)

by increasing the flash energy to such levels. Here the first trace shows the leading edge of the late receptor potential (a-wave) as it would appear if recorded from the most intense flash available from a typical tungsten source. Following the flash, several milliseconds elapse before the late receptor potential develops. The initial events in the excitatory process occur during this interval, but no electrical sign of this activity is present. However, by further increasing the flash energy, it is possible to observe the ERP. It can be seen in Fig. 3 that the amplitude of the ERP increases linearly with flash energy, unlike the amplitude of the late receptor potential, which reaches a maximum value at a somewhat lower flash energy than shown in the first trace.

The linearity of the ERP strikingly distinguishes it from all later electrical responses of the retina, and suggests a direct dependence on visual pigment. The amplitude of the ERP in the eye of the albino rat is shown as a function of flash energy in Fig. 4. The straight line in this figure has a slope of 1, and demonstrates the linearity of the ERP with flash energy up to about 100 times the threshold energy. At higher flash energies, the amplitude approaches a maximum value. Significantly, the saturation of the ERP occurs in parallel with the saturation of the visual pigment by the flash. This is shown by the heavy black line in Fig. 4 which is a theoretical curve depicting the fraction of pigment activated by the flash (for

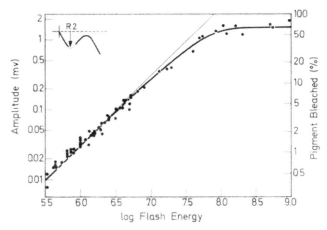

Fig. 4. Amplitude of ERP (R2) as a function of flash energy in the albino rat. Log-log plot. Calculated fraction of pigment bleached by the flash shown to the right. Flash energy calibrated such that log (flash energy) = 0 is the incident beam energy at which one photon is absorbed by the average rod/flash (CONE, 1964)

details, see CONE, 1964). The fit of this curve indicates that the amplitude of the ERP is not proportional to the energy absorbed, but is instead proportional, over the entire observable range, to the number of pigment molecules activated by the flash. That the ERP saturates is significant since it rules out generating mechanisms which depend only on the amount of energy absorbed during the flash. The energy absorbed by the retina does not reach a plateau; it is very nearly proportional to the flash energy throughout the entire range. In squid, the receptor FPV also saturates, the amplitude again being proportional to the fraction of pigment activated

(Hagins and McGaughy, 1967). The linear dependence of the ERP on the number of activated molecules holds true even if much of the visual pigment has been bleached (Cone, 1964). Thus, in the albino rat, for a given incident flash energy, the ERP amplitude is proportional to the amount of rhodopsin present in the eye (Arden et al., 1966a; Cone, 1964).

In contrast to receptor FPV in both vertebrate and squid retinas, FPV observed in other tissues are strictly linear with flash energy (with one partial exception: Brown and Crawford, 1967b), and none have been observed to saturate. By exploiting the saturation property of the visual pigment response in squid, Hagins and McGaughy (1967) identified a small nonsaturating response which is probably generated thermally. Such a thermoelectric voltage (TEV) may also be present in the vertebrate retina but has not yet been identified. However, in the vertebrate eye a large nonsaturating response is produced by melanin-containing cells of the pigment epithelium (Brown, 1965). And a similar but much smaller response can be observed in intact albino eyes (Ebrey and Cone, 1967). It is likely that both these nonsaturating responses are generated thermally. The basis for this suggestion will be discussed later in the article. Approximate flash energies required to produce FPV in the tissues investigated to date can be summarized as follows. To produce a 5 μV (threshold) response, the incident energy from a millisecond flash must be 0.1 to 1 $\times$ 10^{-4} joules/cm^2 for photoreceptors; 0.2 to 1 $\times$ 10^{-2} joules/cm^2 for melinated animal tissues and pigmented plant tissue (leaf); 0.1 to 1 $\times$ 10^{-1} joules/cm^2 for albino animal tissues. The temperature rise during a threshold flash can be roughly calculated to be on the order of 0.01° C for the outer segment of a photoreceptor, and 1° C for melanin-containing cells. For comparison, it is worth noting that in an iron-constantan thermocouple, a temperature rise of about 0.1° C produces a 5 μV response. This indicates that the possible thermal origin of an FPV should not be overlooked.

2. Effects of Temperature

Varying the temperature of the retina dramatically alters the waveform of the ERP, as may be seen in Fig. 5 (Pak and Cone, 1964). Near physiological temperature, R2 is the dominant component (Fig. 5a). As the retina is cooled, R 2 diminishes, and is preceded by the corneal positive component R 1 (Fig. 5b). Near freezing, R 2 is abolished, isolating R 1 (Fig. 5c). These temperature effects gave the first indication that the ERP is generated by two separate processes, the effect of cooling being to suppress and finally abolish the process generating R 2, but not the process generating R 1 (Pak and Cone, 1964). The R 1 component is exceedingly resistant to cooling. It can be elicited from frozen eyes and is still present at $-35°$ C (Pak and Cone, 1964; Pak and Ebrey, 1965). When an eye is frozen and then thawed, however, the ERP is irreversibly abolished, as is the receptor FPV in squid (Hagins and McGaughy, 1967), even though the normal complement of rhodopsin is maintained. Presumably the FPV is abolished by the disruption of the receptor cell structure. Similarly, when the excised eye of a rat is heated for 10 min at temperatures above 48° C, the ERP is irreversibly reduced, and is abolished by temperatures above 59° C (Cone, 1965; Cone and Brown, 1967). Again, the normal complement of rhodopsin is unchanged by this treatment, but the chromophores of the pigment molecules, which are highly oriented in the normal state, become increasingly disoriented as the temperature is increased, and the amplitude of the ERP falls in parallel with the loss of orientation of the chromophores (CONE and BROWN, 1967). Preliminary electron microscopy indicates that the receptor

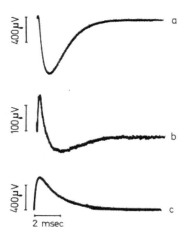

Fig. 5. ERP as a function of temperature in the excised eye of the albino rat. (a) 35° C, (b) 25° C, (c) 0° C, (from PAK and CONE, 1964)

membranes are not disoriented by this treatment (CONE, BROWN, and RICHARDSON, *in preparation*). It appears, therefore, that the ERP depends critically on both the structure of the receptor cell and a high degree of orientation of the visual pigment.

3. Action Spectra

The action spectrum of the ERP in the rod-dominated eye of the albino rat is shown in Fig. 6. In this eye, the action spectrum of the ERP closely matches that of the b-wave of the electroretinogram and also the absorption spectrum of rat rhodopsin (CONE, 1964). Both R 1 and R 2 have the same action spectra, the waveform of the ERP being independent of the stimulus wavelength (PAK and CONE, 1964). In the pure-cone retina of the ground squirrel, the action spectrum of the ERP peaks near 540 nm and is consistent with the absorption spectrum of a cone-type visual pigment, shown by the dashed line in Fig. 6 (PAK and EBREY, 1966). A similar correspondence has been reported for the receptor FPV in both *Limulus* (BROWN et al., 1967) and squid (HAGINS and McGAUGHY, 1967). In contrast to these receptor FPV, in which the action spectra match the absorption spectra of the respective visual pigments, tissue FPV have action spectra which appear to match the absorption spectra of melanins or other broad-banded pigments. Fig. 7 shows action spectra obtained from excised rat and frog eyes for both the ERP and a tissue FPV generated in the pigment epithelium (EBREY and CONE, 1967). The pigment epithelium FPV was isolated by exhaustively bleaching the visual pigments, thereby eliminating the ERP. This figure illustrates striking differences between these two types of FPV. In the pigment epithelium, as well as in other melanin-containing tissues, FPV are photostable, being little affected by the extended exposures to intense light required to bleach the visual pigments. Also, the action spectra of the tissue FPV are nearly flat across the entire visible and near infrared spectrum (Becker and Cone, 1966; Ebrey and Cone, 1967; Brown and Crawford, 1967 b). Furthermore, as can be seen in Fig. 7, tissue FPV are much smaller than the ERP if the stimulus wavelength is well absorbed by the visual

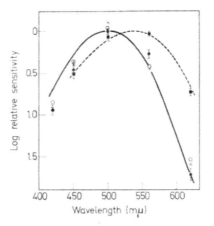

Fig. 6. Action spectra of the ERP and the b-wave of the electro-retinogram in the rod-dominated eye of the albino rat, and the ERP in the pure-cone eye of the Mexican ground squirrel. The action spectrum for R2 is shown by crosses for the ERP of the rat (Cone, 1964), and by small filled squares for the ground squirrel (Pak and Ebrey, 1966). Small filled circles indicate the action spectrum of R1 in the rat (Pak and Cone, 1964), and the action spectrum of the b-wave is shown by open circles (Cone, 1964) and the solid line (Dodt and Echte, 1961). The dashed line indicates the probable absorption spectrum of a visual pigment peaking at 540 nm (from the nomogram by Dartnall, 1953) (from Pak and Ebrey, 1966)

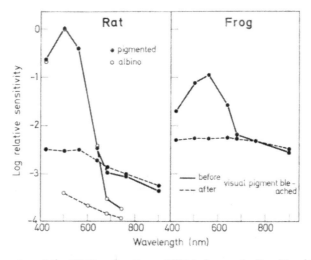

Fig. 7. Action spectra of the ERP and a tissue FPV before and after bleaching the visual pigment in excised eyes of frog, and pigmented and albino rats. Temperature: 12° C frog; 37° C rat. The sensitivity is the reciprocal of the flash energy required to produce a 20 μV response (from Ebrey and Cone, 1967)

pigment. In the eye of the rat, for example, the photostable tissue FPV is some 500 times smaller than the ERP; in the frog; it is about 20 times smaller. FPV can also be observed in albino tissues, and although smaller than in melinated tissue, they are essentially photostable and have broad action spectra (BECKER and CONE, 1966; EBREY and CONE, 1967).

4. Rod and Cone Contribution to the ERP

The wave form of the ERP is nearly the same in both rod- and cone-dominated eyes, and its temperature and flash energy dependence are also similar (PAK and EBREY, 1966). However, in mixed retinas, the ERP appears to be generated primarily, perhaps exclusively, by cones. With intraretinal recordings in the *Cynomolgus* retina, the amplitude of the ERP decreases as the electrode is moved from the fovea to the periphery, paralleling the decrease in density of cones, not the increase in density of rods (BROWN and MURAKAMI, 1964a; BROWN et al., 1965). This finding is given independent support by observations which indicate that the ERP in the isolated retina of the frog is also entirely dominated by cones (GOLDSTEIN, 1967). Frog rods contribute less than 20% of the response (GOLD-STEIN, 1968) even though approximately 90% of the total visual pigment in the frog retina is rhodopsin (LIEBMAN, 1968). As can be seen in Fig. 7, the action spectrum of the frog ERP matches the absorption spectrum of a cone pigment rather than that of rhodopsin (GOLDSTEIN, 1967; EBREY and CONE, 1967). More-over, in an excised frog retina, the ERP recovers rapidly after a bleaching exposure even though there is little or no regeneration of rhodopsin (GOLDSTEIN, 1967). Why the cones should contribute more to the ERP than do the rods is not known, but this difference can probably be attributed to structural differences in the outer sigments of these two types of receptors. In frog cones, for example, all lamellar discs of the outer segments are confluent with the plasma membrane, whereas in frog rods most discs appear to be entirely intracellular (MOODY and ROBERTSON, 1960; NILSSON, 1965).

5. Invertebrate Receptor FPV

The two invertebrate receptor FPV which have been studied in detail have quite different waveforms. In squid, the receptor FPV is a monophasic hyper-polarizing response, the exterior of the outer segment becoming positive (HAGINS and McGAUGHY, 1967; 1968a). But in *Limulus*, the receptor FPV[1] recorded intracellularly in the ventral eye is a biphasic response, an initial depolarizing component followed by a hyperpolarizing component (SMITH and BROWN, 1966; BROWN et al., 1967). The *Limulus* receptor FPV is similar in appearance to the ERP in that it is biphasic, although it has a somewhat slower time course. The ERP has not yet been recorded intracellularly, but the extracellular polarities of its two phases also appear to be the same as those found in the receptor FPV in *Limulus*. Differences in the waveform of the FPV in squid and *Limulus* may be due to differences in visual pigment transitions. The kinetics of transitions concurrent with the receptor FPV have not yet been reported for squid or *Limulus*, but there is a marked difference in later transitions. For example, in squid, retinal always remains

---

[1] SMITH and BROWN have named this response the PEP, for photoelectric potential.

attached to the opsin (Wald and Hubbard, 1957; Hubbard and St. George, 1957/58) but in *Limulus*, retinal eventually dissociates from opsin, as it does in vertebrate pigments (Hubbard and Wald, 1960).

6. Effets of Chemicals

The results of investigations of the effects of ions, fixatives, and other chemicals on the ERP are rather diverse; they support, however, two major conclusions regarding the generating mechanism of the ERP. First, results with ion substitutions and fixatives exclude the possibility that the ERP is generated by passive ionic currents resulting from changes in membrane permeability (Pak, 1965; Brindley and Gardner-Medwin, 1966). Second, R 1 and R 2 often respond in different ways to a particular treatment, which lends support to the conclusion that these two components are generated by separate processes (Pak and Cone, 1964; Brindley and Gardner-Medwin, 1966).

The principle experimental results may be summarized as follows. The replacement of extracellular sodium chloride by foreign salts has little effect on the frog ERP (Pak, 1965; Brindley and Gardner-Medwin, 1966; Pak et al., 1967). Even the immersion of the retina for more than an hour in isotonic KCI fails to abolish the ERP (Pak, 1965; Brindley and Gardner-Medwin, 1966), although all slower electrical responses of the retina are completely abolished by such treatment. The invertebrate receptor FPV is also highly resistant to KCl (Smith and Brown, 1966; Hagins and McGaughy, 1967). On the other hand, isotonic KCl, RbCl, and NH_4Cl solutions selectively enhance the R 2 component of the frog ERP (Pak, 1965; Pak et al., 1967).

Two fixatives, acrolein and formaldehyde, shorten the time course of the ERP (Brindley and Gardner-Medwin, 1966; Arden et al., 1966d; 1968). The R1 component isolated by gluteraldehyde fixation is small in amplitude and displays a rapid time course. Over a remarkably wide range, variation in hydrogen ion concentration has only a small effect on the ERP (Brindley and Gardner-Medwin, 1966; Pak et al., 1967). Changes have been observed, however, at extreme values of pH (Crawford et al., 1967; Pak et al., 1967). In particular, if a frog retina is allowed to equilibrate for about 15 min in Ringer buffered at pH 3.6 or below, selective enhancement of R 1 and reduction of R 2 is observed (Pak et al., 1967).

A reducing agent, sodium bisulphate, has been found to enhance the amplitude of R 2 by about fourfold at 10° C if pH is kept below 7.0 (Arden et al., 1968). On the other hand, other reducing agents tested by these investigators produce little or no effect. Pure oxygen and also the oxidizing agent potassium ferricyanide reduce the size of the ERP, particularly the R 2 component. The thiol-blocking agent, N-ethyl maleimide, and the aldehyde-blocking agent, hydroxylamine, have little effect on the ERP (Brindley and Gardner-Medwin, 1966).

7. Site of Generation of Receptor FPV

In both *Limulus* and squid, convincing evidence has been reported which demonstrates that the receptor FPV is generated within the cell membrane. In the lateral eye of *Limulus*, the receptor FPV reverses polarity between the inside and outside of the retinular cell membrane, and the polarity and amplitude of the FPV are both affected by changes in the polarization of this membrane. As shown in Fig. 8, the dominant phase of this FPV is of opposite polarity when polarization of the membrane is reversed by injecting current into the cell (Smith and Brown, 1966). In squid, the passive spread of FPV current along the receptor outer segments has been investigated by applying flash stimulation locally; the results again indicate that the site of generation is the receptor membrane (Hagins and McGaughy, 1968a; 1968b). Unfortunately, such firm evidence is not yet available for the site of generation of the vertebrate ERP. However, several characteristics of the ERP suggest that it is also generated in the receptor cell membrane. As mentioned earlier, the extracellular amplitude of the ERP is maximal in the neighborhood of the inner segments of the receptors, and the ERP is larger in the fovea than in the periphery of the Cynomolgus retina (BROWN and MURAKAMI, 1964; BROWN et al., 1965).

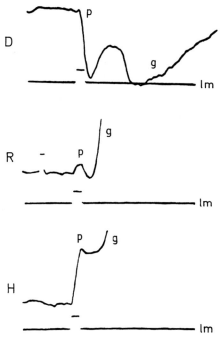

Fig. 8. Receptor FPV (PEP) recorded intracellularly in *Limulus* as a function of membrane potential. The FPV is indicated by "p", the generator potential by "g". Two separate single-barrel microelectrodes in the same retinular cell; one used for passing current, the other for monitoring potential. (*R*) is response in the absence of applied current, (*H*) is response after hyperpolarization, and (*D*) is response with membrane potential forced to a value more positive than a level near zero potential. Calibration pulse in (*R*) is 1 mV, 5 msec. "lm" is light monitor (from SMITH and BROWN, 1966)

Additional suggestive evidence has been reported by ARDEN and IKEDA (1966) in a study of the ERP in dystrophic rats. In such rats the ERP declines in amplitude approximately in parallel with an observable loss in the organization of the receptor outer segments. This loss of organization is caused by a genetic defect which begins to appear after about 25 days of life (DOWLING and SIDMAN, 1962).

8. Comparison of Receptor and Tissue FPV

FPV observed in a variety of tissues such as skin, leaves, pigment epithelium, and iris are in many ways similar to receptor FPV. All require high stimulus flash energies, and there are usually two or three easily separated phases to each response. The earliest phase tends to be the most resistant to low temperatures, to changes in ionic environment, and to various other treatments. The later phases of each

23*

response tend to be ion-dependent, and at least one phase of the pigment epithelium FPV has been shown to originate in the cell membrane (BROWN and CRAWFORD, 1966a). These and other similarities have led some investigators to suggest that tissue FPV may be generated by fundamentally the same process that generates receptor FPV. The similarities of these responses have been thoroughly treated elsewhere (BROWN, 1965; ARDEN et al., 1966b; 1968; BROWN and GAGE, 1966; BROWN and CRAWFORD, 1967a, b). There is, however, evidence to suggest that receptor and tissue FPV may in fact be generated by fundamentally different mechanisms, and it is this evidence which will be discussed here.

Receptor FPV in vertebrates and in squid have been shown to depend directly upon transitions in the visual pigment molecule, and are probably generated at least in part by reversible charge displacements in oriented visual pigment molecules. (Evidence for this will be covered in the following section). In contrast, every characteristic of tissue FPV reported to date is consistent with the possibility that tissue FPV are generated thermally. HAGINS and McGAUGHY (1967) have suggested the name thermoelectric voltage (TEV) for such responses. To say that a response is generated thermally does not specify a particular mechanism. It does assert, however, that the voltage observed is generated as a direct consequence of the conversion of photon energy to thermal energy – the role of the photon being simply to inject thermal energy into the generating structure.

Strong evidence for the thermal origin of an FPV has been reported by HAGINS and McGAUGHY (1967) in the squid retina. They found that if an excised squid retina is placed against a black surface, a large nonsaturating FPV is generated. Similar FPV are generated if the retina is loaded with various dyes. In every case, a response of similar time course and polarity is observed. On the other hand, if the stimulating flash contains wavelengths absorbed well by the visual pigment but not by the dye, a receptor FPV is elicited. These investigators also observed, as mentioned previously, a small nonsaturating component in the normal response of the squid retina which they conclude is generated thermally.

If thermal energy is suddenly injected into a thin plane, such as a cell membrane or a thin layer of cells embedded in a large volume of thermally homogeneous material, after reaching its maximum value, the temperature within the absorbing plane falls as the inverse square root of time following the injection[2]. The falling phase of the TEV reported by HAGINS and McGAUGHY appears to be very nearly proportional to $t^{-\frac{1}{2}}$, as would be appropriate for a response whose voltage is proportional to the temperature of the generating structure. Inspection of published records obtained in other tissues, notably the first phase of the pigment epithelium FPV (BROWN and CRAWFORD, 1967b) and the FPV from mammalian skin (BECKER and CONE, 1966) also shows the falling phase of these responses to be very

---

[2] The temperature is in this case described by the equation for one-dimensional heat flow:

$$\theta \propto t^{-\frac{1}{2}} e^{-x^2/4a^2t}$$

where θ is the change in temperature at a distance x from the absorbing plane, t is the time following the instantaneous injection of heat to this plane, and a^2 is the thermal conductivity of the medium, divided by the product of its density and specific heat. For water, a^2 is about 0.0014 cm² sec⁻¹. θ passes through a maximum when $t = x^2/2a^2$. When $t \gg x^2/4a^2$, $\theta \propto t^{-\frac{1}{2}}$. (For a further discussion of heat flow, see, for example, MARGENAU and MURPHY, 1956).

nearly proportional to $t^{-\frac{1}{2}}$. The thermal origin of tissue FPV is also suggested by the finding that similar FPV occur in both albino and melinated tissues (EBREY and CONE, 1967; BECKER and CONE, 1966). the amplitude of FPV in melinated tissue being on the order of 10 times larger than in albino tissue. Thus melanin may simply augment these responses rather than directly generate them. This result would be expected if the responses were generated thermally since the melanin should increase the heat absorbed in these tissues by roughly a factor of 10.

BROWN and CRAWFORD (1967b) have shown that the pigment epithelium FPV is initiated within a few μseconds of the stimulus flash, and they have raised the question of how melanin granules are able to initiate a transmembrane current in so short a time, since the granules are, for the most part, not in contact with the cell membrane. If it is assumed, however, that the pigment epithelium FPV is produced by the heating of the cell membrane, this question does not arise because heat flows sufficiently rapidly over the distances between melanin granules and the cell membrane to account for the absence of a detectable latency in this response. In the case of an aqueous medium, if the flash energy is delivered instantaneously, and if the distance from the absorbing plane is less than 0.5 μm, the temperature reaches a maximum in less than 1 μsec. Thus in a pigment epithelium cell, even if the flash energy is effectively absorbed only by melanin granules, the temperature of the plasma membrane can reach a maximum in a time of the order of 1 μsec. Moreover, if the cellular dimensions are in the neighborhood of 50 μm, the entire cell will be nearly uniformly heated in less than 10 msec. Indeed, the characteristics of the pigment epithelium FPV look very much as if the response is generated by the heating of the cell membrane, the temperature of the membrane nearest the melanin granules rising somewhat faster than the more distant membrane. An asymmetry in the distribution of melanin granules in the cell could account for the polarity of the response being independent of the direction of the stimulus light, and could also account for the small differences in the response time observed by CRAWFORD et al. (1967) when the direction of the light is changed.

The slower ion-dependent phases of tissue FPV are almost certainly generated by different mechanisms than receptor potentials because there are enormous differences in sensitivity between these two types of response. For example, the third phase of the pigment epithelium response requires on the order of 10^9 times more energy than the late receptor potential for an equal amplitude response in its linear range (CONE, 1965; EBREY and CONE, 1967). Thus although both these responses may arise from changes in membrane permeability, the mechanisms which alter the membrane permeability are probably entirely different.

9. Photoproduct Responses of Receptor FPV

At present, all demonstrations of the dependence of receptor FPV on visual pigment have been based on direct correlations between FPV responses and visual pigment transitions and absorption spectra. A visual pigment molecule is a strongly absorbing pigment in each of its intermediate states. This characteristic has been of primary importance because the course of bleaching can be manipulated by exposing the molecule to light while it is in any of several intermediate states. In particular, photoreversible transitions can occur repeatedly between certain intermediates, and rhodopsin can be photoregenerated from many of them (MATTHEWS et al., 1963; for recent reviews of visual chemistry, see HUBBARD et al., 1965; ABRAHAMSON and OSTROY, 1967; BRIDGES, 1967).

The photoregeneration of the ERP in the rat was first reported by ARDEN and IKEDA (1965), and ARDEN et al. (1966a), who showed that under certain conditions a series of white stimulus flashes can partially photoregenerate rhodopsin, and that the ERP also partially regenerates under these conditions. A demonstration of the

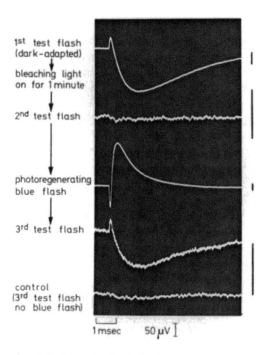

Fig. 9. Photoregeneration of the ERP in the excised eye of the albino rat. Both the test flash and the bleaching light consisted of long wavelengths primarily absorbed by rhodopsin. The blue photoregenerating flash contained wavelengths absorbed by the longer-lived intermediates of the bleaching process. The control trace was obtained from a second eye subjected to the same bleaching exposure and test flashes but without the blue flash. Temperature, 27° C
(CONE, 1967)

photoregeneration of the ERP is shown in Fig. 9 (CONE, 1967). In this figure, the ERP was photoregenerated by presenting an intense blue flash at a time when most of the visual pigment was proceeding through intermediate states which are strongly blue-absorbing. By repeating this procedure, the ERP can be photoregenerated many times, as can rhodopsin in solution (MATTHEWS et al., 1963), and each time the waveform of the ERP is essentially normal. Corresponding results for the photoregeneration of the squid receptor FPV have also been observed (HAGINS and McGAUGHY, 1967). Furthermore, ARDEN et al. (1966a) have demonstrated, in the eye of the rat, that photoregenerated pigment is physiologically active.

One of the most significant findings for research on the ERP was the discovery that when a flash photoregenerates rhodopsin, the waveform of the ERP is dramatically altered. This effect was first reported by ARDEN and IKEDA (1965), and a striking illustration of it is shown in Fig. 9. Here, the response to the photo-regenerating blue flash is biphasic, much like the ERP, but contrasts with the ERP in that its polarity is reversed. In addition, unlike the ERP, it can be fractionated by varying the wavelength of the photoregenerating flash or by presenting the flash at a time when most of the pigment is in only one of the intermediate states in the bleaching process. Three distinct responses can be isolated by these procedures and the rhodopsin intermediate on which each depends has now been identified (CONE, 1967; PAK and BOES, 1967; ARDEN et al., 1966b).

Fig. 10 shows the three photoproduct responses in rat as first identified (CONE, 1967). The responses shown in this figure were obtained by first activating most of the rhodopsin in the eye, and then presenting a stimulus flash after an interval of time sufficient to allow the activated molecules to proceed to a particular

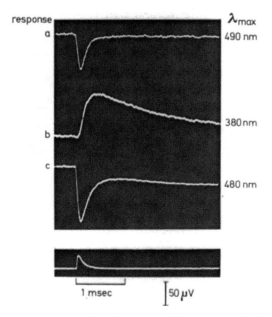

Fig. 10. Photoproduct responses obtained at three different times in the bleaching process in the excised eye of the albino rat. Response a was obtained at 5° C with a white flash delivered 50 msec after a saturating flash had been presented to the dark-adapted eye. Responses b and c were obtained at 27° C following a 30-sec bleaching exposure to tungsten light. Response b was obtained with a narrow-band blue flash with a maximum at 400 nm immediately following the bleaching exposure; response c was obtained with a white flash delivered 3 min after the bleaching exposure. Response c is distorted by the presence of a small amount of rhodopsin regenerated after the bleaching exposure. Wavelength of maximum sensitivity for each response is shown on the right. Flash duration shown by photodiode trace
(CONE, 1967)

intermediate state in the bleaching process. The rhodopsin intermediates on which these responses depend have been identified by their action spectra, their time of appearance after the saturating flash, and their temperature dependence (CONE, 1967; PAK and BOES, 1967; ARDEN et al., 1966b). Photoproduct responses have also been observed in squid photoreceptors (Hagins and McGaughy, 1967), and again each response is found to arise from a known intermediate of squid rhodopsin.

10. Relationship of Receptor FPV Responses to Rhodopsin in Transitions

Not only are receptor FPV generated by the action of light on rhodopsin or rhodopsin intermediates, but it now appears that their waveforms are also directly related to the time course of the pigment transition initiated by the flash. For example, it has been suggested that the time course of R2 in the ERP of the rat is

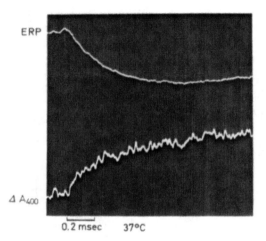

Fig. 11. Time course of ERP and absorbance of retina at 400 nm in the excised eye of the albino rat. Flash duration, 0.1 msec. Total absorbance change, 0.16 A. ERP amplitude, 0.4 mV
(Cone, 1968)

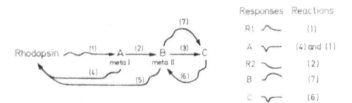

Fig. 12. Photochemical and thermal reactions of rhodopsin, and the corresponding ERP and photoproduct responses in the albino rat. Photochemical reactions are denoted by wavy lines; thermal reactions by straight lines. The five responses so far identified are listed along with sketches to show their polarity and waveform, and reactions on which each depends
(Cone, 1967; 1968)

concurrent with the meta I-meta II (A → B) transition (Cone, 1967). Preliminary measurements which reveal the nature of this correspondence have now been made by observing the meta I-meta II transition directly with a flash photometer, and simultaneously recording the ERP (Cone, in preparation). One result is shown in Fig. 11, in which the time course of the ERP is compared to the concentration of meta II as a function of time after the flash. The concentration of meta II is in this case proportional to the change in absorbance at 400 nm. It is apparent from the similar time course of the two recordings in this figure that the amplitude of R2 closely parallels the concentration of meta II. This implies that the accumulation of the electric charge whose displacement generates R2 must parallel the accumulation of meta II. Similar measurements for responses concurrent with other transitions have also been obtained, and the results are summarized in Fig. 12. In every case, the rising phase of the response parallels the accumulation of the primary photoproduct, provided that this photoproduct forms within less than about 1 msec.

11. Generating Mechanism of Receptor FPV

The striking correspondence between an FPV response and its concurrent rhodopsin transition gives strong support to the hypothesis that these responses are generated by charge displacements produced directly by the activity of rhodopsin molecules (LETTVIN, 1965; CONE, 1965; ARDEN and IKEDA, 1965; BRINDLEY and GARDNER-MEDWIN, 1966). Moreover, the polarity and time course of FPV responses from both rhodopsin and rhodopsin intermediates suggest that these responses are generated by reversible alterations of the charge distribution in the rhodopsin molecule itself (CONE, 1967; HAGINS and McGAUGHY, 1967).

The characteristics of receptor FPV in both *Limulus* and squid imply that the displaced charge must move through, or partially through, the capacitative thickness of the membrane (SMITH and BROWN, 1966; HAGINS and McGAUGHY, 1968a, b). This is an important finding because, as was first pointed out by SMITH and BROWN (1966), it indicates that visual pigment molecules are probably closely incorporated in the structure of the membrane. In addition, HAGINS and McGAUGHY (1967; 1968a) have noted that the characteristics of these receptor FPV also indicate that the pigment molecules are most likely oriented with respect to the cell membrane rather than to the surfaces of the retina, as suggested by BRINDLEY and GARDNER-MEDWIN (1966) in their analysis of the ERP. If the pigment molecules are oriented with respect to the membrane, charge displacements within each pigment molecule can be summed electrically by the capacitance of the membrane. The rising phase of each response would therfore reflect the active charging of the membrane capacitance by light-induced charge displacements, and the falling phase of each response would reflect the resistive discharge of this capacitance. In a preliminary report, HAGINS and McGAUGHY (1968b) state that the receptor FPV in squid is produced by structures with the same electrical cable constants as those generating the ionic membrane current. This indicates that the transretinal FPV is produced by an extracellular flow of charge which is initiated at the outer segment by the active polarization of its membrane, and which flows along the body of the cell, returning intracellularly by passing through more proximal regions of the receptor cell membrane. If the time course of the FPV is short in comparison with the time constant of the receptor cell, most of this current will flow as a displacement current through the capacitance of the membrane, and the decay time of the receptor FPV will be determined by the resistivity of the media surrounding the membrane rather than by the resistance of the cell membrane itself.

In the vertebrate retina, despite the obvious differences in the structure of the receptors, it is also likely that the transretinal ERP is generated by charge from the actively polarized membrane of the outer segment flowing extracellularly to more proximal regions of the receptor membrane. Such a discharge path should cause the ERP to decay exponentially with the same time constant for every response. This is in fact the case. In the rat, the falling phase of the isolated form of R 1, as well as the terminal falling phases of R 2 and the two responses from the B and C intermediates, all decay exponentially with a half-time of about 1 msec. Such a decay time appears reasonable in terms of the resistances and capacitances likely to be involved (Cone, *in preparation*).

12. Magnitude of Charge Displacements

An accurate measurement of the magnitude of the charge displacements which generate receptor FPV has not yet been reported. However, it is possible to show by simple calculations that these responses may well be generated by charge displacements sufficiently small to be confined within a pigment molecule. For example, in the outer segment of a rat rod, there are about 3×10^4 rhodopsin molecules per square micron of membrane. If the specific capacitance of this membrane is 1 μfarad/cm^2, and if one electronic charge for each rhodopsin molecule is displaced through the entire capacitative thickness of the membrane, then the potential developed across the membrane would be 0.5 V. The maximum extracellular amplitude of R 2 is about 0.01 V. Therefore, if it is assumed that the electrodes detect the *full* potential developed across the membrane, and that only one electronic charge is displaced per pigment molecule, this charge need move through no more than 2% (0.01/0.5) of the capacitative thickness of the membrane. For a membrane with a capacitative thickness of 50 A, this amounts to a single charge per pigment molecule moving though 1 A. Obviously, the actual charge displacement must be greater than this since the extracellular electrodes almost certainly detect less than the full membrane potential. An order of magnitude calculation based on data reported by Hagins (1965) indicates, however, that the receptor FPV in squid may also be produced by a single charge per rhodopsin molecule moving through no more than a few Angström units. Though such calculations can be made with little accuracy at this time, they do serve to suggest that a reversible movement of a charge group in the pigment molecule may be sufficient to account for the observed responses.

The type of charge movement involved has yet to be determined. It has been suggested that R 2 may be generated by the binding of a proton from solution since rhodopsin binds such a proton in the meta I-meta II transition (Cone, 1967). In the squid retina, Hagins and McGaughy (1967) have shown that when a photo-steady state is reached by presenting a long series of stimulus flashes, the response to each flash is biphasic with no current flowing through the retina. This indicates that under these conditions the generating charge displacements are exactly photoreversible.

13. Significance of the Receptor FPV

The charge displacements which generate receptor FPV appear to be much too small to act as excitatory electrical signals transmitted to the synaptic end of the receptor (CONE, 1965; HAGINS, 1965). However, the charge displacements are probably large enough to disturb significantly neighboring molecules in the membrane, and they might thereby alter ionic permeablity in the regions surrounding the pigment molecule. HAGINS (1965) has pointed out that in a squid photoreceptor, a "sodium hole" of the order of molecular dimensions can account for the entire ionic current which flows for each pigment molecule activated by light. Hence, it is also possible that a significant change in ionic permeability may be produced by the pigment molecule itself. It is likely, therefore, that the charge displacements which generate receptor FPV may be closely linked to the mechanism by which a pigment molecule initiates the excitatory response of a visual receptor.

Receptor FPV have already been used to investigate electrical and structural properties of squid photoreceptors (HAGINS and McGAUGHY, 1968b) and have helped to determine the association of the pigment molecule with the cell membrane (SMITH and BROWN, 1966; HAGINS and McGAUGHY, 1968a, b; CONE and BROWN, 1967). Since they can reveal properties of the pigment molecule which would be difficult or impossible to observe with previously available methods, it is now evident that receptor FPV will play an important role in future research on photoreceptors.

References

ABRAHAMSON, E. W., OSTROY, S. E.: The photochemical and macromolecular aspects of vision. Progr. Biophys. 17, 179—215 (1967).

ARDEN, G. B., BRIDGES, C. D. B., IKEDA, H.: Isolation of a new fast component of the early receptor potential. Proc. physiol. Soc. (abstr.) C 31 (1966).

— — — SIEGEL, I. M.: Rapid light-induced potentials common to plant and animal tissues. Nature (Lond.) 212, 1235—1236 (1966).

— — — — Mode of generation of the early receptor potential. Vision Res. 8, 3—24 (1968).

— IKEDA, H.: A new property of the early receptor potential of rat retina. Nature (Lond.) 208, 1100—1101 (1965).

— — Effects of hereditary degeneration of the retina on the early receptor potential and the corneo-fundal potential of the rat eye. Vision Res. 6, 171—184 (1966).

— — SIEGEL, I. M.: Effects of light-adaptation on the early receptor potential. Vision Res. 6, 357—371 (1966a).

— — — New components of the mammalian receptor potential and their relation to visual photochemistry. Vision Res. 6, 373—384 (1966b).

BECKER, H. E., CONE, R. A.: Light-stimulated electrical responess from skin. Science 154, 1051—1053 (1966).

BRIDGES, C. D. B.: Biochemistry of visual processes. Comp. Biochem. 27, 31—78 (1967).

BRINDLEY, G. S., GARDNER-MEDWIN, A. R.: The origin of the early receptor potential of the retina. J. Physiol. (Lond.) 182, 185—194 (1966).

BROWN, J. E., MURRAY, J. R., SMITH, T. G.: Photoelectric potential from photoreceptor cells in ventral eye of Limulus. Science 158, 665—666 (1967).

BROWN, K. T.: An early potential evoked by light from the pigment epithelium-choroid complex of the eye of the toad. Nature (Lond.) 207, 1249—1253 (1965).

— CRAWFORD, J. M.: Intracellular recording of rapid light-evoked responses from pigment epithelium cells of the frog eye. Physiologist 9, 146 (1966).

— — Intracellular recording of rapid light-evoked responses from pigment epithelium cells of the frog eye. Vision Res. 7, 149—163 (1967a).

— — Melanin and the rapid light-evoked responses from pigment epithelium cells of the frog eye. Vision Res. 7, 165—178 (1967b).

— GAGE, P. W.: An earlier phase of the light-evoked electrical response from the pigment epithelium-choroid complex of the eye of the toad. Nature (Lond.) 211, 155—158 (1966).

Brown, K. T., Murakami, M.: A new receptor potential of the monkey retina with no detectable latency. Nature (Lond.) **201**, 626—628 (1964a).
— — Biphasic form of the early receptor potential of the monkey retina. Nature (Lond.) **204**, 739—740 (1964b).
— Watanabe, K., Murakawi, M.: The early and late receptor potentials of monkey cones and rods. Cold Spr. Harb. Symp. quant. Biol. **30**, 457—482 (1965).
Cone, R. A.: Early receptor potential of the vertebrate retina. Nature (Lond.) **204**, 736—740 (1964).
— The early receptor potential of the vertebrate eye. Cold Spr. Harb. Symp. quant. Biol. **30**, 483—490 (1965).
— Early receptor potential: photoreversible charge displacement in rhodopsin. Science **155**, 1128—1131 (1967).
— Brown, P. K.: Dependence of the early receptor potential on the orientation of rhodopsin. Science **156**, 536 (1967).
Crawford, J. M., Gage, P. W., Brown, K. T.: Rapid light-evoked potentials at extremes of pH from the frog's retina and pigment epithelium, and from synthetic melanin. Vision Res. **7**, 539—551 (1967).
Dartnall, H. J. A.: The interpretation of spectral sensitivity curves. Brit. med. Bull. **9**, 24 (1953).
Dawson, W. W.: Fast signals in the human visual system. Fed. Proc. **27**, Abstr. 2374 (1968).
Dodt, E., Echte, K.: Dark and light adaptation in pigmented and white rat as measured by electroretinogram threshold. J. Neurophysiol. **24**, 427 (1961).
Dowling, J. E., Sidman, R. L.: Inherited retinal dystrophy in the rat. J. biophys. biochem. Cytol. **14**, 73—109 (1962).
Ebrey, T. G.: Fast light-evoked potential from leaves. Science **155**, 1556—1557 (1967).
— Cone, R. A.: Melanin, a possible pigment for the photostable electrical responses of the eye. Nature (Lond.) **213**, 360—362 (1967).
Galloway, N. R.: Early receptor potential in the human eye. Brit. J. Ophthal. **51**, 261 (1967).
Goldstein, E. B.: Early receptor potential of the isolated frog retina. Vision Res. **7**, 837—845 (1967).
— Visual pigments and the early receptor potential of the isolated frog retina. Vision Res. **8**, 953—964 (1968).
Hagins, W. A.: Electrical signs of information flow in photoreceptors. Cold Spr. Harb. Symp. quant. Biol. **30**, 403—418 (1965).
— McGaughy, R. E.: Molecular and thermal origins of fast photoelectric effects in the squid retina. Science **157**, 813—816 (1967).
— — Membrane origin of the fast photovoltage of squid retina. Science **159**, 213—215 (1968a).
— — Fast photovoltages, receptor currents, and electrical cable constants in squid photoreceptors. Biophys. J. **8**, A-158 (1968b).
Hecht, S., Shlaer, S., Pirenne, M. H.: Energy, quanta, and vision. J. gen. Physiol. **25**, 819—840 (1942).
Hubbard, R., Bownds, D., Yoshizawa, T.: The chemistry of visual photoreception. Cold Spr. Harb. Symp. quant. Biol. **30**, 301—315 (1965).
— George, R. C. C., St.: The rhodopsin system of the squid. J. gen. Physiol. **41**, 501—528 (1957—58).
— Wald, G.: Visual pigment of the horseshoe crab, Limulus polyphemus. Nature (Lond.) **186**, 212—215 (1960).
Lettvin, J. Y.: General discussion; early receptor potential. Cold Spr. Harb. Symp. quant. Biol. **30**, 501—502 (1965).
Liebman, P.: Personal communication to Goldstein, E. B. (Goldstein, 1968).
Margenau, H., Murphy, G. M.: The mathematics of physics and chemistry. Princeton, N. J.: D. Van Nostrand Co., Inc. 1956.
Matthews, R. G., Hubbard, R., Brown, P. K., Wald, G.: Tautomeric forms of metarhodopsin. J. gen. Physiol. **47**, 215—240 (1963).
Moody, M. F., Robertson, J. D.: The fine structure of some retinal receptors. J. biophys. biochem. Cytol. **7**, 87 (1960).
Murakami, M., Pak, W. L.: Unpublished observations (1967).

NILSSON, S. E. G.: The ultrastructure of the receptor outer segments in the retina of the leopard frog (Rana pipiens). J. Ultrastruct. Res. **12**, 207 (1965).

PAK, W. L.: Some properties of the early electrical response in the vertebrate retina. Cold Spr. Harb. Symp. quant. Biol. **30**, 493—499 (1965).

— BOES, R. J.: Rhodopsin: responses from transient intermediates formed during its bleaching. Science **155**, 1131—1133 (1967).

— CONE, R. A.: Isolation and identification of the initial peak of the early receptor potential. Nature (Lond.) **204**, 836—838 (1964).

— EBREY, T. G.: Visual receptor potential observed at sub-zero temperatures. Nature (Lond.) **205**, 484—486 (1965).

— — Early receptor potentials of rods and cones in rodents. J. gen. Physiol. **49**, 1199—1208 (1966).

— ROZZI, V. P., EBREY, T. G.: Effect of changes in the chemical environment of the retina on the two components of the early receptor potential. Nature (Lond.) **214**, 109—110 (1967).

SCHMIDT, W. J.: Polarisationsoptische Analyse eines Eiweiß-Lipoid-Systemes, erläutert am Außenglied der Sehzellen. Kolloid-Z. **85**, 137—148 (1938).

SMITH, T. G., BROWN, J. E.: A photoelectric potential in invertebrate cells. Nature (Lond.) **212**, 1217—1219 (1966).

WALD, G., BROWN, P. K., GIBBONS, I. R.: The problem of visual excitation. J. Opt. Soc. Am. **53**, 20—35 (1963).

— HUBBARD, R.: Visual pigment of a decapod crustacean: the lobster. Nature (Lond.) **180**, 278—280 (1957).

YONEMURA, D., KAWASAKI, K.: The early receptor potential in the human electroretinogram. Jap. J. Physiol. **17**, 235—244 (1967).

— — HASUI, I.: The early receptor potential in the human ERG. Acta Soc. Ophth. Jap. **70**, 120—122 (1966).

Chapter 13

The Nature of the Photoreceptor in Phototaxis

By

Mary Ella Feinleib and George M. Curry, Medford, Mass. (USA)

With 6 Figures

Contents

A. Introduction

It is a common and ancient observation that most organisms change their direction of movement when the pattern of light falling on them is changed. If this change in light pattern induces and determines a new orientation of the *whole* organism, the process is termed "phototaxis". The response is mediated by a change in the motor processes of the organism, and is typified by the swimming of small organisms toward or away from the light. In rooted plants a change in light pattern may determine reorientation of a *part* of the plant with respect to the light; this is called "phototropism", and it is mediated by a change in the pattern of growth.

In the late nineteenth and early twentieth century, many prominent biologists (e.g., Engelmann, Strasburger, Oltmanns, Mast, Loeb) were intrigued by these light responses, particularly by phototaxis. Here was an entire stimulus-response system within a single cell, presumably without a complicated nervous system between receptor and effector. If the mechanism for phototaxis were understood, it might be easier to interpret more complex light responses such as vision. As the years went by, however, our understanding of visual processes increased greatly, while our knowledge of phototropism and phototaxis lagged considerably behind. The photoreceptors have not yet been identified in any of the plants, and we do not know exactly where the "eye" is located. In the case of phototaxis, progress had been hampered until recently by the lack of accurate methods for measuring the response.

There are many phenomena related to phototropism and phototaxis, but distinguished from them by the above definitions. In rooted plants, a symmetrical change in light intensity is often followed by a change in growth rate, but no bending occurs. This is the so-called "light-growth reaction" (BLAAUW's "Lichtwachstumreaktion"); BLAAUW (1914, 1915), and others since, held the view that this is the fundamental phenomenon and phototropism only a special case of it. Similarly, in motile organisms a temporal change in light intensity may initiate a change in swimming rate. This reaction, termed "photokinesis", does not involve a change in swimming direction, but can occasionally lead to apparent phototactic accumulation; if an organism swims rapidly in the dark but very slowly in the light, it will tend to spend more time in the light.

There is another response of motile organisms which is more difficult to classify. It is elicited by an abrupt change in light intensity and typically consists of a brief cessation or reversal of movement, followed by resumption of movement in a new direction. If the organism changes direction when it enters a darkened region but not when it enters an illuminated region, it will tend to remain in the light. This reaction is nonoriented with respect to the light and should not be classified as a taxis *sensu strictu* (FRAENKEL and GUNN, 1961). Nonetheless, since it is a light-induced change in orientation, many workers treat it as a kind of phototaxis, referring to it as "phobotaxis" (PFEFFER, 1904). By contrast, "true" phototaxis, in which orientation of the individual is determined by the light direction, is called "topotaxis".

On the surface, some of these distinctions may seem trivial, but it must be recognized that there is a fundamental difference involved. In the case of an oriented response to light (e.g., phototropism, topotaxis), the organism must have some means of detecting light direction, whereas the other responses (e.g., light-growth reaction, photokinesis, phobotaxis) require only a means of detecting a temporal change in light intensity.

B. Plants in which Phototaxis Occurs

Phototaxis has been studied in the purple photosynthetic bacteria as well as in numerous representatives of all the major groups of algae (BENDIX, 1960). Bacterial phototaxis is strictly phobic (nonoriented). The response is positive over a broad range of light intensities, but reverses to negative at very high intensities (CLAYTON, 1959, 1964). The effectors of phobotactic movement are the whiplike flagella which occur either at one or at both ends of the cell. It has been known since the classical experiments of ENGELMANN (1883), that phobotaxis in bacteria is closely related to photosynthesis; the action spectra for the two phenomena are similar, with peaks in the near infrared and in the blue regions. The photoreceptors for bacterial phobotaxis are evidently the same as those for photosynthesis: principally bacteriochlorophyll and, to some extent, carotenoids. There is no evidence in the bacteria of a localized photoreceptor for phobotaxis (Clayton, 1959, 1964).

The phototactic mechanism in the flagellated algae is quite different from that in the bacteria. Although these algae do show phobic responses under appropriate conditions, the typical response is a topotaxis, with each cell oriented with respect to the light. If a converging beam is used, it becomes clear that the algae swim toward the source of the light and not toward its most intense point (Buder, 1917; Halldal, 1959; Sachs and Mayer, 1961). The ability to orient seems to depend on the presence of a localized photoreceptive region. The response is usually positive at low light intensities and negative at high intensities. It is assumed that phototaxis in the flagellated algae is not directly mediated by the photosynthetic system, since red light, which is absorbed by chlorophyll, is phototactically inactive. The flagellated algae include unicellular green algae (such as *Euglena* and *Chlamydomonas*), colonial green algae (such as *Volvox*), dinoflagellates (such as *Gonyaulax*), and motile cells of multicellular algae (such as *Ulva*). The emphasis in this review will be on phototaxis in these forms, particularly in *Euglena* and *Chlamydomonas*.

The filamentous blue-green algae, the diatoms and the desmids differ from other phototactic forms in that they have no obvious locomotor organelles. Rather they glide slowly over the substratum, probably by secreting a mucilaginous material. There is little precise information concerning the mechanism of phototactic behavior in these organisms. Most of the phototactic blue-green algae and diatoms show phobic as well as topic responses, whereas phototaxis in desmids may be purely topic (Bendix, 1960). Action spectra for phototaxis in blue-green algae suggest that carotenoids and phycobilins are the photoreceptors for topotaxis (Nultsch, 1961), whereas there is evidence that chlorophyll and phycobilins function in phobotaxis (Nultsch, 1962; Nultsch and Richter, 1963). These results suggest that in these organisms, as in the bacteria, there is a direct connection between *phobo*taxis and photosynthesis. Similarly, phobotaxis in diatoms (and topotaxis in desmids) occurs in response to red as well as to blue and green light (Nultsch, 1956; Bendix, 1960). However, Clayton (1964) noted that the responses to red light may be due to secondary nonphototactic effects of the light, particularly to photokinesis.

Finally, there is the interesting case of light-controlled orientation of chloroplasts in certain algae, mosses and angiosperms. These organelle movements are believed to be under the control of a phytochrome system (Haupt, 1966). For a comprehensive review of phototaxis in all of these forms, see Haupt (1959) and Clayton (1959). The subject has also been reviewed in recent years by Bendix (1960), Halldal (1962, 1964), Haupt (1966), Jahn and Bovee (1966), and (with phototropism) by Thimann and Curry (1960). Clayton (1964) has written a penetrating review in which he mentions possible applications of phototactic study to other areas of biological investigation.

C. The Relationship between Phototaxis and Photosynthesis

Indirect Effects of Light: Almost all phototactic plants are also photosynthetic; thus light plays a major and a multiple role in their lives, and its specific role in phototaxis is often difficult to define. Since photosynthesis supplies much of the energy for movement, investigators must be careful to distinguish between general effects on motility and specific effects on phototaxis. The light conditions under which flagellated algae are kept prior to testing influence the threshold (BRUCKER, 1954) and the direction (HALLDAL, 1960) of subsequent phototactic response[1], acting at least in part *via* the photosynthetic system.

HALLDAL (1960) found that exposing *Platymonas* to blue or red light prior to testing reverses its subsequent phototactic response from negative to positive. He suggests that chlorophyll may be one of the photoreceptors for this phenomenon. DIEHN and TOLLIN (1966a, b) reported that when *Euglena* is placed in the dark, the ability to perform positive phototaxis declines, following the same timecourse as the degeneration of the chloroplasts, even though motility remains unimpaired. They concluded that positive phototaxis depends on photosynthesis. Surprisingly, they found that phototaxis totally disappears if the euglenae are grown for 24 hr under continuous light; this is certainly not the case for *Chlamydomonas reinhardi* (FEINLEIB, 1965). BRUCE and PITTENDRIGH (1956, 1958) reported that phototaxis in *Euglena* shows an endogenous diurnal rhythm, which can be shifted in phase by giving the organisms a single 12-hr exposure to light. Some indirect effects of light on phototaxis seem to operate neither *via* the photosynthetic nor *via* the photo-tactic system. HALLDAL (1960) could induce *Platymonas* to change its subsequent phototactic response from positive to negative by illuminating the culture with yellow light prior to testing. Other phototropic and phototactic systems are specifi-cally affected by red light (see Section IV B, below).

The Possible Role of ATP. For the photosynthetic bacteria (where phototaxis and photosynthesis are closely related), CLAYTON (1959, 1964) outlines and cites support for the following hypothesis, presented in its original form by LINKS (1955). A sudden decrease in light intensity on the bacterium results in a decrease in the rate of photosynthetic phosphorylation. This in turn leads to a decrease in the availability of *adenosine triphosphate* (ATP) to the flagella, causing a change in flagellar motion. CLAYTON suggests that phototactic movements in flagellated algae may be governed by a similar mechanism. Experiments with isolated algal flagella (BROKAW, 1962) have shown that ATP is necessary for motility, but it must be established if it is specifically involved in phototactic orientation.

HALLDAL (1957, 1959) found that he could control the sign of phototaxis in *Platymonas* cultures by manipulating the concentrations of Mg^{++}, Ca^{++}, and K^+, which are known to be involved in the utilization of ATP in muscle contraction. Studies with dinitrophenol (DNP), an uncoupler of phosphorylation, yielded conflicting results (HALLDAL, 1958; SACHS and MAYER, 1961; DIEHN and TOLLIN, 1967). However, DIEHN and TOLLIN found that dichlorophenyl dimethylurea (DCMU) and methyl octanoate, inhibitors of photosynthetic phosphorylation, selectively inhibit phototaxis in *Euglena* with little or no effect on swimming speed. It seems to us that if phototaxis depends on changes in the ATP level, then processes other than photosynthetic phosphorylation must influence that level, since red light (absorbed by chlorophyll) is phototactically inactive. SHROPSHIRE

---

[1] It should be noted that in these algae blue light acts as a phototactic stimulus, but red light does not.

and GETTENS (1966) have found a change in ATP level in *Phycomyces* sporangio-phores shortly after stimulation by light and preceding the growth response. This clearly implicates ATP in a light response in a nonphotosynthetic system. It remains to be seen whether light-induced changes in ATP level lie directly in the path between stimulus and response or off to the side.

D. The Relationship between Light-Intensity and Response

1. Studies on Cell Populations

Methods. Many investigators noted that phototactic response increases as a function of light intensity, but because of inadequate measuring techniques few of them were able to describe what sort of function this is. In many of the experiments, cells were permitted to accumulate in one region of a chamber in response to a stimulus light, and the density of the accumulation was gauged by eye (HARTSHORNE, 1953; MAYER and POLJAKOFF-MAYBER, 1959). In only a few studies were measurements made on a response in progress. DESROCHE (1912) and more recently, HARTSHORNE (1953) and HAXO and CLENDENNING (1953) used the technique of timing the movement of a mass of cells swimming toward a stimulus light at various intensities. Since such measurements depend on eye and stopwatch, their accuracy is limited; moreover, they yield little information about the kinetics of the response.

Phototaxis can be measured more accurately by using a photometric technique which follows changes in the optical density of a cell suspension, recording the time course of the response. This type of method has recently been used in intensity-response studies on *Chlamydomonas* (FEINLEIB, 1965; FEINLEIB and CURRY, 1967) and on *Euglena* (DIEHN and TOLLIN, 1966a). In FEINLEIB's system the swimming path is parallel with the stimulus beam; the monitoring beam (phototactically inactive red light) traverses the sample at right angles to the stimulus light and impinges on a pair of photocells. Oriented (topic) movement toward or away from the stimulus is recorded as a difference in output between the two photocells (Fig. 1a). In the apparatus of DIEHN and TOLLIN (a "phototaxigraph", designed by LINDES et al., 1965), the stimulus beam is perpendicular to the swimming path, and a positive response is recorded if the cells accumulate in the illuminated region. Cells may gather in this region topotactically (in directional response to light scattered from other cells in the test beam) or phobotactically (in response to the beam itself). The apparatus probably measures a mixture of these responses.

Results; Adherence to WEBER's Law. From the early work on, it became clear that phototactic systems commonly adhere to the Weber Law over a considerable range of intensities; i.e., the minimal change in intensity which can be "perceived" by the organism (as judged by a just-detectable response) is proportional to the initial intensity prevailing before the change. In HECHT's (1934) terms, "In a word, intensity perception is relative and not absolute." MAINX and WOLF (1939) and HALLDAL (1958) found this to be true in several flagellated algae, and CLAYTON (1953a) found that it holds true for the phototaxis of *Rhodospirillum* over a 10^5-fold range of intensities (threshold $I/I_0 = 0.03$ to 0.05). A similar situation has been found in most cases of phototropism studied in this regard.

Desroche (1912) observed that the rate at which a population of *Chlamydomonas* cells moves toward the light seems to increase linearly with the logarithm of the intensity. This presumably reflects the Fechner extension of the Weber Law; i.e., integration of the Weber expression leads to the formulation that "sensation" (and therefore "response"?) produced by a stimulating agent is proportional to the

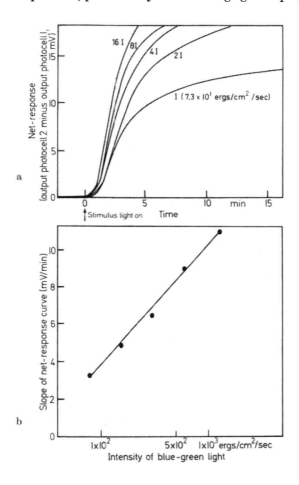

Fig. 1. a) Net topotaxis of a population of *Chlamydomonas reinhardi* cells as a function of time, at five intensities of blue-green light. These curves were traced from recordings made in the measuring system of Feinleib (1965). A fresh sample of cells from the same culture was used for each measurement. b) Positive topotaxis as a function of the intensity of blue-green light. The slope of the linear portion of each curve in Fig. 1a is plotted against intensity (with intensity on a log scale); after Feinleib (1965)

logarithm of its intensity. A logarithmic relationship between stimulus and response has been found in positive phototaxis in *Chlamydomonas* (Feinleib, 1965; see Fig. 1 b), in *Euglena* (Diehn and Tollin, 1966a), and in the so-called "first-positive" range of the dose-response curve in the phototropism of oat and corn seedlings (Thimann and Curry, 1960; Briggs, 1960).

24*

All of these findings are not particularly surprising, since adherence to the Weber Law and a logarithmic connection between stimulus and response are characteristic features of most sensory systems, at least over part of the intensity range. It should be noted that in phototaxis and in phototropism, as in other systems, these "laws" often fail at the intensity extremes.

There have been numerous attempts to account for the relationship between stimulus and response, and in particular for the fact that sensitivity depends on the past history of the system with respect to illumination. This is the problem of "intensity discrimination" in Hecht's (1935) terms, or of "range adjustment" in the terms of Delbrück and Reichardt (1956). Most of the earlier approaches to this problem centered on the notion that the instantaneous concentration of the photoreceptor material itself is the determining factor; i.e., light "bleaches" the photoreceptor which is then reconstituted in a series of dark reactions. Sensitivity is a function of the concentration of "unbleached" photoreceptor at the time of stimulus onset. It has become clear, however, that phototactic and phototropic systems undergo very large changes in sensitivity under conditions in which only negligible "bleaching" of the photoreceptor can occur. Furthermore, the known facts about the kinetics of adaptation are not presently explicable in terms of a "bleaching" system. Most of the recent models have therefore implicated other stages of the systems to account for adaptation; the major features of some of these models will be outlined below.

With regard to phototaxis, the most detailed analysis along this line is that of Clayton (1953b) on the phobotaxis of *Rhodospirillum rubrum*. Clayton used the framework of Rashevsky's theory of excitation, assuming the existence of two substances, one excitatory (s) and one inhibitory (i), both of which are formed at rates proportional to stimulus intensity and dissipated at rates proportional to their excess over the resting values. The rate constants for formation and dissipation of the two materials are different. Excitation occurs when the concentration of excitatory substance exceeds that of inhibitory substance. These premises lead to a set of differential equations whose solutions yield values for s and i as a function of light intensity and time. The derived curves are fitted to the data by experimental evaluation of the arbitrary constants. Clayton shows that the model can account quite neatly for the apparent "all-or-none" nature of the response, for Weber Law adherence, for certain strength-duration relationships, and for summation and accommodation. In Clayton's words, the experimental results show, "on the whole, a fairly good agreement with Rashevsky's theory. The agreement has not been so close, however, as to provide a convincing confirmation of this theory."

In a more recent study of phobotaxis ("stop-response") in *Gyrodinium*, Hand et al. (1967) have returned to the idea that sensitivity is closely related to the concentration of a "photoactive" form of the photoreceptor. The main impetus for this is their discovery that red light rapidly induces "regeneration" of phototactic sensitivity to blue light. They suggest that absorption of blue light shifts a photoreceptor from an active to an inactive state, and that this triggers processes which ultimately result in a motor response. The active state of the photoreceptor is regenerated by a red-light-driven photochemical step. The quantitative aspects of the proposed two-pigment machinery are not formulated.

With regard to phototropism, several models have been developed in some detail in the attempt to account for stimulus-response relationships. Curry (1957) proceeded on the simple idea that light acts by converting a photosensitive molecular species from an "active" form to another form, at a rate proportional to the concentration of the "active" form and to the light intensity. The curvature response was assumed to depend on the amount of conversion produced by a particular light dosage. Integration of the first-order differential equation involved here leads to the expression:

$$C = C_{max} (1 - e^{-kIt})$$

where C_{max} is the maximum curvature obtained as light dosage It is increased. Experimental data obtained with small dosages (in dark-adapted or red-light-adapted plants) can readily be fitted to such an expression; e.g., a portion of the curve is nearly log-linear. The model also accounts for adherence to the reciprocity rule; i.e., response magnitude is determined by the product, $I \times t$, over considerable ranges of I and t. If conversion on both sides (front and back) of the unilaterally illuminated plant is considered, the expression can be modified to:

$$C = C_{max} \left(e^{-kI_b t} - e^{-kI_f t}\right)$$

accounting for decreasing response with increasing dosage as the illuminated side becomes "saturated".

In actuality, the dosage-response curves become very elaborate at high values of $I \times t$, particularly when t is large. There is, in fact, a range of dosages yielding *negative* curvatures in oat seedlings, and then a still higher range yielding so-called "second-positive" responses. Furthermore, phototropically inactive red light substantially decreases the sensitivity of these systems to blue light (as judged by response magnitude for a particular $I \times t$). The simple model accounts for none of these phenomena. Recently, however, ZIMMERMAN and BRIGGS (1963) have greatly extended such a model by assuming three separate and possibly independent mechanisms, each based on a kinetic scheme similar to the above, but each with different rate constants (for light and dark reactions) and differently modified by red light. Under this model they can account for the phototropic dosage-response curves of oat coleoptiles and also for the sensitivity shifts induced by red light. These models do not directly address themselves to the problem of range adjustment.

DELBRÜCK and REICHARDT (1956) chose to study light-growth reactions (see Section I) in *Phycomyces* sporangiophores in a fundamental reexamination of the relations between stimulus and response. This system has a response which is distinctly graded with respect to intensity, and it possesses a clear-cut range adjustment mechanism. (For an elegant description of the *Phycomyces* system and its photoresponses, see SHROPSHIRE, 1963.) DELBRÜCK and REICHARDT defined the level of adaptation A as the intensity with which the specimen will find itself in equilibrium; this is the inverse of "sensitivity" as usually defined. They developed an experimental procedure for determining A and then studied the changes in A resulting from various illumination programs. The most basic finding was that A decreases in the dark exponentially, irrespective of the particular illumination program which established its initial level. The results suggested the following functional relation between $I(t)$ and $A(t)$:

$$dA/dt = (I - A)/b$$

where b is the time constant (ca. 3.8 min) of the system. Noting that stimuli and adaptive levels must be raised proportionally to produce a given "growth output", DELBRÜCK and REICHARDT proposed that the response resulting from any light program is a linear function of the ratio I/A; they called this ratio the "subjective intensity", i. The experimentally measured responses following several illumination programs were in reasonable agreement with those predicted on this assumption.

DELBRÜCK and REICHARDT (1956) postulated the following model as an illustration of the formal relations which they inferred experimentally (see their Fig. 11). The system involves two enzymes, E and E', whose activities are light controlled. The activity of E adjusts slowly to changes in light intensity (time constant of 3.8 min). A precursor material, M, necessary for growth, is supplied at a constant rate, B. E converts M to inactive material, X, in a rapid reaction. The level of activity of E represents the level of adaptation, A, and causes the concentration of M to be proportional to $1/A$. The activity of the second enzyme, E', is assumed to adjust instantaneously to changes in light intensity. E' converts precursor M into wall material, W (thereby, "growth"), in a slow reaction at a rate proportional to its activity (therefore to I), and to the concentration of M (therefore to $1/A$). The growth rate is therefore proportional to the "subjective intensity", I/A. DELBRÜCK and REICHARDT conclude that, "The principal thing that suggests itself from a consideration of such models is a necessity of assuming, in the chain of action starting at the light input, a bifurcation, one branch leading to the setting of the level of adaptation and the other branch utilizing the light input directly so as to compare the instantaneous value of the intensity with the level of adaptation."

CASTLE (1966) constructed an elegant kinetic model for adaptation and the light responses (including phototropism) of *Phycomyces*. His model includes many of the features of those outlined above, but it invokes only a single light catalyzed step in a sequence of processes. The scheme is shown below, using CASTLE's notation:

$$\xrightarrow{c} M \xrightarrow{(k' + k)\,[M]} P \xrightarrow{j\,[P]} \text{wall formation}$$

where c is the constant rate of supply of a material, M, which is used stoichiometrically in growth; $[M]$ is the concentration of material which is subsequently converted to a product, P, by one of two pathways, one light-dependent (rate constant, k') and one light-independent (rate constant, k). The rate of transformation of P, governed by the rate constant j, is assumed to constitute the instantaneous growth rate; i.e., growth velocity, $V = j\,[P]$. All of the steps are assumed to be first-order, irreversible chemical reactions, and changes in the rate constant of the light reaction are considered to be instantaneous. Solution of the differential equations involved, starting from a steady state condition and with values for the constants chosen to approximate the actual responses, leads to the following exponential expression for the time course of the normalized growth velocity after an abrupt change in k' from 0.01 to 0.04 (induced by an intensity increase):

$$V = 1 + 3\,(e^{-0.05\,t} - e^{-0.1\,t}).$$

CASTLE (1966) finds that the major aspects of the cell responses are predicted by the model. As he points out, "the behavior of the model in adaptation is nowhere inconsistent with present empirical knowledge of adaptation in *Phycomyces*." When the model is applied to the asymmetric case (unilateral illumination), it also successfully accounts for all the phototropic phenomena, including, most impressively, the curious phenomenon of phototropic inversion in which a bending cell shows a temporary reversal in bending direction when the unilateral light falling on it undergoes an abrupt increase (or decrease!) in intensity (REICHARDT and VARJU, 1958; DENNISON, 1965).

It will be interesting to see how some of these recent models may apply to phototaxis, which is in many ways related to phototropism (similar action spectra, "adaptation" phenomena, dose-response functions, etc.) and yet is very different with regard to the effector mechanism (change in swimming direction vs. asymmetric change in growth rate). The central idea in the most successful models is the concept of catalytic regulation of the level of a metabolic reservoir by light. In phototaxis, as we have seen, this might involve regulation of the ATP level. Hopefully, further experimentation will reveal the actual components of the mechanisms so strongly implied by these models.

2. Studies on Single Cells

Studies of topotaxis of cell populations are limited in that they fail to distinguish a change in the directness of the swimming path from a change in the swimming rate (photokinesis); i.e., a group of cells may traverse a chamber in less time either because they follow a more direct course or because they swim faster. Several investigators attempted to measure photokinesis during oriented swimming by timing individual organisms as they crossed a microscope field (DESROCHE, 1912; MAST and GOVER, 1922; MAINX and WOLF, 1939). With this technique, however, it is still possible to mistake a straighter path for an increased swimming rate, and, since many algae move rapidly, the investigator's reaction time becomes a limiting factor.

WOLKEN and SHIN (1958) studied photokinesis in *Euglena*, presumably independently of phototaxis. The organisms were placed under a microscope in a flat chamber, with the stimulus light entering at right angles to the plane of the chamber. They timed the movement of the euglenae with a stopwatch and traced some of the swimming paths; from these data they calculated the forward component of the swimming velocity. They found that the swimming rate increases by a factor of 50% between 2 and 40 foot-candles of white light. This effect first becomes evident 10 to 15 min after the light is introduced. The population studies cited earlier demonstrated definite phototactic reactions less than 2 min after stimulus onset in both *Euglena* and *Chlamydomonas* (FEINLEIB, 1965; DIEHN and TOLLIN, 1966a). It would thus appear that the photokinetic effect reported by WOLKEN and SHIN could make only a minor contribution, if any, to intensity-dependent increases in phototactic response. The interpretation of their results is complicated by the possibility that some phototaxis may have occurred either in the vertical plane, in response to the stimulus light, or in the horizontal plane, in response to scattered light from other cells in the stimulus beam. (GERISCH, 1959, has established that swimming rate in *Volvox* undergoes a transient decrease when the light intensity is changed, but a slower swimming rate could not account directly for faster phototactic accumulation.)

FEINLEIB (1965) used a photomicrographic technique to study the topotactic response of *Chlamydcmonas* cells. The stimulus was introduced into one end of a chamber on the microscope stage, and photographs were taken through the microscope using phototactically inactive red light and darkfield optics. The exposure was fixed at 0.2 sec, during which time each cell in the field described a swimming track on the film. Swimming rate was determined by measuring the length of the track, and directness of path was indicated by the angle between the track and the stimulus beam. Such measurements show that swimming rate during positive phototaxis does not increase significantly over a 4-log-unit range of stimulating intensities. Intensity-dependent changes in phototactic response can thus be attributed almost entirely to a change in the directness of the swimming path; this was found to increase as a function of light intensity until the swimming tracks were aligned essentially parallel with the stimulus beam. (In the movement of a population of cells, of course, the number of cells responding also affects the total response measured.)

HAND *et al.* (1967) examined phototaxis in the dinoflagellate *Gyrodinium*, using a "flying-spot" scanning apparatus developed by DAVENPORT *et al.* (1962). In this ingenious system, a light beam generated by an oscilloscope scans the field and falls on a photomultiplier. When an organism intercepts the beam, a signal is received by the photomultiplier; this signal is reported as a potential change, which is ultimately displayed on a readout oscilloscope as an image of the swimming organism (see Section VI). In its present form, this apparatus can only be used to study relatively large phototactic organisms (DAVENPORT, 1968, *personal communication*). HAND and DAVENPORT (1970) and FEINLEIB and CURRY (unpublished) have recently adopted the more versatile technique of monitoring the response of individual cells with a video camera and a video recorder equipped with slow-motion playback.

3. Negative Phototaxis

At high light intensity, the direction of algal phototaxis commonly changes from positive to negative. Since HALLDAL (1959) was unable to observe this phenomenon in short-term experiments (up to 15 min), he concluded that the sign of the response depends only indirectly on the light intensity. It was recently found, however, in *Chlamydomonas*, that changing the light intensity can effect sign reversal within a few seconds (FEINLEIB, 1965).

BÜNNING and TAZAWA (1957) reported negative phototaxis in *Euglena* to be purely phobic. Presumably, the flat swimming chamber used for these observations prevented the algae from responding topically to the light source; it does not seem valid, therefore, to conclude that the algae were incapable of such a response. Definite negative topotaxis has since been observed in several members of the *Volvocales* (HALLDAL, 1958; FEINLEIB, 1965). In *Chlamydomonas*, the negative topic response is often preceded by a phobic reaction, a ca. 1-sec cessation of swimming at the onset of stimulation (FEINLEIB, 1965).

At transitional light intensities between the positive and the negative range, some workers have found an "Indifferenzzone" in which the algae are randomly oriented with respect to the light (BUDER, 1917). However, photomicrographs taken of *Chlamydomonas* during topotactic response indicate that the swimming paths become increasingly well-aligned with the stimulus beam as the light intensity is raised, even when the direction of swimming changes from positive to negative, The net response of the population may thus go through an "indifferent" stage, but the individual cell evidently makes an abrupt switch from a strong positive to a strong negative response.

The reversal of sign is influenced by the age of the culture, the composition of the medium, and the light conditions prior to testing. HALLDAL (1957, 1959) reported that the sign of phototaxis in the marine alga *Platymonas* depends directly on the concentration of cations in the medium. If the ratio of $[Mg^{++}]$ to $[Ca^{++}]$ (at very low$[K^+]$) is 6:1, orientation is random; above that value, response is positive; below, it is negative. In cultures where orientation is close to random, reversal of sign can be induced by illuminating the cells at certain wavelengths prior to testing (HALLDAL, 1960; see Section III above).

MAINX (1929) found that factors which raise the intensity for positive-to-negative reversal in *Pandorina* also lower the threshold intensity for positive response, thereby extending the range of positive phototaxis. He concluded that positive and negative phototaxis are governed by two antagonistic processes which can occur simultaneously in the cell; at the reversal intensity, the negative response overrides the positive. There is no firm evidence for this duality of mechanism. In general, positive and negative phototaxis have the same action spectrum (HALLDAL, 1958, 1961; see Section VI), indicating that they share a common photoreceptor. Nonetheless, in *Chlamydomonas*, negative phototaxis differs from positive in several features which merit further investigation: (a) if the net response of a culture (as recorded photometrically) is strongly positive to blue-green light, it will remain that way for at least 30 min under continuous stimulation, but if the initial net response is negative, it will become increasingly more positive under continuous stimulation (FEINLEIB, 1965; MAYER, 1968); (b) swimming rate may be slightly higher in negative phototaxis (by 20% at the most); and (c) in some strains there may be separate photoreceptors for positive and negative response (FEINLEIB, 1965; see Section F).

E. Morphological Features Related to Algal Phototaxis

The following discussion will concentrate on the morphological features which appear to be related to phototaxis in *Euglena* and *Chlamydomonas*. The description of *Euglena* will refer specifically to *E. granulata*, which has recently been examined by WALNE and ARNOTT (1967). *E. gracilis*, the species most commonly used in phototactic studies, resembles *E. granulata* in most features but is about half its size (GIBBS, 1960; WOLKEN, 1967; LEEDALE, 1967; see Fig. 2a).

E. granulata is a spindle-shaped organism covered with a flexible pellicle which allows the cell to change shape. Two flagella arise in an anterior invagination of the cell; only one of these emerges from the cell and is functional in propelling the organism. *Euglena* generally rotates as it swims, the anterior end tracing a wide circle (LEEDALE, 1967). *E. granulata* contains a dozen or more chloroplasts and a conspicuous orange-red organelle, the stigma or "eyespot."

Euglenoids are unique in that the stigma is independent of the chloroplasts. It is located anteriorly, adjacent to the expanded basal portion (the reservoir) of the anterior invagination. The stigma is roughly cup-shaped, with the inside of the cup facing the reservoir, and is 7 to 8 µ in diameter (2 to 3 µ in *E. gracilis*; Fig. 2b). It consists of a mass of spheroidal or polygonal granules, rather haphazardly

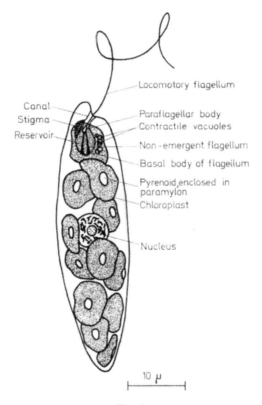

Fig. 2a

Fig. 2a—c. a) Schematic drawing of *Euglena gracilis*, as seen through the light microscope (redrawn from LEEDALE, 1967). b) Electron micrograph of tangential section through stigma of *E. granulata*, showing arrangement of stigma granules into membrane-bounded packets. *SG* = stigma granule. × 27,200. c) Reservoir region of *E. granulata*, showing stigma granules and paraflagellar body, containing crystalline structure. *SG* = stigma granule; *PB* = paraflagellar body. × 52,800. Inset shows detail of this structure. × 128,000. Electron micrographs courtesy of DR. H. J. ARNOTT, Cell Research Institute, University of Texas (for details, see WALNE and ARNOTT, 1967)

Fig. 2b

arranged. In electron micrographs these granules appear as dense osmiophilic bodies, 240 to 1,200 nm in diameter, with some evidence of a reticulate structure. Most of the granules occur in groups of 3 to 5, with each group surrounded by a unit membrane, but there is no membrane bounding the entire stigma. The chloroplasts of *E. granulata* often contain rows of granules or droplets similar to those of the stigma, but smaller. Walne and Arnott (1967) found that the stigma granules exhibit birefringence typical of a spherite and concluded that the pigment molecules may be radially oriented.

Near the base of the emergent flagellum is a swelling known as the parabasal, or paraflagellar, body (Fig. 2c). This structure is often implicated in hypotheses concerning the mechanism of phototaxis. Under the light microscope it appears to be somewhat larger than an eyespot granule, and is not visibly pigmented. Under the electron microscope, Walne and Arnott found the paraflagellar body to consist mainly of an ovoid crystalline structure, with a repeat pattern of about 75 A. The chemical composition is unknown, but the large lattice parameters suggest that it may be partly protein. The stigma and the paraflagellar body in euglenoids almost always occur together, and are closely associated with each other in position (Leedale, 1967). This close association has led some investigators to believe that the two structures constitute a morphological unit (Pringsheim, 1963), but electron micrographs do not reveal a structural connection between

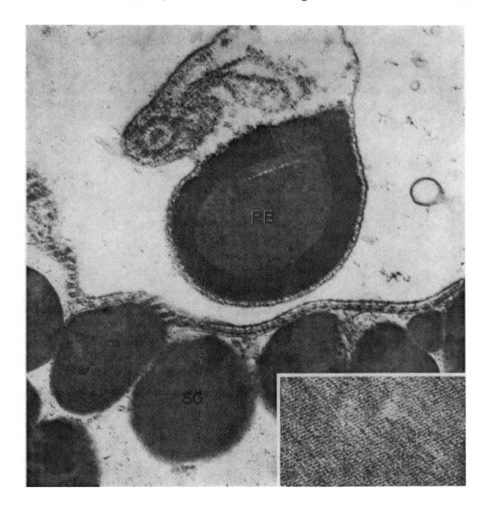

Fig. 2c

them. Moreover, there are mutants of *Euglena* which have lost the stigma but retain a paraflagellar body (VAVRA, 1956; GÖSSEL, 1957).

Chlamydomonas has been extensively used in studies of genetics and photosynthesis, as well as phototaxis. Much of the following description refers specifically to *C. eugametos*, studied by LEMBI and LANG (1965) and by WALNE and ARNOTT (1967). Other species, such as *C. reinhardi*, have a similar structure (SAGER and PALADE, 1957; LEVINE and EBERSOLD, 1960; Fig. 3a). The vegetative cell of *C. eugametos* is ellipsoidal and is enclosed in a rigid cellulose wall. The cell has two flagella at its anterior end and contains a single, large cup-shaped chloroplast. An orange-red stigma is embedded in the peripheral region of the chloroplast, its longitudinal axis lying in a plane with the long axis of the cell and with the flagella. It differs from the stigma of *Euglena* in that it is quite far from the base of the flagella.

The stigma in *C. eugametos* is a curved plate, about 2–3 μ long (Fig. 3b). When viewed in tangential section under the electron microscope, the plate is seen to consist of hexagonally packed osmiophilic granules, 75 to 100 nm in diameter. In longitudinal section the stigma appears as a single row of granules, just inside the chloroplast membrane and subtended by a disc of the chloroplast. In other species of *Chlamydomonas*, there may be several rows of granules, each subtended by a disc of the chloroplast (Sager and Palade, 1957; Lembi and Lang, 1965).

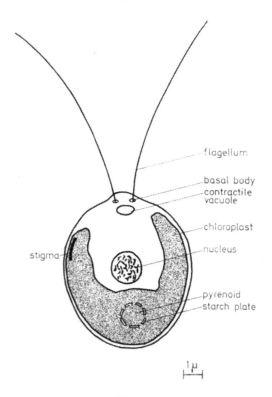

Fig. 3a

Fig. 3a u. b. a) Schematic drawing of a vegetative cell of *Chlamydomonas reinhardi*, as seen through the light microscope (redrawn from Sager and Palade, 1957). b) Electron micrograph of longitudinal section through stigma of *Chlamydomonas eugametos*, showing single layer of granules between chloroplast envelope and chloroplast lamella. *CW* = cell wall; *PM* = plasma membrane; *CE* = chloroplast envelope; *CL* = chloroplast lamella; *S* = stigma granule. × 66,700 (courtesy of Dr. H. J. Arnott)

The stigma granules of *C. eugametos* are smaller than those of *E. granulata*, but they are similar in structure. There is a close and suggestive resemblance between the rows of stigma granules in *Chlamydomonas* and the rows of osmiophilic granules in the chloroplasts of *Euglena*. Similar osmiophilic granules are also scattered, singly or in rows, throughout the chloroplast of *C. eugametos*. The suggestion is that these granules are all essentially alike, and that they simply occur in a more concentrated mass in the stigma.

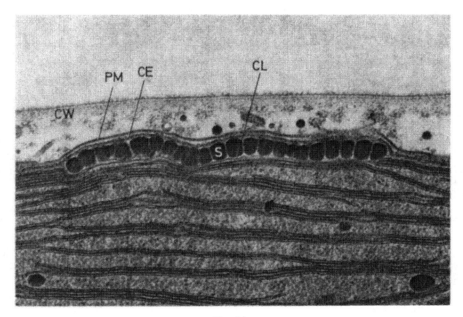

Fig. 3b

In C. *eugametos*, both the chloroplast membrane and the plasma membrane are modified in the region immediately overlying the stigma. Both membranes protrude slightly, conforming to the convex shape of the stigma, and there is a clear area between them. The plasma membrane is slightly thicker here than in the rest of the cell. It is important to note that the flagella of *Chlamydomonas* bear no structure which corresponds to the paraflagellar body of *Euglena*.

F. The Action Spectrum for Phototaxis in Flagellated Algae

As we have seen, the action spectrum for phobotaxis in the bacteria indicates that the photosynthetic pigments are the photoreceptors for phobotaxis in these forms. Spectral studies have not yielded such clear results in the flagellated algae. In most early work, and in some of the recent studies as well, the optical systems have been quite crude; e.g., spectral regions were "isolated" with broad band-pass filters, light intensity was adjusted by methods which changed the spectral distribution of the light or the geometry of the light path, and relative energy determinations, when made at all, were approximate. The index of response was often questionable; e.g., many studies involved threshold measurements, in spite of the fact that topotaxis is a distinctly graded response, so that the "threshold" is defined by the measuring system and is not a characteristic of the phenomenon itself (see the comment of DELBRÜCK and REICHARDT, 1956, on this). Finally, the action spectrum is undoubtedly affected by the presence of screening pigments in these green cells and possibly also by photokinesis (CLAYTON, 1964).

From ENGELMANN's time on, however, it has been clear that the peak of phototactic sensitivity in the flagellated algae lies in the blue or blue-green region.

ENGELMANN (1882) found maximum accumulation of *Euglena* in the 470 to 490 nm region of a projected microspectrum. The numerous action spectra obtained since then are in essential agreement with ENGELMANN's, and most of them have added little new information (e.g., MAST, 1917; LAURENS and HOOKER, 1920; LUNTZ, 1931; MAYER and POLJAKOFF-MAYBER, 1959). For references to the older literature, see HAUPT (1959) and BENDIX (1960). Several of the more recent studies will be discussed in some detail below.

BÜNNING and SCHNEIDERHÖHN (1956) used a threshold method to obtain action spectra for *Euglena*. A flat cuvette containing the suspension was mounted on a microscope stage, and the stimulus beam was sent through the cuvette toward the observer so that a circle of light appeared in the field. At each test wavelength, they determined the minimum intensity needed to produce a just-discernible accumulation of cells in the lighted circle in 10 min. Their spectrum for positive phototaxis had a major peak at 495 nm, a secondary peak at 425 nm, and a shoulder near 475 nm. They further observed that, as the intensity was raised, cells began to leave the lighted circle and to gather in a ring in the dark around the periphery of the circle. At each wavelength they determined the minimum intensity at which the ring formation began; these data were used to construct an action spectrum for negative phototaxis. This spectrum had a maximum at 415 nm, with shoulders at 450, 475, and 510 nm.

It has been pointed out (THIMANN and CURRY 1960) that the methods used in obtaining these two action spectra were not identical. In studying positive phototaxis, BÜNNING and SCHNEIDERHÖHN (1956) determined the "threshold" for response, but in studying negative phototaxis they determined the "threshold" for reversal from positive to negative. Another difficulty results from the optics of their system; orientation parallel with the light beam (topotaxis) was restricted by the flatness of the chamber, and was, in any case, not visible to the observer since his line of vision was parallel with the axis of the stimulating beam. The greatest problem concerning their results was discovered some years later, when BÜNNING and GÖSSEL (1959) found that a major difference between positive and negative spectra occurs only in dark-adapted organisms.

DIEHN and TOLLIN (1966a) used the photometric apparatus described earlier (Section IV A) to obtain an action spectrum for positive phototaxis in *Euglena gracilis* (Fig. 4). They examined 10 regions between 350 and 700 nm, using interference and cut-off filters. Two indices of response were chosen from the recordings: 1. the "initial rate of phototaxis", i.e., the maximum rate of change in optical density as a function of time during the early stages of accumulation; and 2. the final density of accumulation. These two methods produced the same action spectrum, with a broad peak in the region between 450 and 520 nm.

HAND et al. (1967) studied the spectral sensitivity of positive topotaxis in the dinoflagellate *Gyrodinium* in the the region between 430 and 520 nm, using the "flying-spot" scanning apparatus described earlier. They determined the percent increase in the number of cells gathered at the illuminated wall of the swimming chamber, after 1 min of illumination. The stimulus was monochromatic light, adjusted to equal intensities. One part of the action spectrum was reexamined, using the angle between stimulus beam and swimming path as a response index. Both methods gave an action spectrum having a sharp peak at 470 nm.

A series of detailed and elegant spectral studies of algal phototaxis was performed by HALLDAL (1958). He used two-beam null methods to obtain action spectra for positive and negative topotaxis in a number of marine Volvocales and dinoflagellates. In one technique a reference light of fixed intensity was projected on one side of a cuvette, while light from a monochromator was projected on the

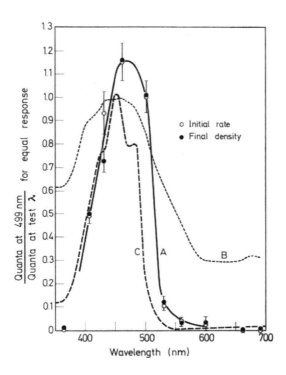

Fig. 4. *Line A;* Action spectrum for positive phototaxis in *Euglena gracilis.* Plotted on the ordinate is the reciprocal of the relative number of quanta required at each wavelength to give the same response, as measured in the system of LINDES *et al.* (1965). Two indices of response are used: "initial rate" of accumulation (the maximum time rate of change in optical density during the initial stages of accumulation) and "final density" of accumulation. *Line B;* Absorption spectrum of a suspension of stigma granules from *E. gracilis. Line C;* Absorption spectrum of an extract of stigma granules in cyclohexane; after DIEHN and TOLLIN (1966a)

other side. The intensity of the test beam was regulated with a variable transformer. The cells swam toward the reference beam or the test beam, whichever was the more effective. When the two beams were equally effective (null), there was random motion; the intensity of the test wavelength at null was used as the index of effectiveness. Some of the experiments were repeated using a reference beam of different intensity and wavelength; these gave the same results.

In an aesthetically pleasing variation of this technique, HALLDAL (1958) illuminated one side of the swimming chamber with a projected spectrum which had an intensity gradient perpendicular to the wavelength gradient. A reference beam of uniform intensity and wavelength was projected on the other side of the chamber. At highly effective wavelengths, the cells accumulated over the entire intensity gradient on the test beam side, whereas at less effective wavelengths they appeared only at the highest intensities. In effect, then, the algae drew their own action spectrum in a single exposure.

In *Platymonas*, *Dunaliella*, and *Stephanoptera* (members of the Volvocales), the action spectrum has a broad maximum, peaking at 495 nm, with a shoulder at 435 nm (Fig. 5). The spectra for positive and negative topotaxis are identical.

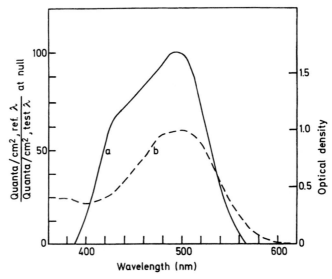

Fig. 5. *Line a;* Action spectrum for positive topotaxis in *Platymonas subcordiformis*. The algae were stimulated from opposite sides with a test beam and a standard reference beam. Plotted on the ordinate is the reciprocal of the number of quanta required at each wavelength to produce random motion of the algae; after HALLDAL (1958). *Line b;* Absorption spectrum of hydroxyechinenone (in chloroform) extracted from *Euglena gracilis*; after KRINSKY and GOLDSMITH (1960)

The peak for positive topotaxis in *Ulva* gametes is at 485 nm with a shoulder at 435 nm, whereas in the dinoflagellates *Peridinium* and *Gonyaulax* it is at 475 nm. In all of these spectra, response drops to zero somewhere between 530 and 570 nm. The one exception is the dinoflagellate *Prorocentrum* which has a peak response in the yellow region of the spectrum at 570 nm. This is one of the rare reports of a definite topic response at wavelengths longer than 550 nm.

Recently, FEINLEIB (1965) and FEINLEIB and CURRY (1967) have observed distinct topotactic responses of *Chlamydomonas reinhardi* to the mercury yellow line (577 nm) isolated with a narrow-band interference filter. In most (but not all) samples of these algae, response was positive at 436 and 546 nm, but negative at 577 nm. The negative response in yellow light could not be reversed to positive by lowering the light intensity. This phenomenon is rather elusive, because, as noted, it has only been observed in certain of the cultures. If it should turn out to be a more general phenomenon, it would indicate that two photoreceptor pigments are involved in some cases of topotaxis. Topotactic response to yellow light should be carefully reexamined in *Chlamydomonas* and other algae.

G. A Comparison of Action Spectra and Absorption Spectra

There is no exact correspondence between the action spectrum for algal phototaxis and the absorption spectrum of any known pigment in the cells. The effectiveness of blue light and the ineffectiveness of red suggests that a yellow (or orange) pigment is involved, presumably a carotenoid or a flavin; action spectra for phototropism point to a similar pigment. One possible reason that the photoreceptor has not been identified may be that it is present only in small amounts; absorption spectra of whole cells support this contention. Although *Peridinium* and *Prorocentrum* have very different action spectra, their in vivo absorption spectra are almost identical (HALLDAL, 1958).

Because of its orange color, the stigma was long considered to be the most obvious candidate for photoreceptor. GÖSSEL (1957), STROTHER and WOLKEN (1960), and WOLKEN (1967) obtained in vivo absorption spectra of the stigma in *Euglena gracilis* using a microspectrophotometer, while BATRA and TOLLIN (1964) measured the absorption spectrum of a suspension of stigma granules isolated from the same organism. All of these spectra show a broad maximum and two or three ill-defined "bumps", between 430 and 530 nm (Fig. 4). The details should probably not be taken too seriously, since there is considerable variation not only from one study to the next but also (in the microspectra) from one stigma to the next. Furthermore, the spectra include absorption peaks of other cellular components, particularly of chlorophyll. Hexane extracts of isolated stigma granules reveal a mixture of carotenoid pigments, with peaks at 420, 450, and 480 nm (BATRA and TOLLIN, 1964; Fig. 4).

The absorption spectra of stigma granules in *Euglena* do resemble, in general, the action spectra for phototaxis in that organism, but the close similarities claimed to exist between certain of the spectra are not obvious to the present authors (see BATRA and TOLLIN, 1964; DIEHN and TOLLIN, 1966a; WOLKEN, 1967). In any event, because of the inaccuracies in both kinds of spectra, detailed comparison is probably not meaningful.

BATRA and TOLLIN (1964) analyzed their stigma extract chromatographically and claimed that its main components are lutein and cryptoxanthin, with some β-carotene. However, upon analyzing the carotenoids of whole cells of *E. gracilis*, KRINSKY and GOLDSMITH (1960) found that these cells contain no lutein at all. They did find, among the minor components of the cell, three orange-red carotenoids which they suggest are possible pigments of the stigma, namely, echinenone, euglenanone, and hydroxyechinenone, all ketoderivatives of β-carotene. Each of these pigments (in chloroform) shows a single broad maximum in the visible range, between 460 and 500 nm (Fig. 5).

KRINSKY and GOLDSMITH (1960) have remarked on the similarity between these absorption spectra and the action spectra for phototaxis, particularly those reported by HALLDAL (1958). They even suggest that *Prorocentrum*, whose action spectrum is shifted toward longer wavelengths, may have a pigment such as hydroxyechinenone as its photoreceptor, but that the pigment may be bound to a protein with a resulting shift in the absorption spectrum. When HALLDAL (1961, 1967) extended his action spectrum for *Platymonas* into the ultraviolet, he reported that it resembles absorption spectra of carotenoid-protein complexes, with peaks at 275, 335 and 400 nm. Phototactic response of green algae to ultraviolet light had earlier been reported by Luntz (1931); further investigation of this response is needed.

More information about the pigments involved in phototaxis might be obtained by comparing the phototactic behavior of normal algae with that of carotenoid-deficient mutants. Since Krinsky and Levine (1964) have recently analyzed the carotenoid contents of several mutants of *Chlamydomonas reinhardi*, such a comparison should be feasible in this species. Finally, it should be noted that poor correspondence between action spectra and absorption spectra might result from the presence in the cell of light-absorbing structures which "shade" the photoreceptor. A hypothesis incorporating this notion is presented in the following section.

H. The Location of the Photoreceptor and the Mechanism of Phototactic Orientation

Phototactic sensitivity in *Euglena* is localized in the anterior part of the cell (Engelmann, 1882). The prevailing view at present is that the photoreceptor is not in the stigma, but in the paraflagellar body; the stigma is thought by many workers to act as a shading body, enabling the organism to detect the direction of light according to the following hypothesis[2] (advanced in its original form by Mast, 1911; see Fig. 6).

Euglena rotates about its longitudinal axis as it swims. If the cell is unilaterally illuminated, the stigma intermittently casts a shadow on the paraflagellar body, producing a signal which causes the cell to turn. The signal stops only when the organism is swimming directly toward or away from the light source. The cell always turns toward its "dorsal" (stigma-containing) side, regardless of the direction of the light. In positive phototaxis, the turning is elicited by a sudden decrease in intensity on the paraflagellar body, whereas in negative phototaxis, it is elicited by a sudden increase in intensity. This "intermittent-shading" hypothesis thus assumes that oriented (topic) response is the result of a series of nonoriented (phobic) responses (Jennings, 1906). The following discussion will treat separately the two major issues involved in the intermittent-shading hypothesis: 1. the location of the photoreceptor; and 2. the mechanism for detecting light direction.

1. Location of the Photoreceptor; the Stigma vs. the Paraflagellar Body

The stigma almost certainly plays some role in phototaxis, since its absorption spectrum resembles the action spectrum for phototaxis and since it is commonly present in phototactic algae, but absent in their nonphototactic relatives (Pringsheim, 1963). In the colonial flagellate *Volvox*, only the stigma-containing (vegetative) cells react to light, those with the largest stigmas reacting most strongly (Gerisch, 1959).

---

[2] This interpretation of the hypothesis is based specifically on the assumption that the paraflagellar body is the photoreceptor and the stigma is the major shading device; for a discussion of other interpretations, see Section H 1 (p. 388/389).

On the other hand, there is no doubt that phototaxis can occur in the absence of a stigma. Phototactic response has been observed in species which never have a stigma — e.g., in certain dinoflagellates (HALLDAL, 1958), and in the chlorophyll-free *Chilomonas* (LUNTZ, 1931) — as well as in mutant forms which have lost the

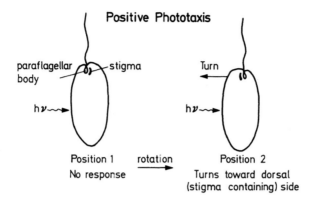

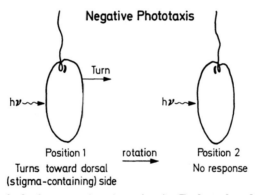

Fig. 6. Possible method of topotactic orientation in *Euglena*, based on the intermittent-shading hypothesis proposed by MAST (1911), with paraflagellar body as photoreceptor and stigma as shading device

stigma — e.g., in a green stigma-less mutant of *Chlamydomonas* (HARTSHORNE, 1953), and in chlorophyll-free stigma-less mutants of *Euglena* (GÖSSEL, 1957; DIEHN and TOLLIN, 1966b). From this we must conclude either that the stigma is not the site of the photoreceptor or that it is one of several sites. The latter view is consistent with the observation that pigmented droplets resembling stigma granules are found in other parts of the cell. It would be important to determine whether these droplets are retained in the stigma-less mutants.

In *Euglena*, the paraflagellar body is an appealing candidate for the site of the phototactic receptor, since it is directly attached to the effector organelle. As evidence that the photoreceptor resides in this structure, many investigators cite the observation of ENGELMANN (1882) that *Euglena* undergoes a shock reaction as soon as the clear protoplasm at the front end of the cell is shaded, even while the stigma is still in the light. Since the distance between the stigma and the paraflagellar body appears to be on the order of 1 μ or less, it is questionable whether Engelmann could shade the one structure without the other. Somewhat more convincing is Mast's (1911) observation that a positively topotactic *Euglena* undergoes a shock reaction only when the stigma is on the side facing the light, i.e., when the stigma is interposed between the light and the base of the flagellum.

On the basis of the following studies, Bünning and his co-workers concluded that the photoreceptor is in the paraflagellar body, and that the stigma operates only in positive phototaxis, as a shading device.

a) Bünning and Schneiderhöhn (1956) claimed that in their action spectra for green *Euglena*, positive phototaxis had its peak at 495 nm, whereas negative response peaked at 415 nm.

b) A chlorophyll-free mutant of *Euglena* which had both the stigma and the paraflagellar body showed only negative phototaxis, with its peak sensitivity at 410 nm (Gössel, 1957).

c) A second chlorophyll-free mutant, which lacked the stigma but retained the paraflagellar body, was positively phototactic over a small range of intensities, and otherwise negative; the peak for both responses lay at 410 nm.

d) *Astasia*, a colorless relative of *Euglena* which lacks both the stigma and the paraflagellar body, showed no phototaxis.

Gössel (1957) argued that since phototaxis can occur in the absence of the stigma but not in the absence of the paraflagellar body, the latter must be the photoreceptor. She noted that the peak of the action spectrum for phototaxis (mostly negative) in the chlorophyll-free mutants and for negative phototaxis in green *Euglena* occurred at about 410 nm, and surmised that this is the peak of the absorption spectrum of the paraflagellar body. The peak at 495 nm for positive phototaxis in green *Euglena* was believed to represent the joint absorption of the paraflagellar body and the stigma. Some of the evidence for this view is open to question; for example, (a) Bünning and Gössel later (1959) showed that under appropriate conditions, the action spectrum for negative phototaxis had its peak not at 410 to 415 nm but at 495 nm. (b) The response of the chlorophyll-free *Euglena* with stigma was negative only; yet Gössel concluded that the stigma acts solely in positive phototaxis. Based on these results of Gössel and on later studies by Diehn and Tollin (1966 b), it seems possible that the negative response in these euglenae is not correlated with the absence of a stigma, but rather with the absence of fully developed chloroplasts.

The only compelling evidence that the paraflagellar body is the photoreceptor in euglenoids is the observation that *Astasia*, which lacks this structure, shows no phototaxis. (It must be noted, however, that *Astasia* also lacks chloroplasts, a stigma, and possibly other components present in its relative *Euglena*.) We feel that it is premature for investigators to refer to the paraflagellar body of *Euglena* as the "photoreceptor" (as many of them have done) since the evidence for this view is inconclusive. It must at least be demonstrated that the (apparently colorless) paraflagellar body contains a pigment whose spectral characteristics are consistent with the action spectra for phototaxis.

For noneuglenoid algae such as *Chlamydomonas*, the intermittent-shading hypothesis would have to be modified. In these cells, there is no organelle resembling the paraflagellar body, and the stigma is relatively far from the flagellar base. WALNE and ARNOTT (1967) have proposed that the clear region of cytoplasm lateral to the stigma is a possible site for the photoreceptor. They suggest that the signal might be conducted from this site to the flagellar base by the cell membrane or by microtubules which extend between these two regions.

Another possibility is that cellular components other than the stigma can function as shading bodies (e.g., the chloroplasts). It has been noted that in algae which lack chloroplasts and stigma, the peak of the action spectrum tends to be shifted toward shorter wavelengths — e.g., 366 nm for *Chilomonas* (LUNTZ, 1931) and 410 nm for mutant euglenae (GÖSSEL, 1957). This shift might occur because the shading is done by violet-absorbing structures such as the cell wall and the cytoplasm (HALLDAL, 1958). Where the stigma is present, it may function as a fine adjustment for orientation; this is consistent with the finding that orientation in stigma-less algae is less precise than in forms with a stigma (HARTSHORNE, 1953; HALLDAL, 1958).

Modifying the intermittent-shading hypothesis by changing the identity of the photoreceptor or the shading body does not challenge its basic premises. A more general form of this hypothesis omits specification of particular sites for these components, but stipulates that at least one of them be localized asymmetrically in the cell. This more general version is evaluated in the following section.

2. A Consideration of Directional "Perception"

Assuming a photosensitive system, it seems to us that the one *necessary* condition for the "perception" of light direction, is that there be a "detectable" difference in the amount of light absorbed by the photoreceptor, depending on the light direction (see also HAUPT, 1966). Let us consider that the photoreceptor is uniformly distributed within a transparent subcompartment of a system, and that light is incident on the system from one direction (unilaterally). In order for a reaction to occur, the photoreceptor molecules must absorb some of the incident light. In a unilateral beam, more light will *necessarily* be absorbed by the photo-receptor molecules in the half of the subcompartment toward the light source than in the half on the other side (neglecting "lens effects" such as those in *Phycomyces*; see CASTLE, 1933). Therefore, there must be a gradient of photochemical action across the subcompartment. If there is a mechanism for detecting the difference in the two halves, the direction of the light will be "perceived." In essence, the direction of light is detected by comparing the light absorbed in two regions of the organism at one point in time.

The above is apparently the means of directional "perception" in most, if not all, phototropic systems, where the sensitive system is essentially stationary with respect to the light at the onset of stimulation. In very small systems, however, the light gradient across the system is undoubtedly very small, and it is difficult to imagine mechanisms capable of detecting the difference. One might argue that this is why *Rhodospirillum* does not have a directional response. (Note that in this case the subcompartment is probably the whole cell.) In small systems where the photoreceptor is distinctly localized (as in most topotactic systems), it is even more unlikely that a gradient can be detected. If there were two or more subcompartments (analagous to a "micro-retina"), the existence of a detectable light gradient would seem more likely, especially if a strongly absorbing "screen" existed between the subcompartments. Although generally ignored in considerations of topotaxis, this possibility is not ruled out by any evidence at hand, and it must be kept in mind.

There is, however, a fundamentally different means of directional "perception" by a free-moving system. The requirement here is that the system assume at least two different orientations with respect to the light at two different times, and that there be a means of detecting the rate of light-absorption by the photoreceptor at any time. If there is a detectable difference, the direction of the light is "perceived;" thus, light direction can be detected by comparing the light absorbed in one region of the organism at two points in time.

There are at least three ways in which such a difference may arise. If the photoreceptor sub-compartment is not spherically (completely) symmetrical, the orientation of the sub-compartment itself may lead to a difference. In *Chlamydomonas*, for instance, we could suppose that the platelike stigma is the photoreceptor; this organelle will obviously absorb very different amounts of light, depending on the orientation of its plane with respect to the light. Secondly, the photoreceptor molecules may be anisotropic or organizationally birefringent, so that (polarized) light may be preferentially absorbed from a particular direction. There is, in fact, evidence of birefringence in the stigma (WALNE and ARNOTT, 1967), and it has been reported that phototactic response varies with the polarization of the stimulus (WOLKEN and SHIN, 1958; BOUND and TOLLIN, 1967). However, since the systems do respond (directionally) to unpolarized light, orientation cannot *depend* on the birefringent properties of the system.

Finally, the difference may arise in a very simple way, if there is light absorption by components other than the photoreceptor, and if the photoreceptor is asymmetrically localized within the system. This is, of course, exactly what is envisioned in the intermittent-shading hypothesis. Using *Chlamydomonas* again as an example, we could suppose that the photoreceptor is just peripheral to the stigma (see WALNE and ARNOTT, Section V above). Clearly, the photoreceptor would absorb *very* different amounts of light, depending on the orientation of the cell with respect to the light. Light absorption by the stigma *and* the chloroplast would contribute to this difference. On this model, light direction might be established by a very small angular change in cell orientation; if the cell were traveling directly toward the light, a very small deviation would be "detectable."

From these considerations we see that the Mast version of the intermittent-shading hypothesis is only one of several reasonable alternatives, and it is not the simplest one. An alga does not have to rotate in order for shading to occur; it must only show random deviations from a straight path. Although most phototactic flagellates do rotate, phototaxis occurs in some which do not (STAHL, 1880; RINGO, 1967). A necessary stipulation of the Mast hypothesis (Fig. 6) is that turning always occurs toward the same side of the organism; otherwise the proposed mechanism aligns the cell parallel with the light beam, but fails to specify whether the cell is directed toward or away from the light source. Apparently, a free-swimming *Euglena* does always turn toward its "dorsal" (stigma-containing) side in a phobic response (JENNINGS, 1906; MAST, 1911), but this may not be true for other organisms. If the model takes into account light-absorbing components located *behind* the photoreceptor, the specificity of turning is no longer required. For instance, if shading by the chloroplasts generated a "turn" signal, a *Euglena* would eventually head toward the light, even if its turns were directed randomly (HALLDAL, 1964).

At the crux of the intermittent-shading hypothesis (in any version) is the assumption that topic response results from one or more phobic responses, initiated by changes in light intensity. The main support for this interpretation is the demonstrable fact that phototactic algae *can* respond to temporal changes in light intensity. In certain algae, it is clear that orientation results from phobic responses (e.g., *Ceratium*; METZNER, 1920), but most cases are still debated. BANCROFT (1913) argued that topotaxis in *Euglena* occurs so swiftly and smoothly that it cannot depend on nonoriented movements. He also reported that in some cultures topotaxis was negative in the same intensity range where phobotaxis was positive. (The cells were scored as positively phobotactic if they displayed a shock reaction when a screen was interposed between them and the stimulus light.)

MAST (1914) criticized these experiments of BANCROFT, pointing out that the shading produced by putting a screen in front of the stimulus light cannot be compared quantitatively with the internal shading resulting from the rotation of the cell. GERISCH (1959) has overcome these difficulties in his recent work on *Volvox*. He shaded the colonies intermittently and found that negative phobic responses (of the individual cells) do occur at the same light intensities as negative topic responses (of the whole colony) if the periods of darkness are very short, lasting about half as long as a single rotation of the colony. Under these conditions, the shading imposed by the experimenter is presumed to be equal in duration to the shading that normally occurs within the organism. Complications arise in trying to apply these observations to *Euglena*. In Volvox, the phobic response consists of a decrease in the frequency of flagellar movement. At intensities in the range for positive topotaxis, an *increase* in intensity elicits the phobic response, whereas at intensities in the negative range, a *decrease* in intensity is the effective stimulus; this is just the opposite of the situation in *Euglena*. GERISCH maintains that his results can best be explained by postulating that the stigma itself contains the photoreceptor. Although these results of GERISCH do indicate that topotaxis in *Volvox* can be resolved into a series of phobic responses, the details are difficult to interpret. We feel that the intermittent-shading hypothesis (in its more general form) is the most attractive one for explaining topotactic orientation, at least in *Euglena*, but it is by no means proven.[3]

The mechanism of phototaxis has been only partly elucidated, yet many intriguing features are already apparent. The phototactic system may well represent the simplest sort of receptor-effector sequence and should offer an excellent opportunity for a thorough analysis from input to output. With the new techniques available for measuring phototaxis, the system can now be examined in detail.

---

[3] There is, in fact, recent evidence that this is not the mechanism of orientation in the dinoflagellate *Gyrodinium* (HAND and DAVENPORT, 1970). At the onset of stimulation, this organism comes to an almost immediate stop. Then, in a single rapid movement, it realigns itself so that its anterior end faces directly toward the stimulus source. This observation suggests that light direction is, in this case, detected by comparing the light absorbed in two regions of the organism at one point in time.

References

Bancroft, F. W.: Heliotropism, differential sensibility, and galvanotropism in *Euglena*. J. Exp. Zool. **15**, 383—428 (1913).

Batra, P. P., Tollin, G.: Phototaxis in *Euglena*. I. Isolation of the eye-spot granules and identification of the eye-spot pigments. Biochim. Biophys. Acta **79**, 371—378 (1964).

Bendix, S.: Phototaxis. Botan. Rev. **26**, 145—208 (1960).

Blaauw, A. H.: Licht und Wachstum. (I) Z. Botan. **6**, 641—703 (1914).

— Licht und Wachstum. (II) Z. Botan. **7**, 465—532 (1915).

Bound, K., Tollin, G.: Phototactic response of *Euglena gracilis* to polarized light. Nature (Lond.) **216**, 1042—1044 (1967).

Briggs, W. R.: Light dosage and the phototropic responses of corn and oat coleoptiles. Plant Physiol. **35**, 951—962 (1960).

Brokaw, C. J.: Flagella. In: Physiology and biochemistry of algae. New York: Academic Press 1962.

Bruce, V. G., Pittendrigh, C. S.: Temperature independence of a unicellular "clock." Proc. nat. Acad. Sci., (Wash.) **42**, 676—682 (1956).

— — Resetting the *Euglena* clock with a single light stimulus. Am. Naturalist **92**, 295—306 (1958).

Brucker, W.: Beiträge zur Kenntnis der Phototaxis grüner Schwärmzellen. I. Die Lichtempfindlichkeit von *Lepocinclis texta* und ihre Abhängigkeit von der Vorbelichtung und vom Kohlensäuregehalt des Mediums. Arch. Protistenk. **99**, 294—327 (1954).

Buder, J.: Zur Kenntnis der phototaktischen Richtungsbewegungen. Jahrb. Wiss. Botan. **58**, 105—220 (1917).

Bünning, E., Gössel, I.: Ergänzende Versuche über die phototaktischen Aktionsspektren von *Euglena*. Arch. Mikrobiol. **32**, 310—321 (1959).

— Schneiderhöhn, G.: Über das Aktionsspektrum der phototaktischen Reaktionen von *Euglena*. Arch. Mikrobiol. **24**, 80—90 (1956).

— Tazawa, M.: Über die negative-phototaktische Reaktion von *Euglena*. Arch. Mikrobiol. **27**, 306—310 (1957).

Castle, E. S.: The physical basis of the positive phototropism of *Phycomyces*. J. gen. Physiol. **17**, 49—62 (1933).

— A kinetic model for adaptation and the light responses of *Phycomyces*. J. gen. Physiol. **49**, 925—935 (1966).

Clayton, R. K.: Studies in the phototaxis of *Rhodospirillum rubrum*. I. Action spectrum, growth in green light, and Weber Law adherence. Arch. Mikrobiol. **19**, 107—124 (1953a).

— Studies in the phototaxis of *Rhodospirillum rubrum*. III. Quantitative relations between stimulus and response. Arch. Mikrobiol. **19**, 141—165 (1953b).

— Phototaxis of purple bacteria. In: Handbuch der Pflanzenphysiologie, 17. Aufl. I. Teil. Berlin-Heidelberg-New York: Springer 1959.

— Phototaxis in microorganisms. In: Photophysiology, Vol. II, 51—77. New York: Academic Press 1964.

Curry, G. M.: Studies on the spectral sensitivity of phototropism. Ph. D. Thesis. Harvard University. Cambridge, Massachusetts (1957).

Davenport, D., Wright, C. A., Causley, D.: Technique for the study of the behavior of motile microorganisms. Science **125**, 1059—1060 (1962).

Delbrück, M., Reichardt, W.: I. System analysis for the light-growth reactions of *Phycomyces*. In: Cellular mechanisms in differentiation and growth. Princeton: Princeton Univ. Press 1956.

Dennison, D. S.: Steady-state phototropism in *Phycomyces*. J. gen. Physiol. **48**, 393—408 (1965).

Desroche, P.: Reactions des *Chlamydomonas* aux agents physiques. Etude de physiologie cellulaire. Paris: Schulz Libraire 1912.

Diehn, B., Tollin, G.: Phototaxis in *Euglena*. II. Physical factors determining the rate of phototactic response. Photochem. Photobiol. **5**, 523—535 (1966a).

— — Phototaxis in *Euglena*. III. Lag phenomena and the overall mechanism of the tactic response to light. Photochem. Photobiol. **5**, 839—844 (1966b).

DIEHN, B., TOLLIN, G.: Phototaxis in *Euglena*. IV. Effect of inhibitors of oxidative and photophosphorylation on the rate of phototaxis. Arch. Biochem. **121**, 169—177 (1967).

ENGELMANN, T. W.: Über Licht- und Farbenperception niederster Organismen. Pflügers Arch. **29**, 387—400 (1882).

— *Bacterium photometricum*. Ein Beitrag zur vergleichenden Physiologie des Licht- und Farbensinnes. Pflügers Arch. ges. Physiol. **30**, 95—124 (1883).

FEINLEIB, M. E.: Studies on phototaxis in *Chlamydomonas reinhardi*. Ph. D. Thesis. Harvard University. Cambridge, Massachusetts 1965.

— CURRY, G. M.: Methods for measuring phototaxis of cell populations and individual. cells Physiol. Plantarum **20**, 1083—1095 (1967).

FRAENKEL, G. S., GUNN, D. L.: The orientation of animals. New York: Dover Public., Inc. 1961.

GERISCH, G.: Die Zelldifferenzierung bei *Pleodorina californica* Shaw und die Organisation der Phytomonadinenkolonien. Arch. Protistenk. **104**, 292—358 (1959).

GIBBS, S. P.: The fine structure of *Euglena gracilis* with special reference to the chloroplasts and pyrenoids. J. Ultrastruct. Res. **4**, 127—148 (1960).

GÖSSEL, I.: Über das Aktionsspektrum der Phototaxis chlorophyllfreier Euglenen und über die Absorption des Augenflecks. Arch. Mikrobiol. **27**, 288—305 (1957).

HALLDAL, P.: Importance of calcium and magnesium ions in phototaxis of motile green algae. Nature (Lond.) **179**, 215—216 (1957).

— Action spectra of phototaxis and related problems in Volvocales, Ulva-gametes and Dinophyceae. Physiol. Plantarum **11**, 118—153 (1958).

— Factors affecting light response in phototactic algae. Physiol. Plantarum **12**, 742—752 (1959).

— Action spectra of induced phototactic response changes in *Platymonas*. Physiol. Plantarum **13**, 726—735 (1960).

— Ultraviolet action spectra of positive and negative phototaxis in *Platymonas subcordiformis*. Physiol. Plantarum **14**, 133—139 (1961).

— Taxes. In: Physiology and biochemistry of algae. New York: Academic Press 1962.

— Phototaxis in protozoa. In: Biochemistry and physiology of protozoa, Vol. 3. New York: Academic Press 1964.

— Ultraviolet action spectra in algology. A review. Photochem. Photobiol. **6**, 445—460 (1967).

HAND, W. G., DAVENPORT, D.: The experimental analysis of phototaxis and photokinesis in flagellates. In: Photophysiology of microorganisms. (P. HALLDAL, Ed.). London: Wiley 1970. In press.

— FORWARD, R., DAVENPORT, D.: Short-term photic regulation of a receptor mechanism in a dinoflagellate. Biol. Bull. **133**, 150—165 (1967).

HARTSHORNE, J. N.: The function of the eyespot in *Chlamydomonas*. New Phytologist **52**, 292—297 (1953).

HAUPT, W.: Die Phototaxis der Algen. In: Handbuch der Pflanzenphysiologie, 17. Aufl., I. Teil.. Berlin-Göttingen-Heidelberg: Springer 1959.

— Phototaxis in plants. Intern. Rev. Cytol. **19**, 267—299 (1966).

HAXO, F. T., CLENDENNING, K. A.: Photosynthesis and phototaxis in *Ulva lactuca* gametes. Biol. Bull. **105**, 103—114 (1953).

HECHT, S.: Vision. II. The nature of the photoreceptor process. In: A handbook of general experimental psychology. Worcester, Mass.: Clark University Press 1934.

— A theory of visual intensity discrimination. J. gen. Physiol. **18**, 767—789 (1935).

JAHN, T. L., BOVEE, E. C.: Motile behavior of protozoa. In: Research in protozoology, Vol. I. Oxford: Pergamon Press 1966.

JENNINGS, H. S.: Behavior of the lower organisms. New York: Columbia University Press 1906.

KRINSKY, N. I., GOLDSMITH, T. H.: The carotenoids of the flagellated alga, *Euglena gracilis*. Arch. Biochem. **91**, 271—279 (1960).

— LEVINE, R. P.: Carotenoids of wild type and mutant strains of the green alga, *Chlamydomonas reinhardi*. Plant Physiol. **39**, 680—687 (1964).

Laurens, H., Hooker, H. D.: Studies on the relative physiological value of spectral lights. II. The sensibility of *Volvox* to wavelengths of equal content. J. Protozool. **30**, 345—368 (1920).

Leedale, G. F.: Euglenoid flagellates. Englewood Cliffs, N. J.: Prentice-Hall, Inc. 1967.

Lembi, C. A., Lang, N. J.: Electron microscopy of *Carteria* and *Chlamydomonas*. Amer. J. Botany **52**, 464—477 (1965).

Levine, R. P., Ebersold, W. T.: The genetics and cytology of *Chlamydomonas*. Ann. Rev. Microbiol. **14**, 197—216 (1960).

Lindes, D., Diehn, D., Tollin, G.: Phototaxigraph; recording instrument for determination of rate of response of phototactic microorganisms to light of controlled intensity and wavelength. Rev. Sci. Instrum. **36**, 1721—1725 (1965).

Links, J.: I. A hypothesis for the mechanism of (phobo-) chemotaxis. II. The carotenoids, steroids, and fatty acids of *Polytoma uvella*. Leiden: Diss. 1955. [Quoted in Clayton (1959, 1964).]

Luntz, A.: Untersuchungen über die Phototaxis. I. Mitteilung: Die absoluten Schwellenwerte und die relative Wirksamkeit von Spektralfarben bei grünen und farblosen Einzelligen. vergl. Physiol. **14**, 68—92 (1931).

Mainx, F.: Untersuchungen über den Einfluß von Außenfaktoren auf die phototaktische Stimmung. Arch. Protistenk. **68**, 105—176 (1929).

— Wolf, H.: Reaktionsintensität und Stimmungsänderung in ihrer Bedeutung für eine Theorie der Phototaxis. Arch. Protistenk. **93**, 105—120 (1939).

Mast, S. O.: Light and the behavior of organisms. New York: John Wiley & Sons, Inc. 1911.

— Orientation in *Euglena* with some remarks on tropisms. Biol. Zbl. **34**, 641—664 (1914).

— The relation between spectral color and stimulation in the lower organisms. J. exp. Zool. **22**, 471—528 (1917).

— Gover, M.: Relation between intensity of light and rate of locomotion in *Phacus pleuronectes* and *Euglena gracilis* and its bearing on orientation. Biol. Bull. **43**, 203—209 (1922).

Mayer, A. M.: *Chlamydomonas*: Adaptation phenomena in phototaxis. Nature (Lond.) **217**, 875—876 (1968).

— Poljakoff-Mayber, A.: The phototactic behavior of *Chlamydomonas snowiae*. Physiol. Plantarum **12**, 8—14 (1959).

Metzner, P.: Bewegungsstudien an Peridineen. Z. Botany **22**, 225—265 (1929).

Nultsch, W.: Studien über die Phototaxis der Diatomeen. Arch. Protistenk. **101**, 1—68 (1956).

— Der Einfluß des Lichtes auf die Bewegung der Cyanophyceen. I. Mitteilung. Phototopotaxis von *Phormidium autumnale*. Planta **56**, 632—647 (1961).

— Der Einfluß des Lichtes auf die Bewegung der Cyanophyceen. III. Mitt. Photophobotaxis von *Phormidium uncinatum*. Planta **58**, 647—663 (1962).

— Richter, G.: Aktionsspektrum des photosynthetischen $^{14}CO_2$-Einbaus von *Phormidium uncinatum*. Arch. Mikrobiol. **47**, 207—213 (1963).

Pfeffer, W.: Pflanzenphysiologie, 2. Aufl., III. Teil. Engl. transl. A. J. Ewart. Oxford: Clarendon Press 1903.

Pringsheim, E. G.: Farblose Algen. Stuttgart: Gustav Fischer 1963.

Reichardt, W., Varju, D.: Eine Inversionsphase der phototropischen Reaktion (Experimente an dem Pilz *Phycomyces blakesleeanus*). Z. Physik. Chem. **15**, 297—320 (1958).

Ringo, D. L.: Flagellar motion and fine structure of the flagellar apparatus in *Chlamydomonas*. J. Cell Biol. **33**, 543—571 (1967).

Sachs, T., Mayer, A. M.: The relationship between metabolism and phototaxis in *Chlamydomonas snowiae*. Phycologia 1, 149—159 (1961).

Sager, R., Palade, G. E.: Structure and development of the chloroplast in *Chlamydomonas*. I. The normal green cell. J. Biophys. Biochem. Cytol. **3**, 463—488 (1957).

Shropshire, W., Jr.: Photoresponse of the fungus, *Phycomyces*. Physiol. Rev. **43**, 38—67 (1963).

— Gettens, R. H.: Light-induced concentration changes of adenosine-triphosphate in *Phycomyces* sporiangiophores. Plant Physiol. **41**, 203—207 (1966).

Stahl, E.: Über den Einfluß von Richtung und Stärke der Beleuchtung auf einige Bewegungserscheinungen im Pflanzenreiche. Bot. Z. **38**, 297—304 (1880).

STROTHER, G. K., WOLKEN, J. J.: Microspectrophotometry of *Euglena* chloroplast and eyespot. Nature (Lond.) **188**, 601—602 (1960).

THIMANN, K. V., CURRY, G. M.: Phototropism and phototaxis. In: Comparative biochemistry, Vol. I. New York: Academic Press 1960.

VÁVRA, J.: Ist der Photoreceptor eine unabhängige Organelle der Eugleniden? Arch. Mikrobiol. **25**, 223—225 (1956).

WALNE, P. L., ARNOTT, H. J.: The comparative ultrastructure and possible function of eyespots: *Euglena granulata* and *Chlamydomonas eugametos*. Planta **77**, 325—353 (1967).

WOLKEN, J. J.: *Euglena*: an experimental organism for biochemical and biophysical studies. New York: Appleton-Century-Crofts 1967.

— SHIN, E.: Photomotion in *Euglena gracilis*. 1. Photokinesis. 2. Phototaxis. J. Protozool. **5**, 39—46 (1958).

ZIMMERMAN, B. K., BRIGGS, W. R.: A kinetic model for phototropic responses of oat coleoptiles. Plant Physiol. **38**, 253—261 (1963).

Chapter 14

Sensory Transduction in Hair Cells

By

Å. Flock, Stockholm (Sweden)

With 30 Figures

Contents

A. Introduction

The neural mechanism that underlies excitation and sensory processing in the inner ear is unlike that of most other mechanoreceptor-nerve preparations in that the receptor is not a part of the sensory neuron but a specialized epithelial cell which excites the sensory neuron by synaptic transmission. The peripheral excitatory processes take place in several sequential stages, and it is difficult to localize the particular stage at which various output characteristics are contributed. This is also because hair cells are used to subserve different functions in the various organs of hearing and equilibrium. It is not intended here to describe the function of separate end-organs as separate entities but to find basic principles of structure and function of hair cells and their nervous connections on a comparative basis.

B. Basic Structure and Function of Hair Cells

The mechanoreceptor cells in the acoustico-lateralis system are all morphologically similar, and they are also homologous in an evolutionary sense. The common

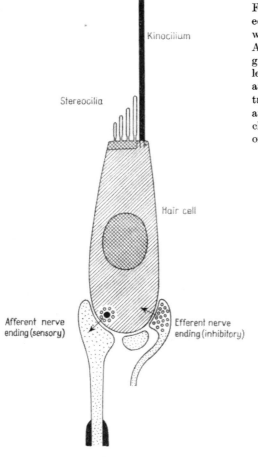

Fig. 1. The common type of hair cell is equipped with a bundle of sensory hairs which receive the mechanical stimulus. A kinocilium is placed at one side of a group of stereocilia which are of increasing length closer to the kinocilium. At the afferent synapse, sensory information is transmitted from the hair cell to the afferent nerve ending, probably by neurochemical transmission. Efferent endings of centrifugal fibers mediate presynaptic inhibition of afferent transmission

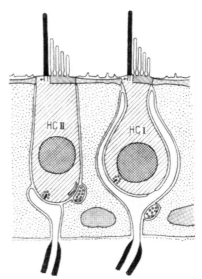

Fig. 1

Fig. 2

Fig. 2. In bird and mammalian vestibular organs, two types of hair cells exist. In addition to common-type hair cells (*HC II*) are amphora-shaped cells enclosed by a large nerve chalice (*HC I*). Efferent nerve endings which contact the nerve chalice probably provide postsynaptic inhibition

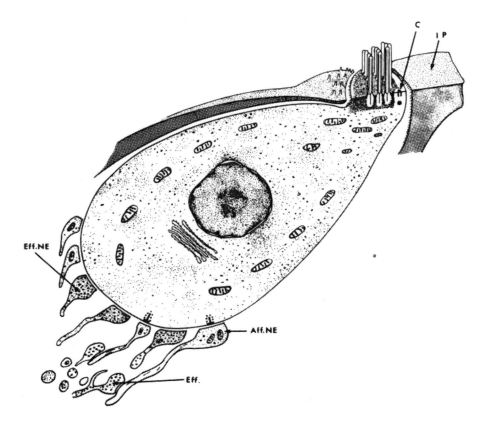

Fig. 3a

Fig. 3a and b. In inner (a) and outer (b) hair cells in the organ of Corti, the extracellular portion of the kinocilium drops off during development, leaving behind its basal body or centriole (*C*) (reproduced from WERSÄLL et al., 1965)

type of hair cell is that present in the inner ear and lateral line of fishes and amphibians and in the vestibular labyrinth of mammals. Details of ultrastructure vary between organs and from one species to another, but the general characteristics are the same (Fig. 1). The cell is cylindrical or flask-shaped and is equipped at its apical end with a bundle of sensory hairs among which one, the kinocilium, is of a different structure and is asymmetrically placed. The hair cell is innervated at its base by afferent endings of sensory nerve fibers and by one or several endings of efferent centrifugal nerve fibers.

Hair cells of higher structural differentiation are present in the inner ear of higher vertebrates. In the vestibular organs of birds and mammals, two types of hair cells are distinguished. Apart from hair cells of the common type (type II of WERSÄLL, 1956), are found amphora-shaped cells (type I hair cells, WERSÄLL, 1956) enclosed up to their neck by a large nerve chalice which may enclose more than one cell (Fig. 2). Outer and inner hair cells in the mammalian organ of Corti lack the kinocilium (Fig. 3).

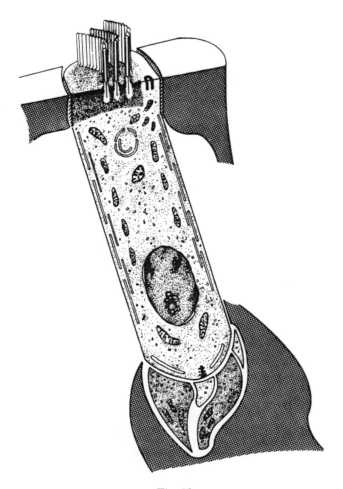

Fig. 3b

Whereas structural differentiation among hair cells is likely to reflect functional specialization, the adequate stimulus for the sensory cell seems to be the same in all organs.

C. Adequate Stimulus for the Hair Cell

In the lateral-line organs, measurements of phase relations between cupular motion and the electrical response from the organ have been done by stroboscopic observation (JIELOF et al., 1952; KUIPER, 1956) or by direct mechanical coupling to the cupula (FLOCK, 1965). The output of the organ is proportional to the degree of displacement of the cupula and not to its velocity of motion. HARRIS and VAN BERGEIJK (1962) found a similar relationship between electrical output and the near-field displacement of vibrating sources in water. In the semicircular canals, the DC response from the crista ampullaris is proportional to the angular displacement of the cupula (TRINCKER, 1957). In behavioral studies, von HOLST

(1950) has shown that the utricle responds to the vector component of gravitational force which is in a plane parallel to the surface of the sensory epithelium and produces shearing displacement of the overlying otolith. Von Békésy (1960) displaced the tectorial membrane with a needle and found the electric output from the organ of Corti to be proportional to displacement and not to its velocity component.

In all these organs, vertical push or pull applied to the cupula, the otolith or the tectorial membrane in a direction perpendicular to the surface of the sensory epithelium is not an effective stimulus; neither is increase or decrease in hydrostatic pressure (Henriksson and Gleisner, 1966) which will cause a response only if it causes distortion of the tissue and subsequent shearing deformation. Whereas the adequate stimulus for the different sense organs may differ, the adequate stimulus for the sensory cell appears to be the same, shearing displacement causing a lateral or angular displacement of the sensory hairs. The specificity to particular types of stimuli is achieved by mechanical coupling to auxiliary structures of different types as illustrated in Fig. 4–8.

D. Adequate Stimulus for the Sensory Organ

The hair cell is basically a directionally sensitive displacement detector, and works on this principle to subserve a multitude of functions. The most primitive

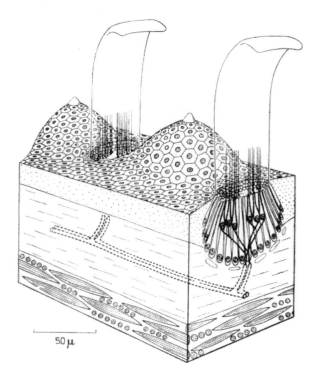

50 μ

Fig. 4. The primitive, epidermal lateral-line organs are situated in the skin with the cupula extending directly out into the surrounding water. Each organ is innervated by branches from two sensory nerve fibers (*Xenopus laevis*) (by courtesy from Görner, 1963)

of acoustico-lateralis organs is the epidermal lateral-line organ which is situated in the skin of fishes and water-living amphibians (Fig. 4). The organ is provided with a gelatinous cupula which projects into the outside water. The organ is activated by cupular displacement which is in turn proportional to some function

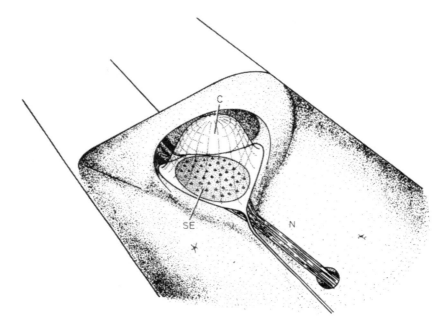

Fig. 5. Lateral-line canals are sunk down below the skin and occur on the skull and along the body of fishes. At regular intervals, sense organs occupy the bottom of the canal. The innervating nerve (N) supplies a disk-shaped sensory epithelium (SE) upon which a gelatinous cupula (C) rests (Lota lota) (from FLOCK, 1965)

of water motion, velocity or displacement, depending on the mechanical characteristics of the coupling between cupula and water. In Xenopus laevis, the flaglike cupula is flattened from two sides and the organ is sensitive to displacement of the cupula along the major axis of its profile. In this direction water motion meets the narrow cupula "edge on" and mainly couples to its flat sides by viscous drag. The resulting cupular displacement, and hence the response of the organ, is consequently a function of water velocity (GÖRNER, 1963; HARRIS and MILNE, 1966).

The closely related lateral-line canal organs in fishes are located in canals sunk down below the skin. These canals are filled with a viscous fluid which communicates with the outside water through narrow pores. A dome-shaped cupula rests on a disk of sensory epithelium and partly occludes the cross-section of the canal (Fig. 5). The organ behaves as a critically damped harmonic oscillator (VRIES, 1956) sensitive to water displacement caused by local water current (DIJKGRAAF, 1963) or by vibrating sources (HARRIS and VAN BERGEIJK, 1962).

The semicircular canals serve as detectors of angular acceleration in the three planes of space. The cupula which rides on the sensory crest in each ampulla responds as a swing-door to volume displacement of canal fluid (Fig. 6). In the utricle and saccule, the hair cells are coupled to a gelatinous membrane containing crystals of calciumcarbonate; this loading by inertial mass make the organs sensitive to linear acceleration and changes of position in the gravitational field (Fig. 7).

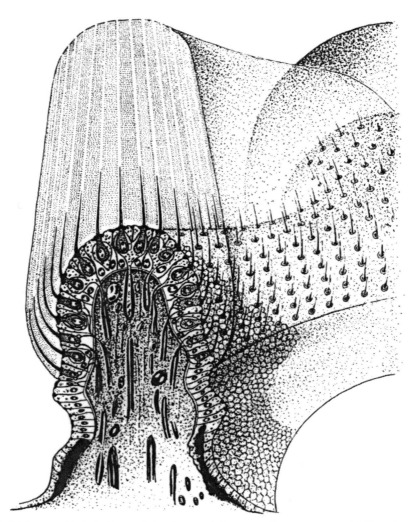

Fig. 6. Crista ampullaris in the semicircular canals is covered by a sensory epithelium housing the receptor cells. The sensory hairs attach to the cupula which acts as a swing door to volume displacement of endolymph (guinea pig) (by courtesy from Wersäll, 1965)

Hair cells serve in tone perception in the organ of Corti in birds and mammals, in the amphibian and basilar papilla in frogs, and in the saccule of ostariophys fishes. In the organ of Corti (Fig. 8), sound frequency is analyzed on the place principle along a vibrating membrane (Békésy, 1960), whereas in the fish saccule other mechanisms are probably involved.

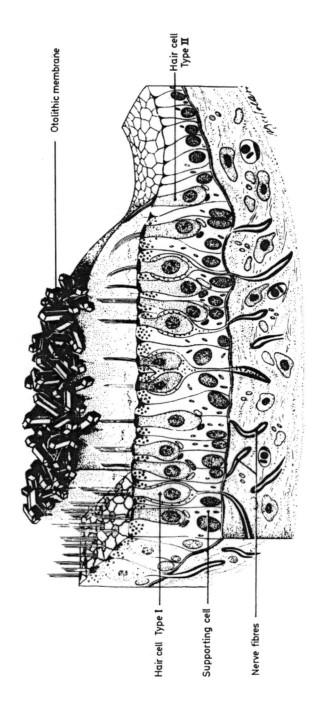

Fig. 7. The sensory epithelium in the utricle and saccule is covered by a gelatinous membrane containing crystals of calcium carbonate. This loading by mass makes the organs sensitive to gravitational forces and linear acceleration (guinea pig) (by courtesy from IURATO, 1967)

These are examples of mechanical systems which provide an input to hair cells. Mechanical energy is transmitted to the hair cell by the sensory hairs via their coupling to these auxiliary structures. The input-output relationship will be influenced also by the mechanical characteristics of this coupling.

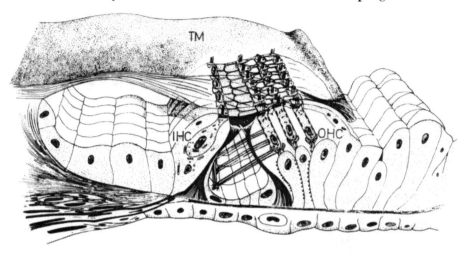

Fig. 8. Three rows of outer hair cells (*OHC*) and one row of inner hair cells (*IHC*) contact the tectorial membrane (*TM*) in the mammalian organ of Corti (from Wersäll et al., 1965)

E. The Sensory Hairs and their Attachment to Auxiliary Structures

1. Sensory Hair Structure

From the apical end of each receptor cell, a bundle of sensory hairs protrude (Figs. 1—3). In lateral-line and vestibular hair cells, each bundle is composed of 30 to 100 stereocilia arranged in a hexagonal pattern behind an asymmetrically located kinocilium (Fig. 9) whereas in cochlear hair cells the stereocilia are differently spaced and the kinocilium is lacking (Fig. 10). The stereocilia consist of a central core of cytoplasm surrounded by a plasma membrane which is continuous with that of the cell. The core contains an assembly of fine parallel fibrils, 30 to 40 A in diameter, which join and extend into the hair cell as a rootlet planted in a granular cuticular plate (Fig. 11). A single dense-core fiber with a central lumen is seen in stereocilia of the mammalian organ of Corti. Systematic variation in the number of cilia and in their arrangement occurs along the coils of the mammalian cochlea (Hawkins, 1965). The diameter of stereocilia varies from one organ to another and between species, ranging from 0.1 to 0.3 µ. Differences in diameter and length do not seem to have functional correlates. The height of the hair bundle is fairly constant at 4 to 10 µ, with the exception of the crista ampullaris where it reaches 40 µ (Wersäll, 1956). Within each bundle, the length of the stereocilia increases stepwise toward the kinocilium (Figs. 1—3). This staircase arrangement of ciliary length is present in all organs. This arrangement was already observed by light microscopists in the nineteenth century (Retzius, 1884) and has received recent attention as one expression of the morphological polarization of the hair cell which is characterized by the asymmetric position of the kinocilium in one end of the sensory hair bundle. This organization may depend on the role of the kinocilium, or the centriole from which it presumably evolves, as a center of development of the hair cells apical end. In the

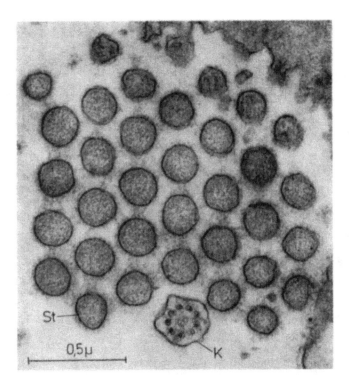

Fig. 9. The hexagonal arrangement of stereocilia (*St*) behind the kinocilium (*K*) is seen in this cross-section through the sensory hair bundle of a common hair cell (lateral-line canal organ)

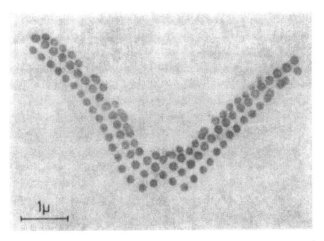

Fig. 10. The stereocilia of outer hair cells in the organ of Corti are spaced in a W pattern

developing organ of Corti (KIKUCHI and HILDING, 1965) and in the fowl embryo otocyst (FRIEDMAN, 1968), immature hair cells possess a rudimentary kinocilium behind which a group of microvilli rise, those close to the kinocilium growing taller than those at the opposite side of the bundle. In the mature mammalian

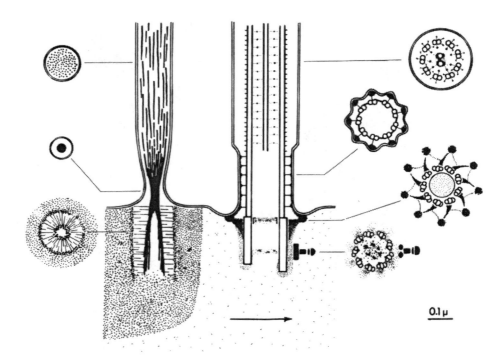

Fig. 11. Schematic drawing illustrating the ultrastructure of the stereocilium with its rootlet and the kinocilium with its basal body. The arrow indicates the direction of excitatory stimulation (from FLOCK, 1965)

organ of Corti, the "extracellular" portion of the kinocilium drops off, leaving behind its intracellular portion, the basal body or centriole. In other acoustico-lateralis organs, the kinocilium is retained. The cytoplasmic core of the kinocilium contains a central pair of tubules surrounded by nine peripheral double-barrelled tubules which connect to triplet tubules comprising the wall of the basal body (Fig. 11).

2. Attachment of Sensory Hairs

The apical ends of the hair cells are locked in a reticular membrane composed of a framework of supporting cells. This surface is rigid to lateral displacements, whereas, at least in the cochlea, it is flexible in the vertical plane. The cupula, otolith membrane or tectorial membrane rests above the sensory epithelium from which it is separated by a low-friction interface, the subcupular or subtectorial space. The cupula and the tectorial membrane were long thought to be composed of acid mucopolysaccharides. Recent chemical analysis by IURATO (1967) shows, however, that the principal constituent is a protein, distinct from collagen or elastin, and that hexuronic acids which are essential components of mucopoly-saccharides are absent. Whereas the chemical composition of the tectorial membrane and the cupula appears to be the same, the submicroscopic organization of the material and the manner of attachment to the sensory hair bundles varies (Fig. 12). The material of the cupula and the otolith membrane is a three-dimensional network of fine fibrils about 30 to 40 A wide, that of the tectorial membrane is a dense amorphous substance, traversed in mammals by oriented filaments.

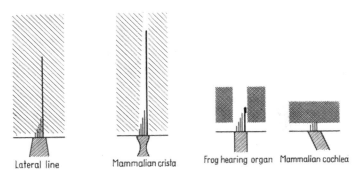

Lateral line   Mammalian crista   Frog hearing organ Mammalian cochlea

Fig. 12. The mode of attachment of the sensory hairs to the auxiliary structure varies between organs. These differences are likely to provide differences in mechanical coupling characteristics relevant in each organ for transmission of adequate stimulus parameters

In lateral-line canal organs, the cupular fibrils interweave between the sensory hairs and attach to their plasma membrane, as seen in Fig. 9. In the mammalian vestibular labyrinth, the sensory hair bundles are lodged in canals in the cupula (WERSÄLL, 1956). In auditory organs, the manner of attachment is different; in the frog amphibian and basilar papilla, the sensory hair bundles project into canals in the tectorial membrane, the subtectorial space being traversed by fine fibrils which attach to microvilli of the supporting cells (Fig. 12). The distal end of the kinocilium possesses a bulbous swelling which contains an amorphous substance. This part of the kinocilium and the first row of stereocilia are in close proximity to the wall of the tectorial membrane canal. Consecutive rows of stereocilia behind the kinocilium attach to one another and to the leading row by fine fibrils. A similar situation exists in the bird organ of Corti (JAHNKE et al., 1968). In the mammalian organ of Corti, the tallest rows of sensory hairs of the outer hair cells attach to the flat lower surface of the tectorial membrane into which the individual stereocilia make imprints (KIMURA, 1966; BREDBERG, 1968). An amorphous substance is occasionally seen to bridge between consecutive rows of sensory hairs in the bundle (KIMURA, 1966).

3. Force Transmission

Little is known about the mechanics of transmission between cupula — tectorial membrane and hair cell. The cupula of lateral-line canal organs is free to slide by shearing displacement on the low-friction subcupular space for a distance no longer than 10 μ (FLOCK, 1967); microphonic potentials saturate at displacement amplitudes of 5 to 10 μ when the cupula is driven by direct mechanical coupling close to the hair cells. In the natural situation, however, displacement amplitudes have been reported not to exceed 1 μ (KUIPER, 1956). At large amplitudes motion is restrained by the attachment of the circumference of the cupula to peripheral supporting cells. A similar function, i.e., to act as strain belts protecting the hair cells from damage at high-intensity stimulation, may be attributed to the subtectorial fibrils in the amphibian papilla and to the attachment of the outer margin of the tectorial membrane in the mammalian organ of Corti (BÉKÉSY, 1960).

During stimulation, lateral force is thus transmitted to the sensory hair bundle. The resulting displacement will be distributed to the sensory hairs and to the apical end of the hair cell according to the distribution of elastic, frictional and inertial elements, and it will give rise to deforming strain of some critical point. Stereocilia appear to be rather stiff whereas the kinocilium is more flexible. ENGSTRÖM et al. (1962) observed that inner and outer hair cell stereocilia do not bend but fan out like stiff rods when a cover glass is pressed against them in the fresh organ of Corti where the tectorial membrane is stripped off. In the crista ampullaris of the guinea pig, the proximal part of the stereocilia is stiff and the distal part is flexible (WERSÄLL, 1956).

If the sensory hairs act as stiff levers, deformation would occur at their base; bending of the hair would cause deformation of the plasma membrane which constitutes the wall of the cilia. If the cuticular plate is laterally displaced with the stereocilia, deformation of the apical cell membrane would occur with concomitant displacement of cytoplasm and cytoplasmic organelles. One cannot choose between these alternatives until the cytological mechanics of the hair cell is better known. A discussion of deformation of this order of magnitude might even be irrelevant since the final mechanical transformer is probably of molecular dimensions; the displacement at the threshold of hearing is in the order of a few Ångström units (JOHNSTON and BOYLE, 1967).

A proper matching of mechanical impedances between the auxiliary structures and the hair cell is important in order to provide the greatest flow of energy to the hair cell. The various kinds of attachment summarized in Fig. 11 may provide mechanical coupling relevant in each organ to meet specific requirements. In lateral-line and vestibular end organs, the hair bundle is held by intertwining cupular fibrils or it is fit into canals of the cupula. Such an arrangement may be suited to maintaining static displacement of the sensory hairs as demanded in tonic receptors. In auditory organs the tips of the tallest sensory hairs contact the tectorial membrane, a coupling which apparently ensures high fidelity at auditory frequencies. The degree of coupling is frequency-dependent because of a combination of small elastic forces and large frictional forces which make the tectorial membrane stiff to vibrations but pliable to slow static displacements. The displacement at threshold is in the neighborhood of thermal noise motion (Brownian motion). The magnitude of thermal noise generated at the receptor depends on the degree of coupling between the hair cell and the tectorial membrane; a rigid coupling improves the ratio of signal to thermal noise (HARRIS, 1966).

F. Receptor Potentials

Local potential changes which vary with the strength of the stimulus have been recorded in several sensory organs. In a number of cases, such receptor potentials (DAVIS, 1961) intervene as a causative link between the various forms of sensory stimuli and the initiation of sensory nerve impulses. In vibration-sensitive organs in the acoustico-lateralis system, vibratory stimulation causes an alternating potential change, the microphonic potential, first discovered by WEVER and BRAY (1930). In vestibular organs (e.g., the crista ampullaris in the semi-circular canals), DC potential shifts accompany sustained static displacement (TRINCKER, 1957). Inversely, microphonic responses can be obtained from the crista by experimental vibratory stimulation and steady potentials from the organ of Corti (BÉKÉSY, 1960). These receptor potentials are superimposed on steady resting potentials which exist within the membranous labyrinth in the absence of stimulation.

1. Resting Potentials

The endolymphatic resting potential is always positive with respect to a distant electrode. In cold-blooded animals and in the mammalian vestibular labyrinth, this potential is < 6 mV (SCHMIDT and FERNANDÉZ, 1962), whereas in the mammalian organ of Corti the endocochlear potential is as high as 80 to 90 mV (BÉKÉSY, 1960). The source of the endocochlear potential is the stria vascularis (TASAKI and SPYROPOLOUS, 1959); that of vestibular resting potentials is unknown.

2. Static Receptor Potentials

Displacement potentials have been recorded by TRINCKER (1959) in the crista ampullaris and the utricle of the guinea pig. The direction of the potential change varies in the negative or positive direction depending on the direction of the displacement. DC potential shifts have been observed in the cochlea when the position of the basilar membrane is shifted by pressure differences between the scala vestibuli and the scala tympani (TASAKI et al., 1954; BUTLER and HONRUBIA, 1963), or when the tectorial membrane is statically displaced with a needle (BÉKÉSY, 1960). The polarity of the potential change is determined by the direction of displacement. The presence of this response depends on the presence of intact hair cells, which are believed to be the responsible generators (BUTLER and HONRUBIA, 1963).

3. Dynamic Receptor Potentials

Microphonic responses have been recorded in most sense organs of the inner ear which are sensitive to vibration, and they have been extensively studied in the mammalian cochlea. VON BÉKÉSY has shown that the electrical energy contained in the cochlear microphonics, as well as in the steady displacement response, is larger than the energy of the input stimulus. The microphonic response is thus not a physical epiphenomenon to the vibration of tissues, but the result of energy amplification in a transducer for which the physical energy acts as a trigger.

The potential is of reverse sign above and below the sensory epithelium (JIELOF et al., 1952; TRINCKER, 1957; BÉKÉSY, 1960). It is well established that the microphonic response depends on the presence of intact hair cells: it disappears when hair cells are destroyed by ototoxic antibiotics (BUTLER and HONRUBIA, 1963) which leave the tectorial membrane and the cupula intact. Its dependence on an intact blood supply and the action of drugs are further arguments against a physical effect as the major contributor.

4. Origin of Receptor Potentials

It is generally believed that the receptor potentials represent the summed output from a large number of hair cells. The question of the origin of these

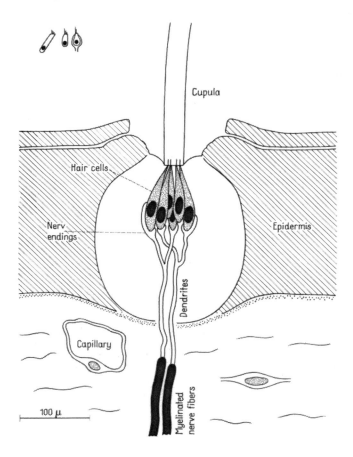

Fig. 13. Lateral-line organ of *Necturus maculosus*. Cochlear and vestibular hair cells are drawn to same scale for size comparison

potentials is a crucial one which is persistently put into debate. Intracellular recordings from single hair cells are decisive in this matter. When traversing the organ of Corti with microelectrodes, negative response potentials in the order of 40 to 90 mV have been recorded (see DAVIS, 1965). These potentials can be

recorded with electrodes with a tip diameter as large as 15 μ (BUTLER, 1965).
Since the diameter of hair cells in the organ of Corti is only 4 to 5 μ, the possibility
is remote that these potentials represent intracellular potentials from hair cells
(GOLDSTEIN, 1968). TRINCKER (1957) recorded similar potentials within the sensory
epithelium of the crista ampullaris in the guinea pig. KATSUKI et al. (1954) reported

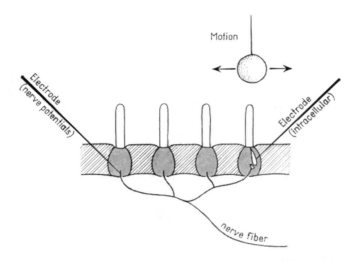

Fig. 14. Illustration of experiment by HARRIS et al. (1970). Each group of organs comprising
a "stich" is innervated by two afferent nerve fibers, one of which is shown in the drawing.
Neural responses evoked by stimulation of one organ were monitored by a microelectrode
inserted into another organ of the same stich. Under visual observation, a second microelectrode penetrated hair cells in the stimulated organ

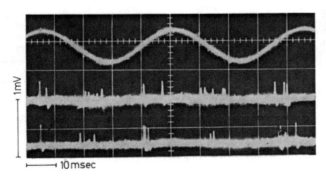

Fig. 15. Neural responses in *Necturus* afferent nerve fibers recorded as shown in Fig. 14.
Two classes of nerve action potentials with different shapes are seen. They belong to the
two afferent nerve fibers and lock to opposite phases of a sinusoidal stimulus. (Original,
from HARRIS et al., 1968)

electrical responses from single hair cells in the fish utricle, but the intracellular
origin of these responses was later disclaimed by the same author (KATSUKI and
USCHIYAMA, 1955). Numerous technical problems accompany such attempts; e.g.,
the approach to the inner ear sense organs is complicated by their enclosure in

the temporal bone and the delicate structure of the membraneous labyrinth. The hair cells are generally quite small, ranging in length from 25 to 40 μ and in diameter from 4 to 10 μ.

It is evident that accurate study of the sensory process in hair cells requires a suitable preparation which is available for experimentation without damage to the sensitive structures and where penetration of single hair cells

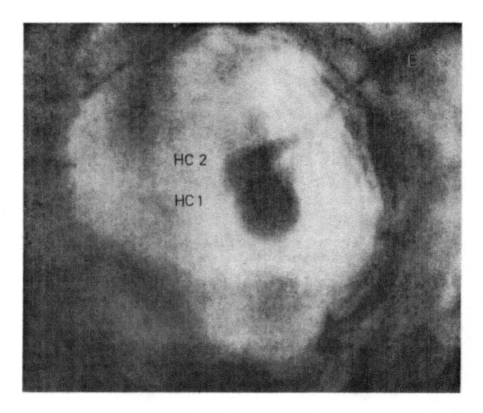

Fig. 16. In the lateral-line organ of *Necturus maculosus*, intracellular recordings from single hair cells can be done under visual control in anesthetized animals. The cell of origin of the recorded potential is identified by vital staining; responses from the two hair cells, *HC 1* and *HC 2* are seen in Fig. 17. *E* is the recording microelectrode. (Original, from Harris *et al.*, 1968)

can be reliably controlled. In other sensory systems, the study of the signi-ficance of the receptor potential and its relation to nerve impulse discharge has depended on preparations which allow recording from individual receptor elements. Famous examples are the compound eye of limulus (Hartline *et al.*, 1952) and the crustacean stretch receptor (Eyzaguirre and Kuffler, 1955). Lateral line organs are easily accessible because of their superficial position in the skin and may provide suitable preparations for such studies in the acoustico-lateralis system. Unique among these organs is that of the salamander *Necturus maculosus* because of the remarkable size of its hair cells (Fig. 13); the size encourages the study of hair cell function on a cellular level. The organs occur on the body and along the tail in small linear groups, called stitches, which run cranio-caudally. Each organ is embedded in the skin and is composed of a central group of 6 to 10 hair cells surrounded and separated by supporting cells. A long and narrow cupula extends from the skin surface several hundred microns into the water. The hair cell in the mudpuppy lateral line is several times larger (60 to 100 μ long and 20 μ wide at the level of the nucleus) than those found in other animals.

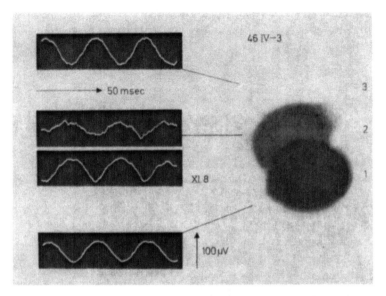

Fig. 17. At the left are intracellular responses from the three hair cells seen at the right (the two lower ones are also seen in Fig. 16). Each record is the averaged response to 128 periods of sinusoidal stimulation. The receptor potentials of cells *1* and *3* are in the same phase, whereas that of cell *2* is 180° out of phase. (Original, from HARRIS *et al.*, 1968)

A bundle of short stereocilia and a single asymmetrically placed kinocilium project from the apical end into the cupular substance. In half the hair cells of each organ the kinocilium is located toward the caudal side of the organ and in the remainder towards the cranial side. Each stitch is innervated by two, large, my-elinated afferent nerve fibers that branch to all its organs. Afferent endings, characterized presynaptically by synaptic bodies surrounded by vesicles, terminate at the base of each hair cell.

In a recent experimental study neural responses evoked from one organ were monitored by a microelectrode inserted into another organ of the same stitch (HARRIS *et al.*, 1968 and 1970; Fig. 14). Two populations of spikes which lock to opposite pha-ses of a sinusoidal displacement were recorded (Fig. 15). Under visual observation, hair cells in the stimulated organ were penetrated by a second pipette (filled with niagara sky blue, KCl or K-citrate). During electric recording, the cell of origin of the recorded potential was identified by injection of dye from the electrode (see KANEKO and HASHIMOTO, 1967). The stained cellular compartment can be identified *in vivo* by high-power water-immersion optics (Fig. 16); the stain also remains during histological processing. In addition to a negative DC shift, a small (800 μV *p-p* max) sinusoidal potential of the same frequency as the stimulus was occasionally seen in hair cells by averaging with computer. This response was of opposite phase in different cells in the same organ (Fig. 17). It is not a motion artifact since it is reversibly abolished by methylene blue while the cupular motion

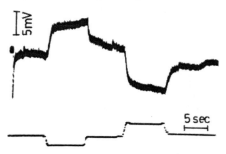

Fig. 18. Static displacement of the cupula shifts the hair cell intracellular potential (upper trace) in the depolarizing or hyperpolarizing direction, depending on the direction of displacement (lower trace). (Necturus maculosus)

and the intracellular DC potential remain the same. We believe this to be the intracellular receptor potential. Also, DC potential shifts can be recorded upon static biasing of the cupula (Fig. 18). Final conclusion about the nature and mechanism of generation awaits further investigation. It is known, however, that the underlying process is intimately related to directional sensitivity.

G. Directional Sensitivity

There is a functional relationship between the electrical response of the hair cell and the structural asymmetry of the hair-bearing end. This correlation was first described by LOWENSTEIN and WERSÄLL (1959) and has since been shown to be valid for the receptor potential as well as for the nerve response in a large number of acoustico-lateralis organs, for which a directional sensitivity based on structural polarization seems to be a pervading functional principle (see WERSÄLL and FLOCK, 1965).

Displacement of the sensory hair bundle in the direction in which the kinocilium is leading is excitatory. It is accompanied by the negative-going phase of the receptor potential as recorded in the endolymph, probably related to depolarization of the hair cell, and it leads to an increased probability of firing of the sensory nerve fiber. Displacement in the opposite direction is inhibitory (or defacilitating), it is accompanied by a positive receptor potential, probably related to hyperpolarization of the hair cell, and it leads to a decreased probability of nerve firing. We shall deal first with the directionality of the receptor potential and return later to the control of nerve discharge rate.

The active sensory epithelium can be looked upon as an array of distributed sources of current which are arranged in a parallel manner across the sensory epithelium. Above and below the sensory epithelium is a volume conductor, the endolymph and the perilymph, into which each source contributes a potential field. An electrode in the endolymph records a potential to which each source adds a value attenuated in proportion to its distance from the electrode. Also, the sign and the amplitude of the potential contributed by each receptor depends on the orientation of that receptor relative to the direction of stimulus displacement as illustrated graphically in Fig. 19. The receptive field of the hair cell

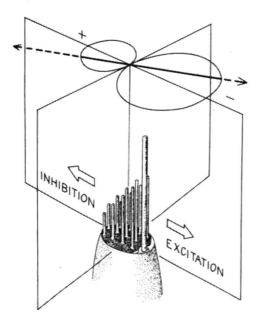

Fig. 19. The directional sensitivity of the hair cell approximates a cosine function of stimulus direction; the output varies as the cosine of the angle between the direction of maximum sensitivity and the applied displacement

approximates a cosine function of the stimulus direction; the output varies as the cosine of the angle between the direction of applied displacement and the direction of maximum sensitivity (organ of Corti, BÉKÉSY, 1960; lateral-line organ, FLOCK, 1965).

It is evident that the study of microphonic potentials as a measure of hair cell function is meaningful only when the spacial arrangement of the contributing receptors is taken into proper consideration. The **crista ampullaris** offers the simplest case (Fig. 20). Angular acceleration causes deflection of the cupula in only one or the opposite direction and affects all hair cells in the same direction simultaneously. All cells face the same way (LOWENSTEIN and WERSÄLL, 1959) and thereby contribute their responses in phase (Fig. 21). Microphonic potentials from fenestrated semicircular canals accordingly follow the frequency of a stimulating tone (VRIES and BLEEKER, 1949), and static displacement potentials are either depolarizing or hyperpolarizing (TRINCKER, 1957). The orientation of the hair cells in the three semicircular canals is such that the horizontal canal is excited by utriculopetal endolymphatic flow, whereas the two vertical canals are excited by utriculofugal endolymphatic flow (Fig. 22). This arrangement provides the basis for EWALD's first law, familiar to otologists since 1892.

In the **lateral line canal organ** of fishes, adjacent hair cells are oriented with their kinocilia facing in opposite directions (Fig. 20) (FLOCK and WERSÄLL, 1962). A similar situation is seen in epidermal lateral line organs in *Necturus* (Fig. 22) and in the toad *Xenopus* (KALMIJN, as quoted by DIJKGRAAF, 1963; FLOCK, 1967). Displacement of the cupula in any direction will cause responses of opposite polarities

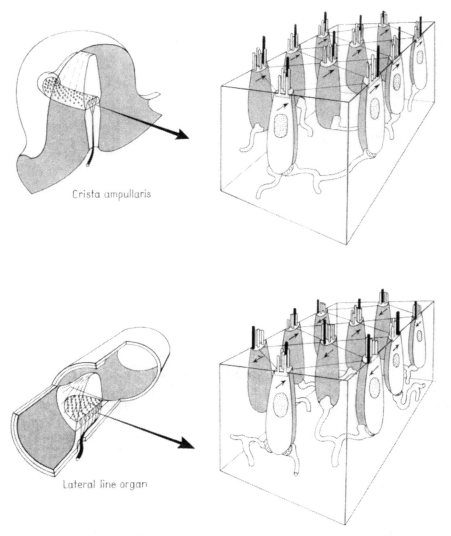

Crista ampullaris

Lateral line organ

Fig. 20. In the crista ampullaris, all cells are oriented with their kinocilia pointing in the same direction, whereas in the lateral-line organ alternating cells face opposite directions (from FLOCK and WERSÄLL, 1962)

from the two sets of cells which work 180° out of phase, as seen in Fig. 15 and 17. When opposing responses are of equal amplitude, no external voltage will be recorded, but when the linear working range of the cell is exceeded, distortion products begin to appear. The microphonic response recorded from the lateral-line canal organ is the second-order harmonic (twice the frequency) of the stimulus frequency (JIELOF *et al.*, 1952). HARRIS and VAN BERGEIJK (1962) have shown that the microphonic response increases as the third power of stimulus intensity as would be expected from a distortion product.

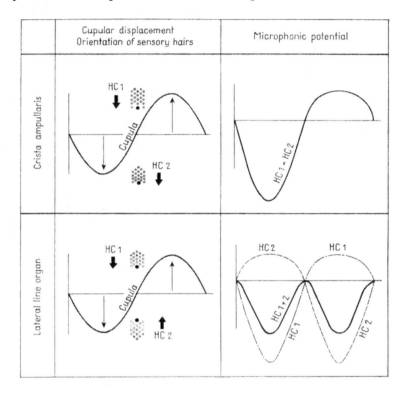

Fig. 21. In the crista ampullaris, the responses from all cells sum in phase. In the lateral-line organ the responses from opposing cells are 180° out of phase (see Fig. 17). The microphonic potential recorded by an external electrode is the distortion product between the two (from FLOCK and WERSÄLL, 1962)

The **saccule** of fishes (LOWENSTEIN *et al.*, 1964; WERSÄLL *et al.*, 1965; HAMA, 1969) and of mammals (SPOENDLIN, 1964) is also equipped with two groups of oppositly orientated hair cells (Fig. 23); here, however, the two groups are not mixed but are confined to different regions, a dorsal and a ventral one, separated by a central line, the striola (WERNER, 1940). In teleostei (bony fish), the otolith is a compact stone (its name derives from its discovery in a species of this subclass of fish by Scarpa in 1789). The two divisions of the sensory epithelium will yield responses 180° out of phase, but now the position of the recording electrode becomes important. This has been demonstrated most elegantly in recent work by FURUKAWA and ISHII (1967a) on gold fish. This fish has hearing, i.e., perception of pressure change rather than near field displacement from vibrating sources (see BERGEIJK, 1967). This is achieved by coupling of the compressive swimbladder by means of a chain of ossicles (Weberian ossicles) to the inner ear. Furukawa and Ishii (1967a) found regional differences in the distribution of saccular microphonics; when recorded over the dorsal part of the macula, the response to the compression

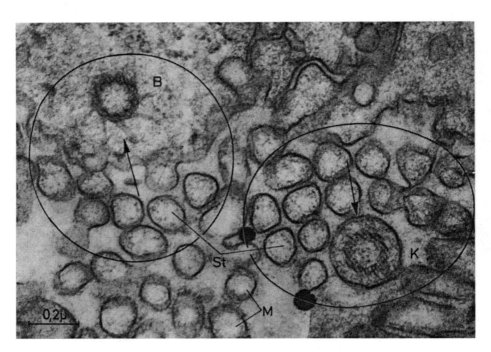

Fig. 22. Adjacent hair cells in *Necturus maculosus* face opposite directions. In one cell the basal body (*B*) is sectioned below the cell membrane, in the other the kinocilium (*K*) is sectioned through. *St*, stereocilia. *M*, microvilli

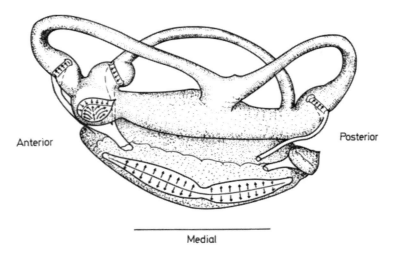

Fig. 23. Each labyrinthine sense organ has a specific pattern of directional sensitivity as determined on the basis of morphological polarization. The drawing is from fish; with modification in topography it applies also to mammals. (From Wersäll et al., 1965)

phase of the sound is predominant over the response to the rarefaction phase. A reverse relationship was recorded over the ventral part, and at an intermediate position, equal responses were obtained to both phases of the sound (double microphonics).

Stimulation at an angle with respect to the hair cell symmetry axis yields a response of a certain magnitude, smaller than that produced by stimulation along the symmetry axis. Stimulation in this oblique direction cannot be distinguished from a stimulus of lower intensity but parallel to the axis of maximal sensitivity. Directional information is provided only when a reference of stimulus intensity is given or by comparison of the output from two hair cells with perpendicular orientations. A single receptor would give a large error in source location since, for small angles, cosine does not change appreciably from unity. Difference detection in an orthogonal pair is more accurate. This is illustrated in the lateral-line system of teleosts with surface-wave perception (SCHWARTZ, 1967). On the head of the fish are several organs sensitive to surface waves produced by flies or other prey dropping onto the water surface. The organs are oriented at an angle with respect to one another, but within each lateral-line organ the hair cells are oriented with parallel sensitivity axes. These fishes orient with high accuracy toward the source of disturbance, but this ability is lost when the animal is left with only a single organ, in spite of the fact that within that organ individual hair cells have directional sensitivity.

The ability to detect stimulus direction is vital in the **utricle** which serves the maintenance of body posture by responding to linear acceleration. The effective stimulus is the shearing displacement of the otolith membrane, which has 360° of freedom and will move in a direction determined by the direction of acceleration. In the utricular macula, the spatial arrangement of hair cell orientation is quite elaborate (Fig. 23). The hair cells are oriented in a "fanning" fashion with a medial zone of cells pointing in anterior-lateral-posterior directions, and a marginal zone of cells pointing in opposite directions. The pattern is similar in fish (FLOCK, 1964) and in mammals (SPOENDLIN, 1964), whereas in the ray the orientation is different (LOWENSTEIN et al., 1964). For any direction of otolith motion, a pair of orthogonal hair cells can be found to supply information about the stimulus direction, presupposing that they have separate input channels to the central nervous system. If, on the other hand, cells with different orientation share a common channel (i.e., are innervated by the same afferent nerve fiber), discrimination of direction is lost but the "opening angle" of directional sensitivity, the receptive field, is increased. This illustrates the importance of the detailed knowledge of peripheral interconnections which is still lacking. One can expect that the composition of the receptor potential in this organ is strongly influenced by the direction of shearing displacement and electrode position since the sign and amplitude of individual contributions from various parts of the sensory epithelium will be different for different stimulus directions.

Outer hair cells in the **organ of Corti** are oriented with their basal bodies facing away from the modiolus, towards the stria vascularis (ENGSTRÖM et al., 1962; FLOCK et al., 1962); this is true also for inner hair cells (DUVALL et al., 1966). The cochlear partition analyzes the periodic fluctuations of acoustic pressure according to frequency; the maximum amplitude of movement is located at a

27*

particular position for each frequency. On either side of this maximum, the phase and amplitude of motion is a function of distance along the basilar membrane (Fig. 24). At each point the local output of the hair cells is proportional to the position of the basilar membrane. The instantaneous microphonic potential thus reproduces the travelling wave as shown by TEAS *et al.* (1962) in multiple electrode recordings at a number of positions along the cochlea. In most recording situations,

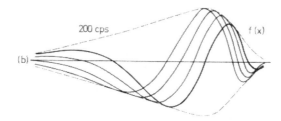

Fig. 24. The travelling wave along the cochlear partition at successive intervals of time. At each instant, different points are in different phase of motion (by courtesy from BÉKÉSY, 1960)

the cochlear microphonic potential that the electrode sees is a spatial average of the travelling wave image, weighted in favor of that part of cochlear partition which is in the vicinity of the electrode. In these cases the electrode does not record from a group of hair cells small enough for them all to be effectively in phase (see WHITFIELD and ROSS, 1965). Cochlear microphonics may therefore be affected by out-of-phase cancellation. Such cancellation would operate only for the AC components, and if nonlinearities exist in the individual generator waveforms, DC potential shifts will accompany the presentation of tones. Such a mechanism has been proposed by WHITFIELD and ROSS (1965) to explain the summating potential (SP) in the cochlea. The alternative theory proposed by DAVIS (1959) is the hitherto most widely accepted; it explains the SP by a steady bending of sensory hairs in proportion to the envelope of the travelling wave — the negative summating potential being generated by inner hair cells, the positive by outer hair cells. It should be pointed out in this context that a negative summating potential is present also in the bird cochlea (STOPP and WHITFIELD, 1964) and in the lateral-line canal organ (KUIPER, 1956), two organs which have only the common type of hair cells. In the lateral-line canal organ, this potential is not caused by a steady displacement of the sensory hairs (FLOCK, 1965). The nature of the summating potential remains as elusive as ever.

H. Fidelity of Mechano-Electric Transduction

Cochlear microphonic potentials increase linearly with increasing stimulus intensity over a remarkable range, about 70 db or a 3,000-fold increase. At higher intensities the linear relationship ceases to exist (DAVIS, 1959). This departure from linearity is not caused by nonlinearities in the mechanical transmission in the middle ear (GUINAN and PEAKE, 1964), or in the motion of the cochlear partition (BÉKÉSY, 1960) which is linear at sound pressure levels as high as 130 db.

The observed nonlinearity, and the ensuing distortion of the output waveform, is likely to originate at the level of transduction from mechanical stimulus to electrical signal.

Nonlinearities in the mechano-electric transducer are readily identified in organs where microphonics are produced by two sets of receptors working 180° out of phase, as in lateral-line organs, the saccular macula or the amphibian papilla. Neighboring receptor cells face opposite directions, but are attached to the cupula, the otolith or the tectorial membrane in the same fashion, and so receive identical input. When the hair cells work within their linear range, the responses from the two groups cancel. When nonlinearities exist in the input-output transfer function, distortion products will emerge as harmonics of the fundamental frequency, as mentioned in the previous section. Second-order harmonics appear at displacement amplitudes of 0.08 μ (HARRIS and VAN BERGEIJK, 1962; FLOCK, 1965) in lateral-line canal organs where KUIPER (1956) measured a maximum cupular displacement of about 1 μ in the intact organ. FURUKAWA and ISHII (1967a) found second-order harmonics in the goldfish saccule at a sound pressure of 60 to 70 db and CAPRANICA et al. (1966) at 45 db in the amphibian papilla in the bullfrog. These values are within the physiological range.

The nonlinearity in the microphonic response of hair cells is such that the "endolymph-negative" response is larger than the "endolymph-positive" response produced by symmetric excursions of the cupula around the resting position. TRINCKER (1959) recorded static displacement potentials in the guinea pig crista ampullaris. He found an S-shaped proportionality between cupular deviation and electric response where the endolymph-negative response was larger than the endolymph-positive response by a factor of two or three. He also states that physiological deviation of the cupula does not exceed 35°, and that within this range the curve is "with good approximation" linear. A shift of position of the basilar membrane toward the scala vestibuli produces a negative static potential which is larger than the positive potential caused by a similar shift in opposite direction (SMITH et al., 1958). These differences in sensitivity to stimulation in opposite directions may have a bearing on changes seen in the response to vibratory stimulation when the sensory hairs are simultaneously given a static bias. A shift in the position of the lateral-line canal cupula causes an increase of those negative peaks in the microphonic response which corresponds to the stimulus phase in that direction and a suppression of the peaks corresponding to the opposite stimulus phase (FLOCK, 1965). A similar effect is caused by superimposing a low-frequency oscillation on a high-frequency vibratory stimulation (KUIPER, 1956). FURUKAWA and ISHII (1967b) showed that microphonic responses in the saccule are changed when the sensory hairs are deviated statically. A pressure increase, causing a shift of the saccular otolith, enhances that phase of the microphonic potential which corresponds to the compression phase of the sound, while the response to the rarefaction phase is suppressed. From these observations, a tentative curve is drawn in Fig. 25 for the input-output transfer function of the single hair cell. This curve plots the amplitude of the receptor potential on the ordinate, as a function of displacement of the sensory hairs on the abscissa. Symmetric vibration of the sensory hairs around the resting position produces a microphonic output which increases linearly with increasing stimulus amplitude when the cell works in the linear range. At higher intensities the relationship is not linear. A static bias given to the sensory hairs gives a DC-potential output and shifts the cells working point along the curve. Suppression of the response occurs at inhibitory bias. The transfer function of Fig. 25 is based mainly on results obtained from the lateral-line organ and the saccule and may be different for hair cells in other organs. For instance, the linear range might be wider in

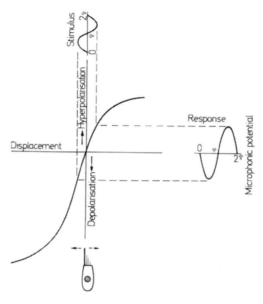

Fig. 25. Hypothetical input-output curve of a hair cell. It plots the amplitude of the receptor potential (ordinate) as a function of displacement of the sensory hairs (abscissa)

the organ of Corti where a true reproduction of stimulus wave shape may be of importance. An important reason for keen interest in the nonlinearity of the transducer mechanism is the fact that enhancement-suppression by static or periodic bias is mirrored by facilitation-defacilitation of firing in the primary afferent neuron. Sand (1936) made the observation that the sensitivity to vibration in the primary sensory fibers from the lateral-line canal organ increases by static displacement of the cupula in one direction and decreases in the other; similar observations have been made by Furukawa and Ishii (1967b) in the fish saccule. Here is a mechanism of peripheral gain control independent of efferent activity. Sharpening of the tuning curve in primary auditory fibers (Kiang et al., 1956), as well as inhibitory interaction between neighboring tones (Frishkopf, 1964), is known to have a peripheral origin independant of efferents.

I. Theories of Hair Cell Transduction

Several theories have been proposed to explain the generation of the microphonic potential and the static displacement potential. It has been established in the previous sections that the graded potential response in acoustico-lateralis organs is a receptor potential resulting from energy amplification by a biological transducer in the hair cell and is not the result of physical effects like piezo-electric potentials or capacitive potentials produced by displacement of charged molecules.

1. Physical Theories

The piezo-electric effect is the production of free electric charge by deformation which is found in some crystals (like those used in microphones and gramophone pick-ups). Such an effect was considered for some time by DE VRIES (1948) to be responsible for the production of microphonic potentials in the labyrinth. This theory was thought especially appealing since the otoliths contain crystals of calcium carbonate. The microphonic potential would then be produced by deformation of the otolith or the cupula. The piezo-effect always has the same frequency as the deformation, which is difficult to reconcile with the presence of double microphonics in some labyrinthine organs. This hypothesis was abandoned after BÉKÉSY (1960) had shown that the energy of the microphonic potential exceeds the energy of the mechanical stimulus. Histochemical investigations by DOHLMAN et al. (1959) and by JENSEN and VILSTRUP (1961) pointed to the presence of acid mucopolysaccharides in the cupula and tectorial membrane. VILSTRUP and JENSEN (1961) studied the electrostatic charges caused by moving such macromolecules in vitro, and found potential differences between the two ends of the advancing gel. When such potentials are produced by oscillating displacement, they resemble microphonic potentials. DOHLMAN (1960) has suggested that a film of polysaccharide molecules covers the sensory hairs and forms the basis for a condensor-like function of an electrostatically charged membrane where the charge is modulated by mechanical displacement of the sensory hairs. The induced potential changes inside and outside the hair cell membrane would act as generator potentials, affecting directly or indirectly impulse activity in the sensory axon.

2. Biological Theories

At some site in the hair cell there is a transducer sensitive to mechanical deformation caused by displacement of the sensory hairs. Recent research points to the basal body of the kinocilium and to the apical cell membrane as possible cites of transducer action.

a) **Mechanosensitivity in Cilia.** The direction sensitivity of the receptor mechanism is somehow related to the asymmetric structure of the hair-bearing end of the hair cell. A morphological polarization is present also on an ultra-structural level by an asymmetric arrangement of the component tubules of the kinocilium and by the presence of a basal foot, which protrudes from the basal body in a direction away from the stereocilia, that is, in the direction of depolarizing stimulation (Fig. 11; FLOCK and DUVALL, 1965). It is interesting to observe the correlation of an analogous morphological polarization of the kinocilium and its basal body to the beating direction of motile cilia (GIBBONS, 1961; AFZELIUS, 1961), a fact that directs our attention to the presence of modified cilia also in other sense organs. Kinocilia act as the sole transducers in hair cells of the squid (BARBER, 1965). It has been suggested by LOWENSTEIN et al. (1964) that the kinocilium may act as a motile cilium in reverse by responding to passive deformation in a certain direction with the initiation of electric change, as proposed for the tympanic organ by GRAY and PUMPHREY (1958). ENGSTRÖM et al. (1962) refer to the basal body as the essential excitable structure. THURM (1967) recently studied the mechanosensitivity of motile mussel cilia. These cilia beat spontaneously but can be brought to rest by increased carbon dioxide pressure and lowered pH value in the surrounding sea-water. Mechanical stimulation then induces a ciliar beat. He found that the locus at which the mechanical stimulus is most effective is at the ciliary base, inside or outside the cell body, in that region which is structurally different from the remaining ciliary shaft (compare Fig. 11). A shearing displacement of the ciliary base parallel to the cell surface is the most effective stimulus. The angular sensitivity varies as a cosine function of the stimulus direction within the plane of the epithelium, with the highest sensitivity in the direction of the active beat; stimulation in the opposite direction does not elicit a stroke, but inhibits spontaneous activity. Remarkable similarities thus exist between the mechanosensitivity of ciliated cells and that of hair cells. THURM (1967) proposes that in mussel cilia the mechanically sensitive structure is not that which is specialized to perform the beat, i.e. the long kinociliary shaft, but instead belongs to a specific structure which is engaged in the control of this movement. He points to the basal body, especially the basal foot as the possible mechanosensitive structure. Ciliated cells whose basal bodies are equipped with a basal foot have a fixed direction of ciliary beat which is related to the orientation of the foot; they are mechanically sensitive and show distinct angular sensitivity in the manner described by THURM. Several ciliated protozooa are capable of changing the direction of ciliary beat according to the direction of preferred locomotion. In several such organisms, the basal foot is absent. Motile cilia of this type in Euplotes do not exhibit angular mechanosensitivity similar to that in mussel cilia (THURM, 1967), a fact that couples directional sensitivity to the presence of basal feet in ciliated cells. However, the presence of a basal foot is not a sine qua non for directional sensitivity in hair cells since cochlear hair cells have full fledged angular sensitivity, as described by BÉKÉSY (1960), in spite of the fact that the basal bodies of outer and inner hair cells lack a basal foot (DUVALL et al., 1965).

The output of the mechanosensitive control mechanism at the ciliary base may be directed in motile cilia to trigger the movement of the cilium, or (in the case of receptor cells) to control a change of membrane permeability (THURM, 1967). In motile ciliated cells, both actions may be simultaneously at work as implied by the finding of HORRIDGE (1965) that a mechanically induced ciliary beat in coomb plate cells of a Ctenophore is accompanied by membrane potential change. HÅKANSSON et al. (1966) observed discrete potential fluctuations related to spontaneous beat activity in respiratory cells in the guinea pig trachea. On the basis of his experience, THURM (1967) concluded that a membrane depolarization is not causative in the ciliary beat mechanism but is rather secondary to this process. In hair cells, however, membrane potential changes may be the causative link between the apical end of the hair cell, where transduction takes place, and the basal end, where synaptic transmission occurs. The following section will deal with electrogenesis in excitable membranes as related to the origin of hair cell receptor potentials.

b) Membrane-Bound Excitability. Receptor potentials in hair cells may be explained in terms common to neurophysiological processes in neurons and several other sensory receptors. It is therefore appropriate first to describe the generation of bioelectric membrane potentials and their general properties.

Electrogenesis in neurons and also in several receptor cells is based on the potential difference between the inside and the outside of the cell. This electrical polarization is established across the semipermeable cell membrane by the different distribution of ions on the two sides of the cell membrane. Potassium is found in high concentration intracellularly where the potential is negative; sodium is generally high extracellularly. There is thus an inward electromotive force for sodium and an outward force for potassium. Chloride is often distributed close to equilibrium and is in higher concentration on the outside. This membrane polarization is maintained by an active transport of ions against their concentration gradients, by a process supported by metabolic energy. Excitation of the membrane causes a conductance change, a selective change in ion permeability to one or several ion species which are allowed to pass across the membrane along the electrochemical potential gradients towards their equilibrium potentials. The associated change in transmembrane potential can be recorded extra- or intra-cellularly and is of opposite polarity on the two sides of the membrane. Sensitive membranes belong to one of two different types depending on the mode of activation; it may be either electrically excitable or electrically inexcitable (GRUNDFEST, 1965). In electrically excitable membranes, membrane conductance is a function of the transmembrane potential, the permeability increases with decreasing membrane potential until, at a critical threshold, total depolarization suddenly occurs. This is the type of membrane where self-sustained, regenerative all-or-nothing impulses are developed, e.g., in nerve fibers. Electrically inexcitable membranes respond only to specific stimuli, i.e., chemical, mechanical, photic or thermal. Obviously these are the adequate stimuli for sensory receptors, except electroreceptors. These membranes respond to a graded stimulus with a graded potential change. There is now considerable evidence that the two membrane types have different locations in receptors. The receptor membrane gives a local response to a local stimulus (LOEWENSTEIN, 1961; TERZUOLO and WASHIZU, 1962). This is of importance since it increases the integrative capacity of these regions by preserving them from explosive depolarization and from successive strong depolarization.

The mechanism of excitation involves a change of state in the molecular structure of the cell membrane. Modern membrane biophysics suggests a number of possible mechanisms. The classical picture of the cell membrane is one of a central core of lipids covered on both sides by a layer of proteins, it appears as two dense lines separated by a lighter zone. Recent evidence indicates that other molecular arrangements can occur in specialized membranes. SJÖSTRAND (1963) described a globular membrane substructure; FERNANDEZ-MORAN (1962) identified elementary particles as constituents of the membrane; and ROBERTSON (1962) observed a hexagonal microstructure. The present trend is to look upon the excitable membrane as possessing functional subunits where molecular transitions provide gating mechanisms for selective ion flow. According to the "sieving" theory (see ECCLES, 1966), such transitions may involve the opening of pores which allow ions to pass according to their size. Fixed positive or negative charges in these channels may provide selectivity toward different ion species. Hodgkin and HUXLEY (1952) suggested that movement of charged particles ("carriers") might regulate ion flow. This theory infers that transmembrane passage involves a temporary binding of the transported material to some component of the membrane, the "carrier", and the passage of this complex across the membrane by diffusion or convection. TASAKI and SINGER (1966) instead assumed a rapidly reversible ion-exchange process involving transitions between two stable states of a macromolecular complex. The structural changes involved would be similar to helix-random coil and oil-water transitions. It seems that the ion-selective mechanisms are functionally different on the two sides of the membrane since the action of drugs on selective ion flow is different depending on whether the drug is applied on the outside or inside of the membrane. The general picture of the receptor membrane is one of a patchwork of membrane units with isolated response capacities. LOEWENSTEIN (1961) has shown that in the denuded Pacinian corpuscle, local responses arise from discrete areas of the membrane and sum spatially with increasing stimulus strength. Similar integration of unit responses occurs in another mechanoreceptor; the crustacean stretch receptor (EYZAGUIRRE and KUFFLER, 1955), and apparently also in photoreceptor cells where discrete potential fluctuations, "quantal bumps", appear as the result of the absorption of photons (ADOLPH, 1964). On illumination, the number of bumps increases until they add to form the graded receptor potential.

It is conceivable that in mechanoreceptor membranes (including the hair cell), mechanical deformation causes changes in the molecular structure of labile membrane subunits acting as gates for ion flow. Associated current flow results in proportional voltage changes, the receptor potentials. In hair cells where the response is highly selective with respect to direction of deformation, the underlying mechanism may involve a change in a steady leakage current as suggested by DAVIS (1965) or a selective conductance change for different species of ions. There is evidence indicating that the generation of labyrinthine receptor potentials does relate to ion flow. The cations sodium and potassium are the two ions mainly engaged in neural electrogenesis. At least potassium is important in the generation of microphonic potentials. Tetraethylammonium is known to block potassium flow across membranes. KATSUKI et al., (1966) found that electrophoretic application of this substance into the scala media will reduce the microphonic potential. Streptomycin is an antibiotic which has a toxic effect on the apical membrane of hair cells (DUVALL and WERSÄLL, 1964). Local application of low concentration of this drug to the lateral-line canal organ causes a partial or total suppression of the microphonic potential which is reversible by rinsing only if sufficient amounts of potassium ions have been added to the rinsing fluid (WERSÄLL and FLOCK, 1964). Sodium ions do not restore the response. Streptomycin combines rapidly and reversibly with the surface component of bacteria (ANAND et al., 1966), a binding which is inhibited by cations. In bacteria, streptomycin affects the membrane permeability to potassium (DUBIN and DAVIS, 1961). The potassium content of the endolymph is unusually high for an extracellular fluid. In the guinea pig it is about 144 mekv/l (SMITH et al., 1954); even the lateral-line canal fluid has a potassium concentration of 40 mekv/l. High external potassium concentration is known to influence the electric constant of polarized membranes (FRANK and FUORTES, 1961). TASAKI (1961) found that a high potassium concentration in the presence of anodal polarization (such as the endocochlear potential) renders the membrane of excitable cells more sensitive to mechanical deformation. The apical ends of the hair cells, with their bundles of sensory hairs, are exposed to the ionic environment of the subcupular or subtectorial space. The ionic composition of this space is unknown. If it is as high in potassium as the endolymph, it must be ionically separated from the lower part of the sensory epithelium because the nerve fibers innervating the hair cells could not function in a high potassium environment (TASAKI et al., 1954). Electron microscopy shows that tight junctions (zonula occludens) between the apical interfaces of supporting and sensory cells can provide such a barrier. It is possible to stain sodium ions specifically with potassium antimonate, which forms an electron dense deposit marking the presence of sodium ions for electron-microscopic identification (KOMNICK and KOMNICK, 1963). Such staining reveals a particularly high concentration of sodium in the cupula and in the subcupular space, but the dense deposit ends abruptly at the intercellular junctions (FLOCK, 1967).

The similarity of the microphonic potential to receptor potentials in other sense organs is indicated by its insensitivity to tetrodotoxin, the puffer fish poison, even in doses which block the nerve action potential (KATSUKI et al., 1966). Tetrodotoxin has a selective action on electrically excitable membranes where it blocks sodium conductance, whereas electrogenesis in receptor membranes remains unaffected as seen in the Pacinian corpuscle and the crayfish stretch receptor (LOEWENSTEIN et al., 1963) and in the insect chemosensory receptor (WOLBARSHT and HANSON, 1965). These findings were in fact predicted by GRUNDFEST (1964) on the basis of his findings that electrically inexcitable synaptic membranes are insensitive to tetrodotoxin.

The microphonic potential in hearing organs is remarkable because of its capability to follow high-frequency sound; echo-locating bats use swept tones of 25 to 100 kHz (GRIFFIN et al., 1960). The conventional cell membrane has a capacitance in parallel with the membrane resistance, and the membrane has the property of a low-pass filter which imposes limits on the high-frequency response. With a high frequency cut off at about 30 kHz, as seen in fast membranes, the response at a frequency one octave higher (60 kHz) would be 6 db down ($= \frac{1}{2}$); so thus, microphonic potentials at high frequency are still compatible with bioelectric membrane potentials.

c) **Conclusion.** Two mechanisms which relate to transduction in hair cells have been discussed. One set of observations points to the presence in the hair cell of a mechanically sensitive point at the ciliary base which controls membrane conductance. If this is the basal body, which is separated from the membrane, the basal body must have some output which is transmitted to the membrane where the conductance change is to take place. The first step in the sensory process must be the absorption of mechanical energy by a sensitive molecular transducer. Molecular excitation does not migrate far from the site of energy absorption by purely physical processes of excitation from molecule to molecule; therefore, molecular excitation is more likely to occur in the structure where the change of state is required to be active. This favors membrane-bound excitability of the type seen in other mechanoreceptors. The functions performed by the basal body and the membrane do not necessarily represent alternative control mechanisms which are mutually exclusive. It is conceivable that mechanosensitivity is located primarily in the membrane and that the labile molecular structure which underlies excitability is actively maintained by, or laid down during development, by the kinocilium with its basal body, which is responsible for the directional polarization of the functional elements on which angular sensitivity depends.

K. Synaptic Transmission in Hair Cells

In primary receptors, the sensory terminal is an integral part of the sensory neuron, and it is well established that the receptor potential is causative in firing the nerve. Hair cells are secondary receptor cells which are contacted by nerve endings of terminal branches of bipolar sensory neurons. Recent ultrastructural and electrophysiological observations have lead to new understanding of the nature of synaptic transmission at this afferent synapse, and may shed light also on the role of the receptor potential in the chain of events that leads to nerve impulse initiation.

Hair cells in vestibular and lateral-line organs of fishes and amphibians are innervated by a number of nerve endings which may be rather small (0.5 to 2 μ) or somewhat larger (5 to 10 μ) contacting a more substancial part of the hair-cell base. Wersäll (1956) described two types of nerve endings in the vestibular labyrinth of mammals. Type II hair cells are innervated as just described, whereas type I hair cells are enclosed by a calyx-like nerve ending completely enclosing the amphora-shaped cell (Fig. 2). Several hair cells may even be enclosed by the same nerve ending. In the organ of Corti, afferent nerve endings are small; generally only one of them contacts each hair cell and it is surrounded by several large efferent nerve endings. The nerve ending cytoplasm is generally clear, containing neurofilaments, mitochondria and a few vesicles of varying diameters ranging from 300 to 700 A. The pre- and post-synaptic membranes are separated by a synaptic cleft with a fairly regular width of about 200 A. At points along the contact, regions of ultrastructural specialization occur which resemble axodendritic synapses in the central nervous system. These areas are characterized by accumulation of vesicles about 350 A in diameter which surround a dense synaptic body inside the hair cell, and by an increased density and thickness of the subsynaptic membrane in the adjacent area. Such a structural specialization was first described in the guinea pig cochlea by Smith and Sjöstrand (1961) and has since been described in all labyrinthine and lateral-line organs studied (see Wersäll et al., 1965; Spoendlin, 1965). The details of ultrastructure vary somewhat from one organ to another and from species to species. Fig. 26 shows how synaptic vesicles are lined up along the presynaptic membrane in the salamander *Necturus maculosus*. The number of specialized sites varies with the size of the nerve ending and with the type of hair cell. The size and appearance of these vesicles are like synaptic vesicles in presynaptic terminals where neurochemical transmission occurs. According to the now generally accepted theory of the quantal nature of neurochemical synaptic transmission (see Martin,

1966), the synaptic vesicles contain fixed amounts of transmitter substance which on excitation is released into the synaptic space, which it passes by diffusion, to act at the postsynaptic membrane. As the result of the impact of each transmitter package, local postsynaptic potentials are generated. The transmitter is

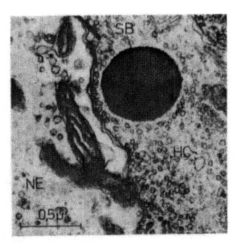

Fig. 26. The afferent synapse between a hair cell (*HC*) and a nerve ending (*NE*) in *Necturus maculosus* is characterized by a dense synaptic body (*SB*) surrounded by vesicles which line up along the presynaptic membrane. Area corresponding to frame in Fig. 27

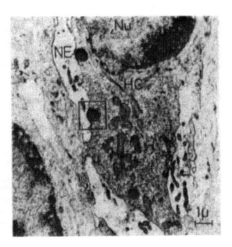

Fig. 27. Large afferent nerve endings (*NE*) contact the base of a hair cell (*HC*) in *Necturus maculosus*. Area corresponding to frame is seen in Fig. 26

inactivated after its absorption in the postsynaptic membrane, and its effect decays with a certain time constant, the relase of each synaptic vesicle thus contributing a miniature postsynaptic potential to the total postsynaptic response. When the postsynaptic potential is large enough to exceed a threshold level, it triggers a nerve action potential in the electrically excitable part of the neuron.

In central neurons this point is probably at the axon hillock; in peripheral sensory axons it is probably at the point where myelinization of the axon starts. Recent electrophysiological findings suggest that synaptic transmission of this type occurs at the junction between hair cells and afferent nerve endings.

Afferent nerve endings in the salamander *Necturus maculosus* are unusually large (Fig. 27). Two types of spontaneous electric potentials have been observed

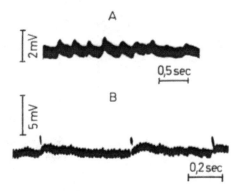

Fig. 28. Two types of spontaneous potentials can be recorded with microelectrodes from the base of hair cells in *Necturus maculosus*. Slow potentials are seen in record A; in record B, fast potentials occur in the rising phase of slow potentials

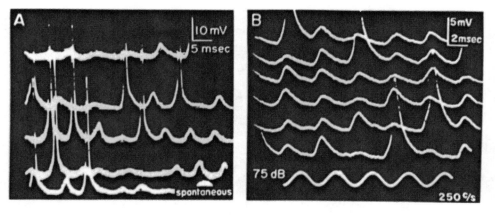

Fig. 29. Postsynaptic potentials recorded from afferent nerve fibers in the saccule. (A) shows spontaneous activity; (B) shows synchronization to a periodic stimulus (by courtesy from FURUKAWA and ISHII, 1967a)

when recording from the lower part of the hair cell with microelectrodes. Slow potentials, with a wave shape similar to postsynaptic potentials, are seen in record A of Fig. 28. In record B, fast potentials occur in the rising phase of the slow potentials. The slow potentials are interpreted as excitatory postsynaptic potentials generated in the sensory nerve endings by secretion from the hair cell of one or more quanta of neurochemical transmitter substance contained in the synaptic vesicles. The fast potential is interpreted as a nerve action potential,

attenuated by electrotonic spread from its point of origin below the basement
membrane where the myelinization of the sensory axon starts. When recording
intracellularly from primary sensory fibers in the goldfish saccular nerve, FURU-
KAWA and ISHII (1967a) observed spontaneous potential fluctuations of varying
amplitude which occured in irregular series. Nerve action potentials were elicited
by potential fluctuations large enough to reach a certain level (Fig. 29A). When
a tone was presented, these potentials were synchronized to the tone frequency
(Fig. 29B) and their amplitude increased with increasing tone intensity. The

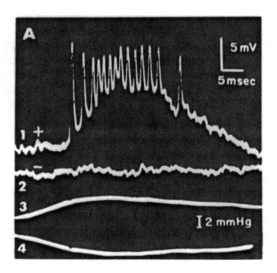

Fig. 30. Recording from an afferent nerve fiber in the saccule. Displacement of the otolith
in one direction (*3*) caused depolarization of the nerve fiber and discharge of nerve impulses
(*1*), whereas a shift in the opposite direction (*4*) had no effect (*2*) (by courtesy from FURUKAWA
and ISHII, 1967b)

quantal fluctuations in potential size, the augmentation of the potentials by
anodal polarization, and the presence of a delay of 0.6 to 0.8 msec between the
microphonic potential and these potentials were taken as evidence that excitatory
postsynaptic potentials were recorded. In some fibers the postsynaptic potentials
were synchronized to the compression phase of the stimulating sound, in others
to the rarefaction phase, whereas still other fibers responded to both phases.
FURUKAWA and ISHII propose that these fibers are connected to hair cells in the
dorsal, the ventral, or both halves of the saccular macula respectively, where hair
cells are oriented with opposite directions of excitatory sensitivity (Fig. 23).
A pressure increase caused a slow depolarization in fibers sensitive to the compression phase of a tone, whereas a pressure decrease had no effect (Fig. 30).
In fibers sensitive to the rarefaction phase, the depolarizing response was evoked
by a pressure decrease, whereas an increase in pressure had no effect.

This is strong evidence that the hair cell releases an excitatory chemical transmitter substance at the synaptic junction with the sensory nerve ending. Lateral
displacement of the sensory hairs in the excitatory direction cause an increased

rate of release and an increased probability of firing, whereas inhibitory displacement reduces transmitter release rate and decreases the probability of firing.

A fundamentally different type of synaptic contact is seen in vestibular type I hair cells in mammals. At points along the extensive synaptic junction between the hair cell and the nerve chalice, the pre- and postsynaptic membranes closely appose each other (SPOENDLIN, 1965; ADES and ENGSTRÖM, 1965; WERSÄLL and LUNDQUIST, 1966). SPOENDLIN (1966) showed that at these sites the adjacent cell membranes may fuse to form a tight junction, of the type characteristically seen where electric transmission occurs. It has to be decided by electrophysiological methods if these tight junctions really represent sites of electric (ephaptic) transmission (GRUNDFEST, 1959). Where such transmission exists, presynaptic potential fluctuations transfer with neglible delay and little attenuation to the postsynaptic cell. This type of transmission is clearly different from the neurohumoral transmission previously discussed.

Hair cells are also contacted by the endings of efferent centrifugal nerve fibers. They are characterized by the presence of a large number of vesicles inside the nerve ending and by a subsynaptic sac closely apposed to the postsynaptic membrane. Such endings, of varying shape and abundance, are present in all acousticolateralis organs thus far examined (see WERSÄLL et al., 1965), except in the basilar papilla of the frog where efferent innervation is absent (ROBBINS et al., 1967; FRISHKOPF and FLOCK, 1967). Stimulation of the efferent system is inhibitory on the afferent discharge (FEX, 1962), the action probably being *pre*synaptic inhibition on the hair cell. In vestibular hair cells of type I, efferent endings terminate on the nerve calyx and here *post*synaptic inhibition is indicated although it has not been demonstrated physiologically. FEX (1967) has recorded electric potentials in the organ of Corti in response to efferent stimulation which he interprets as inhibitory postsynaptic potentials. This action is probably related to the release of acetylcholine since specific acetylcholin-esterase is found in relation to efferent pathways and at the efferent nerve endings (SCHUKNECHT et al., 1959; HILDING and WERSÄLL, 1962).

L. Receptor Potentials and Synaptic Transmission

The possible role of the microphonic potential, or the displacement potential, in nerve impulse initiation in the acoustico-lateralis system has long been debated. Evidence now accumulates in favor of a causative relationship as maintained by DAVIS (1965). The opposite view is taken by GRUNDFEST (1965) who looks on the microphonic potential as an epiphenomenon, secondary to the secretory action of hair cells. Recent pertinent experiments have been performed in muscle motor end plates (KATZ and MILEDI, 1967) and on the squid giant synapse (see KUSANO and LIVENGOOD, 1967), where electrical excitability was blocked by tetrodotoxin. Presynaptic depolarization by electric current was shown to initiate transmitter release evidenced by depolarizing postsynaptic potentials. The amplitude of the postsynaptic potential was proportional to the magnitude of presynaptic depolarization, and it was concluded that the potential level in the presynaptic terminal controls the rate of release of transmitter substance at the synapse and that intracellular potential changes (or the associated currents) are instrumental in controlling this release rate, in positive or negative direction depending on the direction of potential change. Intracellular potential changes do occur in hair cells as a result of mechanical stimulation (HARRIS et al., 1970). Postsynaptic potentials of a type strongly suggesting neurochemical transmission of quantal nature have been recorded (FURUKAWA and ISHII, 1967a), but final proof of causative dependence requires experimental evidence that intracellular depolarization or hyperpolarization by current, in the absence of mechanical stimulation, controls the frequency of miniature postsynaptic potentials in a predicted manner.

In vestibular type I hair cells, the intracellular receptor potential may communicate directly with the afferent terminal via an electrical synapse.

The mechanism which couples presynaptic depolarization to liberation of synaptic transmitter in neurochemical synapses is unknown. In motor end-plates and in the squid giant synapse, an inward flow of calcium ions in the terminal membrane seems to be involved (KATZ and MILEDI, 1967; MILEDI and SLATER, 1966), although permeability changes to sodium and potassium do not appear to be causative (KUSANO et al., 1967). Cations also determine the degree of electric coupling across junctional membranes between epithelial cells (LOEWENSTEIN, 1966).

M. Excitation and Integration in Afferent Terminals

A background of continuous spike discharge is commonly seen in acoustico-lateralis organs. The origin of this spontaneity is not in the nerve fiber but in the process in the hair cell which causes release of synaptic transmitter, or at an earlier stage which controls this process. Ambient acoustic noise is not the causative factor in the ear (KIANG et al., 1965) nor is it spontaneous water motion in the lateral line (HARRIS and FLOCK, 1965). Each nerve fiber innervates several hair cells, so there are multiple sites of origin of generator potentials in different hair cells and even in the same nerve ending where several synaptic sites exist (Fig. 27). The nature of signal transmission in the peripheral terminals is unknown; the possibility that they conduct all-or-nothing impulses cannot be excluded, although it is more likely that individual hair cells innervated by the same branch give graded contributions to the probability of an impulse arising at a point where their actions summate. On the basis of such integrating capacity, multiple innervation may serve to lower the threshold in response to simultaneous subliminal excitation of several synapses (DAVIS, 1961). In the case of random transmitter release in the absence of a synchronizing stimulus, postsynaptic potentials from individual sites arrive at the electrically excitable point at random times where their summed action is proportional to the square root of the number of active synapses. In the presence of a periodic stimulus, the postsynaptic responses from all receptors have a constant phase relation and their summed action is proportional to the number of active synapses. Signal to noise ratio is thus improved. The quantal fluctuation of generator potential size observed by FURUKAWA and ISHII (1967a) is suggestive of synchronous release of a varying number of synaptic transmitter packages, rather than a random Poisson-like distribution in time as in muscle motor end-plate (GAGE and HUBBARD, 1965). The nature of the process which synchronizes transmitter release is obscure.

N. Coding in Primary Afferent Neurons

Coding of sensory information in the inner ear and lateral line is done on two principles; temporal coding is based on the time pattern of responding units, whereas spatial coding involves excitation of different units when some stimulus parameter is changed.

In some organs, repetitive nerve firing occurs at regular intervals which change as a function of the stimulus; the classical example is the isolated ray labyrinth (Lowenstein, 1956). Coding in such a system, [for example, coding of angular acceleration in the ray semicircular canal (Lowenstein and Sand, 1940)] is by frequency modulation of ongoing activity. In this situation, spontaneous discharge is not simply due to pick up of noise by a sensitive receptor working close to threshold, but constitutes the reference level for bidirectional sensitivity. More often, however, nerve firing is of a probabilistic nature, spontaneous and evoked activity being characterized by variability of interspike intervals in primary auditory neurons (see Kiang, 1965), in utricular units in the frog (Gualtierotti and Alltucker, 1966), or in the epidermal lateral-line organ (Harris and Milne, 1966). The statistical nature of spontaneous activity is closely modeled by a Poisson process (Rodieck et al., 1962); after a brief refractory period the occurence of an impulse is equally likely at any time. Two origins of variability of firing are seen, one in a random distribution of generator potentials in time, and one in the quantal nature of generator potential size. In Fig. 29, the fluctuation of generator potential amplitude is the more impressive of the two. From the point of view of higher neural centers, fluctuation of nerve impulse intervals must decrease the sensitivity to threshold stimulation since this is determined by the smallest significant alteration in the output. It must also decrease temporal resolution since integration must take place over some time (Bullock and Diecke, 1956). The summing of activity in independent input channels is probably an important mechanism for gaining reliability from noisy signals.

Static displacement or low-frequency stimulation results in an increased average rate of nerve firing during the excitatory phase, and a decreased firing rate during the inhibitory phase. This is true not only for lateral-line and vestibular end organs of equilibrium, but also in the organ of Corti of the cat where Kiang et al. (1965) found that base-line activity is increased or depressed by the rarefaction and condensation phases of low-frequency sound. The response to periodic stimulation at higher frequencies is probabilistic; spikes tend to follow in the excitatory phase of the stimulus (Rose et al., 1967). Firing does not occur at every period but tends to occur at some integral multiple of the stimulus, i.e., not at every second or third cycle as in simple frequency division but in a probabilistic way (Tasaki, 1954). As frequency is increased, a limit is reached above which the neural response is no longer phase-locked to the stimulus; it fires as soon as its recovery process allows (Kiang et al., 1965). The demonstration of chemical synaptic transmission at the hair-cell nerve-ending junction supports the suggestion of Kiang et al. (1965) that the following-capacity of a unit is associated with the decay time of the excitatory synaptic process, the time resolution being limited by the rate of destruction of transmitter substance. Pooling of the excitatory transmitter may provide rectification necessary for high-frequency discrimination. Generator potentials in the goldfish reach peak amplitude in about 1 msec (FURUKAWA and ISHII, 1967a), whereas those in the salamander are very slow, requiring about 100 msec (Fig. 28). These times are influenced by the recording situation; the delay of the peak increases with increasing distance of the recording electrode from the site of potential generation.

In animals lacking an organ which performs frequency analysis on the place principle, information of sound frequency is likely to be conveyed by neurons following the stimulus frequency (SCHWARTZKOPFF, 1962). In the cochlea, the spectral content of sound is analyzed according to the frequency selectivity of the basilar membrane (BÉKÉSY, 1960), but temporal coding is used as well. Psychoacoustic experiments (SHOUTEN, 1960) as well as neurophysiological findings (MÖLLER, 1968) show that pitch may be determined by the periodicity of sound, and temporal coding is a prerequisite to binaural localization of low-frequency tones which relies on information pertaining to phase (DETHRAGE, 1966).

Analysis of sound frequency in the cochlea is done on the place principle. Other forms of spatial coding are seen in the vestibular organs. BÉKÉSY (1967) points to the finding of LAGRANGE (1867—1892) that any rotation of a body in space can be described as the combination of three rotations parallel to the axis of a coordinate system. A natural application is the system of three semicircular canals with axes along three perpendicular coordinates. Comparison of these three outputs by higher neural centers provides the necessary information. Primary afferent neurons in the utricle exhibit specific directional sensitivity (LOWENSTEIN and ROBERTS, 1949) which correlates with the orientation of receptor cells in the end organ (see page 11). The sensory elements manifest trigonometric resolution of applied linear acceleration (MILSUN and JONES, 1967); the response amplitude varies with the angle between the directions of applied acceleration and that of maximum sensitivity for the cell. Different regions, innervated by separate nerve fibers, respond to specific directions of linear acceleration and provide spatial coding of directional information. Experiments by HIBBARD (1964) on grafted labyrinths in amphibia show that the peripheral spatial pattern has a central projection and that apropriate topic correspondence is a prerequisite to the integrity of rightening reflexes.

FURUKAWA and ISHII (1967) found two major types of response characteristics in primary sensory nerve fibers in the fish saccule. The so-called type S_1 are large fibers which have no spontaneous activity, have a rather high threshold, and adapt quickly to tonal stimuli. The fast adaptation rate does not seem to be due to adaptation of the microphonic response but rather to decline of the post-synaptic potentials with time. So-called type S_2 fibers are smaller; they show spontaneous discharge, have a lower threshold and no marked adaptation to tones. S_1 fibers are, in all probability, associated with the Mauthner cell, a giant neuron found in the medulla in ostariophys fishes. The endings of large myelinated fibers terminate on this neuron with electrically excitatory synapses (FURSHPAN, 1964) where rapid transmission occurs. The axon of the Mauthner cell crosses the midline of the medulla, bends caudally and runs down the side of the body to innervate muscles which flip the tail towards this side. S_1 fibers and the Mauthner cell constitute a fast afferent reflex arc which provides the flight reaction of the fish. In general, electric synapses seem to occur where sure-fire transmission is vital, and may be characteristic of the executive type of neuronal net activity rather than the analyzer-integrator type (SCHMITT, 1966). It is interesting to recall the pattern of innervation of type I and type II hair cells in the vestibular labyrinth (WERSÄLL, 1956). Type II hair cells are similar to those found in lower animals; they are innervated by rather thin fibers (2 to 5 µ) which spread diffusely to innervate a considerable number of hair cells. Type I hair cells are innervated by thick myelinated fibers (5 to 8 µ) which innervate one or a few hair cells where synaptic contact regions suggesting electric transmission are seen. It is tempting to speculate that these structural analogies suggest a similar functional differentiation. A subdivision of primary neurons into quickly adapting on-off units and slowly adapting tonic units has been observed in most organs belonging to the acoustico-lateralis system. Possibly these systems supply different types of sensory information, the tonic response serving perception, the dynamic response serving discrimination of rate of change used in higher orders of neural organization. In sensory neurophysiology, interest in recent years has shifted from static to dynamic properties of sensory systems. Today great interest is taken in "feature extraction" processes by lateral interconnection between neurons at different levels in the central nervous system which enhance significant information in the sensory input. HUBEL and WIESEL (1962) and others identify cells in the visual system which respond selectively to gradients and rate of change of brightness, orientation of contour or motion, etc. In the acoustico-lateralis system, feature extraction is less well documented although selective sensitivity to swept tones is one example of advanced information processing (WHITFIELD and EVANS, 1965). What is required in the future is a study of how various neurons respond to a wide variety of signals, including stimuli modulated with respect to time, amplitude and direction.

Acknowledgement

This work has been supported by the swedish Medical Research Council and from funds of the Kardinska Institute, Stockholm.

References

ADES, H. W., ENGSTRÖM, H.: Form and innervation of the vestibular epithelia. In: First symposium on the role of the vestibular organs in the exploration of space. NASA report SP-77. Washington: U. S. G. P. O. 1965.

ADOLPH, A. R.: Spontaneous slow potential fluctuations in the *Limulus* photoreceptor. J. Gen. Physiol. 48, 297—322 (1964).

AFZELIUS, B. A.: Flimmer flagellum of the sponge. Nature (Lond.) 191, 1318 (1961).

ANAND, N., DAVIS, B. D., ARMITAGE, A. K.: The effect of streptomycin on *Escherichia coli* — uptake of streptomycin by *Escherichia coli*. Nature (Lond.) 185, 22—23 (1964).

BARBER, U. C.: Preliminary observations on the fine structure of the octopus statocyst. J. Microscopie 4, 547—550 (1965).

BÉKÉSY, G. von: Experiments in hearing (Research articles from 1928 to 1958). New York: McGraw Hill Book Co., Inc. 1960.

— Some similarities in sensory perception of fish and man. In: Lateral line detectors. Bloomington: Indiana Univ. Press 1967.

BERGEIJK, W.: The evolution of vertebrate hearing. In: Contributions to sensory physiology, Vol. 2. New York: Academic Press Inc. 1967.

BREDBERG, G.: Cellular pattern and nerve supply of the human organ of Corti. Acta otolaryng. (Stockh.) Suppl. 236, 1—135 (1968).

BULLOCK, F. H., DIECKE, F. P. J.: Properties of an infrared receptor. J. Physiol. 134, 47—87 (1956).

BUTLER, R. A.: Some experimental observations on the DC resting potentials in the guinea pig cochlea. J. Acoust. Soc. Amer. **37**, 429—433 (1965).
— HONRUBIA, V.: Responses of cochlear potentials to changes in hydrostatic pressure. J. Acoust. Soc. Amer. **35**, 1188—1192 (1963).
CAPRANICA, R., FLOCK, Å., FRISHKOPF, L. S.: Microphonic response from the inner ear of the bullfrog. J. Acoust. Soc. Amer. **40**, 1262 (1966).
DAVIS, H.: Excitation in auditory receptors. In: Handbook of Physiology. Section I: Neurophysiology. Washington: Amer. Physiol. Soc. 1959.
— Some principles of sensory receptor action. Physiol. Rev. **41**, 391—416 (1961).
— A model for transducer action in the cochlea. Cold Spr. Harb. Symp. quant. Biol. **30**, 181—190 (1965).
DETHRAGE, B. H.: Examination of binaural interaction. J. Acoust. Soc. Amer. **39**, 232—249 (1966).
DIJKGRAAF, S.: The functioning and significance of the lateral-line organs. Biol. Rev. **38**, 51—105 (1963).
DOHLMAN, G.: Histochemical studies of vestibular mechanisms. In: Neural mechanisms of the auditory and vestibular system. Springfield, III.: Charles C. Thomas, Publ. 1960.
— ORMEROD, F. C., McLAY, K.: The secretory epithelium of the internal ear. Acta oto-laryng. (Stockh.) **50**, 243—249 (1959).
DUBIN, D. T., DAVIS, B. D.: The effect of streptomycin on potassium flux in *Escherichia coli*. Biochem. biophys. Acta (Amst.) **52**, 400—402 (1961).
DUVALL, J., FLOCK, Å., WERSÄLL, J.: The ultrastructure of the sensory hairs and associated organelles of the cochlear inner hair cells, with reference to directional sensitivity. J. Cell. Biol. **29**, 497—505 (1966).
DUVALL, A. J., WERSÄLL, J.: Site of action of streptomycin upon inner ear sensory cells. Acta oto-laryng. (Stockh.) **57**, 581—598 (1964).
ECCLES, J. C.: The ion mechanisms of excitatory and inhibitory synaptic action. Ann. N. Y. Acad. Sci. **137**, 473—494 (1966).
ENGSTRÖM, H., ADES, H. W., HAWKINS, J. E.: Structure and functions of the sensory hairs of the inner ear. J. Acoust. Soc. Amer. **34**, 1356—1363 (1962).
EWALD, J. R.: Physiologische Untersuchungen über das Endorgan des Nervus octavus. Wiesbaden: Bergmann 1892.
EYZAGUIRRE, C., KUFFLER, S. W.: Processes of excitation in the dendrites and in the soma of single isolated sensory nerve cells of the lobster and crayfish. J. Gen. Physiol. **39**, 87—119 (1955).
FERNANDEZ-MORAN, H.: Cell membrane ultrastructure, low- temperature electron microscopy and x-ray diffraction studies of lipoprotein components in lammellar systems. Circulation **26**, 1039—1065 (1962).
FEX, J.: Auditory activity in centrifugal and centripetal fibers in cat, a study of a feedback system. Acta physiol. scand. Suppl. **189**, 1—68 (1962).
— Efferent inhibition in the cochlea related to hair cell DC activity of the crossed olivo-cochlear fibers in the cat. J. Acoust. Soc. Amer. **41**, 666—675 (1967).
FLOCK, Å.: Structure and function of the macula utriculi with special reference to directional interplay of sensory responses as revealed by morphological polarization. J. Cell Biol. **22**, 413—431 (1964).
— Electron microscopic and electrophysiological studies on the lateral line canal organ. Acta oto-laryng. (Stockh.) Suppl. **199**, 1—90 (1965).
— Ultrastructure and function in the lateral line organs. In: Lateral line detectors. Bloomington: Indiana Univ. Press 1967.
— DUVALL, A. J.: The ultrastructure of the kinocilium of the sensory cells in the inner ear and lateral line organs. J. Cell Biol. **25**, 1—8 (1965).
— KIMURA, R., LUNDQUIST, P.-G., WERSÄLL, J.: Morphological basis of directional sensitivity of the outer hair cells in the organ of Corti. J. Acoust. Soc. Amer. **34**, 1351—1355 (1962).
— WERSÄLL, J.: A study of the orientation of the sensory hairs of the receptor cells in the lateral line organ of a fish with special reference to the function of the receptors. J. Cell Biol. **15**, 19—27 (1962).

FRANK, K., FUORTES, M.: Excitation and conduction. Ann. Rev. Physiol. **23**, 357—386 (1961).

FRIEDMANN, I.: The chick embryo otocyst: A model ear. J. Laryngol. Otol. **82**, 185—202 (1968).

FRISHKOPF, L. S.: Excitation and inhibition of primary auditory neurons in the little brown bat. J. Acoust. Soc. Amer. **36**, 1016 (1964).

— FLOCK, Å.: Ultrastructure of the basilar papilla in the bullfrog. J. Acoust. Soc. Amer. **41**, 1578 (1967).

FURSHPAN, E. J.: Electrical transmission at an excitatory synapse in a vertebrate brain. Science **144**, 878—880 (1964).

FURUKAWA, T., ISHII, Y.: Neurophysiological studies on hearing in gold fish. J. Neurophysiol. **30**, 1377—1403 (1967a).

— — Effects of static bending of sensory hairs on sound reception in the gold fish. Jap. J. Physiol. **17**, 572—588 (1967b).

GAGE, P. W., HUBBARD, J. I.: Evidence for a poisson distribution of miniature end plate potentials and some implications. Nature (Lond.) **208**, 395—396 (1965).

GIBBONS, I. R.: The relationship between the fine structure and direction of beat in gill cilia of a lamellibranch mollusc. J. biophys. biochem. Cytol. **11**, 179—205 (1961).

GOLDSTEIN, M.: The auditory periphery. In: Medical Physiology. St. Louis: Mosby 1968. **2**, 1465—1498.

GÖRNER, P.: Untersuchungen zur Morphologie und Elektrophysiologie des Seitenlinieorgans vom Krallenfrosch (Xenopus laevis Daudin). Z. vergl. Physiol. **47**, 316—338 (1963).

GRAY, E. G., PUMPHREY, R. J.: Ultrastructure of the insect ear. Nature (Lond.) **181**, 618 (1958).

GRIFFIN, D. R., WEBSTER, F. A., MICHAEL, C. R.: The echolocation of flying insects by bats. Anim. Behav. **8**, 141—154 (1960).

GRUNDFEST, H.: Synaptic and ephatic transmission. In: Handbook of physiology. Section I: Neurophysiology. Washington: Amer. physiol. soc. 1959.

— Effects of drugs on the central nervous system. Ann. Rev. Pharmacol. **4**, 341—364 (1964).

— Electrophysiology and pharmacology of different components of bioelectric transducers. Cold Spr. Harb. Symp. quant. Biol. **30**, 1—14 (1965).

GUALTIEROTTI, T., ALLTUCKER, D.: The relationship between the unit activity of the utricle-saccule of the frog and information transfer. In: Second symp. on the role of the vestibular organs in space exploration. NASA report SP-115. Washington: U. S. G. P. O. 1966.

GUINAN, J. J., Jr., PEAKE, W. T.: Middle ear characteristics of anesthetized cats. J. Acoust. Soc. Amer. **41**, 1237—1261 (1967).

HAMA, K.: A study of the fine structure of the saccular macula of the gold fish. Z. Zellforsch. **94**, 155—171 (1969).

HÅKANSSON, C. H., and TOREMALM, N. G.: Studies on the physiology of the trachea. III. Electrical activity of the ciliary cell layer. Amer. Otol. Rhinol. Laryngol. **75**, 1007—1019 (1966).

HARRIS, G. G.: Brownian motion in the cochlear partition. J. Acoust. Soc. Amer. **40**, 1264 (1966).

— BERGEIJK, VAN, W. A.: Evidence that the lateral-line organ responds to the nearfield displacements of sound sources in water. J. Acoust. Soc. Amer. **34**, 1831—1841 (1962).

— MILNE, D. C.: Input-output characteristics of the lateral-line organs of Xenopus laevis. J. Acoust. Soc. Amer. **40**, 32—42 (1966).

— FLOCK, Å.: Spontaneous and evoked activity from the Xenopus laevis lateral line. In: Lateral Line Detectors. Bloomington: Indiana Univ. Press 1967.

— FRISHKOPF, L., FLOCK, Å.: Receptor potentials in the hair cells of mudpuppy lateral line. J. Acoust. Soc. Amer. **45**, 300—301 (1969).

— — — Receptor potentials from hair cells of the lateral line. Science **167**, 76—79 (1970).

HARTLINE, H. K., WAGNER, H. G., MacNICHOL, E. J.: The peripheral origin of nervous activity in the visual system. Cold Spr. Harb. Symp. quant. Biol. **17**, 125—141 (1952).

HAWKINS, J.: Cytoarchitectural basis of the cochlear transducer. Cold Spr. Harb. Symp. quant. Biol. **30**, 147—157 (1965).

HENRIKSSON, N. G., GLEISNER, L.: Vestibular activity of experimental variation of labyrinthine pressure. Acta oto-laryng. (Stockh.) **61**, 380—386 (1966).

HIBBARD, E.: Selective innervation and reciprocal functional suppression from grafted extra labyrinths in amphibians. Expl. Neurol. **10**, 271—283 (1964).

HILDING, D. A., WERSÄLL, J.: Cholinesterase and its relation to the nerve endings in the inner ear. Acta oto-laryng. (Stockh.) **55**, 205—217 (1962).

HODGKIN, A. L., HUXLEY, A. F.: A quantitative description of membrane current and its application to conduction and excitation in nerve. J. Physiol. **117**, 500—544 (1952).

HOLST, E. VON: Die arbeitsweise des Statolithenapparates bei Fischen. Z. vergl. Physiol. **32**, 60—120 (1950).

HORRIDGE, G. A.: Intracellular action potentials associated with the beating of the cilia in ctenophore comb plate cells. Nature (Lond.) **205**, 602 (1965).

HUBEL, D. H., WIESEL, T. N.: Receptive fields, binocular interaction and functional architecture in the cat's visual cortex. J. Physiol. **160**, 106—154 (1962).

IURATO, S.: Tectorial membrane. In: Submicroscopie structure of the Inner Ear. Oxford: Pergamon Press 1967.

JAHNKE, V., LUNDQUIST, P.-G., WERSÄLL, J.: Some morphological aspects on sound perception in birds. Acta Oto-laryng. **67**, 583—601 (1969).

JENSEN, C. E., VILSTRUP, T.: On the chemistry of human cupulae. Acta Oto-laryng. **52**, 383 (1960).

JIELOF, R., SPOOR, A., DE FRIES, H.: The microphonic activity of the lateral line. J. Physiol. **116**, 137—157 (1952).

JOHNSTON, B. M., BOYLE, A. J. F.: Basilar membrane vibration examined with the Mössbauer Technique. Science **158**, 389—390 (1967).

KANEKO, A., HASHIMOTO, H.: Recording site of single cone response determined by an electrode marking technique. Vision Res. **7**, 847—851 (1967).

KATSUKI, Y., USCHIYAMA, H., TOTSUKA, G.: Electrical responses of the single hair cell in the ear of fish. Proc. Japan. Acad. **30**, 248—255 (1954).

— — Note on the hair cell potential of the ear of fish. Proc. Japan. Acad. **31**, 99 (1955).

— YANAGISAWA, K., KANZAKI, J.: Tetraethylammonium and tetrodotoxin: effects on cochlear potentials. Science **151**, 1544—1545 (1966).

KATZ, B., MILEDI, R.: Tetrodotoxin and neuromuscular transmission. Proc. Roy. Soc. (Lond.) Ser. B. **167**, 8—22 (1967).

KIANG, N. Y.-S., WATANABE, T., THOMAS, E. C., CLARK, L. F.: Discharge patterns of single nerve fibers in the cat's auditory nerve. Research monograph No. 35. The MIT Press, Cambridge, Massachusetts (1965).

KIKUCHI, K., HILDING, D. A.: The development of the organ of Corti in the mouse. Acta oto-laryng. (Stockh.) **60**, 207—222 (1965).

KIMURA, R.: Hairs of the cochlear sensory cells and their attachment to the tectorial membrane. Acta oto-laryng. (Stockh.) **61**, 55—72 (1966).

KOMNICK, H., KOMNICK, U.: Elektronenmikroskopische Untersuchungen zur Funktionellen Morphologie des Ionentransportes in der Salzdrüse von *Larus argentatus*. Z. Zellforsch. **60**, 163—203 (1963).

KUIPER, J. W.: The microphonic effect of the lateral line organ. Publ. Biophys. Group „Natuurkundig laboratorium", Groningen, 1—159 (1956).

KUSANO, K., LIVENGOOD, D. R., WERMAN, R.: Correlation of transmitter release with properties membrane of the presynaptic fiber of the squid giant synapse. J. gen. Physiol. **50**, 2579—2601 (1967).

LAGRANGE, J. L.: Oevres de Langrage. Paris: Serret et Darboux 1867—1892.

LOEWENSTEIN, W. R.: Excitation and inactivation in a receptor membrane. Ann. N. Y. Acad. Sci. **94**, 510—534 (1961).

— Permeability of membrane junctions. Ann. N. Y. Acad. Sci. **137**, 441—472 (1966).

— TERZUOLO, C. A., WASHIZU, Y.: Seperation of transducer and impulse-generating processes in sensory receptors. Science **142**, 1180—1181 (1963).

LOWENSTEIN, O.: Comparative physiology of the otolith organs. Brit. Med. Bull. **12**, 110—114 (1956).

— OSBORNE, M. P., WERSÄLL, J.: Structure and innervation of the sensory epithelia of the labyrinth in the Thornback ray (*Raja clavata*). Proc. Roy. Soc. Biol. **160**, 1—12 (1964).

— ROBERTS, T. O. M.: The equilibrium function of the otolith organs of the thornback ray (Raja clavata). J. Physiol. **110**, 392—415 (1949).

LOWENSTEIN, O., SAND, A.: The individual and integrated activity of the semicircular canals of the elasmobranch labyrinth. J. Physiol. **99**, 89—101 (1940).

— WERSÄLL, J.: A functional interpretation of the electron microscopic structure of the sensory hairs in the cristae of the eleasmobranch *Raja clavata* in terms of directional sensitivity. Nature (Lond.) **184**, 1807—1810 (1959).

MARTIN, A. R.: Quantal nature of synaptic transmission. Physiol. Rev. **46**, 51—66 (1966).

MILEDI, R., SLATER, C. R.: The action of calcium on neuronal synapses in the squid. J. Physiol. **184**, 473—498 (1966).

MILSUN, J. H., JONES, G. M.,: Trigonometric resolution of neural response from the vestibular otolith organ. In: Digest of the 7th Int. Conf. on Medical and Biological Engineering. Stockholm: Almqvist & Wiksell, 1967.

MÖLLER, A. R.: Unit responses in the rat cochlear nucleus to repetitive transient sounds. Acta physiol. scand. **75**, 542—551 (1969).

RETZIUS, G.: Das Gehörorgan der Wirbeltiere. II. Das Gehörorgan der Reptilien, der Vögel und der Säugetiere. Stockholm: Die Centraldruckerei 1884.

ROBBINS, R. G., BAUMKNIGHT, R. S., HONRUBIA, V.: Anatomical distribution of efferent fibers in the 8th cranial nerve of the bullfrog (*Rana catesbeiana*). J. Acoust. Soc. Amer. **41**, 1581 (1967).

ROBERTSON, J. D.: The occurence of a subunit pattern in the unit membranes of club endings in Mauthner cell synapses in goldfish brains. J. Cell Biol. **19**, 201—221 (1963).

RODIECK, R. W., KIANG, N. Y.-S., GERSTEIN, G. L.: Some quantitative methods for the study of spontaneous activity of single neurons. Biophys. J. **2**, 351—368 (1962).

ROSE, J. E., BRUGGE, J. F., ANDERSON, D. J., HIND, J. E.: Phase locked response to low frequency tones in single auditory nerve fibers of the squirrel monkey. J. Neurophysiol. **30**, 769—793 (1967).

SAND, A.: The mechanism of lateral sense organs of fishes. Proc. Roy. Soc. Biol. **123**, 472—495 (1937).

SCARPA, A.: Anatomische Untersuchungen des Gehörs und Geruchs. Aus dem Lateinischen original: Anatomicae disquisitiones de audit et olfactu. Ticini 1789. Nürnberg: Kaspeschen Buchhandlung 1800.

SCHMIDT, R. S., FERNANDEZ, C.: Labyrinthine DC potentials in representative vertebrates. J. Cell Comp. Physiol. **59**, 311—322 (1962).

SCHMITT, F. O.: Molecular and ultrastructural correlates of function in neurons, neuronal nets, and the brain. Naturwissenschaften **53**, 71—79 (1966).

SCHUKNECHT, H. F., CHURCHILL, J. A., DORAN, R.: The localization of acetylcholinesterase in the cochlea. Arch. oto-laryng. (Stockh.) **69**, 549—559 (1959).

SCHWARTZ, E.: Analysis of surface wave perception in some teleosts. In: Lateral line detectors. Bloomington: Indiana Univ. Press 1967.

SCHWARTZKOPFF, J.: Vergleichende Physiologie des Gehörs und der Lautäußerungen. Forschr. Zool. **15**, 213—336 (1962).

SHOUTEN, J. F., RITSMA, R. L., LOPES CARDOZO, B.: Pitch of the residue. J. Acoust. Soc. Amer. **34**, 1418—1424 (1962).

SJÖSTRAND, F.: A new ultrastructural element of the membranes in mitochondria and of some cytoplasmic membranes. J. Ultrastruct. Res. **9**, 340—361 (1963).

SMITH, C. A., DAVIS, H., DEATHERAGE, B. H., GESSERT, C. F.: DC potentials of the membraneous labyrinth. Amer. J. Physiol. **193**, 203—206 (1958).

— LOWRY, O. H., WU, M. L.: The electrolytes of the labyrinthine fluids. Laryngoscope **64**, 141—153 (1954).

— SJÖSTRAND, F. S.: A synaptic structure in the hair cells of the guinea pig cochlea. J. Ultrastruct. Res. **5**, 184—192 (1961).

SPOENDLIN, H.: Organization of the sensory hairs in the gravity receptors in utricule and saccule of the squirrel monkey. Z. Zellforsch. **62**, 701—716 (1964).

— Ultrastructural studies of the labyrinth in sequirel monkeys. In: First symposium on the role of the vestibular organs in the exploration of space. NASA report SP-77. Washington U. S. G. P. O.

SPOENDLIN, H.: Some morphofunctional and pathological aspects of the vestibular sensory epithelia. In: Second symposium on the role of the vestibular organs in the exploration of space. NASA repott SO-115. Washington: U. S. G. P. O. 1966.

STOPP, P. E., WHITFIELD, I. C.: Summating potentials in the avian cochlea. J. Physiol. **175**, 45—46 (1964).

TASAKI, I.: Nerve impulses in individual auditory nerve fibers of guinea pig. J. Neurophysiol. **17**, 97—122 (1954).

— Afferent impulses in auditory nerve fibers and the mechanism of impulse initiation in the cochlea. In: Neural mechanisms of the auditory and vestibular system. Springfield, III.: Charles C. Thomas, Publ. 1961.

— DAVIS, H., ELDREDGE, D. H.: Exploration of the cochlear potentials in the guinea pig with microelectrodes. J. Acoust. Soc. Amer. **26**, 765—773 (1954).

— SINGER, I.: Membrane macromolecules and nerve excitability: A physico-chemical interpretation of excitation in squid giant axons. Ann. N. Y. Acad. Sci. **137**, 792—806 (1966).

— SPYROPOLOUS, C. S.: Stria vascularis as a source of endocochlear potential. J. Neurophysiol. **22**, 149—155 (1959).

TEAS, D. C., ELDREDGE, D. H., DAVIS, H.: Cochlear responses to acoustic transients: an interpretation of whole nerve action potentials. J. Acoust. Soc. Amer. **34**, 1438—1459 (1962).

TERZUOLO, C. A., WASHIZU, Y.: Relation between stimulus strength, generator potential and impulse frequency in strech receptor of crustacea. J. Neurophysiol. **25**, 56—66 (1962).

THURM, U.: Steps in the transducer process of mechanoreceptors. Symp. zool. Soc. Lond. **23**, 199—216 (1968).

TRINCKER, D.: Bestandspotentiale im Bogengangssystem des Meerschweinchens und ihre Änderungen bei experimentellen Cupula-Ablenkungen. Pflügers Arch. ges. Physiol. **264**, 351—382 (1957).

— Neuere Untersuchungen zur Elektrophysiologie des Vestibular-Apparates. Naturwissenschaften **46**, 344—350 (1959).

VRIES, H.: Die Reizschwelle der Sinnesorgane als physiologisches Problem. Experientia (Basel) **4**, 205—240 (1948).

— BLEEKER, I. D.: The microphonic activity of the labyrinth of the pigeon. — II. The responce of the cristae in the semicircular canals. Acta oto-laryng. (Stockh.) **37**, 298—306 (1949).

WERNER, C. F.: Das Labyrinth. Leipzig: Thieme 1940.

WERSÄLL, J.: Studies on the structure and innervation of the sensory epithelium of the cristae ampullares in the guinea pig. Acta oto-laryng. (Stockh.) Suppl. **126**, 1—85 (1956).

— FLOCK, Å.: Suppression and restoration of the microphonic output from the lateral line organ after local application of streptomycin. Life Sci. **3**, 1151—1155 (1964).

— — Functional anatomy of the vestibular and lateral line organs. In: Contributions to Sensory Physiology, Vol. I. New York: Academic Press Inc. 1965.

— — LUNDQUIST, P.-G.: Structural basis for directional sensitivity in cochlear and vestibular sensory receptors. Cold Spr. Harb. Symp. quant. Biol. **30**, 115—145 (1965).

— LUNDQUIST, P.-G.: Morphological polarization of the mechanoreceptors of the vestibular and acoustic systems. In: Second symp. on the role of the vestibular organs in the exploration of space. NASA report SP-115. Washington: U. S. G. P. O. 1966.

WEVER, E. G., BRAY, C. W.: Action currents in the auditory nerve in response to acoustical stimulation. Proc. nat. Acad. Sci. (Wash.) **16**, 344—350 (1930).

WHITFIELD, I. C., EVANS, E. F.: Responses of auditory cortical neurons to stimuli of changing frequency. J. Neurophysiol. **28**, 655—672 (1965).

— ROSS, H. F.: Cochlear-microphonic and summating potentials and the outputs of individual hair cell generators. J. Acoust. Soc. Amer. **38**, 126—131 (1965).

WOLBARSHT, M. L., HANSON, I. E.: Electrical activity in the chemoreceptors of the blowfly. III. Dendritic action potentials. J. gen. Physiol. **48**, 673—683 (1965).

Chapter 15

Transducer Properties and Integrative Mechanisms in the Frog's Muscle Spindle

By

D. Ottoson, Stockholm (Sweden), and G. M. Shepherd, New Haven, Conn. (USA)

With 39 Figures

Contents

I. Introduction

Since the classical studies by ADRIAN and ZOTTERMAN in 1926, the function of the muscle spindle has attracted a continuously increasing interest. Research in this field has been focused mainly on the reflex functions and the motor regulation of the spindle; comparatively little has been done on the action of the spindle as a mechano-electrical transducer.

Work on the receptor functions of the spindle goes back to MATTHEWS' study in 1931 in which the basic relations between the stimulus and the spindle response were disclosed. The preparation employed consisted of a frog's toe muscle; in most cases it was found to contain one end organ, which by staining could be identified as a muscle spindle. When the muscle was loaded with weights, MATTHEWS found that the spindle discharged with an initial high rate which declined to a steady level during maintained constant stretch. He suggested that the rapid fall to the steady level was due to differences in viscosity between different regions of the spindle. The static discharge was characterized by a striking regularity at firing frequencies above 20 per sec, and it increased linearly in rate with the logarithm of the load. There was a depression of the excitability of the spindle after loading, and this "adaptation remainder" was related to the total amount of previous stimulation i.e., the magnitude and duration of the previous stretch.

An important step forward in the understanding of the mechanisms underlying impulse generation in the spindle was made with the discovery of the receptor potential by KATZ in 1950. In recording sensory impulses from a frog's spindle he observed that the propagated impulse discharge was superimposed upon a slow depolarization which increased in amplitude with the intensity of the stimulus. This depolarization was not the result of summation of after-potentials of the action potentials since it often began before the first impulse. Furthermore, the slow potential could be obtained in isolation after the conducted activity was eliminated by bathing the preparation with procaine. KATZ concluded that this potential represented the activity of the sensory terminals and that it was a link between the mechanical stimulus and the afferent impulse discharge. The receptor potential exhibited two components, an initial one associated with dynamic stretching and a later one associated with maintained stretch. After release of stretch, there was a transient potential change of opposite polarity. KATZ (1950b) discussed various alternative explanations for the production of the receptor potential (e.g., a capacitance change, chemical changes or permeability changes caused by the deformation of the membrane), but he concluded that the experimental evidence did not provide any direct evidence for any of these hypotheses.

The process of encoding the characteristic features of the mechanical stimulus into an afferent impulse message involves a sequence of mechanical and electrical events. One of the primary and functionally most important steps in this process is the conversion of the mechanical deformation into a depolarization of the sensory endings. Another important step is the conversion of this depolarization into a propagated impulse message. Katz' study disclosed a close relationship between the intensity of the local depolarization of the endings and the frequency of the afferent impulse discharge. It also focused attention on the challenging problems involved in the function of the spindle as a mechano-electrical transducer.

It has been the aim of the studies which will be reviewed in this article to gain insight into the mechanisms of the encoding process of the muscle spindle. For this purpose, the frequency characteristics of the impulse response to controlled stretches of various parameters have been analyzed and related to the underlying receptor potential. Studies have also been carried out on the ionic mechanisms involved in the different stages of the receptor activity. All of the results described are derived from experiments on single frog spindles. Isolation of the preparation was found to be a necessary prerequisite for precise study of the electrical changes taking place in the sensory terminals. The isolated preparation also offers several other advantages compared with the in situ spindle. Since the spindle can be observed under the microscope, the experimental conditions throughout an experiment can be easily controlled. The mechanical stimulus can be applied directly to the polar ends of the spindle; distortion of the stimulus caused by transmitting structures in the bulk of extrafusal muscle tissue is therefore reduced to a minimum. Furthermore, since motor activity is abolished in the isolated spindle, the complications of muscular activity which otherwise might affect the sensory discharge are eliminated. The results to be described therefore relate only to the sensory function of the spindle but it is hoped that they may indirectly provide information about the action of the spindle under various conditions of intrafusal motor activity.

II. Structure of the Spindle

It was established early in light microscopic work (see review of Regaud and Favre, 1904/05) that there are great variations in structure of the muscle spindle from one species to another and also that within the same species there may be different types of spindles. The anatomically simplest spindle has been found in reptiles. In these species the spindle is unifascicular, i.e., it contains only a single intrafusal muscle fibre and usually each spindle has only one sensory fibre. In general the mammalian spindle is considerably more complex; it often contains two types of sensory endings and several muscle fibers. The amphibian muscle spindle (Fig. 1) takes an intermediate position between the complex mammalian

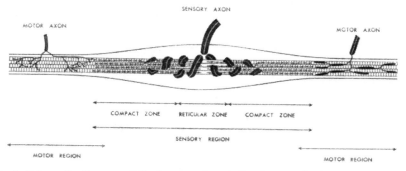

Fig. 1. Schematic diagram of the frog muscle spindle. See text for further description

spindle and the simple spindle in reptiles (see CAJAL, 1888; GRAY, 1957). It receives only one sensory fiber which branches repeatedly and coils around the intrafusal muscle fibers. The branches lose their myelin sheaths and terminate as beaded chains which run alongside the intrafusal muscle fibers. A frog's spindle contains 3 to 12 intrafusal muscle fibres and has an overall diameter of about 100 μ. Striation of the fibers is absent for a distance of about 100 μ in the central equatorial region of the spindle. The muscle fibers are innervated with large and small motor fibers, which terminate as "en plaques" end plates and "en grappes" end plates, respectively.

Recent electron microscopic work has revealed features in the spindle structure of considerable interest. Particularly important from the point of view of the sensory function of the spindle is the fine structure of the terminal branches and their relation to the intrafusal muscle fibers. KATZ (1961) found that the sensory terminals are formed by thin cylindrical tubes (about 0.15 μ in diameter) which connect varicose swellings in series. Many of these bulbs are seated in sockets in the surface of the muscle and appear to be anchored to the muscle by thin filaments. KARLSSON, ANDERSSON-CEDERGREN, and OTTOSON (1966) confirmed that there is an intimate contact between the bulbs and the intrafusal muscle fibers but they found very few connecting filaments. In the reticular zone they found that the bulbs were located close to regularly spaced muscular processes containing one sarcomere of contractile material (Fig. 2). The distribution of the sensory endings is bimodal with maxima at the borders between the reticular and the sensory compact zones. It is estimated that the total number of bulbs is about 10,000 (KATZ, 1961). To account for this great number of sensory structures, there must be an immense ramification of the single afferent nerve fiber. KARLSSON et al., (1966) estimated that up to sixth orders of dichotomous branching are required to give the number of cross-sectioned structures found in the transition zone.

In the spindle there is also a great number of satellite cells which are closely associated with the surface of the intrafusal muscle fibers and with the sensory endings. They appear as long fusiform cells oriented along the longitudinal axis of the muscle fiber. The nucleus is about 20 μ in length and occupies the larger part of the cell body. The cytoplasm of the cell contains few mitochondria in contrast to the sensory endings, but has a variety of vesicular bodies. The majority of the cells appear to be located in the sensory compact zones and they are relatively scarce in the reticular zone. At present there is no information on the function of these cells.

In the central equatorial zone of the spindle, the myofilaments of the intrafusal fibres are reduced in number. KATZ (1961) proposed the name "reticular zone" for this portion and "compact zones" for the polar portions of the intrafusal fibres where they have the same appearance as in the extracapsular regions. The sensory innervated region may consequently be divided into a sensory reticular zone and two sensory compact zones. In the reticular zone the muscle fibers lose about 85% of their content of myofilaments and split up into a complex network of fine connective tissue fibrils. Some fibers appear, however, to escape this differentiation and pass through the reticular zone with only a slight reduction in their myofilament content. The intrafusal fibers contain a peculiar ladderlike struc-

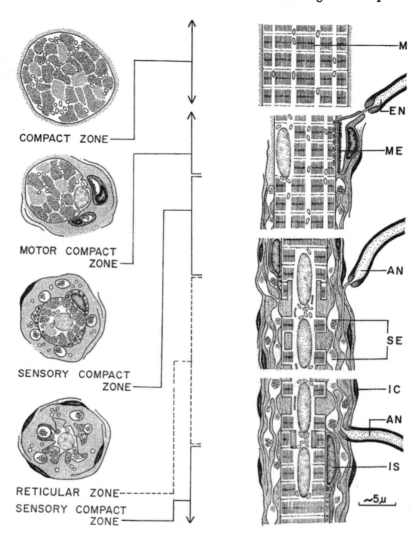

Fig. 2. Schematic presentation of the fine structure of the frog muscle spindle. Longitudinal (right) and transverse (left) aspects of sensory and motor innervation of different zones of one intrafusal muscle fibre (*M*). In the reticular zone the sensory nerve endings (*SE*) of the afferent fibre (*AN*) are in close contact with regularly spaced muscular processes. Motor nerve endings (*ME*) in the compact zones are characterized by an accumulation of small vesicles. Intrafusal satellite cells (*IS*) with dark nuclei are seen in intimate contact with the muscle fibres (from Karlsson, Andersson-Cedergren and Ottoson, 1966). Copyright Academic Press

ture within the region of sensory contacts, characterized by filamentous material in the interior of the intrafusal fibres. This material is longitudinally oriented and interrupted at regular intervals by discs of high electron opacity. Katz has suggested that these ladders may form intracellular "microtendons" between myofilaments and certain sensory contacts.

It appears likely that the structural differences between different zones of the spindle are functionally significant. MATTHEWS (1931) suggested that mechanical properties of different areas of sensory innervation might account for the adaptation of the impulse discharge during maintained stretch. KATZ (1961) arrived at a somewhat similar conclusion on the basis of electron microscopic observations. He suggested that the typical mechanical structure of the reticular zone may be responsible for the dynamic phase of the response, the static response being produced mainly by the compact zones.

The spindle is enclosed in a capsule which is a direct continuation of the connective tissue sheath surrounding the sensory axon. The capsule is composed of relatively thick lamellae and exhibits a basement membrane coating. It tapers off toward the polar end of the spindle and adheres closely to the surface of the muscle fibers. It would appear, therefore, that there is little communication between the space enclosed by the capsule and the extrafusal fluid. SHERRINGTON (1894) succeeded in injecting the spindle space through the lymphatics and suggested that it was a lymph space. More recent experiments (VON BRZEZINSKI, 1961) have failed to confirm this finding. An interesting observation is that the spindle fluid contains a relatively high amount of hyaluronic acids, which are probably formed by the capsule. It is possible that the hyaluronic acids by virtue of their high viscosity provide the spindle with a protective fluid cushion. It has recently been suggested that the intracapsular fluid may play a role in distributing pressure within the spindle (BRIDGMAN and ELDRED, 1964).

In addition to the outer capsule, there is also an inner capsule which encloses the individual intrafusal fibers in the reticular region. In the compact zones, the enclosure of the fibers is less complete and there is a free communication between the main outer space and the inner space formed by the investment of the individual fibers. The characteristic encapsulation of the fibers in the reticular zone may simply provide a mechanical protection for the muscle fibers. Furthermore, the individual capsules might have some other functions related to the sensory innervation. It is of interest that in this region there is a typical localization of the sensory bulbs to the sarcomeres of the intrafusal fibers.

III. Experimental Methods

All experiments described in this article have been carried out on spindles isolated from the *Musculus extensor digitorum longus IV* of the frog (*Rana temporaria*). The spindle and about 2 mm of its sensory axon were dissected free from the muscle and transferred to a small perspex chamber containing Ringer solution. It was then mounted between two fine nylon rods, one of which was fixed and the other attached to a loudspeaker coil. Stretches of varying velocities and amplitudes were obtained by feeding electrical pulses of different waveforms into the loudspeaker. The mechanical stimulus was monitored directly on one oscilloscope beam by a high-sensitivity capacitance meter (HAAPANEN, 1962) which measured the displacement of the spindle. Recordings were made with calomel half-cell electrodes connected to the preparation through Ringer-agar bridges. Amplification was in most cases made with a direct coupled amplifier.

IV. General Properties of the Spindle Response
A. Impulse Response

When the isolated spindle is held at resting length, it usually fires at a low and irregular rate. This spontaneous discharge is also present before isolation, i.e., when the spindle is in situ (see MATTHEWS, 1931; KATZ, 1950 a), and is therefore not to be regarded as due to any damage during the dissection procedure. The discharge goes on for hours with very little change in frequency and seems to be a sensitive index of the functional state of the spindle. For instance, the effects of drugs or changes in the ionic composition of the external solution are usually revealed by a change in the spontaneous firing long before any significant effects on the response to stretch can be observed (OTTOSON, 1961). In addition to the impulses, small abortive spikes of different amplitudes often appear. These seem to occur quite independently of the large impulses and usually do not propagate into the afferent fiber. KATZ (1950 a) who first observed these potentials, suggested that they represented action potentials in the terminal branches of the afferent fiber and that several such spikes might build up to a propagated impulse.

When the spindle is stretched there is an initial dynamic discharge of impulses followed by a decline in frequency to a steady static level during maintained stretch. At the transition from dynamic to static stretch, a pause in the firing often occurs. Aften cessation of stretch, the spindle is silent for a period before the spontaneous firing reappears. These features of the impulse response change in a regular way with changes of the parameters of stretching. Thus the discharge during dynamic stretching starts earlier and reaches higher frequencies when the amplitude or velocity of stretching is increased. The static response on the other hand is independent of the velocity of the applied stretch and only dependent on its absolute amount. In the following account the characteristics of the different phases of the impulse discharge and their relation to the applied stimulus will be described in more detail.

1. Relation between Stimulus Strength and Impulse Frequency

When the spindle is subjected to a rapid stretch, only one spike occurs during the period of extension unless the stretch is very strong. To study the change in the dynamic response during extension, relatively slow stretches must be used so that several spikes are produced. The general appearance of the discharge during extension at a rate of 5 mm/sec to three different amplitudes is illustrated in Fig. 3.

A typical feature of the discharge is the linear increase in frequency with increasing lengthening of the spindle (SHEPHERD and OTTOSON, 1965). For a given velocity of stretch, the peak frequency of dynamic discharge consequently is a direct function of the magnitude of extension. The static discharge exhibits a similar linear relation to spindle length but, in contrast to the dynamic discharge, the static firing rate is independent of the velocity of stretching. In the diagram in Fig. 4, the dynamic and static frequencies of the response are plotted versus extension. Both the dynamic and the static frequencies rise in direct relation to stretch. Since in this case a slowly rising stretch was used, the difference in slope between the lines is small; with greater velocity of stretching, the line for the dynamic discharge would have become steeper, the static discharge remaining unchanged.

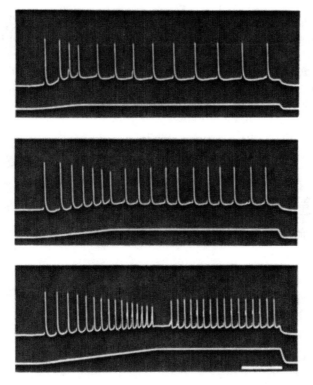

Fig. 3. Characteristic response of muscle spindle to linearly rising stretch to three different amplitudes. Lower trace in this and following records indicates extension. Time bar: 50 msec

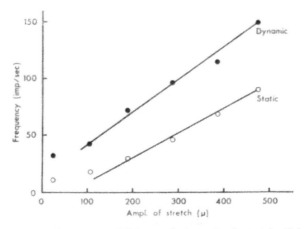

Fig. 4. Relation between frequency of firing and strength of stretch. Values for dynamic discharge obtained by measuring spike intervals of response to linearly rising stretch at 2.6 mm/sec; static values obtained by corresponding measurement 200 msec after onset of sustained stretch to different lengths

An interesting property of the spindle is that the dynamic discharge reaches its maximum firing rate at lower levels of extension than the static response. With a fast rising stretch the maximum dynamic firing is usually obtained at an extension of about 30% of the spindle, whereas to elicit the maximum static frequency the spindle has to be lengthened by 100% or more. The static firing may then reach the same value as the dynamic discharge. However, such stretches produce irreversible damage to the spindle as evidenced by the finding that the firing of the spindle goes on at a high rate for several seconds after release of stretch. After the spindle is subjected to several such strong stretches, the static response usually disappears and only the dynamic discharge remains.

2. Effect of Changes in Velocity of Stretch

As described above, the length of the spindle at any moment during stretching is signalled by the frequency of discharge at that time. Another important function of the spindle is to convey information about the velocity of the stimulus. Studies of the response of the spindle to linearly rising controlled stretches shows that the

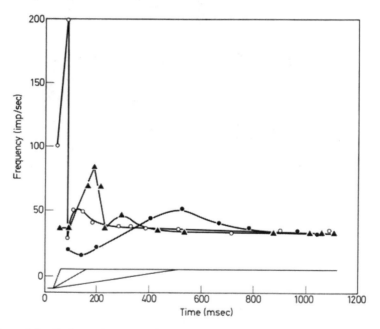

Fig. 5. Plots of the discharge frequency during stretch of a spindle at three different velocities (13, 2.6 and 0.7 mm/sec) to the same amplitude of maintained extension. Stretches are indicated at the bottom of the diagram. Frequency values obtained by measuring spike intervals and plotting their reciprocals along the time axis as the spikes occur (from Shepherd and Ottoson, 1965)

frequency of firing is directly related to the rate of stretch (as illustrated in Fig. 5), and that this holds true at all lengths of the spindle during dynamic stretching. In this respect the spindle behaves closely similar to touch receptors in the toad's skin (Lindblom, 1962, 1965). It thus appears that for a given velocity of stretching the amount of extension is signalled by the frequency at that moment, whereas the velocity is signalled by the rate of increase in firing. As a result of the close relationship between the frequency characteristics of the impulse response and the parameters of stretch, the spindle is able to convey precise information about the strength and velocity of the stimulus at any moment during dynamic extension.

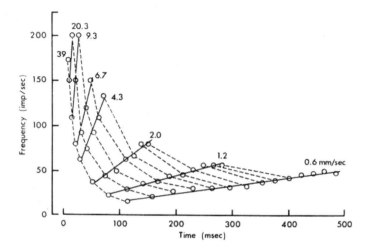

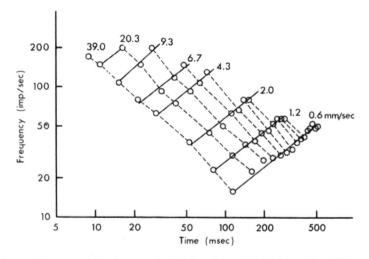

Fig. 6. Upper diagram: spike frequencies during dynamic discharge for different rates of stretching. Final extension about 300 μ for all stretches. Linear scales. Lower diagram: same data plotted on logarithmic scales. Resting spindle length: 1.0 mm (from SHEPHERD and OTTOSON, 1965)

A closer analysis of the dynamic discharge over a wide range of velocities of stretching brings out several other interesting features. The extreme regularity of the discharge is one of these. In Fig. 6 the frequency of each impulse has been

29*

taken as the reciprocal of the preceding spike interval and plotted against time, the initial spike of each response being omitted. The spike frequencies for any given rate of stretch fall along a straight line. It may also be noted that the dotted lines connecting corresponding spikes in different responses form a family of curves. When the values are plotted on double logarithmic scales (lower diagram), the frequencies for any given rate of stretch fall on an approximately straight line and the lines for different rates are parallel; furthermore, the dotted lines connecting corresponding spikes in the different responses are also parallel. The peak frequency of the dynamic discharge decreases along a straight line with decreasing velocity of stretch; the slope of this line, however, is different from that of the lines connecting the corresponding spikes. An interesting consequence of the regular behavior of the spindle is that if the dynamic discharges to two stretches of different velocities are known, it is possible to predict the frequency characteristics for any other rate of stretch within the physiological range.

An analysis of the response with respect to the level at which the spikes occur at different velocities of stretching gives further evidence of the faithfulness whereby the properties of the stimulus are encoded in the impulse response. In the diagram in Fig. 7, the levels of stretch at which the spikes arise are plotted for the

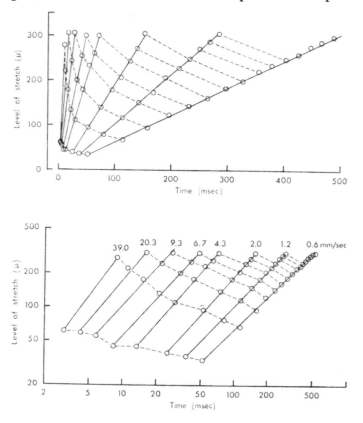

Fig. 7. Upper diagram: levels of extension for onset of spikes in dynamic discharge for different rates of stretching. Linear scales. Lower diagram: same data plotted on logarithmic scales. Resting spindle length: 1.0 mm (from Shepherd and Ottoson, 1965)

same velocities of stretching as shown in Fig. 6. Each linearly rising stretch is indicated by an upward sloping line, and corresponding spikes are again connected with dotted lines. The diagram shows that the corresponding spikes arise at successively higher levels as the velocity of stretching is increased. It is likely that this change reflects in part the increasing influence of nerve refractoriness as the intervals between the spikes become smaller. Plotting of the data on double logarithmic scales gives a family of parallel lines. Here again it is possible to predict the characteristics of the discharge with respect to the level of onset of the spikes at any rate of stretch, if the response patterns for two different stretches are known.

Another example of the great regularity of the response may be seen when an extra impulse is interjected during the discharge to stretch. Such an impulse may be elicited by an electrical shock to the nerve or by a brief stretch superimposed upon the stretch giving a sustained discharge. As illustrated in Fig. 8, the interjected spike causes a shift in the entire spike train. It can be seen that in general

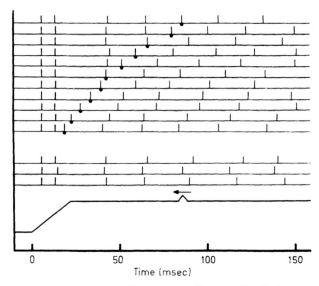

Fig. 8. Effect of an interjected extra spike on succeeding impulse discharge. Spikes indicated diagramatically by vertical bars. Successive runs with extra spike (marked with dot) elicited at different times by brief stretch superimposed upon steady stretch. Three runs without interjected spike shown below (from OTTOSON, McREYNOLDS, and SHEPHERD, 1969)

the timing of the succeeding discharge is merely reset by the extra spike. An exception to this rule is observed when the extra spike closely follows a previous spike; in such a case, the shift in the succeeding train is small or absent. These findings suggest that the spike frequency of the discharge is mainly determined by the time course of the generative process. Since the depolarization during maintained stretch is relatively constant, it would appear that the interjected spike delays or transiently wipes out the currents leading up to the generation of the succeeding spike, thereby resetting the entire impulse generating process. During dynamic stretching, an interjected spike has similar effects. It appears that with weak or moderately strong stretches the firing rate is not determined by refractoriness of the nerve, whereas with strong stretch and high rates of firing refractoriness is relatively more important (see ARVANITAKI, 1938; HODGKIN, 1948; EYZAGUIRRE and KUFFLER, 1955; LOEWENSTEIN, 1958; CHAPMAN, 1966).

3. The Transitional Interval

MATTHEWS (1931) observed that a slight pause in the discharge often appeared when the muscle spindle was rapidly loaded. A similar firing gap was later noted by KATZ (1950a) and may also be discerned in records from the mammalian spindle (JANSEN and MATTHEWS, 1962; JANSEN and RUDJORD, 1963).

The studies on the isolated spindle have established that the pause is a regularly occurring phenomenon (Fig. 9) and that its relative duration is closely related to the velocity and strength of the stimulus (SHEPHERD and OTTOSON, 1965). Analysis of the time characteristics of the discharge of the spindle during the period of

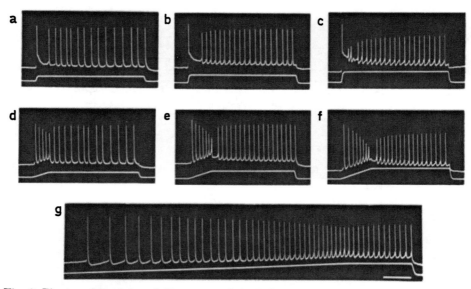

Fig. 9. The transition interval. Responses of the isolated spindle to increasing amplitudes of extension at different velocities. a-c, at 130 mm/sec; d-f, at 13 mm/sec; g, at 1.3 mm/sec. Amplitude of extension the same in a and d, b and e, c, f and g. Note development of "pause" between dynamic and static discharges with higher velocities and levels of stretching. Time bar: 50 msec (from SHEPHERD and OTTOSON, 1965)

transition from dynamic to static stretching shows that with a moderately strong stretch the firing gap usually begins to appear at velocities of stretching above 2 to 3 mm/sec (Fig.10). With further increase in velocity, the pause becomes more and more marked; the last dynamic interval decreases, however, and the interval between the first two static spikes remains constant in duration. For a given velocity, there is a given strength of the stretch below which the pause is not present. This level is in general lower for higher velocities of stretch. With stretches below about 2 mm/sec, the pause is usually not present at any strength of stretch. The development of the pause with lengthening of the spindle is illustrated in the dia-

gram in Fig. 11 which shows the relation between the duration of the pause and
the strength of stretch for three different velocities. Since the static discharge
increases with increasing amplitude, the transitional interval in the diagram is
plotted on the ordinate as a percentage of the sum of the duration of the pause and
the first static spike interval.

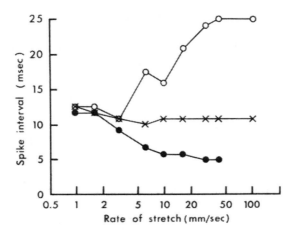

Fig. 10. Diagram showing duration of interval between last two dynamic spikes (filled circles),
first two static spikes (crosses) and transition interval between last dynamic and first static
spike (open circles), at different rates of stretch. Abscissa: logarithmic scale; ordinate: linear
scale (from SHEPHERD and OTTOSON, 1965)

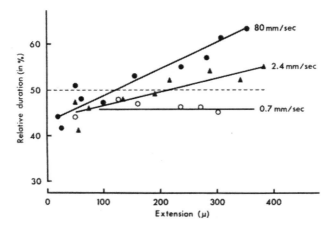

Fig. 11. Relative duration of transition interval for three different velocities of stretch.
Transitional interval given in per cent of sum of transition interval and first static spike
interval. Resting spindle length: 1.2 mm (from SHEPHERD and OTTOSON, 1965)

These findings clearly demonstrate that the relative duration of the transitional
interval is closely related to the properties of the stretch. It appears as a gap in the
discharge only within a given range of strengths and velocities of stretching. Within this range, however, the interval is more marked as the stretch is faster and

stronger. But, above a given strength of stretch, spikes may begin to appear in the pause, and with overstretch the dynamic and static discharge becomes continuous.

Several explanations may be offered for the appearance of the firing gap. One possibility is that the pause is the result of mechanical events in the spindle and is related to differences in the visco-elastic properties of the reticular zone and the compact zones. Another possibility is that the pause is related to the dynamic fall of the receptor potential. As will be described in Section VI, the responsiveness of the sensory endings undergoes characteristic changes during the transition from dynamic to static stretch, changes which are closely related to the properties of the stimulus. It appears in this light that the pause reflects the relative repolarization of the endings as represented by the dynamic decay of the receptor potential.

Whatever the mechanism underlying the pause, its presence as a break between the dynamic and static discharge may have important functional significance. In separating the dynamic and static firing, the pause would appear to represent a sharpening of mechanical contrast which would provide the central nervous system with information about the change in rate of stretching at the transition from dynamic to static stretch. It is interesting to note that a similar pause also appears in recordings from the *Limulus* eye (see Dowling, 1968). It would appear therefore that the pause is a general property of the responses of slowly adapting sense organs. In the encoding process the presence of a silent period would represent a differentiation of the dynamic and the static phases of a stimulus.

4. Latency of Response

With a fast rising stretch, the first spike of the impulse response usually appears within about 0.7 to 0.8 msec. As the velocity of stretch is decreased, the latency increases; with very slow rates of stretching the first spike may not occur until about 200 msec after the onset of stretch. When the latency is plotted against rate of stretching on logarithmic scales, the values fall on an approximately straight line with a slope of -1 over most of the range of velocities. Usually the spike is preceded by a prepotential which with a fast rising stretch appears 0.3 to 0.4 msec earlier than the spike. These values may be regarded as typical for the spindle kept at resting length. If the spindle is slightly prestretched before the test stretch is applied, values less than 0.3 msec for the onset of the prepotential may be obtained. Since transmission of the stimulus to the sensory endings must take a given time, the true latency of the response must be shorter than 0.3 msec. It is of interest to note that other mechanoreceptors appear to have similarly short latencies (Gray and Sato, 1953; Thurm, 1964). The rapid onset of the response suggests that there are no intermediate processes, such as the release of a chemical transmitter, involved in the initiation of this activity. It would therefore appear that the depolarization of the endings is the direct result of the mechanical deformation of the sensory transducer membrane.

5. Threshold

The minimum extension required to elicit an impulse from the spindle varies with several factors, the principal one being the initial resting length of the spindle.

The absolute value obtained in threshold measurements will therefore greatly depend on the criteria used to determine the resting length of the spindle. If the spindle is kept slightly relaxed the values will be high; if it is kept slightly stretched they will be correspondingly lower.

In most of the experiments described here the spindle was mounted in a relaxed state and then slowly stretched until an increase in the spontaneous firing was noted; it was thereafter left either in this position or slightly relaxed. The threshold values obtained with a fast rising brief testing stretch then usually varied from 10 to 40 μ. Since the average length of the spindle in most experiments was about 1 mm or slightly less, the threshold extension would be about 1 to 5% of the spindle length. When the spindle was pre-extended the threshold fell in direct relation to the amount of extension (see Fig. 25), and with an initial length which exceeded the normal resting length by 50% stretches of a few microns were sufficient to elicit an impulse. Thus the spindle is an exquisitely sensitive receptor organ.

As described above, the latency of the onset of the dynamic discharge increases in direct relation with decreasing velocity of stretch. This implies that there exists a corresponding regular relationship between the level of onset of the discharge and velocity of stretch. An analysis of this relation shows that the level of onset of the electrical activity is relatively constant over a wide range of rate of extension. A closely similar relation between stimulus gradient and threshold has been demonstrated for touch receptors in the frogs skin (GRAY and MALCOLM, 1951; HÖGLUND and LINDBLOM, 1961). With the fastest stretches there is a decrease, and with slow stretches there is a tendency toward increased values. It therefore appears that within an intermediate range the spindle fires at approximately the same level whether the stretch is slow or fast. This would imply that there is relatively little accommodation of the nerve fiber at threshold levels.

B. Receptor Potential

It is generally recognized that the afferent nerve discharge from sense organs is initiated by a local depolarization of the receptor elements. This potential is a nonpropagated event and forms the link between the stimulus and the conducted impulse response. Such local potentials have been described for different types of sense organs (see GRANIT, 1955; GRAY, 1959). In 1942, BERNHARD, GRANIT and SKOGLUND introduced the term "generator potential" for these potentials. DAVIS (1961) has suggested that this term should be reserved for potentials which trigger the nerve response, whereas potentials which are produced by special receptor elements should be called receptor potentials. However, this distinction has not been generally adopted. One reason for this may be that the exact structural origin of the potential cannot always be established. In the present study the terms receptor potential and generator potential are used interchangeably to designate the graded response obtained after blocking of the conducted activity.

KATZ (1950b) was the first to describe the receptor potential of the muscle spindle. He found that the potential was characterized by an initial peak which occurred during dynamic lengthening and by a later static component during maintained stretch. At release of stretch, a transient potential change occurred in the opposite direction. Katz also showed that there is a close relation between the intensity of depolarization of the endings and the frequency of impulses in the afferent nerve. Since Katz' original investigations in 1950 no further study has been reported on the characteristics of the receptor potential of the spindle. We shall therefore describe in more detail the typical properties of this potential in the isolated preparation.

1. Site of Mechano-electrical Transduction

It is well known that sensitivity to mechanical stimulation is an inherent property of a variety of excitable tissues. It was demonstrated by Tigerstedt as early as 1880 that a nerve can be excited by mechanical stimulation. Recent studies by Goldman (1964) have shown that a mechanical pulse applied to a giant axon of the lobster produces a depolarization which increases in amplitude with the stimulus and that when threshold is reached an action potential is elicited. With respect to the spindle, it is generally assumed that the production of the generator potential takes place in the unmyelinated terminals of the afferent fiber. However, at the present time there is no direct evidence for this hypothesis and it cannot be a priori excluded that the myelinated branches participate in the transducer action. Direct recording from single endings appears extremely difficult because their diameter is of the order of a few microns. Microstimulation of different regions of the spindle with the tip of a glass rod (tip diameter about 5 μ) mounted on a piezo-crystal shows that the stem fiber and the myelinated branches are relatively insensitive to mechanical stimulation, whereas the polar regions of the spindle are very sensitive (Ottoson, *unpublished observations*). There is an interesting similarity in this respect between the muscle spindle and the Pacinian corpuscle. As demonstrated by Loewenstein (1961), the nerve ending of the Pacinian corpuscle is specifically sensitive to mechanical stimulation and responds to compression by less than 1 μ. It appears likely that the terminals are responsible for the transducer action of the spindle. It might well be, however, that stretching in addition causes a small depolarization of the branches and that this effect contributes to the initiation of the discharge.

The nerve terminals form long chains of extremely thin strands with numerous varicose swellings. Electron microscopic studies (Katz, 1961; Karlsson et al., 1966) have shown that many of these bulbs are closely attached to the muscle. Katz (1961) suggested that the conversion of the mechanical stimulus into an electrical change is localized in the sensory bulbs, the thin tubes between them serving as connecting lines. As pointed out by Katz, the sensory bulbs would provide a suitably large contact surface with the intrafusal muscle fibers. A depolarization of the bulbs would be relatively little attenuated when conducted passively by the connecting tubes which, because of their small diameter, would serve as a high-impedance line. A potential generated by the thin tubes would, on the other hand, be shunted by the bulbs.

At present there appears to be no means for a precise identification of the membrane responsible for the transducer action. It might well be that the transduction of the stimulus takes place in the bulbs, but it is also possible that the connecting thin tubes participate and that the bulbs mainly serve other functions, such as energy stores. The great number of mitochondria in them supports this view. Other

structures may also be indirectly involved in the excitatory process and the main-
tainance of ionic concentration gradients. The dense accumulation of satellite cells
in the sensory innervated region may, for instance, be of importance in relation to
the transduction process.

The small diameter of the terminal branches has prevented recording directly
from them, but since their activity spreads electrotonically to the nerve it is
possible to obtain a great deal of information by recording from the stem fiber.
During this spread, the potential produced by the endings is bound to suffer some
attenuation. Since the conductance distances are short, it appears likely that the
attenuation is comparatively small. It may be concluded, therefore, that the potential recorded from the stem fiber would give a relatively true picture of the
potential generated in the terminals. However, the absolute value of the potential
when recorded from the afferent fiber may vary a great deal from one preparation
to another, depending on how successfully the fiber has been isolated in the dissec-
ting procedure.

2. Isolation

From studies on different sense organs it has been established that treatment
of a sensory ending with local anaesthetics at concentrations sufficiently strong to
block propagated activity leaves the receptor potential relatively unaffected. By

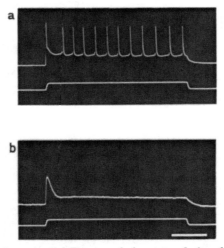

Fig. 12. Isolation of receptor potential. Response before (a) and after (b) blocking of conducted
activity with 0.2% lignocaine. Time bar: 50 msec (from OTTOSON and SHEPHERD, 1965)

this procedure the activity of the sensory receptors, which are otherwise difficult
to study because of their anatomical structure, may be isolated and analyzed. The
records in Fig. 12 show an example of the response of the isolated spindle to a
stretch of 400-msec duration before (a) and after (b) the treatment of the preparation with 0.2% lignocaine solution. The response obtained after blockage exhibits
properties typical for the receptor potential; i.e., an initial dynamic phase (KATZ,
1950b) developed during the period of lengthening and an ensuing static phase
developed during maintained stretch. The "off" response described by KATZ

(1950b) is not seen in this recording; in the isolated spindle it usually appears only after strong stretches. Its exact relation to the parameters of the stretch awaits further study.

3. Relation to Amplitude of Stretch

As demonstrated by Ottoson and Shepherd (1965), changes in the strength of the applied stretch are accompanied by characteristic changes of the receptor potential. A general view of these changes is given in Fig. 13 which shows tracings

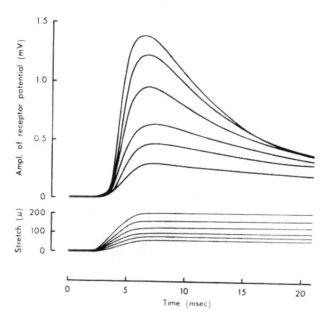

Fig. 13. Characteristic features of dynamic phase of receptor potential. Responses to steplike stretching of spindle to different lengths after blocking of conducted activity with 0.2% lignocaine. Resting spindle length: 1.0 mm

of receptor potentials obtained by applying steplike stretches of increasing amplitudes. With a weak stretch the difference in height between the dynamic and the static phase is very small. As the magnitude of the stretch is increased, the dynamic phase increases more rapidly than the static component, and the potential acquires its characteristic appearance with an initial dynamic peak which declines to a steady static level during maintained stretch. With high velocity of stretch, the maximum dynamic peak is usually obtained at a lengthening of about 30% of the resting length whereas the static potential level continues to rise with extensions above this value. This difference between the two components is clearly brought out by the diagram in Fig. 14. In this experiment the dynamic peak reached its maximum value at a lengthening of about 200 μ, whereas the static potential still increased as extension was increased further. This raises the question at what level of stretch does the static component reach its maximum. Experiments with very strong stretches show that the static potential continues to rise in height until it approaches the amplitude of the dynamic peak. The strength of stretching needed to obtain such large static responses usually represented an extension of more than 100 % of the resting length and resulted in irreversible damages to the spindle.

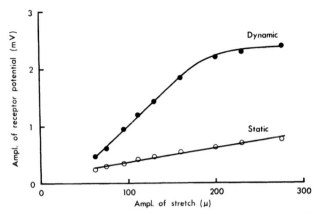

Fig. 14. Diagram illustrating relation between strength of stimulus and amplitude of dynamic and static phase of receptor potential. Resting spindle length: 0.9 mm

4. Relation to Velocity of Stretch

A characteristic property of the spindle response is the close relation between the frequency pattern of the dynamic discharge and the velocity of the stretch. It might therefore be expected that the depolarization of the endings would show a similar close relationship. This is also borne out by studies on the isolated receptor potential (KATZ, 1950b; OTTOSON and SHEPHERD, 1965). The general characteristics of the dynamic phase of the receptor potential with different rates of stretching are illustrated in Fig. 15. The responses shown were obtained by stretching the spindle at different velocities to the same final level. Stretch was then released in each case so that there was no phase of static extension. In this way, purely dynamic responses were obtained. With the fastest rate of stretching the receptor potential has a rapidly rising initial phase. At lower velocities of stretching the onset of the response is more gradual, the potential rises more slowly, and the peak amplitude of the potential decreases. Measurements of the change in height with velocity of stretching show that the amplitude of the response increases almost linearly with the rate of stretching, when the values are plotted on logarithmic scales.

How faithfully the receptor potential reproduces the properties of the stimulus is clearly brought out by the fact that the rate of rise of the potential is a direct function of the velocity of stretch. Since the discharge of the afferent nerve is triggered by the receptor potential, it follows that the dynamic impulse response must be closely related in frequency to the velocity of stretching. This is also evidenced by measurements of the peak frequencies of the impulse discharge at different velocities of stretching. The precision with which information of the velocity of the stimulus is encoded in the afferent impulse message is thus to be attributed primarily to the typical transducer properties of the endings.

The differentiation of the receptor potential into two phases with different characteristics brings up the question of the origin of the two components of the spindle response. MATTHEWS (1931) was the first to suggest that different regions of the spindle by virtue of their specific visco-elastic properties might account for the dynamic and static phases of the discharge. This idea has later gained a great

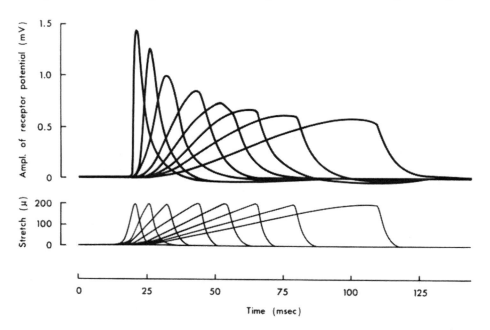

Fig. 15. Dynamic receptor potentials for different velocities of stretching. No static phase of stretch (from OTTOSON and SHEPHERD, 1970d)

deal of support from electron microscopic studies (KATZ, 1961; KARLSSON et al., 1966) which have shown that there are two distinct zones of sensory innervation within the spindle. In the central "reticular" region, the endings lie on fibers which are almost devoid of myofibrils. In this zone there is also a relatively dense network of supporting connective tissue fibrils. The other "compact" zone is represented by the polar regions in which the muscle fibers contain the normal amount of myofibrils. It appears likely therefore that the compact zones are more viscous than the reticular zone. KATZ (1961) has suggested that the sensory endings located in the reticular zone generate the dynamic response whereas the static response would be produced mainly by endings in the compact zones. In the mammalian spindle the dynamic sensitivity of the primary endings may be explained by an analogous mechanical differentiation, as indicated by observations on intrafusal fibers of the rat (SMITH, 1966).

A number of experiments have been carried out to isolate the activity of the reticular region by selective elimination of the response of the endings in the compact zones (OTTOSON and SHEPHERD, unpublished observations). This isolation was accomplished by crushing increasingly greater parts of the spindle with microforceps starting from the polar ends of the spindle. One would expect that only the dynamic response should be left after the entire two compact zones were treated in this way. In all experiments, however, the two phases were still obtained although they were smaller in amplitude than before. If the reticular zone was then also gradually pinched starting from the borders of the compact zones, the remaining response declined rapidly and then suddenly disappeared. This apparently occurred when the larger myelinated branches were pinched. However, even when only a small portion of the reticular zone was still responding, the two components of the response were present. These observations would suggest that the reticular zone provides at least part of the static response. However, the evidence is not conclusive since the borderlines between the reticular and compact zones are not easily determined in the living preparation.

As described above, the dynamic response reaches its maximum when the spindle is extended with a rapid stretch to about 30% of its resting length. It is interesting to note that the number of endings located in the reticular zone comprises about onethird of the total number of endings (see KATZ, 1961; KARLSSON et al., 1966). If it is assumed that the dynamic response is produced in the reticular zone, it appears that all the reticular endings are excited when the spindle is extended by about onethird of its resting length; of the endings generating the static response, only a portion would be active as judged from the height of the receptor potential. To activate all the endings the spindle must be subjected to an extensive lengthening. It can be presumed that during physiological conditions the spindle would most likely not be exposed to stimuli beyond the maximum capacity of the endings responsible for the dynamic response and definitely not for that of the endings generating the static component.

In conclusion there appears at the present time to be no conclusive evidence that the velocity response of the spindle derives from the central reticular region only (see also OTTOSON and SHEPHERD, 1968, 1970a). Even if the dynamic response originates mainly from the reticular zone as suggested by the structure of this region, it seems likely that the reticular endings also contribute to the static response. Likewise it appears that part of the dynamic response may come from endings in the compact zones. The conclusion at this time would therefore be that the two zones of sensory innervation contribute to both the phases of the response.

5. Relation between Tension and Spindle Activity

In studies on the stimulus-response relationship of mechanoreceptors such as the stretch receptor or the muscle spindle, the activity changes are generally related to extension. As will be discussed further below, there are reasons for thinking that tension may be a more adequate representation of the stimulus. In recording the discharge of the single in situ spindle, MATTHEWS (1931) found that the static firing rate was proportional to the logarithm of the load of the muscle. A similar relationship between load and impulse activity of the stretch receptor has been reported by TERZUOLO and WASHIZU (1962). Other authors (KRNJEVIC and VAN GELDER, 1961; WENDLER and BURKHARDT, 1961; BROWN and STEIN, 1966) have reported varying degrees of departure from the linear relationship.

Recent experiments (HUSMARK and OTTOSON, 1970a) on the isolated spindle have shown that there is a close relation between the tension developed during stretch and the output of the receptor in terms of the receptor potential and the impulse response (Fig. 16). With a steplike stretch tension rises to a peak and then gradually falls to a more or less constant level during maintained stretch. The amount of tension developed during phasic stetching as well as under steady stretch is a logarithmic function of the amount of lengthening. As would be expected from the linear relation between the response and lengthening, both the dynamic and the static phase of the response are linear functions of the logarithm of the corresponding tension. The relation between the decline of tension during

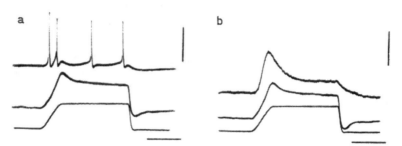

Fig. 16. Tension changes in isolated spindle during stretch. Upper trace: response of spindle; middle trace: tension; lower trace: stretch. a, impulse response; b, receptor potential obtained after blocking of conducted activity with 0.2% lignocaine. Vertical bar in a, 2 mV; in b, 1 mV. Time bars: 20 msec (from Husmark and Ottoson, 1970a)

steady stretch and the response of the spindle is of particular interest with respect to the mechanisms underlying the adaptation of the spindle. The fact that the time constant of fall in tension in the early period of maintained stretch is approximately the same as that for the decay of the receptor potential in the corresponding period would suggest that adaptation is related to tension changes. This would also appear to support the conclusion that tension may be regarded as a more direct representation of the actual stimulus to the sensory endings than the applied stretch, as has been suggested for the crustacean stretch receptor (Krnjevic and Van Gelder, 1961; Brown and Stein, 1966). However, the fall in tension is quantitatively always less than the decline of the receptor potential. This suggests that adaptation can not be entirely of mechanical origin. This is also borne out by the observation (Husmark and Ottoson, 1970b) that the receptor potential still decays when tension is kept constant, as will be described below. The actual stimulus to the sensory endings is most likely the structural deformation of the mechanosensitive membrane. It is unlikely that there is a simple relation between this deformation and lengthening or tension. On the basis of present evidence therefore, neither lengthening nor tension can be taken as a direct representation of the stimulus to which the endings are exposed.

6. Decay of Receptor Potential

On release of stretch, the sensory endings repolarize as indicated by the decay of the receptor potential. During maintained stretch, there is a phase of relative repolarization as represented by the decline of the potential from the dynamic peak to the static level. It can be assumed that the time characteristics of these different decay phases of the receptor potential as illustrated in Fig. 17 represent functionally important properties of the transducer action. Thus, adaptation in terms of the decline in firing rate at the onset of maintained stretching may be attributed to the fall of the receptor potential from the dynamic peak to the static

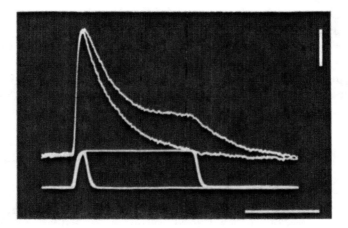

Fig. 17. Decay characteristics of receptor potential. Superimposed records of response to brief dynamic and maintained stretch to the same length. Time bar: 25 msec. Vertical bar: 0.5 mV (from OTTOSON and SHEPHERD, 1970a)

level. It is also likely that the time course of fall after release of dynamic or steady stretch plays an essential role in the recovery of the spindle. Information about the characteristic decay properties of the receptor potential may therefore be useful for the understanding of the function of the spindle.

The decline of the potential elicited by a brief purely dynamic stretch is characterized by an exponential fall towards base line, except for responses to either very weak or supramaximal stretches. The time constant for decay of the response to a moderately strong stretch is 8 to 12 msec and becomes slightly shorter with increasing strength of the stimulus, but it remains relatively unaffected by changes in velocity of stretching. Because of the comparatively slow decline, the receptor potential outlasts the stretch for a considerable time; for instance, a brief dynamic stretch with a duration of 3 to 4 msec may give rise to a receptor potential which does not return to base line until 40 msec after termination of stretch.

The decline of the potential on release of sustained stretch exhibits similar properties: it is exponential; its time constant is very close to or the same as that of the dynamic potential; it diminishes slightly with increasing strength of stretch; and it is relatively unaffected by the duration of the stimulus.

The return of the spindle to resting conditions is most likely brought about by the elastic elements. It can be assumed that at release of stretch the elastic elements which have been extended during stretch pull the internal structures of the spindle back to their original resting position. The fall of the potential would, according to this view, reflect the internal mechanical restitution, and the time course of the fall would be determined by the mechanical properties of the spindle.

However, several other factors, such as pressure changes within the spindle and the electrical constants of the membranes of the sensory endings and the nerve branches, may also influence the time course of the decay.

When stretch is maintained, the potential falls to a steady level, the amplitude of which is dependent on the strength of the stimulus. This decay is also exponential. With strong stretches, its time constant is slightly shorter than that of the purely dynamic potential. With decreasing velocity of change, there is a tendency towards a shortening. For a given amplitude and velocity of stretching, the time constant of the dynamic fall to static level may therefore be the same as that for the dynamic fall to zero.

The decay of the spindle potential might be explained as the result of a passive leakage of charge in the sensory terminals; its time constant would consequently be determined by the electrical constants of the sensory membrane. Alternatively, it could be suggested that the decay of the potential reflects the dissipation of deformation of the sensory membrane. In this latter case, the mechanical properties of the spindle would be assumed to determine the time course of the fall.

7. Adaptation

As already described it is a typical feature of the impulse response of the muscle spindle that the frequency falls from the peak value reached at the end of dynamic stretch toward a lower static level during steady stretch. The muscle spindle is therefore usually classified as a slowly adapting receptor. However, there seems to be no general agreement as to the actual definition of the term adaptation. The adaptive fall occurs in two phases: an initial phase characterized by a rapid decline and a later phase of slow gradual decline. The term adaptation is sometimes used to designate the first stage, sometimes the second phase. It would appear that adaptation in a broad sense includes the initial rapid fall as well as the prolonged later period. The greater part of adaptive fall occurs in the first stage. In the following the term adaptation will be used for the activity changes taking place in the first phase only.

In studies of the adaptation of the spindle the decline in impulse frequency is usually taken as an index of the adaptation. However, the receptor potential is a more direct representation of the activity of the sensory endings. In relating the amount of adaptation to the properties of the stimulus the analysis of the changes of the receptor potential gives a more detailed picture than a corresponding study of the impulse response. In terms of fall of the receptor potential, adaptation increases with increasing amount of lengthening of the spindle for stretches up to 25—30 % of its resting length; it then decreases with still stronger stretches. Adaptation also changes with the velocity of stretch; it decreases with decreasing velocity and may be zero when stretch is sufficiently slow. Thus, the amount of adaptation is not a constant or given functional feature of the spindle but varies within wide limits with the properties of the applied stimulus.

There is little doubt that the visco-elastic properties of transmitting stuctures are of great importance for the adaptive properties of mechano-receptors. An elegant demonstration of the functional importance of the mechanical coupling between the receptor and the mechanical stimulus is given by Loewenstein and Mendelson (1965) in the Pacinian corpuscle. They removed the lamellae of the receptor by careful dissection so that the stimulus could be applied directly to the ending. The generator potential of the decapsulated ending was identical to that of slowly adapting mechano-receptors. However, in spite of the generator potential being sustained, the nerve fired only one impulse and steady polarizing currents also gave only a few impulses, as had earlier been observed by GRAY and MATTHEWS (1951). This suggests that the nerve of the Pacinian corpuscle is rapidly accommodating. This is in contrast to the spindle nerve which gives a sustained discharge under similar conditions. It would appear, therefore, that the difference in adaptation between the Pacinian corpuscle and the spindle is related to differences in mechanical factors as well as to differences in the accommodation properties of the axons.

Accommodation would appear to play a relatively small role for the adaptation of the spindle as evidenced by the finding that there is no decline in firing rate during constant depolarization (EDWARDS, 1955; LIPPOLD et al., 1960). With strong stretch, however, accommodation may be a contributing factor as indicated by the finding of WENDLER and BURKHARDT (1961) that the discharge of the crayfish stretch receptor decreased when strong depolarizing currents were used.

Although the structural properties of the spindle may be of importance for adaptation there is strong evidence that other factors may contribute and perhaps be even more important than the mechanical ones. In studies of the spindle by stroboscopic photomicroscopy it has been found (OTTOSON and SHEPHERD, 1968, 1970a) that the length changes of the central reticular zone closely follow the applied stretch in time course and magnitude. No relative shortening corresponding to the decline of the receptor potential could be demonstrated. This suggests that the early adaptive fall of the spindle response is not due to differential length changes as would be expected if adaptation was caused by a sliding movement of visco-elastic elements. However, the finding that tension decays during maintained stretch indicates that there are mechanical alterations related to the viscosities taking place in the early period of steady stretch. The fact that these changes are not revealed as gross length changes suggests that they occur at the ultra-structural level.

Studies of the receptor potential evoked when the spindle is kept under constant tension appear to provide a clue to the contribution of mechanical factors to adaptation (HUSMARK and OTTOSON, 1970b). Under constant tension there is a fall of the receptor potential from the dynamic peak toward the steady level closely similar to the fall under constant length. However, under constant tension the fall is less than that under constant length. This would suggest that adaptation may be only partly attributed to mechanical properties. This raises the question as to the origin of the remaining adaptation. One possibility would be that the relative fall is related to specific conductance changes in the sensory membrane and associated with movements of ions. This is supported by the finding that changes in the potassium concentration in the external fluid have profound influence on the adaptation of the receptor potential (HUSMARK and OTTOSON, unpublished observations).

In conclusion it would appear that adaptation of the spindle is the result of a complex interplay of factors, some of which are mechanical and some of which are non-mechanical. It is important to reiterate that the receptor potential of the

30*

spindle is closely like the receptor potential of other slowly adapting receptors. It is probable therefore that the characteristic features of the spindle response reflect basic response characteristics of the sensory membrane of slowly adapting end organs. In the spindle these properties may be modified by the specific characteristics of the transmitting structures.

C. Stimulus-response Relationship

The analysis of the response of the muscle spindle suggests that the characteristics of the stimulus are encoded into the impulse message with a high degree of precision. It would appear that in one phase of stretch the afferent discharge does not convey faithful information about the stimulus. This is the early period of maintained stretch when the frequency falls although stretch is constant; i.e., the spindle "adapts". However, if the static discharge is regarded as a separate message independent of the dynamic discharge, the absolute value of the frequency would be of little importance. The only requirement would be that different amounts of steady stretch are signalled by different frequencies, which is precisely what the spindle does as we have shown above. The fact that the spindle uses a lower frequency range in signalling the amount of steady stretch may be regarded as a rational way of transmitting information over prolonged periods of time. If static stretches were signalled with the same frequencies as corresponding levels of dynamic stretch, the spindle would run the risk of becoming fatigued. By shifting to a lower frequency range during static stretch, however, the spindle avoids this risk.

The relation between the transducer action and the afferent impulse pattern is clearly brought out by a direct comparison of the response before and after blocking of the impulse activity. A study of the records in Fig. 18 shows that several of the features of the complicated potential patterns of the unblocked response may be attributed to the interaction between the conducted activity of the nerve and the depolarization of the endings. By comparative analysis of the impulse activity and the receptor potential, the particular features of the different stages in the development of the response may be identified.

The linear relation between the stimulus and the afferent impulse message (see Fig. 4) reveals an important functional property of the spindle. The same relation has also been described for the vertebrate spindle (Eldred et al., 1953) and for the crustacean stretch receptor (Burkhardt, 1959; Terzuolo and Washizu, 1962; Brown and Stein, 1966). This typical behavior implies that during dynamic or sustained stretching the spindle fires at a rate which increases proportionally with extension. The receptor potential initiating the afferent discharge exhibits analogous characteristics (see Fig. 14); the dynamic peak as well as the static level of the potential are directly related to length. Thus it follows that the firing rate of the afferent fiber must be a direct function of the level of depolarization of the endings. An analysis of the relation between the frequency characteristics of the discharge and the level of receptor potential provides conclusive evidence for the linear transformation of the dynamic as well as the static components of the response. In the diagram in Fig. 19, the frequency of the dynamic discharge during a relatively slowly rising stretch is plotted versus the level of the receptor potential at which the individual spikes took off. The values for the static relationship derive from the same spindle and are obtained by subjecting the spindle to sustained stretches of different strengths. For both the dynamic and the static discharges

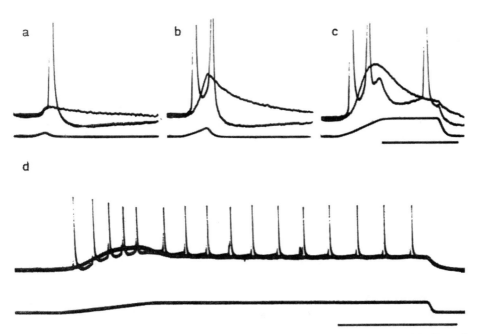

Fig. 18. Relation between impulse response and underlying receptor potential. Records obtained before and after blocking of conducted activity with 0.2% lignocaine. a-c, response to brief stretches showing development of dynamic impulse response and transitional interval in relation to receptor potential. d, similar recording to more slowly rising prolonged stretch. Records photographically superimposed. Time bars: in c, 25 msec; in d, 100 msec (from OTTOSON and SHEPHERD, 1970b)

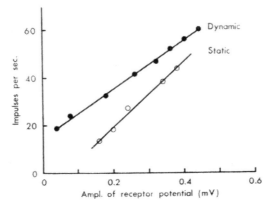

Fig. 19. Relation between amplitude of receptor potential and frequency of firing during dynamic and static phase of stretching. Dynamic values obtained from response to linearly increasing stretch at 1.2 mm/sec. Static values from the same spindle subjected to steady stretches of different lengths (from OTTOSON and SHEPHERD, 1970b)

the frequency increases proportionally with depolarization of the endings. This is in close agreement with the observations of Katz (1950a). The less steep slope of the dynamic line compared to that of the static line in Fig. 19 is due to the relatively slow velocity of dynamic stretching used in this particular experiment. With faster stretching the dynamic slope is steeper, whereas the static slope remains constant.

V. Recovery after Stretch

Following release of stretch, there is a period during which the sensitivity of the spindle to a succeeding stretch is depressed. Several hypotheses may be suggested to explain this depression. It could be due to electrical events associated with the repolarization of the sensory endings or to accumulated refractoriness of the afferent fiber and its branches. Another possibility is that the depression is caused by mechanical alterations in the spindle; in this case, the recovery would reflect the mechanical restitution of elements which are extended during stretch. To assess the contribution of the various factors which might be responsible for the depression, it is essential to know the recovery cycle of the afferent fiber as well as the alterations in responsiveness of the sensory endings in the aftermath of stretching.

1. Recovery Cycle of the Afferent Fibre

Information about the functional characteristics of the afferent fiber may be obtained by classical methods of testing the refractoriness of the fiber with electrical shocks following a single impulse. It is then found that the afferent nerve is absolutely refractory for about 2 msec after a single impulse evoked by an electrical shock (Ottoson and Shepherd, 1969). The ensuing relative refractory period has a duration of 4 to 5 msec and is followed by a supernormal period lasting for 20 to 30 msec; there then follows a period of slightly lowered excitability lasting for 60 msec or more. It is clear from this that the duration of the absolute refractoriness will not be a limiting factor for the spindle to respond to strong stretches since it would permit high frequency discharges. However, the relative refractory period and the ensuing subnormal period might be of significance for the frequency coding of stretches with specific parameters.

An interesting observation is that the excitability of the afferent nerve is affected by stretching of the spindle. In general the effect is characterized by a shortening of the refractory period and a lowering of the threshold of the fiber when tested with electrical shocks. With increasing strength of stretch, the period of supernormality increases in duration. It has proved difficult to characterize quantitatively this change with changes in the parameters of stretch. It appears that the duration of the period of raised excitability increases in relation to the increase in amplitude of the stretch, although it seems to be relatively independent of the duration of stretch. The lowering of the threshold of the nerve may be explained as a result of the depolarization spreading from the endings. This increased excitability of the afferent nerve in the aftermath of stretch as tested with electrical shocks stands in striking contrast to the raised threshold of the spindle to mechanical stimulation. When tested with a brief test stretch, the sensitivity of the spindle is reduced in direct relation to the strength and amplitude of the preceding stretch (OTTOSON *et al.*, 1969).

The mechanical changes left after a stretch may affect not only the sensitivity of the spindle to stretch but also the pattern of discharge to a succeeding stimulus. An example of such aftereffects is shown in Fig. 20. A brief stretch which only

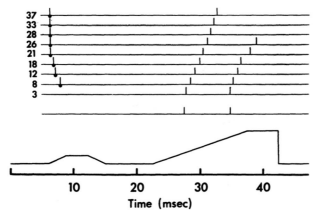

Fig. 20. Mechanical after-effects. Successive runs of response to brief conditioning stretch followed by slowly rising stretch. Stretch indicated by first line from below. Conditioning stretch increased in strength for each run, second stretch being kept constant. Numbers to the left of each line give amplitude of conditioning stretch in arbitrary units. First run shows response to second stretch alone; amplitude 3: conditioning stretch below threshold for giving a spike. Spikes indicated diagrammatically with vertical bars; spike elicited by conditioning stretch marked with dot (from OTTOSON, McREYNOLDS, and SHEPHERD, 1969)

elicited one spike was introduced in front of a linearly rising stretch which gave two dynamic spikes. These two spikes appeared later and later with increasing amplitude of the preceding brief stretch until the second spike dropped out. Since the first stretch evoked only one single impulse in all the runs, the delaying effect must be attributed to mechanical changes taking place in the spindle after stretch.

The finding that the mechanical sensitivity of the spindle is lowered in the aftermath of stretch even though the electrical test shock reveals a raised excitability of the afferent fiber would at first appear contradictory. The change in excitability of the nerve may be explained by the effect of the depolarization spreading from the endings. When stretch is made gradually stronger, the receptor potential increases and the threshold of the fiber falls in direct relation to the amount of depolarization. The fact that the sensitivity to stretch is reduced in spite of the lowered threshold of the fiber could be attributed to aftereffects associated with the mechanical reorganization of the spindle or to depressant effects on the sensory terminals, such as K^+ accumulation. It appears likely that the recovery of the spindle is determined by the time course of the recovery cycle of the fiber only when stretch is at threshold and elicits only a single impulse. With increasing strength of the stimulus, the mechanical aftereffects and the depolarization due to the receptor potential will become increasingly dominant, whereas the refractoriness of the fiber remains relatively constant provided that only a single impulse is produced.

In most situations, stretch gives rise to a discharge of varying frequency and duration. It can be assumed, therefore, that there may be an accumulation of refractoriness related to the impulse discharge. It is difficult to discriminate between the mechanical aftereffects and those which may be attributed to summated refractoriness since changes in the parameters of stretch are almost always associated with corresponding variations in firing. Indirect experimental evidence suggests, however, that compared with the mechanical aftereffects refractoriness is responsible for a relatively small fraction of the depression following stretch.

2. Depression of Receptor Potential

Recordings of the receptor potential have provided further insight into the mechanisms governing the time course of recovery of the spindle. It has been found that the response of the sensory endings is depressed in the aftermath of stretch and that this depression is a function of the duration and strength of the preceding stretch (OTTOSON et al., 1969). As shown in the records in Fig. 21, the depolarization produced by a given test stretch becomes smaller when the interval between two stretches is made successively shorter. It may be noted that the peak of the response reaches the same level at all intervals. The results illustrated in Fig. 21 are typical for experiments in which the first conditioning stretch is weak or moderately strong. The time course of depression then appears to be closely related to the decay of the preceding response; when the interval between the

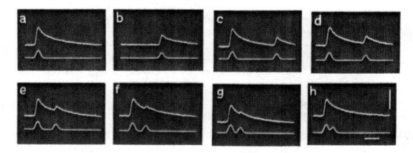

Fig. 21. Depression of receptor potential following brief stretch. a, response to conditioning stretch; b, to test stretch; c-h, to test stretch at different intervals after conditioning stretch. Vertical bar: 1 mV. Time bar: 10 msec (from OTTOSON, McREYNOLDS, and SHEPHERD, 1969)

two stretches is long enough, so that the response to the conditioning stimulus has returned to baseline before the second stretch is applied, there is little depression left. This would suggest that the depression is a function of the amount of depolarization of the endings. However, depression after strong stretches outlasts the decay of the response. It is possible, therefore, that mechanical deformation outlasts the electrical depolarization following release of stretch. A depressant effect at the sensory membrane itself cannot be ruled out. Similar observations have been made on other mechanoreceptors such as the Pacinian corpuscle (GRAY and SATO, 1953; DIAMOND et al., 1958; LOEWENSTEIN and ALTAMIRANO-ORREGO, 1958; LOEWENSTEIN and MENDELSON, 1965) and on insect mechanoreceptors (THURM, 1965).

It can be assumed that the restitution in the aftermath of stretch is the result of a springlike action of elastic elements which have been extended during stretch. The effect of a stretch when applied before these structures have returned to their normal resting length will be reduced by an amount which is directly related to the differences between the actual length and the length after complete restitution i.e., the normal resting length of the spindle. The behavior of the spindle would in this respect resemble that of a plastic cord with rheoscopic properties. The time course of mechanical recovery would consequently be determined by the mechanical properties of the intrafusal fibers and the connective tissues within the spindle. In the spindle in situ it is likely that besides the tissues within the spindle itself the extrafusal fibers and connective tissues may influence the rate of mechanical restitution and consequently also the time course of recovery of the spindle.

When the duration of stretch is increased, the depression of the response to a test stretch becomes more profound. An illustrative example of this is shown in Fig. 22. After a brief dynamic stretch, the response to the test stretch was only slightly reduced, whereas it became progressively smaller when the conditioning stretch was prolonged, and was almost abolished when the duration of the conditioning stretch was 600 msec. This depressant effect could be due to mechanical

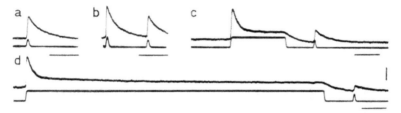

Fig. 22. Depression of receptor potential with increasing duration of preceding stretch. a, response to test stretch alone; b-d at constant delay after conditioning stretches of increasing durations. Vertical bar: 2 mV. Time bar: 50 msec (from OTTOSON, McREYNOLDS, and SHEPHERD, 1969)

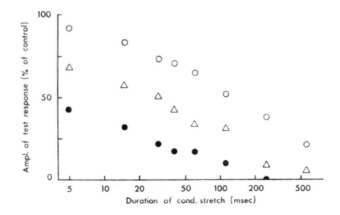

Fig. 23. Depression of receptor potential in relation to duration of preceding stretch. Interval from end of preceding stretch to beginning of test stretch: 10 msec (filled circles), 20 msec (triangles) and 50 msec (open circles). Amplitude of receptor potential to test stretch expressed in per cent of unconditioned value (from OTTOSON, McREYNOLDS, and SHEPHERD, 1969)

changes which develop progressively during a maintained stretch and are similarly time-dependent in their restitution. Alternatively, the depressant effect could be due to recovery processes in the nerve terminals themselves. A comprehensive analysis of the depression after stretches of different durations and amplitudes provides further evidence for the close relation between the mechanical parameters of the stimulus and the amount of the depression, as illustrated in Fig. 23.

In conclusion, the responsiveness of the spindle after release of stretch is determined by the interaction of mechanical events related to the structural reorganization of the spindle and electrical events associated with depolarization and recovery of the endings and refractoriness of the afferent fiber and its myelinated branches. The mechanical factors and the recovery of the endings together with refractoriness tend to depress responsiveness; this depressive effect is counteracted by the lowering of the nerve threshold caused by the receptor potential.

VI. Sensitivity and Extension

1. Changes in Impulse Threshold with Changes in Length

The muscle spindle in situ is exposed to considerable variations in length and tension as a result of the activity of fusimotor fibers or activity of the extrafusal

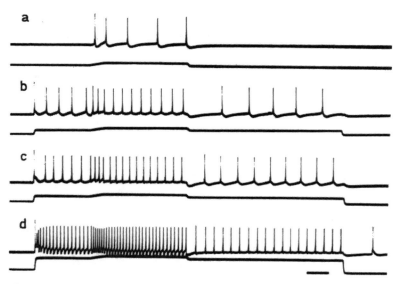

Fig. 24. Increase in responsiveness to stretch with increasing lengthening of the spindle. Test stretch (a) superimposed upon steplike maintained stretches to different lengths (b-d). Time bar: 50 msec (from Ottoson, McReynolds, and Shepherd, 1969)

part of the muscle. It seems clear, therefore, that the responsiveness of the spindle may vary a great deal depending on its actual mechanical state. In general, a lengthening of the spindle is accompanied by an increased responsiveness (Ottoson et al., 1969). An example of this is illustrated in Fig. 24. A small stretch which evokes only a very weak discharge when the spindle is relaxed becomes increasingly more effective when the spindle is lengthened. Both the dynamic and the static discharge appear to increase. Inspection of the response to the underlying stretch with and without the superimposed briefer stretch, however, reveals that the static discharge evoked by the test stretch remains relatively constant, whereas the dynamic phase increases.

Measurements of threshold changes with changes in length reveal that the sensitivity of the spindle in terms of the amount of stretch required to elicit an impulse increases directly with lengthening. This is demonstrated in Fig. 25 which shows results obtained from three spindles. The threshold was measured with brief

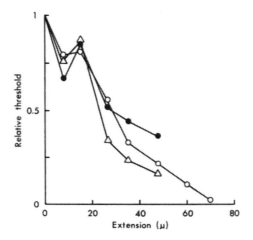

Fig. 25. Decreasing threshold of spindle with increasing extension. Values from three different spindles. Threshold measured with brief stretches at approximately 30 msec after prestretching spindle to different lengths (from OTTOSON, McREYNOLDS, and SHEPHERD, 1969)

test stretches applied about 30 msec after the spindle had been prestretched. With increase of the amount of prestretching, the threshold fell progressively. The early break in the curve is related to the appearance of the dynamic spike evoked by the steplike prestretch and is a measure of the refractoriness following this impulse. No additional dynamic impulse was evoked by the background stretch with further lengthening, and the influence of refractoriness therefore remained constant. The fall in threshold consequently reflects the effects of lengthening.

A similar direct relation between the amount of extension and the mechanical sensitivity of the spindle is found during the dynamic phase of linearly rising stretch. At the transition from dynamic to static extension, the spindle shifts over from recording velocity and amplitude of the stimulus to signalling only steady displacement. As described above, there occurs within a given range of velocities and amplitudes of stretch a pause in the discharge during this period. It would appear likely that the appearance of this pause is related to a relative reduction in sensitivity of the spindle. Tests with brief stretches show that the impulse threshold in this period varies in a regular way with the parameters of stretch. In general, the interval between the last dynamic and the first static spike is characterized by a comparatively high threshold. For certain amplitudes and velocities of stretch there is, however, an early brief period of increased excitability related to the peak of the underlying receptor potential. It thus appears that the transitional interval is a period of complex interaction of different factors, some of which depress and some of which tend to augment the sensitivity of the spindle. There is an interesting resemblance between the muscle spindle and touch receptors in the toad skin with respect to the excitability changes taking place during mechanical stimulation. As demonstrated by Lindblom (1963), the excitability of the touch receptor increases in direct relation to the stimulus during linearly rising mechanical deformation of the skin. At the transition from dynamic to static stimulation, the sensitivity of the touch receptor falls, the magnitude and time course of this change being dependent on the parameters of the mechanical deformation.

It may be concluded that the variations in responsiveness of the spindle during different phases of stretching are associated with excitability changes in the afferent fiber and its branches. The influence of such changes has been assessed by testing the afferent fiber with electrical shocks applied to the nerve in the interval between the impulse elicited by dynamic and static stretch. The results show that the threshold of the fiber is lowered during stretching in direct relation to the amplitude of the stretch (Ottoson and Shepherd, 1969). The lowering of the threshold of the afferent nerve during stretch is most likely to be attributed to the depolarization caused by electrotonic spread of the receptor potential.

In general, the impulse frequency during steady stretch appears to be a function of two main events, the amount of depolarization of the terminals and the time course of recovery of the fiber. Since each impulse is initiated by the production of local currents of the nerve, the recovery will also be dependent on the interaction of refractoriness and the development of the local potential initiating the impulse. At low levels of sustained stretch, refractoriness may be relatively unimportant. With increasing stretch and particularly during fast dynamic stretches, refractoriness becomes more dominant and may be a limiting factor with respect to the ability of the spindle to respond to fast transients of the stimulus. At the same time the depolarizing action of the sensory terminals becomes more powerful and will tend to lower the threshold of the fiber and shorten its recovery period. However, there is also evidence that the sensitivity of the spindle at any moment during stretch is not a simple function of the depolarization level and refractoriness. This is shown, for instance, by the observation that the threshold just after release of a sustained stretch (i.e., during the decay of the static part of the receptor potential), has reached resting level. Furthermore, the excitability is higher during the rising part of the dynamic receptor potential than during its fall to the static level.

2. Changes of Responsiveness of the Sensory Endings with Changes in Length

It is obvious that the increase in mechanical sensitivity of the spindle during stretch cannot be explained only by the lowered threshold of the fiber. It is likely that the responsiveness of the sensory endings is also altered when the spindle is stretched. An analysis of the responses of the endings to brief test stretches applied during lengthening of the spindle provides evidence that the increase in mechanical sensitivity as described above is mainly due to an increased responsiveness of the endings.

An illustrative example of the variations of the isolated receptor response of the endings during lengthening and after release of stretch is shown in Fig. 26. A brief stretch which elicited a relatively small response when applied to the relaxed spindle produced a large potential when superimposed upon a steady maintained stretch. When applied in the aftermath of this stretch, it gave no response

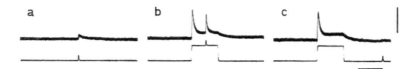

Fig. 26. Receptor potential evoked by brief dynamic stretch. a, at resting length; b, during static phase of steady stretch; c, in the aftermath of steady stretch. Time bar: 100 msec. Vertical bar: 3 mV

at all. A quantitative measure of the change in responsiveness of the endings with increasing lengthening of the spindle can be obtained by applying a given brief test stretch to a spindle which is prestretched to different lengths. When this is done, it is found that the response to the superimposed test stretch increases up to a maximum with increasing length of the spindle and then decreases with further elongation (Fig. 27). It is of interest that the maximum response is obtained when the spindle is prestretched by about 30% of its resting length.

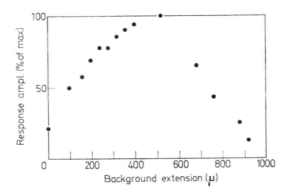

Fig. 27. Dynamic responsiveness of sensory endings at different lengths of spindle. Amplitude of dynamic receptor potential elicited by brief test stretch (of constant strength) superimposed upon steady background extension of spindle to different lengths. Response amplitude given in per cent of height of potential obtained from spindle at resting length (background extension zero). Resting spindle length: 970 μ (from OTTOSON, McREYNOLDS, and SHEPHERD, 1969)

Closer analysis of the response to brief test stretches superimposed upon a sustained stretch shows that not only does the amplitude of the test response increase with increasing lengthening of the spindle but also the rate of rise of the response becomes faster. Similar tests during the dynamic phase of stretch (Fig. 28) show that for a given velocity of stretching the response to the test stretch increases in amplitude in direct relation to the underlying lengthening of the spindle.

In the early period of static stretch corresponding to the dynamic decay of the receptor potential, there is a reduction of the responsiveness of the endings in terms of the increment in depolarization added by a brief test stretch. This reduction is particularly pronounced with strong and fast velocities of the underlying stretch.

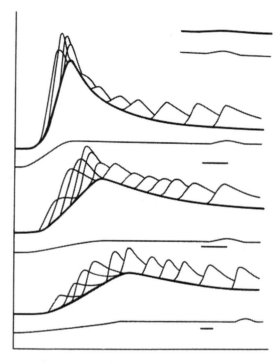

Fig. 28. Responsiveness of sensory endings during stretching. Superimposed tracings of responses to constant test stretch at various times during dynamic and static extension at three different velocities (to the same final amplitude). Stretch indicated by line below each set of tracings. Inset at upper right shows response to test stretch alone (lower line stretch). Time bar: 10 msec; note slower time base in lowest tracing (from Ottoson, McReynolds, and Shepherd, 1969)

These observations suggest that the sensitivity of the spindle under different conditions of lengthening is closely related to the responsiveness of the endings. The changes in responsiveness, in turn, arise most likely as a result of mechanical changes in the spindle. Thus the enhancement of the receptor response during stretch may be attributed to stretching of the elastic elements within the spindle. By prestretching the spindle, slack is taken up and an applied second stretch becomes more effective. An interesting point is that the increasing responsiveness with increasing length of the spindle seems to suggest a nonlinear characteristic for the production of the dynamic response. This behavior could partly be accounted for by the assumption that elastic elements are arranged parallel to the spindle. Slack in these elements may be taken up gradually with increasing extension, and an additional stretch would consequently become more and more effective.

This also directs attention to the differences in structure between different regions of the spindle. It would appear likely that the sensory endings throughout the spindle are not affected to the same extent by changes in length of the spindle. Also, variations in the velocity of lengthening would affect the sensitivity in the reticular and compact zones differently. All this implies that the activity within different regions of the spindle is bound to vary during stretch. The general effect, however, is characterized by an increased sensitivity with increasing lengthening of the spindle and a depression in the aftermath of stretch.

In conclusion, whereas the response of the spindle to a given stretch is extremely constant under resting conditions, as was first noted by MATTHEWS (1931), it may be modified by a number of factors. These factors may be of mechanical origin and related to changes in tension, or they may be related to electrical changes in the nerve and in the endings. The mechanical effects are revealed by an enhancement of the response of the sensory endings with extension. The electrical factors influencing the discharge pattern are related to depolarization and recovery of the endings and the recovery cycle of the nerve fiber as has been described above.

In the experiments described above, the entire bundle of intrafusal muscle fibers was extended during lengthening of the spindle, and the sensory endings on the different fibers were consequently affected more or less to the same extent. This would resemble the situation in situ when the whole muscle is extended. A similar situation might also occur during fusimotor activation.

Anatomical studies (BOYD, 1962; BARKER, 1967) have demonstrated that the efferent fibers innervating the intrafusal muscle fibers can be divided into several different types. Functionally the motor fibers may be separated into two distinct groups, usually called dynamic and static fusimotor fibers (MATTHEWS, 1963, 1964). In the mammalian spindle, stimulation of the dynamic fibers increases the response of the primary endings to dynamic stretching whereas the static fibers have no effect or cause a reduction of the dynamic sensitivity of the spindle. It has been suggested that this functional arrangement provides the basis for a control of the dynamic and static response of the spindle (JANSEN and MATTHEWS, 1962).

In the amphibian spindle, two kinds of intrafusal muscles with separate motor innervation have been described. Activation of each of these kinds of fibers suggests a functional differentiation with respect to the effect on the response to dynamic stimuli (MATTHEWS and WESTBURY, 1965). The results on the isolated spindle provide information about the effect of a simultaneous and equal extension of all the intrafusal fibers comparable to activation of both types of fusimotor fibers. The differentiation of these two types in the isolated preparation awaits future study.

In the diagram in Fig. 29, an attempt has been made to summarize the excitability changes of the nerve and the sensory endings during a moderately strong stretch which rises linearly to a maintained level. The composite response exhibits the characteristic dynamic discharge of increasing frequency followed by a pause during the transition to steady stretch and the ensuing static discharge. The underlying receptor potential is seen below the impulse response. The third diagram from the top gives a schematic representation of the changes in responsiveness of the endings as tested with brief stretches. There is a linear increase during dynamic stretch followed by a rapid fall in the transitional interval and a later gradual rise to the static level. After release of stretch, the sensitivity drops below the resting value. The changes in excitability of the nerve are illustrated in the bottom diagram. The overall change is an increase in excitability related to the depolarization represented by the receptor potential. This change is interrupted by the changes in the aftermath of each of the impulses. It may be noted that the recovery of the nerve gradually becomes faster with increasing level of dynamic stretching.

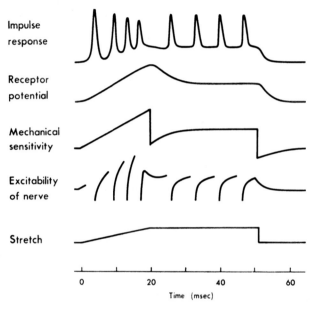

Fig. 29. Schematical diagram summarizing functional changes in muscle spindle during and after stretch (from Ottoson and Shepherd, 1969)

The recovery after the last dynamic spike is modified by the fall of the receptor potential, and the threshold for firing an impulse is not reached in this particular situation. During the static phase of stretch, the recovery cycles following successive spikes are more or less identical since the level of underlying depolarization remains constant. After release of stretch, the excitability of the nerve decays slowly in relation to the fall of the receptor potential and in contrast to the abrupt reduction in mechanical sensitivity.

VII. Impulse Initiation

1. General Mechanisms

The impulse response recorded from the afferent nerve is the ultimate result of a sequence of events taking place in the spindle during stretch. These events involve the mechanical transmission of the stimulus to the endings, the transduction of the deformation of the sensory membrane into a depolarization, and the conversion of this depolarization into an impulse message. The different stages involved in the electrical processes of impulse initiation can be identified by a detailed analysis of the response obtained to a given stimulus before and after blocking of the conducted activity. However, several important features associated with the generation of impulses may be discerned in recordings of the composite response to critically adjusted stretches.

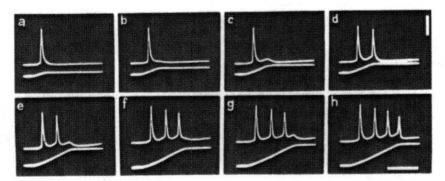

Fig. 30. Development of dynamic discharge. Spindle response to linearly increasing stretch to increasing lengths. Velocity of stretch about 11 mm/sec. Time bar: 20 msec. Vertical bar: 2 mV (from OTTOSON and SHEPHERD, 1965)

The records in Fig. 30 illustrate an experiment in which the development of the dynamic discharge was studied in this way. The first record in this series shows the response to a stretch which was just strong enough to elicit a spike. With a somewhat stronger stretch, a small hump appeared in the aftermath of the spike. This potential hump increased when stretch was allowed to proceed further, and at a critical strength a second spike arose. With further extension, the same sequence of events was repeated, leading to the building up of the dynamic discharge.

The depolarization of the endings and the development of the afferent impulse represent the two principal stages in the excitatory process of the spindle during stretch. However, interposed between these two steps there are also potential changes related to activity in the myelinated branches and to activity at the first node of the afferent fiber (KATZ, 1950a). These potentials can be presumed to play an important function in the development of the characteristic pattern of discharge to a given stretch. A detailed analysis of the response under various experimental conditions provides interesting information about the sequence of processes leading to the appearance of the impulse discharge in the afferent fiber.

2. Activity of the Myelinated Nerve Branches

It appears likely that the impulses are initiated at the first node of the fiber and that they arise as a direct result of the depolarization which spreads electrotonically from the endings. However, the spindle contains a varying number of myelinated branches which converge toward the stem fiber. The question may therefore be raised if conducted activity takes place in these branches or if the branches serve only as passive conducting lines for the spread of the receptor potential to the first node. If they do conduct impulses, which appears very likely, the implication would be that there are three main steps involved in the impulse generating process: 1) the production of the receptor potential; 2) initiation of conducted activity in the myelinated branches and the confluence of these impulses towards the first node; and 3) the initiation of the impulses in the stem fiber. This would also imply that the initiation of the impulses in the stem fiber may be the result of depolarization caused by the confluence of impulses in the branches towards the first node. The function of the receptor potential would then be to initiate conducted activity in the branches.

The functional role of the branches in the initiation of the afferent discharge is difficult to assess because of their small diameters. The fact that they are enclosed in the spindle capsule also makes it difficult to record from them. However, it is possible by various experimental methods to gain information about their activity. One way to do this is to subject the spindle to repeated stretches. With a critical strength, velocity and frequency of stimulation, the single impulse will then split up into smaller spikes. An example of this is shown in Fig. 31. The

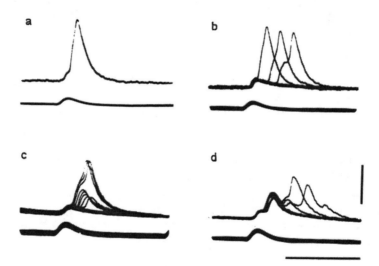

Fig. 31. Progressive breakdown of impulse transmission during repetitive stretch. a, response to single stretch; b-d, superimposed tracings responses in course of repetitive stretches at 50/sec (b), 70/sec (c) and 100/sec (d). Time bar: 10 msec. Vertical bar: 1 mV (from Ottoson and Shepherd, 1970c)

spindle in this experiment was subjected to a brief stretch giving rise to one impulse. When the stretch was repeated at a frequency of 50 to 100 per sec, there occurred a series of changes leading to a decomposition of the impulse into smaller subunits which with prolonged stimulation dropped out in an all-or-nothing manner, finally leaving the underlying receptor potential alone. A similar decomposition of the single impulse also appears when the spindle is treated with tetrodotoxin or procaine in low concentrations (Albuquerque et al., 1969).

The small spikes which are isolated by repetitive stimulation or by treatment of the spindle with tetrodotoxin appear to correspond to the spontaneous abortive spikes which are often seen in recording from the resting spindle. All of the experimental evidence suggests that these spikes represent the activity of the myelinated branches (see KATZ, 1950a). The demonstration that these spikes form a part of the response has important implications for interpretation of the processes underlying the initiation of the impulse response in the afferent fiber. The obvious conclusion is that the receptor potential initiates conducted activity in the myelinated branches and that these impulses propagate toward the first node of the stem fiber. The fact that the activity of the separate branches is not seen in the response to stretch, except under certain conditions, suggests that normally the activity is synchronous. This view is also supported by the fact that the conducting distances are short. Differences in time course of the activity arising in branches from different parts of the spindle therefore tend to be cancelled out. The decomposition of the impulse during repetitive stretch or under the action of tetrodotoxin would be explained as the result of a desynchronization of the activity in the branches. The direct cause of this desynchronization is probably the development of blockage in different branches.

The demonstration that the branches conduct impulses raises the question of to what extent this activity is responsible for the initiation of the discharge of the stem fiber. The observation that the randomly occurring abortive spikes which are seen in the resting spindle do not propagate into the stem fiber (see KATZ, 1950a) suggests that activity in one branch alone is not enough to produce an afferent impulse. When several branches are active synchronously, the depolarization of the node would reach threshold for initiating an impulse in the stem fiber. It would appear likely, therefore, that the discharge of the spindle is governed mainly by the activity of the branches when stretch is weak. With increasing intensity of the stimulus, the receptor potential may become increasingly dominant with respect to the development of the afferent discharge. This view is also supported by the observation of the regularity of the discharge and the close relationship between the receptor potential and the impulse pattern.

3. Local Response of the Afferent Nerve

It is obvious from the above that the events taking place at the first node play a critical role in the development of the characteristic features of the pattern of the afferent discharge. One important step in the initiation of the impulses is the production of local currents in the stem fibre. This local response can be discerned as a prepotential (see EYZAGUIRRE and KUFFLER, 1955) preceding the spikes in the response to stretch. During dynamic stretching, the prepotentials build up in rapid succession with increasing velocity and amplitude of stretch, whereas during

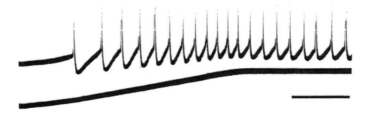

Fig. 32. Increase in rate of rise of prepotentials with increasing level of stretching. Time bar: 100 msec (from OTTOSON and SHEPHERD, 1970b)

maintained stretch their rate of rise is more or less constant and directly related to the amplitude of the stretch (Fig. 32). Measurements of the time characteristics of the prepotentials show that there is a linear relation between the frequency of firing and the velocity of rise of the prepotentials during dynamic stretching (Fig. 33a). The slope of the line relating the two parameters varies with the velocity of stretching. Since the prepotentials arise as a result of the depolarization

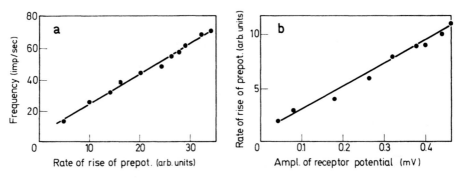

Fig. 33. a, relation between frequency of dynamic firing and rate of rise of prepotentials. b, rate of rise of prepotentials related to amplitude of depolarization of sensory endings (from Ottoson and Shepherd, 1970b)

of the endings, one might expect to find a similar close relation between their rates of rise and the magnitude of the receptor potential. This is also borne out by the analysis of responses to a given stretch before and after blocking of the conducted activity. When the individual prepotentials are related to the underlying receptor potential, it is found that their rates of rise are a direct function of the amount of depolarization of the endings (Fig. 33b).

The prepotentials are graded events, the development of which can be analyzed by subtraction of the receptor potential from the response obtained before blocking. By critical adjustment of the parameters of linearly rising stretches, response patterns suitable for disclosing different stages in the production of the prepotentials can be chosen. One example of the results obtained in this way is illustrated in Fig. 34. The isolated receptor potential is shown by the dotted line and the column to the right shows the derived curves obtained after subtraction of the receptor potential from the composite responses. It may be noted that the dynamic spikes are followed by marked undershoots and are considerably briefer than the static spikes. The potential changes leading to the building up of the dynamic discharge cannot be analyzed in detail in this record since the sweep is not fast enough. However, the development of the static discharge with increasing stretch can be followed more closely. As can be seen, the spikes are preceded by slowly rising potentials. These potentials, which are absent after blocking of the conducted response, represent the local current in the nerve associated with the initiation of the impulse (see Hodgkin, 1948). With increasing stretch and increasing depolarization of the endings, these prepotentials become steeper leading to an increased firing frequency. The level of onset of the impulses during a given discharge seems to remain relatively constant.

Hence, preceding the appearance of the impulse response in the afferent fiber, there is a sequence of electrical changes in the sensory system of the spindle. The first stage in this process is the production of the generator potential by the transducer membrane, the structural identity of which still remains unsettled.

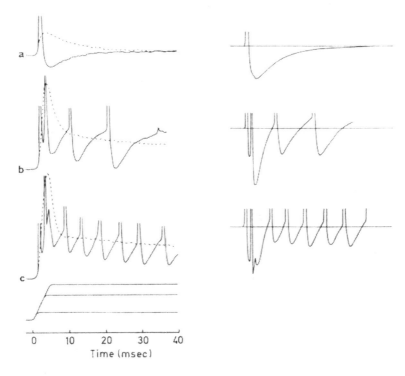

Fig. 34. Component analysis of spindle response. Left column: recordings before (solid line) and after (dashed line) blocking of conducted response to steplike stretches of increasing strengths. Right column: conducted activity derived by subtraction of receptor potential from composite response

The second stage in the encoding process is the initiation of conducted activity in the myelinated branches and the propagation of these spikes toward the first node. Owing to the short conductance distances, the spread of the impulses from the preterminals to the first node of the trunk fiber can be estimated to be only fractions of a millisecond. Under normal conditions, the transducer system therefore works in perfect synchrony and the contribution of the individual branches cannot be discerned. The impulse discharge of the trunk fiber represents the total integrated effect of the sensory system of the spindle.

VIII. Mechanisms of Excitation

1. Transmittance of the Stimulus

As discussed in Section IV, it is likely that the generator potential of the spindle arises as a result of deformation of the sensory terminals. In 1961, KATZ suggested that the transducer action is localized to the bulbous expansions of the terminals. Many of these bulbs are closely attached to the surface of the muscle fiber but a considerable number of them are lying free in the lymph space. At the present time nothing is known about the mechanical deformation of the endings during stretching. Since the bulbs vary in their structural relation to the muscle fiber as well as in their size and configuration, it seems most likely that elongation of the spindle must give rise to a diversified pattern of deformation in different bulbs. The structural differences between the intrafusal fibers in the reticular zone and in the compact zones probably also give rise to regional differences with respect to the distortion of the sensory endings, whether the transducers are the bulbs or the connecting links or both of these structures.

The structural organization of the spindle makes it probable that the externally applied stimulus must undergo a distortion before the sensory endings are reached. It would appear, therefore, that the parameters of the externally applied stimulus in terms of lengthening may represent an inaccurate measure of the characteristics of the actual stimulus to which the sensory membrane is exposed. As long as the mechanical deformation of the endings cannot be characterized, there would appear to be no means of determining the precise relation between the stimulus and the transducer action.

2. Osmotic Excitation

One way to produce a deformation of the sensory endings with little or no intervention of the structure which transmits the stimulus during stretching is to expose the spindle to changes in tonicity of the external solution. Observations on the effect of anisotonic solutions (Ottoson, 1964) are also of interest in relation to the mechanisms underlying the differentiation of the response into a dynamic and a static phase. As discussed above, it is possible that this differentiation is partly of mechanical origin and related to the differences in visco-elastic properties between the reticular zone and the two compact zones. It can be assumed that exposure of the spindle to anisotonic solutions would cause changes in volume of the sensory bulbs and that these changes would be relatively uniform throughout the different regions of the spindle. Furthermore, the deformation induced by the volume change can be postulated to be directly related quantitatively to the changes in osmolarity.

Studies on the isolated spindle have demonstrated that changes in the tonicity of the bathing solution have profound effects on the activity of the spindle. Increase of the osmotic pressure produces a discharge which in moderately strong solutions may go on for several minutes (Fig. 35 A). With stronger solutions, the firing becomes more intense and also more short-lived. In these cases, the firing usually stops abruptly and the spindle remains silent as long as it is immersed in the hypertonic solution. Measurements of the frequency characteristics of the discharge show that the frequency increases linearly with the logarithm of the osmotic strength for concentrations up to four to five times that of Ringer's solution. The typical differentiation of the response into a dynamic and an ensuing static phase is never seen. Either the firing goes on for several minutes with a slow gradual decline in frequency or it stops abruptly as is the case when the spindle is exposed to strong solutions.

Exposure of the spindle to hypotonic solutions also causes a discharge. However, the hypotonic excitatory action is typically of short duration; generally it lasts only for a few minutes and then stops abruptly. Like the hypertonic effect, the discharge induced by hypotonic solutions increases in direct relation with the

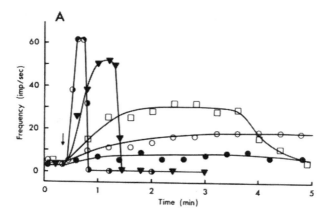

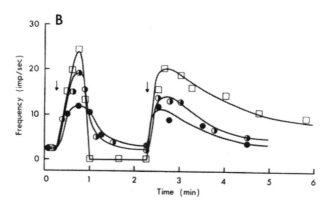

Fig. 35. Osmotic excitation. A: effect of hypertonic NaCl-Ringer's solution. Change from isotonic to hypertonic solution marked by arrow. Concentrations expressed relative to Ringer's solution: $1.5 \times R$ (filled circles), $2 \times R$ (open circles), $3 \times R$ (triangles) and $5 \times R$ (squares). B: effect of hypotonic solutions. Change from isotonic to hypotonic solution marked by first arrow from left, return to Ringer's solution by second arrow. Concentrations relative to Ringer's solution: $0.25 \times R$ (half-filled circles), $0.1 \times R$ (squares) (from OTTOSON, 1965)

changes in osmolarity (Fig. 35B). For obvious reasons the effect of low tonicity can only be studied within a small range of changes in tonicity. A comparison with the hypertonic effect is also restricted by the fact that with low tonicity there is the complicating effect of lack of sodium ions. An interesting finding is that on return to Ringer's solution there is regularly a prolonged discharge resembling the hypertonic effect.

Exposure of the spindle to anisotonic solutions also produces characteristic effects on the responsiveness of the spindle to stretching. Increased osmotic pres-

sure causes a reduced sensitivity to stretch, whereas reduced pressure gives rise
to a transient increase followed by a decrease in responsiveness. The curves in
Fig. 36 summarize these effects. Curve *a* shows the excitatory action of different
osmotic strengths in the range from four times Ringer to 0.25 times Ringer; curve
b illustrates the corresponding changes in sensitivity to stretch.

Electron microscopic studies (Karlsson and Ottoson, 1970) have shown that
changes in tonicity of the bathing fluid are accompanied by pronounced effects in

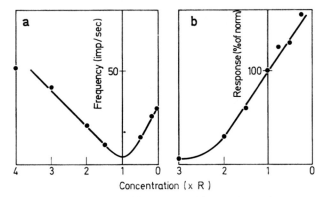

Fig. 36. Diagram summarizing osmotic effects on spindle. a, relation between osmotic pressure
and excitatory action. b, responsiveness of spindle at different osmotic pressures. Response
to test stretch expressed in per cent of that obtained in Ringer's solution. Abscissa in a and b:
osmotic pressure of bathing fluid in units relative to Ringer's solution (from Ottoson, 1965)

terms of volume changes of the sensory nerve endings whereas other cell components within the spindle are little affected at first (Fig. 37). This suggests that the
excitatory effect is the result of a direct action of the anisotonic solutions on the
endings and not a result of volume changes of the intrafusal fibers or other cells.
It would appear from the osmotic effects that the sensory endings behave as
simple osmotic sacs within a limited range of osmotic strengths. The fact that both
increase and decrease in osmotic strength have an excitatory action suggests that
the membrane is sensitive to distension as well as to compression.

The observations on the osmotic effects also appear to have implications with
respect to the interpretation of the mechanisms responsible for the characteristic
adaptation properties of the spindle. The finding that moderately strong solutions
produce a long-lasting effect with very little decline clearly shows that the endings
are able to maintain a constant activity when exposed to a constant stimulus. The
abrupt cessation of the discharge with strong hypertonic solutions appears to be
due to overstimulation in a manner somewhat resembling the effect of overstretch.
The osmotic effect is most likely due to a direct action on the endings as indicated
by the electron microscopic findings and can be assumed to remain constant as
long as the solutions are not changed. The effect is also most likely the same in
the reticular region as in the compact zones.

The osmotic stimulating effects on the spindle are of interest in relation to the
electrokinetic model of a mechanoreceptor developed by Teorell (1962, 1966). In
this model it is suggested that excitation is produced by a sequence of changes
in the gradients of osmotic pressure, ionic concentration and electrical potential

across the transducer membrane. TEORELL attributes particular importance to the electro-osmotic flow of water. It is possible and even likely that excitation of the muscle spindle is associated with a flow of water through the sensory membrane.

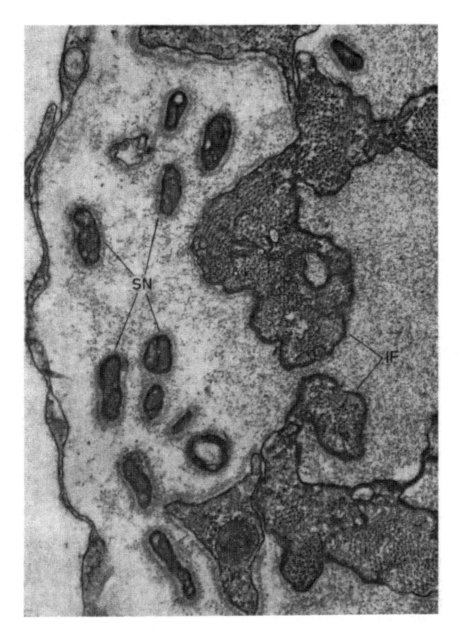

Fig. 37. Osmotic effects on the muscle spindle. Cross section of part of a seemingly normal intrafusal muscle fibre in its reticular zone (*IF*). To the left several shrunken sensory nerve endings (*SN*) embedded in filamentous extracellular material. Magnification: about × **38,000** (from KARLSSON and OTTOSON. In course of publication, 1970)

It is still too early for deciding to what extent the proposed functions of the electro-hydraulic analogues of Teorell are applicable to the events which take place in the sensory end bulbs.

3. Mechanisms of Mechano-electrical Conversion

a) **Chemical Transmitter.** Several hypotheses have been suggested to explain the conversion of the mechanical stimulus into an electrical depolarization of the sensory terminals. One possibility is that stretching leads to the release of a chemical transmitter which affects the membrane of the sensory endings. This idea gains some support from the fact that the receptor potentials in general are closely similar to end-plate potentials and postsynaptic potentials. It has also been reported that acetylcholine has an excitatory action on different types of end organs (see Gray and Diamond, 1957). Studies on the isolated spindle provide no evidence for the assumption that acetylcholine or any related substances are involved as chemical intermediaries in the initiation of the receptor potential. Exposure of the spindle to acetylcholine at increasing concentrations up to 1×10^{-4} mM has no excitatory effect on the spindle. At higher concentrations the drug causes a decrease in the spontaneous discharge. Other cholinesters such as butyrylcholine, propionyl-choline and succinylcholine are equally ineffective. The failure of succinylcholine is of interest since it is well known that this substance when injected intravenously produces a vigorous discharge of muscle spindles in the cat (Granit et al., 1953; Henatsch and Schulte, 1958). It has been suggested that this effect is due to a direct action on the sensory terminals of the spindle. The experiments on the isolated spindles as well as later studies on the cat (Smith and Eldred, 1961; Fehr, 1965) clearly show, however, that succinylcholine has no direct excitatory action on the sensory endings of the spindle.

In view of the relative ineffectiveness of the cholinesters, one would not expect anticholinesters to affect the spindle. However, with the exception of prostigmine, several anticholinesters have been found to exert an excitatory action on the isolated spindle (Ottoson, 1961). This action may not be considered as evidence for a cholinergic mechanism at the sensory endings, since the excitatory effect is only obtained with relatively high concentrations of the anticholinesterases. It is possible, therefore, that the action is unspecific and not related to blockage of cholinesterases. Whatever the explanation of the excitatory effect obtained with the inhibitors may be, the findings suggest the existence of enzymatic systems involved in the processes maintaining the membrane potential of the endings.

b) **Deformation and Membrane Permeability.** The most probable explanation of the transducer action appears to be that mechanical deformation of the endings leads to changes in the permeability properties of their membranes and thereby to a depolarization. Stretching of the spindle would open up channels which are closed in the spindle at resting length and would permit ions to pass through the membrane along their concentration gradients. It seems obvious that the amount of deformation of the sensory membrane produced by a given stretch must be dependent on how closely the sensory terminals are attached to the intrafusal fibers and also on the mechanical properties of the transmitting structures. Since the reticular zone differs from the compact zones, it is likely that membrane deformation and thereby depolarization has different characteristics in the different areas. On release of stretching, the membrane would return to its previous state and the channels close. The time course of the restitution of the membrane and the closing of the channels must be dependent on the mechanical properties of the sensory membrane and of structures attached to the endings. Since the mechanical properties of the spindle are different in different regions, the time course of mechanical restitution is probably not the same throughout the branching tree of the terminal endings.

At the present time there is very little direct evidence about the mechanical changes which take place during and after stretching. Studies at the ultrastructural level of the effect of changes in osmotic pressure (KARLSSON and OTTOSON, 1970) suggest that comparatively small amounts of deformation of the sensory bulbs might be sufficient to cause an excitatory effect. This is also consistent with the high mechanical sensitivity of the spindle.

4. Ionic Mechanisms

In view of the role played by sodium in synaptic transmission and impulse generation, it would be reasonable to expect that this ion also would be involved in the transfer of charges across the transducer membrane of the spindle endings. The early studies by KATZ (1950 b), however, did not provide evidence supporting this view. KATZ observed that the receptor potential of the spindle could still be obtained when sodium in the external solution was replaced by choline chloride. Observations by GRAY and SATO (1953) on the Pacinian corpuscle also failed at first to provide evidence that sodium ions are essential for the transducer action. Later perfusion experiments (DIAMOND et al., 1958) demonstrated clearly, however, that the amplitude and rate of rise of the receptor potential were closely related to the concentration of sodium in the perfusion fluid. At present there is also evidence that sodium ions play an essential role in the production of the receptor potential of the spindle (OTTOSON, 1964).

The immediate effect of withdrawal of sodium on the resting spindle is a decrease and abolition of the spontaneous discharge. This is followed by a diminution of the impulse responses evoked by stretch, and after immersion for 3 to 10 min all signs of conducted activity are abolished. At this stage the receptor potential is relatively little affected. With prolonged exposure to the sodium-free solution, the receptor potential also begins to diminish. However, even after soaking in a sodium-free solution for 30 min, about 20 to 30 % of the original receptor potential can still be obtained. This finding is consistent with observations on the Pacinian corpuscle (DIAMOND et al., 1958) and suggests that ions other than sodium participate in the production of the receptor potential.

A likely ion would be potassium. Observations on the effect of varying concentrations of potassium provide evidence that this ion is also involved in the transport of charges across the membrane (HUSMARK and OTTOSON, 1970c). It is also plausible that calcium affects the permeability of the membrane and thereby indirectly the amount of depolarization of the transducer elements (OTTOSON, 1965).

The finding that ions other than sodium participate in the production of the receptor potential suggests that the conductance change produced in the sensory membrane by the mechanical stimulus is not sodium-selective. If this is so, the permeability changes taking place in the spindle membrane would be similar to those underlying the development of end-plate potentials (FATT and KATZ, 1951) and postsynaptic potentials (see ECCLES, 1964). The idea of a nonselective permeability change also appears more attractive if it is assumed that pores are opened up in the membrane when the spindle is stretched. Such a mechanism would imply that the number of pores opened as well as their size would vary with the amount of stretch. A mechanical pore hypothesis would therefore be difficult to reconcile with the assumption of a selective permeability change.

It is an old observation (OVERTON, 1902) that lithium can replace sodium in the processes underlying the production of the action potential of a muscle fiber. Lithium also appears to be able to replace sodium in the generation of receptor potentials. It has been demonstrated (ALBUQUERQUE et al., 1969) in the isolated spindle that the receptor potential obtained after blockage of the impulse activity with tetrodotoxin remains unchanged when sodium in the external solution is replaced by lithium. When lithium is subsequently removed, the potential declines by about the same amount as when the spindle is soaked in a sodium-free solution.

It has been established that tetrodotoxin blocks the action potentials of nerve and muscle fibers by causing a blockage of the sodium-carrying system. Since sodium also is responsible for the production of the greater part of the receptor potential in mechanoreceptors such as the crustacean stretch receptor, the muscle spindle and the Pacinian corpuscle, one would expect that this fraction of the response should also be blocked. LOEWENSTEIN and his co-workers (1963), who were the first to study this problem, found that the toxin affected the receptor potential of neither the crustacean stretch receptor nor that of the Pacinian corpuscle. Later, OZEKI and SATO (1965) and NISHI and SATO (1966) reported that the toxin caused a 40% reduction in the receptor potential of the Pacinian corpuscle and attributed this effect to a blockage of the sodium channels in the sensory membrane.

The problem of whether tetrodotoxin blocks the sodium-carrying system of the sensory membrane has interesting implications. If the toxin causes a block, this would suggest that basically similar mechanisms control the sodium gate of the receptor membrane and of the membrane producing conducted action potentials. A failure of tetrodotoxin to affect the receptor potential, on the other hand, would suggest that the sodium-carrying system of the transducer membrane is different from the membrane of the afferent nerve. Recent studies on the effect of tetrodotoxin on the isolated muscle spindle (ALBUQUERQUE et al., 1969) appear to demonstrate quite definitely that tetrodotoxin has no appreciable action on the receptor potential in this preparation (Fig. 38, record b). Analysis of the response before and after blockage of the impulse activity provides no evidence for reduction of the receptor potential even at very high concentrations of tetrotoxin. Removal of sodium from the external solution after blockage of the conducted activity is accompanied by a reduction of the receptor potential by about 50% (Fig. 38, record c). This effect is reversible and the potential can be restored to its original value by return to normal sodium concentration in the presence of tetrodotoxin. This clearly demonstrates that tetrodotoxin does not block the sodium channels of the sensory membrane of the spindle. There is also an interesting interaction between tetrodotoxin and calcium: increase in concentration of calcium reduces and retards the blocking action of the toxin whereas removal of calcium has the opposite effect.

The observation that tetrodotoxin does not affect the receptor potential leads to the obvious conclusion that the sodium-carrying system of the transducer membrane of the muscle spindle is different from that involved in the production of

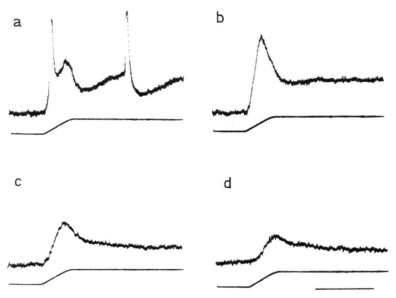

Fig. 38. Effect of removal of sodium and calcium on receptor potential. a, response in Ringers fluid; b, after blocking of conducted activity with tetrodotoxin at a concentration of 100 ng/ml; c, after removal of sodium, NaCl being replaced by choline chloride; d, after subsequent removal of calcium. Time bar: 30 msec (from ALBUQUERQUE, CHUNG, and OTTOSON, 1969)

conducted impulses. However, the failure of tetrodotoxin to affect the sensory endings could be due to a failure of the toxin to gain access to the gates controlling the flow of sodium ions. Studies on the perfused giant axon have shown that the toxin is without effect when applied to the internal surface of the axon membrane. This finding has been taken as evidence that in the squid axon the sodium gate is located on the outer surface of the membrane and that tetrodotoxin is unable to penetrate the membrane. A tempting interpretation of the failure of tetrodotoxin to block the receptor potential would be that the gate of the sodium channel is located on the inner surface of the receptor membrane. In the muscle spindle there seems to be no way of testing this hypothesis. In other receptor cells such as the crustacean stretch receptor which is large enough to permit internal application it would appear to be possible.

IX. Concluding Remarks

Looking through the microscope at an isolated muscle spindle under darkfield illumination reveals a structure of fascinating beauty such as only nature can provide. The afferent nerve is clearly discerned as it penetrates the capsule and breaks up into myelinated branches which wind around the intrafusal fibers like brightly shining coils. In proceeding toward the ends of the spindle, the branches abruptly disappear, losing their myelin sheath, and in the living preparation one can only imagine their continuations into the fine terminal chains. The structural beauty of this little organ is rivalled only by its remarkable functional properties.

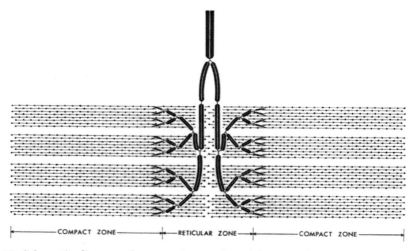

Fig. 39. Schematic diagram of sensory innervation of frog muscle spindle to illustrate its functional organization. Only four intrafusal muscle fibres shown, each with eight rows of sensory chains arising from preterminal nerve branches at the transition between reticular zone and compact zone of the intrafusal muscle fibres. Stem fibre gives rise — by dichotomous division — to five orders of myelinated branches with nodes at the branch points. Longitudinal extent of sensory innervated region 500 μ. Diagram drawn to scale after data from KATZ (1961) and KARLSSON, ANDERSSON-CEDERGREN and OTTOSON (1966) (from OTTOSON and SHEPHERD, 1970c)

The functional task which the spindle must fulfill is quite delicate. It has to provide the central nervous system with information about the mechanical situation in the muscle over a wide range of phasic and static conditions of stretch. The only way this information can be conveyed is in the form of a train of impulses. The experimental data clearly show that the report delivered by the spindle is highly accurate. The firing rate at any moment during stretch provides the central nervous system with information about the actual length of the spindle while the velocity of the stimulus is signalled by the rate of increase in firing frequency. In addition to this, the presence of a pause at the end of dynamic stretching reports the transition from dynamic to static stretch.

The studies described in the present article have given us some insight into the mechanism of the encoding process and have permitted us to glance a bit into the wonderful and concealed world of this complex organ. What we have perceived has helped us to some extent toward understanding the function of the spindle, but there are still many things which remain enigmatic.

Today we know that sodium ions are necessary for the development of the transducer function of the endings. In addition, potassium appears to be involved and possibly other ions. It is conceivable, for instance, that chloride ions contribute to the potential by passing into the terminals during stretch.

The flow of ions can only occur as a result of conductance changes in the transducer membrane. The most likely explanation is that these changes arise as a result of mechanical deformation of the membrane. The obvious implication of this assumption is that the ionic flow is quantitatively related to the amount of deformation. Such a mechanism would explain the close congruity between the parameters of the stimulus and the electrical changes induced in the sensory endings. Opening of ionic channels by deformation of the membrane is also indicated by findings which suggest that the permeability changes induced by stretch are nonselective. An interesting observation emerging from the studies on the effect of tetrodotoxin is that the sodium gate control of the sensory membrane appears to be different from that of the membrane producing conducted activity.

Ever since the classical study of MATTHEWS (1931), it has been clear that the function of the spindle has to be explained in relation to its structure. Recent experiments suggest, however, that the structural changes related to the typical features of the response have to be sought for at the ultrastructural level rather than in gross mechanical differences between the central reticular region and the polar zones. There is also evidence that besides the mechanical factors the electrical properties of sensory membrane may be responsible for part of the adaptation of the spindle. Furthermore there is strong evidence of the preponderant importance of the elastic elements for the return of the spindle to its resting position after release of stretch.

Repolarization of the endings would therefore be governed by the intrafusal mechanical restitution and modified by the electrical constants of the membrane. It also seems obvious that the transmission of the stimulus in terms of lengthening is greatly dependent on the elastic properties of the intrafusal fibers and connective tissues of the spindle. A particularly significant function may be attributed to the elements which attach the sensory endings to the muscle fibers. Many of the bulbs are closely associated with the muscle fibers, but a great number of endings are either entirely free or loosely linked to the muscle fibers. Stretching would therefore appear to produce a nonuniform excitation pattern within the spindle. However, differences in activity in different regions of the spindle appear to be smoothed out as the activity is fed into a single conducting line represented by the afferent fiber.

Although we have a great deal of information at the present time about the functional properties of the transducer membrane, we still know very little about its structural identity. Indirect experimental evidence suggests that the conversion of the stimulus takes place in the sensory terminals, and there also appears to be reason to believe that the transducer action may be localized to the varicose swellings of the terminals. In a single frog spindle there are several thousand bulbs which, by virtue of their number, provide a considerable receptive membrane area (Fig. 39). The current produced by this membrane feeds into the stem fiber and acts on the nodal membrane which has an area of only a few microns. This structural arrangement provides a high safety factor and contributes to the synchronization of activity within the spindle. Hence, the muscle spindle may be considered to be a sensory system in miniature. Through the channels represented by the branching tree of a sensory fiber, information is collected from thousands of transducer elements. The discharge of the afferent nerve represents the total integrated activity of the individual subunits of this sensory system.

In the foregoing we have tried to summarize the knowledge gained about the function of the frog spindle as a mechano-electrical transducer. In our studies we have asked the spindle many questions. In some few cases it has answered us in a clear and straightforward manner, in many cases the answer has been impossible to interpret, and in still other cases the answer has been a question. The basic mechanisms underlying the transducer process are still kept by the spindle as a well-guarded secret of its own.

Acknowledgement

This work has been supported by grants from the Swedish Medical Research Council (Projects B67-14X-43-03, B68-14X-43-04, B69-14X-43-05A, B70-14X-43-06B).

The authors are greatly indebted to Dr. I. Husmark and Mr. L. Ström for valuable discussions during the preparation of this manuscript. We wish to thank Miss Olga Popoff for technical assistance.

References

Adrian, E. D., Zotterman, Y.: The impulse produced by sensory nerve endings. Part 2. The response of a single end-organ. J. Physiol. (Lond.) 61, 151—171 (1926).

Albuquerque, E., Chung, S. H., Ottoson, D.: Impulse generation in the isolated muscle spindle under the action of tetrodotoxin. Acta Physiol. Scand. 75, 301—312 (1969)

Arvanitaki, A.: Les variations graduées de la polarisation des systèmes excitables. Paris: Hermann 1938.

Barker, D.: The innervation of mammalian skeletal muscle. In: Ciba: Foundation Symp. on Myotatic and Vestibular Mechanisms. pp. 3—15. London: J. & A. Churchill, Ltd. 1967.

Bernhard, C. G., Granit, R., Skoglund, C. R.: The breakdown of accommodation — nerve as model sense-organ. J. Neurophysiol. 5, 55—66 (1942).

Boyd, I. A.: The structure and innervation of the nuclear bag muscle fibre system and the nuclear chain muscle fibre system in mammalian muscle spindle. Phil. Trans. R. Soc. (B) 245, 81—136 (1962).

Bridgman, C. F., Eldred, E.: Hypothesis for a pressure sensitive mechanism in muscle spindles. Science 143, 481—482 (1964).

Brown, M. C., Stein, R. B.: Quantitative studies on the slowly adapting strech receptor of the crayfish. Kybernetik 3, 175—185 (1966).

Brzezinski, D. K. Von: Untersuchungen zur Histochemie der Muskelspindeln. II. Mitteilung. Zur Topochemie und Funktion des Spindelraumes und der Spindelkapsel. Acta Histochem. 12, 277—288 (1961).

Burkhardt, D.: Die Erregungsvorgänge sensibler Gangliezellen in Abhängigkeit von der Temperatur. Biol. Zbl. 78, 22—62 (1959).

Cajal, S. R.: Terminaciones en los husos musculares de la rana. Rev. trim. Histol. norm. Patol. no. 1 (1888).

Chapman, R. A.: The repetitive responses of isolated axons from the crab, Carcinus maenas. J. exp. Biol. 45, 475—488 (1966).

Davis, H.: Some principles of sensory receptor action. Physiol. Rev. 41, 391—416 (1961).

Diamond, J., Gray, J. A. B., Inman, D. R.: The relation between receptor potentials and the concentration of sodium ions. J. Physiol. (Lond.) 142, 382—394 (1958).

Dowling, J. A.: Discrete potentials in the dark-adapted eye of the crab Limulus. Nature (Lond.) 217, 28—31 (1968).

Eccles, J. C.: The Physiology of Synapses. Berlin-Heidelberg-New York: Springer 1964.

Edwards, C.: Changes in the discharge from a muscle spindle produced by electrotonus in the sensory nerve. J. Physiol., (Lond.) 127, 636—640 (1955).

ELDRED, E., GRANIT, R., MERTON, P. A.: Supraspinal control of the muscle spindles and its significance. J. Physiol., (Lond.) **122**, 498—523 (1953).

EYZAGUIRRE, C., KUFFLER, S. W.: Processes of excitation in the dendrites and in the soma of single isolated sensory nerve cells of the lobster and crayfish. J. Gen. Physiol. **39**, 87—119 (1955).

FATT, P., KATZ, B.: An analysis of the end-plate potential recorded with an intracellular electrode. J. Physiol., (Lond.) **115**, 320—370 (1951).

FEHR, H. U.: Activation by suxamethonium of primary and secondary endings of the same de-efferented muscle spindle during static stretch. J. Physiol., (Lond.) **178**, 98—110 (1965).

GOLDMAN, L.: The effects of strech on cable and spike parameters of single nerve fibres; some implications for the theory of impulse propagation. J. Physiol., (Lond.) **175**, 425—444 (1964).

GRANIT, R.: Receptors and sensory perception. New Haven: Yale Univ. Press (1955).

— SKOGLUND, S., THESLEFF, S.: Activation of muscle spindles by succinylcholine and decamethonium. The effects of curare. Acta physiol. scand. **28**, 134—151 (1953).

GRAY, J. A. B.: The spindle and extrafusal innervation of a frog muscle. Proc. Roy. Soc. (B) **146**, 416—430 (1957).

— Mechanical into electrical energy in certain mechanoreceptors. Progr. Biophys. Biophys. Chem. **9**, 285—324 (1959).

— DIAMOND, J.: Pharmacological properties of sensory receptors and their relation to those of the autonomic nervous system. Br. med. Bull. **13**, 185—188 (1957).

— MALCOLM, J. L.: The excitation of touch receptors in frog's skin. J. Physiol., (Lond.) **115**, 1—15 (1951).

— MATTHEWS, P. B. C.: A comparison of the adaption of the Pacinian corpuscle with the accommodation of its own axon. J. Physiol., (Lond.) **114**, 454—464 (1951).

— SATO, M.: Properties of the receptor potential in Pacinian corpuscles. J. Physiol., (Lond.) **122**, 610—636 (1953).

HAAPANEN, L. E.: A high-sensitivity capacitance-meter. Electron. Engng. **34**, 183—185 (1962).

HENATSCH, H. D., SCHULTE, J. J.: Wirkungsmechanismus von Acetylcholin und Succinyl-cholin auf die Muskelspindeln des Frosches. Pflügers Arch. Ges. Physiol. **265**, 440—456 (1959).

HODGKIN, A. L.: The local electric changes associated with repetitive action in a non-medul-lated axon. J. Physiol., (Lond.) **107**, 165—181 (1948).

HÖGLUND, G., LINDBLOM, U.: The discharge in single touch receptors elicited by defined mechanical stimuli. Acta physiol. scand. **52**, 108—119 (1961).

HUSMARK, I., OTTOSON, D.: Relation between tension and sensory response of the isolated frog muscle spindle during stretch. Acta physiol. scand. In press (1970a).

— — The contribution of mechanical factors to the early adaptation of the spindle response. Acta physiol. scand. In press (1970b).

— — The effect of potassium on adaptation of the muscle spindle. In course of publication (1970c).

JANSEN, J. K., MATTHEWS, P. B. C.: The central control of the dynamic response of muscle spindle receptors. J. Physiol. (Lond.) **161**, 375—378 (1962).

— RUDJORD, T.: The silent period during twitch contraction of the soleus of the decerebrate cat. Acta physiol. scand. **59**, Suppl. **213**, 69—70 (1963).

KARLSSON, U., ANDERSON-CEDERGREN, E., OTTOSON, D.: Cellular organization of the frog muscle spindle as revealed by serial sections for electronmicroscopy. J. Ultrastruct. Res. **14**, 1—35 (1966).

— OTTOSON, D.: Electronmicroscopical studies on the effect of anisotonic solutions on the muscle spindle. In course of publication (1970).

KATZ, B.: Action potentials from a sensory nerve ending. J. Physiol., (Lond.) **111**, 248—260 (1950a).

— Depolarization of sensory terminals and the initiation of impulses in the muscle spindle. J. Physiol., (Lond.) **111**, 261—282 (1950b).

— The termination of the afferent nerve fibre in the muscle spindle of the frog. Phil. Trans. R. Soc. **243**, 221—240 (1961).

Krnjevic, K., Gelder, N. M. Vav: Tension changes in crayfish stretch receptors. J. Physiol. (Lond.) 159, 310—325 (1961).

Lindblom, U.: The relation between stimulus and discharge in a rapidly adapting touch receptor. Acta physiol. scand. 56, 349—361 (1962).

— Phasic and static excitability of touch receptors in toad skin. Acta physiol. scand. 59, 410—423 (1963).

— Properties of touch receptors in distal glabrous skin of the monkey. J. Neurophysiol. 28, 966—985 (1965).

Lippold, O. C. J., Nicholls, J. G., Redfearn, J. W. T.: Electrical and mechanical factors in the adaption of a mammalian muscle spindle. J. Physiol., (Lond.) 153, 209—217 (1960).

Loewenstein, W. R.: Excitation and changes in adaptation by stretch of mechanoreceptors. J. Physiol., (Lond.) 133, 588—602 (1956).

— Generator processes of repetitive activity in a Pacinian corpuscle. J. gen. Physiol. 41, 825—845 (1958).

— On the "specificity" of a sensory receptor. J. Neurophysiol. 24, 150—158 (1961).

— Altamirano-Orrego, R.: The refractory state of the generator and propagated potentials in a Pacinian corpuscle. J. gen. Physiol. 41, 805—824 (1958).

— Mendelson, M.: Components of receptor adaptation in a Pacinian corpuscle. J. Physiol. (Lond.) 177, 377—397 (1965).

— Terzuolo, C. A., Washizu, Y.: Separation of transducer and impulse-generating processes in sensory receptors. Science 142, 1180—1181 (1963).

Matthews, B. H. C.: The response of a single end organ. J. Physiol., (Lond.) 71, 64—109 (1931).

Matthews, P. B. C.: The response of de-efferented muscle spindle receptors to stretching at different velocities. J. Physiol., (Lond.) 168, 660—678 (1963).

— Muscle spindles and their motor control. Physiol. Rev. 44, 219—288 (1964).

— Westbury, D. R.: Some effects of fast and slow motor fibres on muscle spindles of the frog. J. Physiol., (Lond.) 178, 178—192 (1965).

Nishi, K., Sato, M.: Blocking of the impulse and depression of the receptor potential by tetrodotoxin in non-myelinated nerve terminals in Pacinian corpuscles. J. Physiol., (Lond.) 184, 376—386 (1966).

Ottoson, D.: The effect of acetylcholine and related substances on the isolated muscle spindle. Acta physiol. scand. 53, 276—287 (1961).

— The effect of sodium deficiency on the response of the isolated muscle spindle. J. Physiol., (Lond.) 171, 109—118 (1964).

— The action of calcium on the frog's isolated muscle spindle. J. Physiol. (Lond.) 178, 68—79 (1965).

— The effect of osmotic pressure changes on the isolated muscle spindle. Acta physiol. scand. 64, 93—105 (1965).

— McReynolds, J. S., Shepherd, G. M.: Sensitivity of isolated frog muscle spindle during and after stretching. J. Neurophysiol. 32, 24—34 (1969).

— Shepherd, G. M.: Receptor potentials and impulse generation in the isolated spindle during controlled extension. Cold Spr. Harb. Symp. quant. Biol. 30, 105—114 (1965).

— — Changes of length within the frog muscle spindle during stretch as shown by stroboscopic photomicroscopy. Nature (Lond). 220, 912—914 (1968).

— — Relation of afferent nerve excitability to impulse generation in the frog muscle spindle. Acta physiol. scand. 75, 49—63 (1969).

— — Length changes within isolated frog muscle spindle during and after stretch. J. Physiol. (Lond.) 207, 747—759 (1970a).

— — Steps in impulse generation in the isolated muscle spindle. Acta physiol. scand. In press (1970b).

— — Synchronization of activity in afferent nerve branches within the frog's muscle spindle. In course of publication (1970c).

— — Transducer characteristics of the muscle spindle as revealed by its receptor potential. In course of publication (1970d).

OVERTON, E.: Beiträge zur allgemeinen Muskel- und Nervenphysiologie. II. Über die Unentbehrlichkeit von Natrium-(oder Lithium-) Ionen für den Kontraktionsakt des Muskels. Pflügers Arch. ges. Physiol. **92**, 346—386 (1902).

OZEKI, M., SATO, M.: Changes in the membrane potential and the membrane conductance associated with a sustained compression of the nonmyelinated nerve terminal in Pacinian corpuscles. J. Physiol., (Lond.) **180**, 186—208 (1965).

REGAUD, C., FAVRE, M.: Les terminaisons nerveuses et les organes nerveux sensitifs de l'appareil locomoteur. Rev. gen. Histol. **1**, (1904—05).

SHEPHERD, G. M., OTTOSON, D.: Responses of the isolated muscle spindle to different rates of stretching. Cold Spring Harbor Symp. quant. Biol. **30**, 95—103 (1965).

SHERRINGTON, C. S.: On the anatomical constitution of nerves of skeletal muscles, with remarks on recurrent fibres in the ventral spinal nerve-root. J. Physiol. (Lond.) **17**, 211—258 (1894).

SMITH, C. M., ELDRED, E.: Mode of action of succinylcholine on sensory endings of mammalian muscle spindles. J. Pharmacol. **131**, 237—242 (1961).

SMITH, R. S.: Properties of intrafusal muscle fibres. Nobel Symp. I. Muscular Afferents and Motor Control. pp. 69—80. Stockholm: Almqvist & Wiksell 1966.

TEORELL, T.: Excitability phenomena in artificial membranes. Biophys. J. **2**, 27—52 (1962).

— Electrokinetic considerations of mechanoelectrical transduction. Ann. N. Y. Acad. Sci. **137**, 950—966 (1966).

TERZUOLO, C. A., WASHIZU, Y.: Relation between stimulus strength, generator potential and impulse frequency in stretch receptor of *Crustacea*. J. Neurophysiol. **25**, 56—66 (1962).

THURM, U.: Das Receptorpotential einzelner mechano-receptorischer Zellen von Bienen. Z. vergl. Physiol. **48**, 131—156 (1964).

— An insect mechanoreceptor. I. Fine structure and adequate stimulus. Cold Spring. Harbor Symp. quant. Biol. **30**, 75—82 (1965).

— An insect mechanooreceptr. II. Receptor potentials. Cold Spring Harbor Symp. quant. Biol. **30**, 83—94 (1965).

TIGERSTEDT, R.: Studien über die mechanische Nervenreizung. Finneschen Litteratur Gesell. Helsingfors 574—659 (1880).

WENDLER, L., BURKHARDT, D.: Zeitlich abklingende Vorgänge in der Wirkungskette zwischen Reiz und Erregung. Versuche an abdominalen Streckreceptoren dekapoder Krebse. Z. Naturf. **16**, 464—469 (1961).

— Über die Wirkungskette zwischen Reiz und Erregung. Versuche an den abdominalen Streckreceptoren des Flußkrebses. Z. vergl. Physiol. **47**, 279—315 (1963).

Chapter 16

Static and Dynamic Behavior
of the Stretch Receptor Organ of Crustacea

By

C. A. Terzuolo and C. K. Knox, Minneapolis, Minnesota (USA)

With 19 Figures

Contents

A. Introduction

The word transducer has become commonplace in receptor physiology. It is most frequently used to emphasize the transformation of the energy applied to the receptor organ into a different form of energy by the properties of the elements which constitute the system (Bullock, 1959). However, other features of transducer machines are found in receptor systems. For instance, the places of input and output are specified if the biological transducer is taken to be composed of all those elements (or parameters) which are interposed between the point of application of the physiological stimulus and the generator potential, which in turn is defined as the transducer output.

To interpret the meaning of the word transducer as applied to receptor organs, it may be useful to consider its general features in different receptor systems. In some of these systems, the transducer includes certain parameters whereby the input energy induces changes which in turn are capable of affecting the behavior of the neuronal element of the system, this being insensitive, per se, to the input energy. Therefore, these elements are responsible for conferring to the receptor its specific sensitivity to a particular type of energy. Additionally, they in part define the operational range of the receptor system by selecting a restricted range of the specific energy, by selecting the lower and upper limits of input intensity which may be detectable, or by both. A filtering action is also present in several receptor systems in which the neuronal element may be sensitive per se to the specific energy e.g., in mechanoreceptors (see LOEWENSTEIN and SKALAK, 1966). In the stretch receptors, for instance, specific anatomical structures permit the sensory neuron to detect displacement information in a manner determined wholly by their mechanical properties and geometry. In still other receptor systems, the nature of the eventual restrictions which may be imposed by certain anatomical arrangements is not self-evident, as in the majority of corpuscles, or no such structures are present, the receptor being a free nerve ending. In these cases, the problem of explaining receptor specificity is a difficult one, unless the presence of special properties in the neuronal element itself is postulated, as in some of the hypotheses which have been proposed to account for the ability of thermoreceptors to detect temperature changes.

Even in the simplest of the above cases, the transducer would include several elements since the membrane of the sensory neuron alone is characterized by more than one parameter. Then, if one represents the transducer by a black box which contains all these parameters

Input	Transducer	Output
$x(t)$		$y(t)$
$\xrightarrow{\hspace{1cm}}$ applied energy	operator, O	generator potential $\xrightarrow{\hspace{1cm}}$.

The output $y(t)$ will be determined by the input $x(t)$ through the system operator, O, according to the following expression

$$y(t) = O\{a_1(t), a_2(t) \dots a_n(t)\} x(t)$$

where $y(t)$ is a vector since O depends on $a_1, a_2 \dots a_n$ parameters, which may or may not be time variant.

Although this input-output relationship might be mathematically definable in favorable cases, knowledge of the behavior of the operator parameters is essential for defining the actual transduction processes. In this context, the stretch receptor organ of *Crustacea* is almost unique among mechanoreceptors, since its morphological features not only afford the possibility of an accurate quantitation of both input and output variables, but also the changes occuring in some of the operator parameters during the transduction process can be measured.

a) System Analysis Approach. Although the conceptual framework for most of the experiments to be reviewed on the static properties of the transducer and its mechanisms are certainly familiar to all readers, a few general considerations underlying the experiments on the dynamic behavior may be briefly mentioned.

First of all, if applicable, the approach outlined above allows prediction of the input-output relation within the limits of linear behavior of the system. Although these limits are likely to be modest with respect to the operational and functional range of the receptor, modeling of the system may become possible on other than a purely empirical basis; if this is so, an approach to the analysis of nonlinear behavior becomes available. In any event, if specific mechanisms are proposed for the transformations taking place in the transducer, their static and dynamic behavior must be compatible with the data obtained from system analysis if these mechanisms are to be accepted as plausible. The data and the conclusions to be presented here about the crustacean stretch receptor will be a clear illustration of the latter statement.

The second point concerns the requirement of stationarity. Biological transducers, and more generally receptor systems, also conform to the behavior of transducer machines (see Asby, 1963) in that reproducible states are found at the output which can be recognized as dependent on a given set of input variables, or simply input. However, the output states to a "constant" input are not single-valued and therefore can be defined only probabilistically (presence of noise in the system). Nevertheless, if the distribution of output states remains constant over a prolonged period of observation, the system can still considered to be stationary. This requirement of system analysis is fulfilled even when the behavior of the system depends on the state of the operator at the time that the input is applied, provided that for a given set of initial conditions the behavior is reproducible.

The third general consideration underlying the study of the dynamic behavior of a receptor organ by system analysis is that the receptor must satisfy the requirement of continuity over the range of input states used. That is, the output in response to a known input must be defined (single-valued, bounded) and must vary continuously with the input so that its derivative exists wherever the derivative of the input does. In receptor systems, the experimental verification of this continuity may be hindered by the limits of resolution imposed by the properties of the system and the recording equipment (in particular noise). In practice then, the input states leading to identifiable output states must be widely spaced. Nevertheless, a reasonable assurance of continuity can be obtained by using an input which varies continuously with time, as in frequency analysis. Before considering this technique in some detail, two other equivalent procedures shall be mentioned.

The impulse response is only of theoretical interest in our context since the adequate input for determining this function (Dirac delta function) is defined to be zero everywhere except at time $t = o$, where it is of infinite value subject to $\int_{-\infty}^{\infty} \delta(t) dt = 1$. This is mentioned only because the term "transfer function", which will be frequently used here, is defined as the Laplace transform of the response to a delta function.

An input function more easily applicable experimentally is the step or Heaviside function. It is defined as equal to zero for $t < 0, 1/2$ for $t = 0$, and 1 for $t > 0$. Its Laplace transform is $1/s$. Being the integral of the delta function, the response of a linear system to this input is the integral of the impulse response.

Frequency analysis, the method of choice as argued above, is based on the fact that a large group of functions, which include all those dealing with a linear system, can be regarded as the sum of sinusoidal functions of various periods and magnitudes (Fourier integral). The transfer function of a linear system can therefore be synthesized from a knowledge of its response to sinusoidal inputs of different amplitudes and frequencies. To obtain these data, the gain of the system (ratio between the magnitude of the sinusoidal output and that of the input) and the phase differences between output and input are measured. These quantities can then be related to the transfer function when the independent variable of the Laplace transform is substituted with $j\omega$ (j being the imaginary $\sqrt{-1}$, and ω the frequency in radians/sec). This leads to a function which contains both real and imaginary parts. The measured gain and phase are then related to this function as follows

$$G = \sqrt{R(F)^2 + I(F)^2} \text{ and } \tan\theta = \frac{I(F)}{R(F)}$$

where G is the gain, θ the phase angle, $R(F)$ the real part and $I(F)$ the imaginary part of the function. The system analysis approach also permits one to define, in suitable cases, the transducer transfer function without directly measuring its output (see below).

b) The Preparation. The stretch receptor organs of *Crustacea* were originally described in the lobster by ALEXANDROWICZ (1951, 1952). Their morphological details in the crayfish were then studied by FLOREY and FLOREY (1955). Two electron-microscope studies are also available (PETERSON and POPE, 1961; BODIAN and BERGMAN, 1962).

As shown in Fig. 1, each receptor (three are two in each side of the abdominals and last two thoracic segments) consists of: (i) a bundle of interconnected muscle

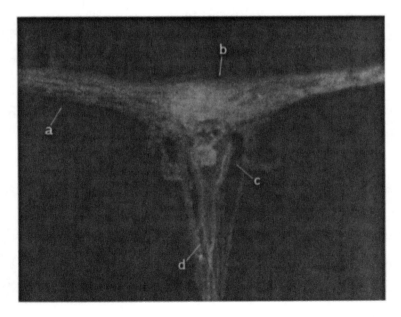

Fig. 1. Photomicrograph of the tonic stretch receptor organ. *a* muscle bundle; *b* region where the dendrites are mostly concentrated; *c* cell body with nucleus; *d* axon. (From TERZUOLO et al., 1967)

fibers to which the dendrites of the sensory neuron are firmly attached, and (ii) the sensory neuron, whose cell body is big enough to permit penetration with one or two microelectrodes without apparent damage.

The preparation can be easily isolated (Fig. 2), in which case it is an open loop system. Demonstration of its stationarity under this condition rests on the experimental finding that its behavior does not change in the course of the experiments (up to 18 hr in the lobster), provided that the temperature is kept constant.

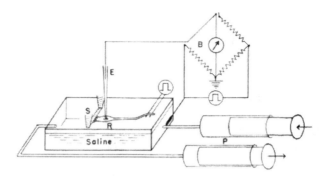

Fig. 2. Experimental arrangements to study the behavior of the isolated preparation. The preparation (R) is held by forceps (S) connected to a mechanical stretching device. A microelectrode (E) is shown impaling the soma and mounted in a bridge circuit to allow simultaneous stimulation and recording. Wire electrodes on the axon can be used for either stimulation or recording. The reciprocal pump system allows substitution of the extracellular medium after impalement of the cell. Two intracellular microelectrodes can also be used

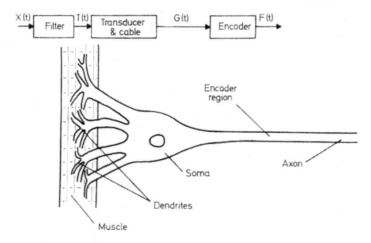

Fig. 3. Diagram of the preparation and the systems which constitute the receptor organ

Wiersma et al. (1952), by recording the impulse activity from the axon of the sensory neuron with a wire electrode, were the first to demonstrate that the receptor converts displacement information into neuronal information. They also demonstrated that one receptor of each pair is slowly adapting (i.e., it responds to a sustained stretch with a sustained impulse activity) and that the other receptor is fast adapting. These two types are readily identifiable because of gross differences in the size of the muscle bundle and other details.

Proceeding from this work, EYZAGUIRRE and KUFFLER (1955a, b) recorded from the soma of the sensory neuron using an intracellularly placed microelectrode. With this technique, they were able to demonstrate that when the muscle bundle is stretched a depolarization is produced, the so-called generator potential, which is in turn responsible for impulse initiation. This observation opened the way to all subsequent studies on the behavior of the receptor organ.

Before analyzing this behavior, it is helpful to consider schematically the systems in which the stretch receptor organ of *Crustacea* can be conceptually and experimentally subdivided. As shown in Fig. 3, these are: (i) a mechanical filter, whose components are the physical parameters of the muscle bundle to which the input force is applied (this filter, according to the definition of transducer given above, is actually a part of this system, but insofar as it can be analyzed separately in the present preparation, it will be considered independently); (ii) the transducer, which tentatively can be said to convert the filter output into the voltage change known as generator potential; and (iii) an encoder, which converts the information contained in the generator potential into repetitive impulse activity. This originates in the axon of the cell (EDWARDS and OTTOSON, 1958; WASHIZU and TERZUOLO, 1966), several hundred micra away from the place of output of the transducer. It is postulated that the three systems are placed in series. However, a structure with cable-like properties is interposed between the transducer, assumed to be located in the dendrites (EYZAGUIRRE and KUFFLER, 1955a, b; TERZUOLO and WASHIZU, 1962), and the encoder.

Although this point will be taken up later, the facts supporting the view that the transducer and encoder systems are connected in series shall be briefly summarized here for the sake of clarity. These facts are: (i) the two systems appear to be located in separate regions of the cell (TERZUOLO and WASHIZU, 1962; WASHIZU and TERZUOLO, 1966), being connected by a cable; (ii) one can block selectively the operation of the encoder system without affecting the behavior of the transducer (see p. 507); and (iii) the generator potential can be substituted by a voltage drop across the membrane resistance, as by applying a current across the cell membrane, without altering the encoder behavior (see p. 510).

Points (ii) and (iii) imply the possibility that the transducer and encoder outputs can be measured separately. Since the mechanical filter can also be analyzed separately, it becomes possible to study individually each of the systems into which the receptor organ can be conceptually subdivided. This provides the opportunity to test if the series arrangement postulated above is correct. Indeed, given two or more independent linear systems with operators $O_1, O_2 \ldots O_n$, it can be easily shown that their combined behavior is given by the product of the individual operators.

Our analysis will be limited to the static and dynamic behavior of the slowly adapting stretch receptor organ. Proceeding from these data, the mechanisms of the transduction process will be discussed.

B. Static Behavior

The overall behavior of both crayfish (*Procambarus alleni* and *Orconentes virilis*) and lobster (*Homarus*) preparations is shown in Fig. 4. A linear relationship is present between length of the muscle bundle (displacement) and impulse frequency under static conditions (Terzuolo and Washizu, 1962). In these experiments, impulse activity was recorded from the axon, and the maximum length of the muscle bundle at which no sustained impulse activity was present was taken as resting length or unity. No appreciable hysteresis was found for displacements as large as 1.4 times resting length. This range of input amplitudes actually covers the entire operational range of the receptor under physiological conditions.

Fig. 4. Receptor output to a slowly varying stretch (Lobster preparation). Upper trace: generator potential and impulse activity recorded intracellularly from the soma (the interrupted line provides a reference). Lower trace: output of a linear potentiometer measuring the amount of applied stretch. Calibrations: 50 mV and 0.5 sec

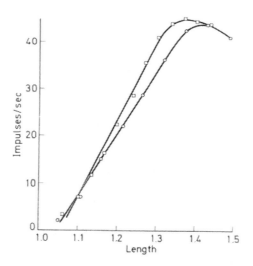

Fig. 5. Relation between impulse frequency and length of the muscle bundle in slowly adapting stretch receptors of the lobster. The results from two preparations are shown and are fully explained in the text. (From Terzuolo and Washizu, 1962)

The measurements reported in the plots of Fig. 5 were made 40 sec after the new level of stretch was slowly attained using a ramp input. The standard deviation of impulse frequency, under these conditions, is between 0.1 and 0.3 impulses per second within the frequency range considered (Firth, 1966; Terzuolo et al., 1968). The temperature dependency of this overall input-output relation is quite high in the lobster, within the temperature range of 11 to 17° C

(TERZUOLO et al., 1968). Data for the crayfish stretch receptor organ are also available (BURKHARDT, 1959).

The above linear relationship between displacement and impulse frequency results from the fact that the relations between muscle length and generator potential amplitude and between generator potential amplitude and impulse frequency are both linear, at least within the input and output ranges considered here. Data concerning the first of these two relationships are provided by two independent sets of experiments.

In the first set, the cell body of crayfish sensory neuron was impaled with a microelectrode and a steady hyperpolarizing current was applied to prevent the initiation of impulse activity by the encoder. The amplitude of the generator potential as a function of displacement was then measured (TERZUOLO and WASHIZU, 1962). In the second set of experiments, impulse activity was suppressed by adding tetrodotoxin to the bathing solution. In the crayfish preparation, concentrations of the order of 1 to 5×10^{-6} (wt/vol) reversibly block the physicochemical events

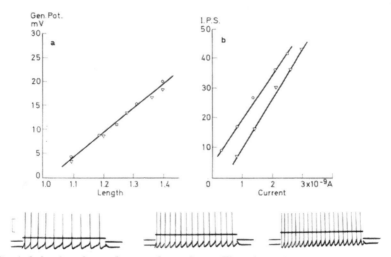

Fig. 6. Static behavior of transducer and encoder. a) The relation between generator potential and length of the muscle bundle for two crayfish preparations. b) The relation between current and impulse frequency. Three records from one experiment are shown. The current applied through the impaling microelectrode (see Fig. 2) is measured on the lower beam. Calibrations: 35 mV, and 1×10^{-9} A; 0.5 sec

responsible for the impulse activity (LOEWENSTEIN et al., 1963). However, only a small decrease in membrane resistance occurs, and the amplitude of the generator potential is unaffected. Hence, it is argued that the transduction processes are unaltered by the toxin. Under these conditions, a linear relationship is found between displacement and transducer output (Fig. 6).

The encoder behavior under stationery conditions is described by the results of experiments (TERZUOLO and WASHIZU, 1962) in which depolarizing current pulses of increasing strength were applied across the cell membrane, the same microelectrode being used to record the impulse activity (Fig. 3). The data for two cells using currents adequate to produce impulse frequencies comparable to those measured for displacements up to 1.4 times resting length are summarized in the plots of Fig. 6b. Notice that the relationship is linear at any one degree of sustained stretch (Terzuolo and Washizu, 1962). Moreover, the voltage changes across the membrane caused by currents of the same order of magnitude are also linearly related to current strength, at least in the majority of cells in which these measurements were made under adequate conditions (see p. 518).

C. Dynamic Behavior

a) Receptor Transfer Function. To relate displacement information to neuronal information in a way adequate to obtain the gain and phase measurements required by the sinusoidal analysis technique, a choice shall be made regarding the output variable to be quantitated. When individual impulses are considered, their instantaneous frequency in a train is given by the reciprocal of the interval between each impulse and the one preceding it. Although the choice of this variable — instantaneous impulse frequency — is arbitrary and other variables could

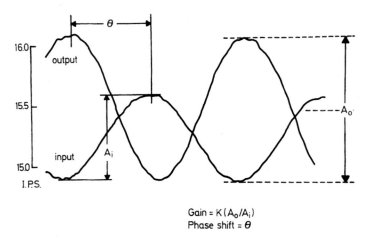

Gain = K (A$_0$/A$_i$)
Phase shift = θ

Fig. 7. Measurements of gain and phase. These data are obtained from the record (output of the averaging computer as displayed by an x–y plotter) in the way indicated in this figure

be considered, it does not introduce undue assumptions about the behavior of the systems to be studied. The procedures used in processing the input and output variables are fully described by Borsellino et al. (1965) and by Terzuolo et al. (1968). Fig. 7 is an actual record of the averaging computer output, as displayed by an x–y plotter. From such records, the gain and phase as a function of frequency can be measured and Bode diagrams are made. By subtracting from those diagrams the transfer function of the processing device[1], the receptor transfer function is obtained.

---

[1] This device was a zero-order-hold. Its transfer function is given by $H(s) = \dfrac{1-e^{T_o s}}{s}$, where $T_0 = 1/\bar{\nu}$ is the average carrier rate. Its behavior, therefore, is frequency-dependent, the gain being zero at $\bar{\nu}$ and its multiples. The sharp fall in gain of such a device at frequencies higher than $\bar{\nu}/2$ limits the accuracy of measurements when the value of $\bar{\nu}$ is approached. Therefore, data for frequencies above $\bar{\nu}/2$ are not given.

In this work (BORSELLINO *et al.*, 1965), it was found that input amplitudes producing changes in instantaneous frequency of 10% or less of the mean steady rate $\bar{v}$, produced a sinusoidally modulated output train without altering the value of $\bar{v}$. Moreover, the gain and phase shift at any one frequency were independent of such amplitudes. Hence, the criteria for linear behavior are satisfied.

Fig. 8 summarizes the data obtained from one set of these experiments (TERZUOLO *et al.*, 1968). By normalizing the input frequency with respect to the mean carrier rate $\bar{v}$, and by superimposing the curve (relative gain), the results from all experiments can be represented on a single plot.

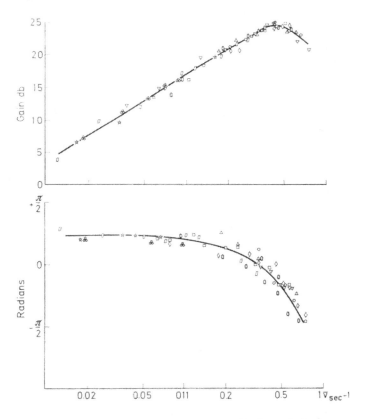

Fig. 8. Receptor transfer function (lobster preparation). Each symbol represents a different preparation. (From TERZUOLO *et al.*, 1968, after subtraction of the transfer function of the demodulating device)

The Bode diagram shows that the overall gain of the receptor has a flat maximum near $\bar{v}/2$. The low frequency side of the gain curve has a slope of 3 db/octave, and the phase shift approaches 35° at low frequencies, suggesting an $s^{1/2}$ term in the transfer function with a pole close to the origin in a pole-zero diagram. A superimposed second-order transfer function with a damping factor of $\gamma = 0.7$ would yield the flat maximum and high frequency cut-off observed. Thus, the overall transfer function can be approximated by

$$h(s) = \frac{s^{1/2}}{s^2 + 2\gamma\, ps + p^2} = \frac{s^{1/2}}{s^2 + 1.4\, ps + p^2}$$

where $p = 2\pi\,\bar{\nu}$ and s is the Laplacian operator. Similar results were obtained by Brown and Stein (1966), but their analysis was limited to a smaller range of frequencies.

Although mathematically descriptive, the above empirically derived relationship does not provide much insight about ways in which the individual compartments of the receptor contribute to its overall behavior. Only the fact that the gain decreases as the input frequency approaches that of the carrier rate can be construed to indicate the presence of some general feature of a holding system (Borsellino et al., 1965).

The relationship between maximum gain and carrier frequency has one more implication. If this behavior is confirmed in other receptor systems, it can imply that the central nervous system may tune tonic receptor organs via its efferent control so as to receive selected information.

b) **Encoder Transfer Function.** The dynamic behavior of the encoder was determined by applying sinusoidally modulated currents through a microelectrode placed in the soma. In these experiments (Terzuolo et al., 1968), a steady carrier frequency was obtained by stretching the muscle bundle, by means of a steady depolarizing current, or by both. Impulse activity was processed as described in the preceding section. The transfer function obtained is that between the applied current and the resulting impulse frequency changes. Because the membrane behaves electrically as if it were made up of parallel resistive and capacitive elements, it can be expected that, whereas the voltage and current will be in phase at very low frequencies and for small currents, a drop in gain (voltage/current) and a phase lag would appear as the input frequency increases. Unpublished data by Purple and Handelman indicate that in the slowly adapting stretch receptor neuron, the gain has fallen by 3 db and the voltage is lagging the current by 45° at frequencies between 15 and 30 Hz. Therefore, frequencies in excess of 10 Hz were not used. Notice that the voltage changes measured in these experiments are those which develop across the soma membrane. This is also the case when the generator potential is measured (although the pattern of current flow is different). Therefore, it can be contended that the generator potential is adequately substituted by applying a current through the impaling microelectrode, the input to the encoder being the same. The Bode plots of Fig. 9 describe the encoder behavior. This was found to be linear for input amplitudes which modulate the instantaneous frequency up to 75% of the carrier rate.

These data conclusively prove that the dependency of the frequency function on the carrier rate is a property of the encoder. This feature is also observed in relaxation oscillators, the gain of these devices being zero at the carrier frequency and its harmonics. This results from the fact that the information is sampled in a discontinuous way, with sampling interval equal to $1/\bar{\nu}$. Actually, the transfer function of a relaxation oscillator is the same of that of a zero-order hold (see footnote 1). The encoder properties, however, differ significantly from such behavior. Proceeding from these data, the contribution of several membrane features to the dynamic properties of the encoder were analyzed by BAYLY (1968). He considered a relaxation oscillator with a linear charging function whose parameters were chosen so as to correspond to those of the sensory neuron. The model allows one to test the influence of various parameters on the encoder behavior.

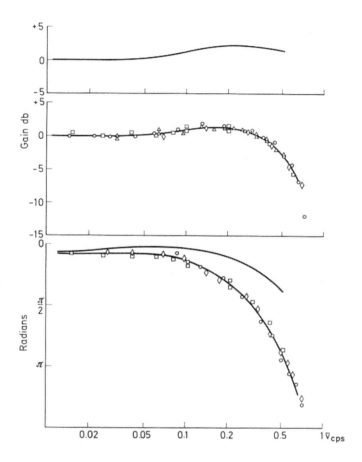

Fig. 9. Encoder transfer function. The experimental data points are shown. The curves describe the phase and the gain after subtraction of the transfer function of the demodulating without dato points device. (From TERZUOLO et al., 1968)

These parameters involve the resetting of the membrane voltage to the resting level by the spike as well as geometrical features such as the axon diameter, the distance from the soma of the site of impulse origin and the passive electrical characteristics of the membrane.

c) **The Transducer Behavior.** Assuming that the encoder and transducer systems are placed in series, as in the diagram of Fig. 3, the dynamic input-output relation between displacement and generator potential can be derived by subtracting from the Bode plots which describe the receptor transfer function those describing the transfer function of the encoder. The validity of this approach can then be tested experimentally by measuring directly the generator potential after suppressing the encoder with tetrodotoxin.

Owing to the difficulty of penetrating lobster neurons, these experiments were performed in crayfish preparations (HANDELMAN, 1968). The agreement between calculated and measured values was found to be good. Fig. 10 shows the data

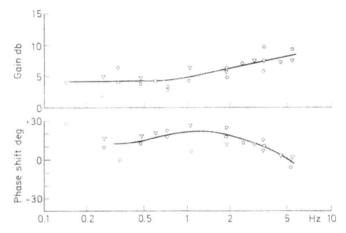

Fig. 10. Transfer function of the crayfish transducer as determined in one preparation by subtraction of the encoder from the overall transfer function (circles) and by direct measurement of the generator potential after suppression of impulse activity by tetrodotoxin (triangles)

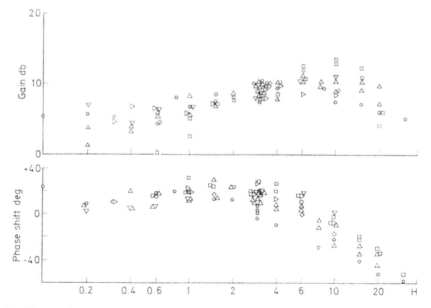

Fig. 11. Measured transfer function of the transducer in crayfish preparations. The generator potential was measured after application of tetrodotoxin. Each symbol represents a different preparation. Higher input frequencies could be used after suppressing the encoder, the limits imposed by the combined properties of the encoder and the demodulator (zero-order hold) being eliminated. (HANDELMAN, University of Minnesota, M. S. Thesis)

for one of the experiments, and Fig. 11 is a composite of all experiments in crayfish preparations. These experiments also showed that the maximum transducer output is a function of muscle length, about a twofold increase of the generator potential amplitude being found for a length increase from 1.1 to 1.33. Fig. 12 illustrates the transducer behavior for the lobster stretch receptor,

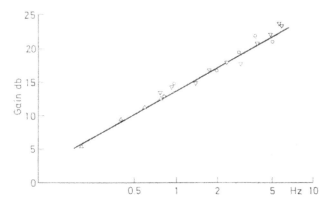

Fig. 12. Gain curve of the transducer of lobster stretch receptors. Different symbols indicate different preparations

obtained by subtracting the encoder transfer function from the receptor transfer function (TERZUOLO et al., 1968). In this case, the transducer output has a slope of approximately 3 db octave, which is steeper than that of the crayfish. A phase lead is present at the lowest frequencies. An interpretation of this behavior will be attempted when discussing the transducer mechanism.

d) Properties of the Mechanical Filter. The viscoelastic properties of the muscle bundle of the receptor organ are likely to affect its dynamic behavior, since under physiological conditions the force is applied at the ends of the muscle bundle, and the dendrites of the neuron, where transduction takes place, occupy only the central 0.5 mm of the muscle bundle (ALEXANDROWICZ, 1951). To examine these properties, the length of the muscle bundle of the lobster receptor organ was modulated sinusoidally at different lengths.

A vibrator (GOODMAN, 30 ohm) was used and tension was measured with an RCA-5,734 mechano-electrical transducer fitted with a tapered, 2.5 cm glass stylus. The excursions of the displacement vibrator were sensed by semiconductor strain gages bonded to a thin cantilever beam which followed the vibrator rod movements. The analog voltages, displacement and tension, were amplified and averaged on the computer to improve the signal to noise ratio. The dynamic stiffness, S, of the muscle was computed from the input (displacement) and output (tension) sinusoids by taking the ratio between tension amplitude and displacement amplitude to determine the modulus, S_0, and by measuring the phase shift, θ, $(S = S_0/\theta)$. Thus, the dynamic stiffness is a mechanical impedance and is a measure of the force produced in the muscle in response to a unit displacement. Since the steady state tension-length curve shown in Fig. 13 is nonlinear the value of S at zero frequency, given by the slope of the curve in Fig. 13, increases with length. Thus, S can be expected to show a dependency on length. The value of S at zero frequency is denoted as K_s, the steady state elasticity, and is shown plotted against length in Fig. 14 for three receptor muscles. For any constant length above rest length ($L_n = 1.0$ at the length where tension begins to increase from

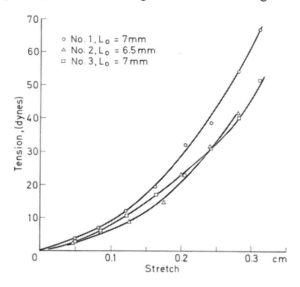

Fig. 13. Static tension-length curves in lobster stretch receptor organs. The data from three preparations are shown. Since the relationship is nonlinear, the elasticity of the muscle is a function of length

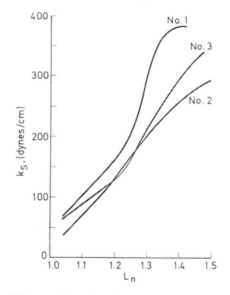

Fig. 14. Steady state elasticity as a function of muscle length. Data from the three preparations shown in Fig. 12. The values are computed as the slope of the tension-length curve and determine the dynamic stiffness, S, at very low frequencies. Resting length ($L_n = 1.0$) is defined as zero tension length and is generally 10% less than resting length defined on the basis of impulse activity

zero), measurement of S is confined to linear changes in tension with length. For $L_n = 1.1$ to 1.5, the tension amplitude increases linearly with displacement amplitudes of up to 100μ peak to peak. Moreover, the phase angle remains constant at any one frequency over the range of 0.3 to 100 Hz. Therefore, a plot of S_0 and θ against frequency represents the linear behavior of the muscle at a particular length. Fig. 15 A is a representative plot from one lobster receptor muscle. What is immediately apparent is that S_0 is a power function of frequency with an

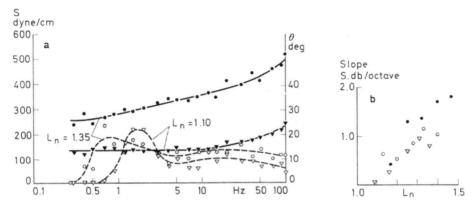

Fig. 15. Dynamic behavior of the receptor muscle bundle in lobster. a Comparison of the results obtained at two different lengths in one muscle, $L_n = 1.10$ (modulus, filled triangles; phase, open triangles) and $L_n = 1.35$ (modulus, filled circles; phase, open circles). b The increase in the modulus of S as a function of length (for frequencies up to 10 Hz). Notice that the slope of S is much less than that seen in the transducer, indicating that tension alone does not determine transducer behavior (see Fig. 11)

exponent which increases with length. Also, tension leads displacement over the entire frequency range. This is to be expected if the muscle is behaving as a passive viscoelastic element. Fig. 15b shows the increases of S_0 with length; it is of considerable interest that this increase is much less than that seen in the transducer. This point will be taken up later.

Proceeding from these data the behavior of the muscle bundle was modeled on an analog computer (KNOX, 1968). The results indicate that the viscoelastic behavior of the muscle can be approximately described by a three-compartment model which includes a series elasticity representing the steady state elasticity K_s and two Maxwell elements (an elasticity and viscosity in series) which are in parallel with K_s (see Fig. 16). Since all elements are functions of muscle length, the model is nonlinear. The time constant C_L/K_L is inversely proportional to length whereas C_H/K_H is proportional to length, the empirical relationships being, $C_L/K_L (L_n - 1.0) = 0.4$ sec, $(1.0 \leq L_n \leq 1.5)$, and $C_H/K_H = 10(L_n - 1.0)$ msec, $(1.0 \leq L_n \leq 1.5)$. K_L and K_H increase linearly with L_n, the empirical relationship being $K_L = 1,100(L_n - 1.1)$ dynes/cm $(1.1 \leq L_n \leq 1.5)$, and $K_H = 500(L_n - 1.0)$ dynes/cm $(1.0 \leq L_n \leq 1.5)$. Correlation of the data with the anatomy of the muscle suggests that the elements K_L and C_L are associated with myofilament interactions, and K_H and C_H with connective tissue and myofibril interactions. This

33*

inference is drawn in part by considering that damping arises in the muscle as a result of viscous coupling between elastic elements mediated by a fluid which is free to move. In this case, the damping coefficients are proportional to the surface area of the interacting elements and inversely proportional to their separation. Since the area separation ratio is quite large for the myofilaments as compared to myofibrils and connective tissue, C_L is assigned to this compartment. This assignment is also supported by the facts that overlap between thick and thin filaments

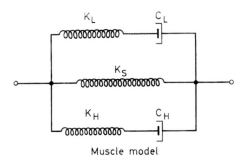

Muscle model

Fig. 16. Viscoelastic model of the muscle bundle. The model is nonlinear since the values of the elements are functions of length as discussed in the text

changes little with the amounts of stretch used and that filament spacing does not appear to change; hence, C_L/K_L would be inversely proportional to length as is found. It is also found that for values of L_n less than 1.15, the myofibrils are kinked, as if the connective tissue sheath that surrounds the muscle bundle is shortening them. The inflection in the K_S curves in Fig. 14 occurs at approximately the length where myofibrils would begin to contribute to elasticity of the muscle, and indeed, the slope of K_S below the inflection corresponds closely to that of K_H, whereas above the inflection it corresponds to that of K_L. This further supports the assignment of the elements to the aforementioned compartments.

D. Transducer Mechanisms

The data presented in the preceding sections which are pertinent to a discussion on the transducer mechanisms can be summarized as follows:

(1) Although the relationship between length of the muscle bundle and magnitude of the generator potential is linear under static conditions over the entire operational range of the receptor, the dynamic input-output relation is linear for displacements of only a few micra, that is, a very small fraction of the operational range.

(2) The absolute output voltage of the transducer is related to the muscle length, since the amplitude of the generator potential for a constant dynamic displacement increases with increasing length of the muscle bundle.

(3) The gain curve describing the transducer output of the lobster preparation has a constant 3 db/octave slope at all lengths. The slope of the modulus of dynamic stiffness of the muscle bundle is considerably less and is dependent on length.

Point (1) above indicates that the transducer is highly rate-sensitive under conditions in which work is done on the muscle fiber. (See LOEWENSTEIN and FUORTES, this volume, for comparison with other receptors). Actually, changes in resting length of 0.6 to 2% are required under static conditions to elicit equivalent changes in impulse frequency at the lower and upper limits of stretch. Furthermore, point (2) suggests that either the transducer input or some parameter of the transducer is a function of muscle length (see also BROWN, 1967).

One question then is if the filter output, i.e., tension, can directly account for the observed proportionality between generator potential amplitude and length under dynamic conditions. This possibility is suggested by the fact that similar proportionality is also present between tension amplitude and length of the muscle bundle. In particular, the increase in the ratio between generator potential amplitude and displacement amplitude compares well with the elasticity increase of the muscle bundle (KRNJEVIC and VAN GELDER, 1961) in the crayfish preparation. The results in the lobster receptor organ are similar, although obtained indirectly by comparing changes in impulse frequency with changes in elasticity. Another finding which bears on this point is the following: the amplitude of modulation which produces the maximum linear change in impulse frequency at a given length is also the maximum amplitude of modulation compatible with linear dynamic behavior at a much smaller length[2]. The linear range for dynamic behavior of the transducer is far less than that of the filter output, however.

Point (3) above shows that the dynamic stiffness of the muscle bundle alone cannot account for the gain curve of the transducer. This curve and the phase lead of the generator potential indicate that an atypical differentiation is performed either by some parameter of the transducer or by some unmeasured input variable. Similar behavior is also found in mammalian primary endings of muscle spindles where the gain curve has a slope of 10 db/decade for frequencies between 0.1 and 10 Hz, the phase leading by approximately 20° at the lowest frequency (POPPELE and TERZUOLO, 1968). Concerning this point, it should be understood that the rate of changes in the generator voltage, as recorded in the soma, can only be attenuated by the cable properties of the dendrites. No accurate estimate can however, be made of the frequency response characteristics of this cable. Notice also that the phase lead is present at the lowest frequency although at these frequencies the tension does not lead the displacement. At higher frequencies, the contributions of the tension to the transducer slope may go unnoticed since these changes are small compared to those introduced by the transduction process.

All the above considerations, developed from the three points stated at the beginning of this section, should be taken into account when attempting to analyze the mechanisms of the transduction process. To this end, we feel it is useful to focus initially upon a set of data obtained under static conditions. These data indicate: (i) a decrease of membrane resistance as a function of stretch; (ii) a relationship between the amplitude of the generator potential and the membrane voltage; and (iii) a dependency of the generator potential on extracellular Na^+. Each of these points will now be considered in some detail.

---

[2] These data are from one experiment only. The two lengths in this experiment produced 12 and 5.5 impulses/sec, resepctively.

Measurements of membrane resistance using one or two intracellular micro-electrodes yielded values ranging between 5.8 and 2.2 MΩ, in the absence of impulse activity (Terzuolo and Washizu, 1962; Washizu and Terzuolo, 1966). The current-voltage relation was found to be linear at all levels of hyperpolarization, whereas depolarizing currents (following substitution of external Na$^+$ by choline or of NaCl by sucrose) indicated the presence of delayed rectification (Edwards et al., 1963). Although this property varied considerably among cells, only in a few instances was it pronounced at levels of depolarization equivalent to those attained by stretch. This linear membrane behavior supports the validity of two sets of experiments. In the first one, the membrane resistance was measured at different levels of sustained stretch, the generation of impulse activity being pre-vented either by a background hyperpolarizing current or by tetrodotoxin (Terzuolo and Washizu, 1962). In four cells, a decrease of membrane resistance was measured and found to be approximately 2/3 of the resting value for amounts of stretch normally leading to impulse frequencies up to 38 impulses sec. In the

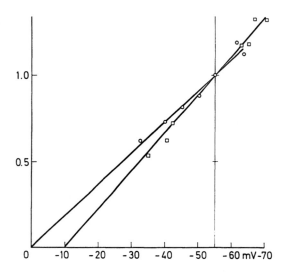

Fig. 17. Equilibrium potential of the generator potential. Ordinate: amplitude of the generator potential for a constant amount of stretch in arbitrary units (the amplitude of the generator potential before changing the membrane voltage is taken as unity). Data from two prepara-tions are shown. (From Terzuolo and Washizu, 1962)

second set of experiments, the amplitude of the generator potential to a "constant" amount of stretch was measured at different membrane voltages. The data from two of these experiments (plotted in Fig. 17) show that the amplitude of the generator potential is linearly related to the membrane potential. Moreover, the lines passing through the experimental points suggest an equilibrium potential near to zero membrane potential. If so, and because of the resistance measure-ments reported above, one must consider the possibility that an increase in mem-brane permeability is produced by the adequate input to the receptor (stretch) which is not likely to be restricted to any given ionic species (Terzuolo and Washizu, 1962). Experiments were, therefore, performed in which the extracellular ionic environment was changed after the cell was penetrated, and control measurements were made (EDWARDS *et al.*, 1963). It was found that the generator potential was markedly reduced when both Na^+ and Cl^- were replaced by sucrose and acetate (Fig. 18). Replacement of Na^+ by choline also reduced the magnitude of the generator potential. Replacement of Cl^- alone with propionate ions, on

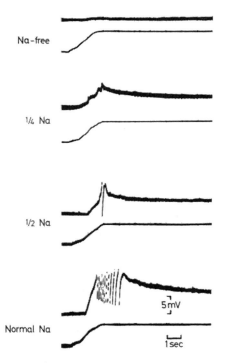

Fig. 18. Effect of removal of Na^+ on the generator potential. Upper beam: intracellular recording. Lower beam: applied stretch. (From EDWARDS *et al.*, 1963)

the other hand, did not lead to a change in the generator potential. Removal of Ca^{++} or an increase of external K^+ did cause a reduction of the generator potential. However, since the membrane resistance under these conditions is reduced, these results do not necessarily indicate a role either for Ca^{++} or K^+. Also, a fivefold increase of the Ca^{++} level did not appreciably alter the amplitude of the generator potential.

The results of this set of experiments on the effect of extracellular ionic substitution, therefore, indicate a role only for Na^+. However, since the equilibrium potential for the generator potential was found to be at about zero membrane potential, it is likely that the permeability to K^+ or Cl^-, or both, is also increased. Most of the current, however, would be carried by Na^+ (see OTTOSON, this volume, for a comparison with muscle spindle).

An equivalent electrical model, suggested by the above analysis of the mechanisms of the transduction process under steadystate conditions, is illustrated by the diagram of Fig. 19. It should be obvious, however that this model cannot account for the dynamic behavior of the transducer (TERZUOLO *et al.*, 1968). In particular, the phase lead and the increase in gain as a function of frequency, as well as the very limited range of linear dynamic behavior, are unaccounted for by the model. Therefore, the results obtained by system analysis oblige us to reconsider the transducer mechanisms.

From an empirical viewpoint, a differention operation could be introduced in the model by inserting a reactive element. However, no such typical operation

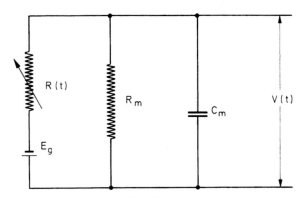

Fig. 19. An incomplete analogue model for the transduction process in stretch receptors

takes place in the transducer since at the lowest input frequencies, where the cable properties of the dendrites should not significantly affect the measurement, only a 30 to 40° phase lead is present in lobster preparations. Moreover, the slope of the gain curve is only half of that expected from a differentiation operation.

Alternatively, a change in membrane capacity could be considered (KATZ, 1950). Such a change has not been found, however, in axons submitted to a mechanical stimulus. Only resistive changes were observed under this condition, which were maintained throughout a sustained stimulus (GOLDMAN, 1965).

Another possibility is that of supplementing the model with features taken from the model of electrokinetic membrane processes proposed by TEORELL (1959, 1966), in which the thermodynamic relationship between hydrostatic pressure and electrochemical potentials are considered. Given a tight mechanical coupling between the dendrites and the muscle bundle, the longitudinal tension which develops as a consequence of stretch can produce instantaneous pressure changes, assuming that the cytoplasm is not redistributed instantaneously owing to its viscosity. If so, adequately large pressure changes could develop in the smaller dendrites. Consequently, the transducer behavior could result from more than one process, namely, the transformation from tension to pressure and that between pressure and voltage across the membrane. Both processes can introduce a phase lead. Moreover, the second one can be responsible for the very limited range of linear behavior of the transducer, since it is described by a second-order nonlinear differential equation (TEORELL, 1959). An analogue model which includes the features just described can indeed exhibit a behavior qualitatively similar to that of the transducer of the crustacean stretch receptor (KNOX and TERZUOLO, *unpublished data*). Its adequacy, however, remains to be tested. This will depend most crucially on the ability of the model to duplicate the nonlinear features of the transduction processes.

References

ALEXANDROWICZ, T. S.: Muscle receptor organs in the abdomen of *Homarus vulgaris* and *Palinurus vulgaris.* Quart. J. micr. Sci. **92**, 163—199 (1951).
— Receptor elements in the thoracic muscles of *Homarus* vulgaris and Palinurus vulgaris. Quart. J. micr. Sci. **93**, 315—346 (1952).
ASBY, W. R.: An introduction to cybernetics. New York: J. Wiley & Sons 1963.
BAYLY, E.: Ph. D. Thesis University of Minnesota (1967).
BODIAN, D., BERGMAN, R. A.: Muscle receptor organs of crayfish: Functional anatomical correlations. Johns Hopk. Hosp. Bull. **110**, 78—106, (1962).
BORSELLINO, A., POPPELE, R. E., TERZUOLO, C. A.: Transfer functions of the slowly adapting stretch receptor organ of Crustacea. Cold Spr. Harb. Symp. quant. Biol. **30**, 581—586 (1965).
BROWN, M. C.: Some effects of receptor muscle contraction on the responses of slowly adapting abdominal stretch receptors of the crayfish. J. Exp. Biol. **47**, 445—458 (1967).
— STEIN, R. B.: Quantitative studies on the slowly adapting stretch receptor of the crayfish. Kybernetic **3**, 175—185 (1966).
BULLOCK, T.: Initiation of nerve impulses in receptors and central neurons. Rev. mod. Phys. **31**, 504—514 (1959).
BURKHARDT, D.: Die Erregungsvorgänge sensibler Ganglienzellen in Abhängigkeit von der Temperatur (Untersuchungen an den abdominalen Streckrezeptoren des Sumpfkrebses, Astucus leptodactylus). Biol. Zbl. **78**, 22 (1959).
EDWARDS, C., OTTOSON, D.: The site of impulse inition in a nerve cell of a crustacean stretch receptor. J. Physiol. (Lond.) **143**, 138—148 (1958).
— TERZUOLO, C. A., WASHIZU, Y.: The effect of changes of the ionic environment upon an isolated crustacean sensory neuron. J. Neurophysiol. **26**, 948—957 (1963).
EYZAGUIRRE, C., KUFFLER, S. W.: Processes of excitation in dendrites and in the soma of single isolated sensory cells of the lobster and crayfish. J. genet Physiol. **39**, 87—119 (1955a) (1955a).
— — Further study of soma, dendrite and axon excitation in single neurons. J. genet Physiol. **39**, 121—153 (1955b).
FIRTH, D. R.: Interspike interval fluctuations in the crayfish stretch receptor. Biophys. J. **6**, 201—215 (1966).
FLOREY, E., FLOREY, E.: Microanatomy of the abdominal stretch receptors of the crayfish (*Astacus flunatilis L.*) J. genet Physiol. **39**, 69—85 (1955).
GOLDMAN, D. E.: The transducer action of mechanoreceptor membranes. Cold Spring Harb. Symp. quant. Biol. **30**, 59—67 (1965).
HANDELMAN, E.: M. S. Thesis, University of Minnesota, (1969).
KATZ, B.: Depolarization of sensory terminals and the initiation of impulses in the muscle spindle. J. Physiol. (Lond.) **111**, 261—282 (1950).
KNOX, C. K.: Simulation of the transfer function of a *Crustacean* muscle bundle. Simulation. **10**, 93—96 (1968).
KRNJEVIC, K., VAN GELDER, N. M.: Tension changes in crayfish stretch receptors. J. Physiol. (Lond.) **159**, 310—325 (1961).
LOEWENSTEIN, W. R., SKALAK, R.: Mechanical transmission in a Pacinian corpuscle. An analysis and a theory. J. Physiol. (Lond.) **182**, 346—378 (1966).
— TERZUOLO, C. A., WASHIZU, Y.: Seperation of transducer and impulse-generating processes in sensory receptors. Science **142**, 1180—1181 (1963).
PETERSON, R. P., POPE, F. A.: The five structure of inhibitory synapses in the crayfish. J. biophys. biochem. Cytol. **11**, 157—169 (1961).
POPPELE, R. E., TERZUOLO, C. A.: Myotatic reflex: its input-output relation. Science **159**, 743—745 (1968).

TEORELL, T.: Electrokinetic membrane processes in relation to properties of excitable tissues. J. genet. Physiol. **42**, 847—863 (1959).

—' Electrokinetic considerations of mechanoelectrical transduction. *Ann.* N. Y. Acad. Sci. **137**, 950—966 (1966).

TERZUOLO, C. A., HANDELMAN, E., ROSSINI, L.: An isolated crustacean neuron preparation for metabolic and pharmacological studies. In: Invertebrate Nervous System. Wiersma, C. A. G. Ed. Chicago: Univ. Chicago Press 1967.

— PURPLE, R. L., BAYLY, E., HANDELMAN, E.: Post synaptic inhibition. Its action upon the transducer and encoder systems of neurons. In: Structure and Function of Inhibitory Neuronal Mechanisms. E. J. C. von Euler, Ed. Oxford: Pergamon Press p. 261—275 (1968).

— WASHIZU, Y.: Relation between stimulus strength, generator potential and impulse frequency in stretch receptor of *Crustacea*. J. Neurophysiol. **25**, 56—66 (1962).

WASHIZU, Y., TERZUOLO, C. A.: Impulse activity in the crayfish stretch receptor neuron. Arch. ital. Biol. **104**, 181—194 (1966).

WIERSMA, C. A. G., FURSHPAN, E., FLOREY, E.: Physiological and pharmacological observations on muscle receptor organs of the crayfish, *Cambarus larkii*. J. exp. Biol. **30**, 136—150 (1953).

Chapter 17

Patterns of Organization of
Peripheral Sensory Receptors*

By

Bryce L. Munger, Hershey, Pa. (USA)

With 11 Figures

Contents

A major obstacle to successful understanding of the sequence of events involved in somatic sensibility is the accurate identification and characterization of modality-specific receptors. This problem of identification has plagued the scientific community for over 100 years (Boeke, 1932, 1940; Stöhr, 1928; Weddell et al., 1955, 1962; Winkelmann, 1960, 1967). Recent studies utilizing refined cytologic technics have not provided a solution to the general problem. In a few instances, questions raised over the past century have been answered with the use of electron microscopy; however, the ultrastructural approach has raised as many questions as it has answered.

The purpose of this discussion is to review the basic patterns of organization of known sensory receptors in an attempt to correlate some of the vast previous literature with current biologic studies. I have arbitrarily organized the material into as small a number of discrete entities as possible, admittedly biased by personal prejudices. Those receptors characterized as "corpuscles" have been classed as one group. Those receptors which are associated with epithelial secretory cells are designated as epithelial cell-neurite complexes, and they are discussed as a separate entity.

The general properties of dermal innervation will be considered first because this information is necessary for considering the remainder of the topics. The next logical subject is the innervation of hair, which deals both with the dermal neural network and specific receptors. As pointed out by Boeke (1940), "We do not even know the nature of the most efficient sensory mechanism in the skin — pain." This statement still stands today. As a result, I will say little about this most fascinating aspect of sensory physiology and anatomy.

* This study was supported in part by U. S. Publik Health Service Research Grants GM-10102 and AM-11407.

A. Dermal Neural Network

The dermis of the skin contains a confusing array of neural elements (Stöhr, 1928; Miller et al., 1960; Weddell et al., 1955, 1962). Some of these elements represent the sensory interface between our external environment and central nervous system, and therein lies our interest. However, the dermis also carries motor, sympathetic, and parasympathetic elements, distributing them to specific terminals. The ultimate problem is to characterize a single specific element in this confusing array.

We may now pose the question, "What is the body of factual information pertaining to the organization of subcutaneous and dermal components of the peripheral nervous system?" At the level of light microscopy, an overlapping arborization of fibers can be defined in methylene blue- or silver-stained pre-

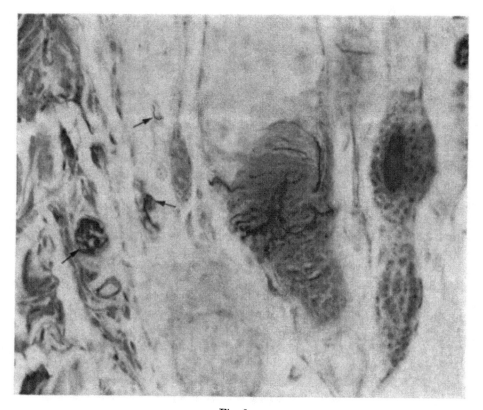

Fig. 1a

Fig. 1a and b. a) Hair and surrounding connective tissue of rat skin [silver method of Sevier and Munger (1965)]. The neurites circularly disposed around the hair follicle turn abruptly to end in lanceolate (palisade) endings arranged parallel with the shaft of the hair. In the connective tissue around the hair follicle, isolated neurites and small bundles of neurites can be seen (arrows). A delicate connective tissue sheath (sheath of Henle) surrounds the fibers and fiber bundles. × 512. b) Human dermis. A small cluster of unmyelinated neurites and Schwann cell investment are enclosed by a delicate process of a fibroblast (H) – – a sheath of Henle. A single myelinated axon is free in the connective tissue. × 3,400

parations. Our studies utilize a modification of Richardson's (1960) silver technique (SEVIER and MUNGER, 1965, 1968). Regional differences in this terminal arborization have been described (MILLER et al., 1960; WEDDELL et al., 1955). Unfortunately, the final termination of a specific fiber is not easily defined unless this fiber terminates in a defined nerve ending (i.e., Pacinian corpuscle) or on a specific structure (i.e., hair; Fig. 1a). As a result, the so-called "dermal nerve net" is defined by the absence of *easily* characterized terminals. This "wastebasket nature" of a presumably important component of the nervous system is most unfortunate. The studies of WEDDELL and his co-workers (1945, 1955, and 1960) and MILLER and his co-workers (1960) have clearly shown that nerves do distribute themselves widely in the connective tissue of the dermis. The definition of a specific ending is more difficult. WEDDELL (1955) is of the opinion that varicosities along the fiber represent sensitive terminals and that these terminals in turn constitute free nerve endings. Many of these neural elements will also send a branch to a hair, as discussed below.

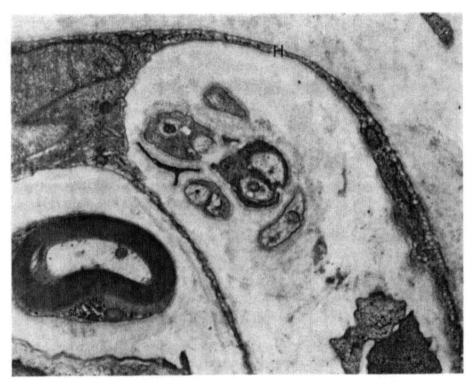

Fig. 1 b

In electron micrographs our confusion as to the nature of free nerve endings and the dermal nerve net is compounded. In sections of skin from primates, including man, isolated myelinated nerve fibers can be found scattered in the connective tissue. A delicate investment of fibroblast cytoplasm (a sheath of Henle; Fig. 1b) accompanies the nerve fiber. Unmyelinated axons are also at

least partially surrounded by a sheath of Henle, and usually two or three neurites
are found in a single Schwann cell. Isolated Schwann cells and single unmyelinated axons are found only infrequently, and naked axons are difficult − if not
impossible − to identify. A terminal, in the strict sense of the definition, cannot
be characterized accurately (STÖHR, 1928). The presence of a somewhat enlarged
portion of a nerve containing a few mitochondria, as seen in the electron microscope, is only speculatively a terminal. Although electron microscopic characterization of a terminal is difficult (CAUNA, 1969), both MILLER (MILLER et al., 1958,
1960) and WEDDELL (WEDDELL, 1945, 1960; and WEDDELL et al., 1955, 1962)
have characterized, at the level of the light microscope, terminals of sensory
nerves. In both glabrous and hairy skin, nerve fibers arborize and terminate
either with or without an expanded tip (DOGIEL, 1903). Nerve terminals ending
on hairs, Merkel cells, or in corpuscular endings are discussed separately below.
The problem confronting any interpretation of electron micrographs (CAUNA,
1969; CAUNA et al., 1969) or the careful studies of MILLER or WEDDELL relative
to free nerve endings is the difficulty in characterizing any ending in a functional
sense. We thus identify terminals as receptors by inference rather than by positive
identification. Which of these endings could fulfill the criteria as a specific
sensory receptor (and transducer) remains a problem for future study.

B. Innervation of Hair

Many hairs are exquisitely sensitive to mechanical displacement (WERNER
and MOUNTCASTLE, 1965). Two types of hairs − sinus hairs (vibrissae) and
somatic sensory hairs − are innervated by a somewhat different and complex
manner. Vibrissae are lacking in man, but are present in many primates and in
most other mammals.

Conventional hairs are innervated by 10 to 15 individual nerve fibers (WED-
DELL, 1945) forming a collar of terminals arranged parallel to the shaft of the
hair and located immediately above the enlargment of the hair bulb (Fig. 2a and c;
MONTAGNA and ELLIS, 1965). The collar of parallel, or palisade, endings (MILLER,
1960) is surrounded by circumferentially disposed nerve fibers (Fig. 2c; RETZIUS,
1892; SYMONOWICZ, 1909). These terminals have been identified in the electron
microscope by YAMAMOTO (1966), ANDRES (1966), and ORFANOS (1967). ANDRES
(1966) has termed these endings "lanceolate" endings, a term denoting the shape
of the terminal neurite (Fig. 2b). The terminal has the shape of a flattened cylinder
and contains numerous mitochondria. The terminal is encased on its two flat
sides by flattened discs of Schwann cell cytoplasm (Fig. 2b and d). There is no
Schwann cell cytoplasm directly apposed to the external root sheath; the neurite
itself is in direct contiguity with the fused basal lamina of the hair follicle and
Schwann cell. The enveloping Schwann cell shows considerable pinocytotic
activity toward both the connective tissue and the neurite terminal. The accumulation of mitochondria in the neurite terminal is the only distinguishing cytologic
feature of this region of the fiber, and this feature is seen in virtually all sensory
terminals discussed below. There remains some uncertainty as to how the electron
microscopy description corresponds to that obtained by light microscopy, specifically the relationship between the longitudinal and circular elements. MILLER

(1960) characterizes two types of terminals on hairs by light microscopy, but only one type has been identified in electron micrographs to date.

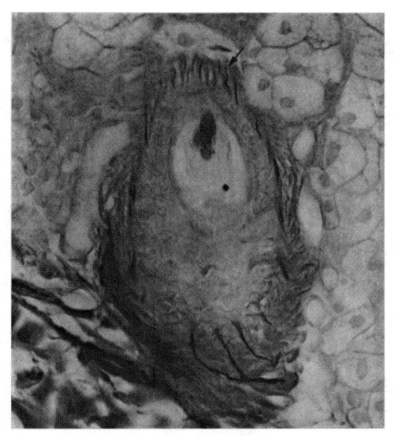

Fig. 2a

Fig. 2a—d. a) Oblique section of shaft of a hair of baboon skin [silver method of SEVIER and MUNGER (1965)]. The lanceolate (palisade) endings (arrow) are disposed parallel to the shaft of the hair. At the lower portion of the micrograph, circular elements can be seen giving rise to longitudinally positioned lanceolate endings. A portion of the sebaceous gland envelops the majority of the hair shaft. × 512. b) Sensory hair from a cinomolgous monkey. The lanceolate endings and hair shaft are cut in cross section. The neurites (arrows) are flattened cylinders containing numerous mitochondria; the Schwann cells on each side are also flattened. Numerous pinocytotic vesicles are present in the Schwann membrane both toward and away from the neurite. As the endings abut the basal lamina of the external root sheath (E), no intervening Schwann cell cytoplasm is present and the neurite is contiguous with the basal lamina × 10,250. c) Idealized drawing of a sensory hair. One or more small bundles of neurites approach the hair follicle and pursue a circumferential course around the shaft. Fibers then turn abruptly and ascend or descend along the hair shaft. Fig. 2b is a cross section of such longitudinally disposed fibers; Fig. 2c illustrates the relationship of the Schwann cells to sensory neurites as seen in Fig. 2b. d) Lanceolate ending on a sensory hair The neurite is flattened and cupped between two flattened expanses of Schwann cell cytoplasm. The sensory neurite contains numerous mitochondria. Pinocytotic vesicles are frequently seen in the surrounding Schwann cell

The pattern described above is an important fundamental framework of reference for tactile transducing systems. It has been observed by electron microscopy in the sensory somatic hairs of mouse, rat, monkey, and man. The basic pattern of similarity is a flattened neurite, sandwiched between flattened

Fig. 2b

portions of Schwann cell cytoplasm, arranged perpendicular to the hair shaft (Fig. 2b and d). Schwann cell cytoplasm is absent on the thin edges of the sensory neurite (i.e., toward and away from the hair follicle basement membrane). Units of this nature ring the follicle and presumably are deformed during mechanical dislocation of the hair. The events of transduction are most likely localized to this complex, and they represent rapidly adapting receptors (ADRIAN and ZOTTER-MAN, 1926; IGGO, 1963b; WERNER and MOUNTCASTLE, 1965).

Vibrissae, or sinus hairs, possess a complex pattern of innervation (TRETJA-KOFF, 1911; VINCENT, 1913; ANDRES, 1966). These receptors have an innervation

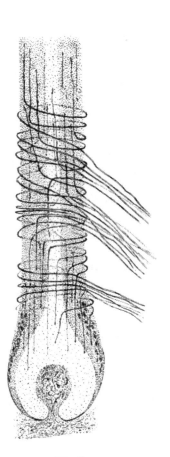

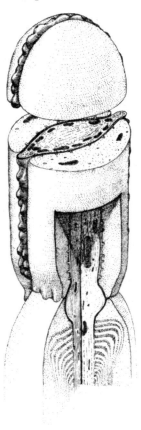

Fig. 2c   Fig. 2d

typical of somatic sensory hairs as well as Merkel cell-neurite complexes (KADA-NOFF, 1924) and occasional corpuscular endings. Vibrissae are complicated structures which, for purposes of simplicity, can be regarded as hairs embedded in erectile tissue; the whole complex is surrounded by a dense connective tissue capsule. The nerve bundle containing over 100 nerve fibers (VINCENT, 1913; ZUCKER and WELKER, 1969) penetrates the capsule and passes through the erectile tissue to the hair shaft. Nerve fibers terminating in the upper portion of the shaft are associated with Merkel cells (PATRIZI and MUNGER, 1966; ANDRES, 1966). The neurite with its enveloping Schwann cell abuts the basement membrane of the hair follicle, and the neurite passes through the basement membrane and between the cells of the external root sheath. It then bends and courses parallel to the long axis of the hair shaft. In this position, it becomes intimately associated with a Merkel cell (TRETJAKOFF, 1902). The Merkel cell contains numerous secretory granules polarized in the cell towards the neurite (as in Fig. 3b). The complex can be termed a *"Merkel cell-neurite complex"* and is discussed in more detail subsequently.

The second type of sensory innervation of a sinus hair is identical in all respects to that seen around the shaft of conventional hairs (VINCENT, 1913; ANDRES, 1966). These lanceolate endings bear a striking resemblance to the individual units comprising a Meissner corpuscle. This similarity is discussed below.

ANDRES (1966) has also described corpuscular endings and unmyelinated terminals in sinus hairs. The function of these components is not known. ZUCKER and WELKER (1969) have studied in detail single-unit recordings of the output of rat vibrissae at the level of the trigeminal ganglion. Both slowly and rapidly adapting fibers can be found in a single vibrissae. Since the lanceolate endings on conventional hairs are rapidly adapting, and Merkel cell-neurite complexes are considered to be slowly adapting (IGGO, 1963a, b and MUNGER, 1965), this double pattern of innervation is consistent with the functional capacity of the unit. ZUCKER and WELKER (1969) have also found a complex pattern of directional sensitivity which lacks a morphologic explanation at this time. This observation deserves future study.

C. Epithelial Cell-neurite Complexes

The controversy over the reality or fiction of intraepidermal innervation has been somewhat resolved after a century of scientific dispute (DOGIEL, 1903; STÖHR, 1928; MUNGER, 1965). Nerves can be found passing into the epidermis in certain areas of skin of most species studied to date, e.g., rat, cat, dog, opossum, monkey, and man (MUNGER, 1965). Differences in distribution within a given species are marked (MERKEL, 1880; RANVIER, 1880; RETZIUS, 1892; BOTEZAT, 1908; KADANOFF, 1928; DUTHIE and GAIRNS, 1960). In most instances, the neurite enters the epidermis at the base of the rete pegs (RANVIER, 1880; MILLER, 1960) and becomes associated with a Merkel cell (MUNGER, 1965). MERKEL'S description (1880) of a specialized cell associated with some sensory neurites has been confirmed to an amazing degree, considering the widespread scientific neglect of his thesis for most of the twentieth century. His description is not correct in all instances, but his exhaustive study remains a veritable gold mine of suggestions for further study. I have proposed that this cell be called a *Merkel cell*, as is also suggested by TRETJAKOFF (1902), rather than a *"Tastzellen"* cell as suggested by MERKEL. The latter term denotes a functional capacity for which there is no basis at the present time.

The Merkel cell is found immediately above the basal layer of the epidermis (Fig. 3a) and is easily distinguished from other epidermal cells at the ultrastructural level by the accumulation of secretory granules in the cytoplasm subjacent to the neurite (Fig. 3b; MUNGER, 1965). The cell is oval and has a lobulated relatively pale nucleus. The Golgi apparatus and associated prosecretory granules are found on the opposite side of the nucleus from the accumulation of secretory granules. The neurite in apposition to the Merkel cell is somewhat enlarged and characteristically contains numerous mitochondria. Desmosomes are present between the Merkel cell and adjacent epidermal cells (MUNGER, 1965). ANDRES (1966) has described thickenings of the plasma membrane of the neurite as it abuts onto the Merkel cell. Such thickenings have not been observed in the material prepared in our laboratory.

In some areas of epidermis, the neurite will course beyond the Merkel cell and ascend to the level of the stratum granulosum where vesiculation and fragmentation of the neurites are seen. As the neurites course through the epidermis,

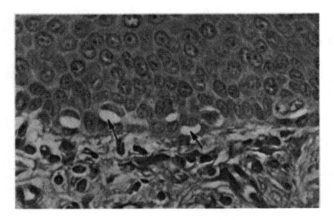

Fig. 3a

Fig. 3a and b. a) Opossum snout skin (formalin fixation and paraffin embedding, *H* and *E*). The emptyappearing cells at the base of the rete peg are Merkel cells (arrows). This fixation artifact is consistently seen in paraffin sections. × 400. b) Opossum snout skin. The central cell is a Merkel cell possessing the characteristic lobulated nucleus (*M*). Numerous electron-opague secretory granules are present in the cytoplasm adjacent to the neurite (*N*). Glycogen particles are scattered throughout the cytoplasm of the Merkel cell. A desmosome (*D*) is present between the Merkel cell and the adjacent epidermal cell. The dermis is indicated beneath the figure number. × 24,000

they are enveloped by epidermal cells in a manner entirely analogous to that of a Schwann cell enveloping an unmyelinated axon (MUNGER, 1965). The neurites thus appear to course *through* the epidermal cells, as described correctly by BOEKE (1932). The neurites in this location lack the accumulation of mitochondria seen at the level of the Merkel cell.

Accumulations of these Merkel cell-neurite complexes (SMITH, 1967, 1970) in hairy skin comprise the so-called "touch domes" of IGGO (1963a, b; IGGO and MUIR, 1963) or the *Haarscheiben* of PINKUS (1902, 1905; Fig. 4). PINKUS described the distribution of these epidermal elevations in considerable detail. They are associated in a regular fashion with hairs, thus the designation *Haarscheibe*. Investigators redescribing these structures have given them a variety of terms, including *tylotrich pad* (STRAILE, 1960; MANN and STRAILE, 1963). I, along with SMITH (1967, 1970), prefer the original term of Pinkus, Haarscheiben.

In cats, Haarscheiben are found cephalad to a hair, whereas in most other mammals they are found caudal to a hair (MANN and STRAILE, 1963). The geo-

34*

graphy of these two receptors determines which direction a hair must be moved in order to fire the Haarscheiben. Numerous studies have demonstrated that Haarscheiben are slowly adapting mechanoreceptors (Iggo, 1963a, b; Mann and Straile, 1963; Werner and Mountcastle, 1965; Tapper, 1965; Smith, 1967).

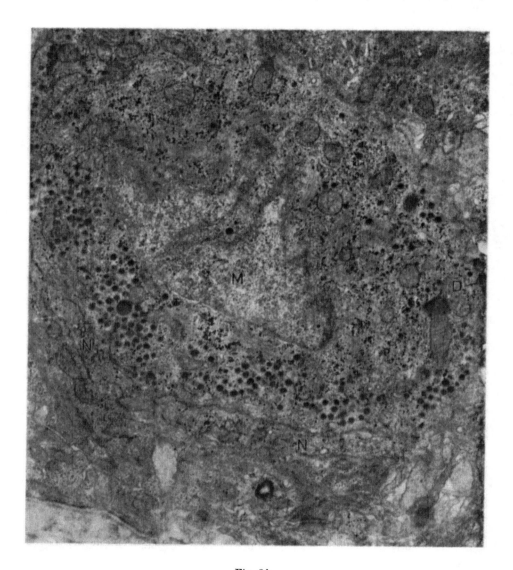

Fig. 3b

These studies, together with those cited previously, lead to the conclusion that hairs are composite mechano-sensitive complexes. Movement of a hair can result in firing of lanceolate endings on the shaft (rapidly adapting) or, in some cases, of specialized groups of epithelial cell-neurite complexes (slowly adapting) located on the hair (vibrissae) or skin surface (Haarscheiben). The fact that only some hairs are associated with Haarscheibe has led STRAILE (1960) to term these follicles *tylotrich follicles* and the Haarscheiben *tylotrich pads*. It appears as though different nerve fibers innervate the lanceolate endings on the hair shaft and the Haarscheiben in the epidermis. Movement of a hair could thus produce rapidly or slowly adapting responses in single fibers, depending on which neurite was under study.

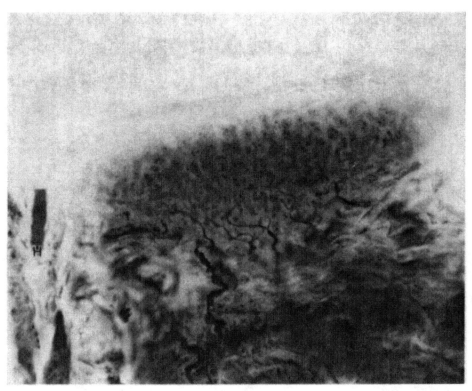

Fig. 4. *Haarscheibe* from baboon skin [silver method of SEVIER and MUNGER (1965)]. An elevated flattened dome of epidermal cells is found adjacent to a hair follicle (*H*). Numerous neurites enter the base of the epidermis, and considerable branching of neurites can be seen.
× 512

Slowly adapting mechanoreceptors are found in the glabrous snout skin of animals (BARKER and WELKER, 1969). This is the system in which Merkel cell-neurite complexes were first defined by electron microscopy (MUNGER, 1965). Wherever Merkel cell-neurite complexes are found, slowly adapting receptors can also be defined. The suggestion is plausible that Merkel cells modulate the activity of sensory nerves, making them slowly adapting. The exact location of the events of transduction in this system cannot be determined precisely, for the secretory product of the Merkel cells has defied characterization to date. Histochemical studies of the secretory product of Merkel cells (MUNGER, 1965) have been negative except for a moderate periodic acid-Schiff positivity. The detailed pharmacologic studies of Smith and Creech (1967) also provide no clue as to the nature of the secretory product. Thus, a neurotransmitter cannot be implicated at the present time. The suggestion has been offered that the Merkel cells might be tropic or trophic in nature (Munger, 1965; Smith and Creech, 1967). This conclusion is speculative and without any supporting data at present.

The Grandry corpuscle found in the skin of birds (e.g., in the duck bill) is an epithelial cell-neurite complex somewhat different from the Merkel cell-neurite complexes described above. These epithelial cell-neurite complexes were described by Grandry (1869) and considered by Merkel (1878) as another example of a receptor containing the *Tastzellen* described by him. The Grandry corpuscle is a large unit approximately 60 μ in diameter. The bulk of the receptor consists of two large cells, each shaped like half a grapefruit and facing each other on the flattened side. A flattened, branched neurite terminal is found between the opposed cells. The entire complex is invested by small satellite cells (Fig. 5a).

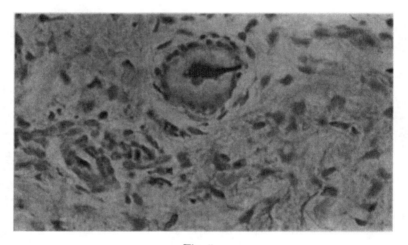

Fig. 5a

Fig. 5a and b. a) Grandry corpuscle from duck bill skin [silver method of Sevier and Munger (1965)]. The blackened neurite exhibits considerable variation in pattern at the terminal (? branching). One epithelial cell is present above and the nucleus of the other can be seen below the neurite. The small nuclei of satellite cells are visible surrounding this epithelial cell-neurite complex. The skin surface is to the top of the micrograph. × 512. b) Grandry corpuscle from duck bill skin. The central neurite (*N*) contains numerous mitochondria. This neurite is enclosed on two sides (top and bottom) by secretory cells containing many electron-opaque secretory granules. They are not polarized towards the neurite as in Merkel cell-neurite complexes (Fig. 3b). Satellite cells (*S*) intimately surround the complex. The dense connective tissue of the dermis occupies the remainder of the micrograph. × 4,500

Based on electron microscopic studies of Grandry corpuscles (Munger, *unpublished observations;* De Iraldi and Rodriguez-Pérez, 1961; Quilliam, 1966 Andersen and Nafstad, 1968), these receptors consist of two large secretory cells and an enclosed neurite ter- minal (Fig. 5b). Numerous electron-opaque secretory granules are present in the cytoplasm of the secretory cells. These granules resemble those found in Merkel cells in many respects; they are not, however, distributed in the same manner in the cell. The secretory granules in the Grandry cells are distributed at random and are not polarized toward the neurite as they are in Merkel cells. Histochemical methods have failed to reveal the nature of the secretory granules, as is also the case in Merkel cells. The cytoplasm of these cells is filled with dispersed ribosomes and granular endoplasmic reticulum. A prominent Golgi apparatus with associated prosecretory

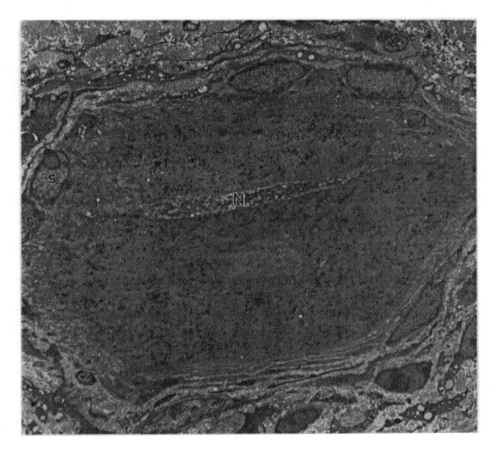

Fig. 5b

granules is also present. These cells may indeed be identical functionally to Merkel cells of the epidermis and hair; however, we have no evidence on this point at the present time. The neurite is flattened into a disclike shape and contains numerous mitochondria, as do nerve terminals seen elsewhere. Some branching of the terminal occurs but not to the extent depicted by the drawings of BOEKE (1932); in this regard, the silver impregnations (Fig. 5a) are somewhat misleading. Junctional complexes are not observed between the Grandry cells and sensory neurite.

Satellite cells envelop the complex and interdigitate with the Grandry cells and with each other; also, junctional complexes are present between these apposing cells. A basal laminais present at the satellite cell-connective tissue interface; no basal lamina is present between the satellite cell and Grandry cell. Occasional pockets of connective tissue are trapped between the complicated interdigitations of the satellite cell, but these do not represent an exception to the generalization made above that the satellite cell separates the Grandry cells from the connective tissue space.

The Grandry corpuscles always have a definite orientation with respect to the surface of the skin of the duck bill; the flattened neurite terminal and the flattened side of the Grandry cell are always exactly parallel to the skin surface. In some instances, two or three such complexes are stacked one upon the other. In such cases, we have a sequence of Grandry cell-neurite-Grandry cell-neurite-Grandry cell. The central Grandry cell is flattened on two sides, i.e., shaped like a doughnut. The stacked Grandry corpuscles still maintain the parallel relationship to the skin surface.

Both Merkel cell-neurite complexes and Grandry corpuscles are examples of what I prefer to regard as a receptor complex of neurite terminal and associated specialized secretory cell. I thus propose that they be designated epithelial cell-neurite complexes.

The origin of the specialized cell is uncertain. Merkel cells have only been observed in stratified squamous epithelium (epidermis and hair), and they possess desmosomes attaching them to adjacent squamous cells. This fact suggests that the Merkel cell is a specialized epidermal cell. Grandry corpuscles, on the other hand, are dermal, and the cell origin of the specialized secretory cell could more plausibly be considered as the Schwann cell. Both complexes suggest a modulating role for the epithelial cell. Since Merkel cell-neurite complexes are slowly adapting mechanoreceptors, we can make the generalization that when a secretory cell is associated with a sensory neurite, the complex is slowly adapting in a physiologic sense. The important problem remaining to be determined is the nature of the secretory product of the specialized cell and its role in the function of the sensory neurite.

D. Corpuscular Nerve Endings

From a morphologic point of view, the corpuscular nerve endings can be defined with ease, and, as a result, they have received the most detailed study over the past century. I define corpuscular endings as those large globular circumscribed endings enveloped by a distinct capsule. The nineteenth century cytologists easily recognized the importance of these specialized receptors because large-diameter nerve fibers could be traced to their termination in the receptors. The extrapolation that these units were sensory in nature was made with ease. Controversy, of course, was rampant, for many divergent opinions were expressed concening the function of the units, but this controversy soon gave way to agreement, to some extent, that these units were sensory endings. Agreement on how they were organized, however, was another matter. Only recently, with the aid of the electron microscope, have some details of the internal structure and organization been accurately elucidated. From a functional point of view, the important element of transduction at a cellular level appears to be the membrane of the neurite (LOEWENSTEIN and RATHKAMP, 1958; OZEKI and SATO, 1964). The associated lamellar cells of the core remain a questionable entity in terms of function. The intermediate lamellae and capsule of Pacinian corpuscles have been characterized as the responsible components for the off response (LOEWENSTEIN and SKALAK, 1966) and for the adaptive characteristics of Pacinian corpuscles (LOEWENSTEIN and MENDELSON, 1965).

Among corpuscular receptors, certain similarities in the pattern of organization are apparent which may assist our comprehension of the function of the various components. The model system usually chosen in a discussion of corpuscular nerve endings is the Pacinian corpuscle. Three distinct cellular zones can be defined: (1) the capsule; (2) the middle zone (lamellae); and (3) the neurite and closely associated lamellar cells (core). Other corpuscular endings evidence some of these same components but in differing proportions.

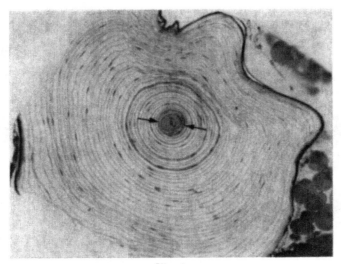

Fig. 6a

Fig. 6a—d. a) Pacinian corpuscle from cat mesentery [2 μ-thick section of tissue prepared for electron microscopy; trichrome method (SEVIER and MUNGER, 1968)]. Part of the corpuscle (upper and lower left) was removed in trimming the araldite block. The arrows delimit the core. The neurite profile is seen in cross section in the center of the core. The cleft in the core (plane of bilateral symmetry) is from top to bottom. Compare with Fig. 7a and note difference in magnification. × 160. b) Pacinian corpuscle from cat mesentery. The neurite is cut in cross section with a similar orientation to Fig. 6b. The arrow denotes the edge of the core (as in Fig. 6a). Stacked cytoplasmic lamellae are derived from lamellar cells, the nuclei of which (L) are scattered (unlike the Herbst corpuscle in Fig. 7b). The cleft (C) in the cytoplasmic lamellae is distinct. The appearance of the cells outside the core (to the right of the arrow), comprising the cellular lamellae of the intermediate zone, are more electron-opaque than the cells of the core. × 7,000. c) Same section as in Fig. 6b at higher magnification. The central neurite contains numerous mitochondria. Collagen fibrils (arrow) are present between individual lamellae. Junctional complexes (J) are present between successive cytoplasmic lamellae, and between lamellae and central neurite. × 20,000. d) Idealized drawing of a Pacinian corpuscle. The sensory neurite at this magnification is the slender filament at the arrow. Surrounding the neurite are the densely packed cellular lamellae of the core. The cleft in the core faces the viewer. The remainder of the corpuscle represents the successive intermediate cellular lamellae and intervening connective tissue matrix

The capsule of a Pacinian corpuscle is a continuous sheet of simple squamous *epithelium* and associated connective tissue. SHANTHAVEERAPPA and BOURNE (1955, 1963) have greatly increased our understanding of this layer and have defined it as a *perineural epithelium*. This term emphasizes the epithelial nature of the capsule which completely segregates the contents of the corpuscle from the connective tissue space surrounding it. A dense mesh of collagen envelops the whole corpuscle.

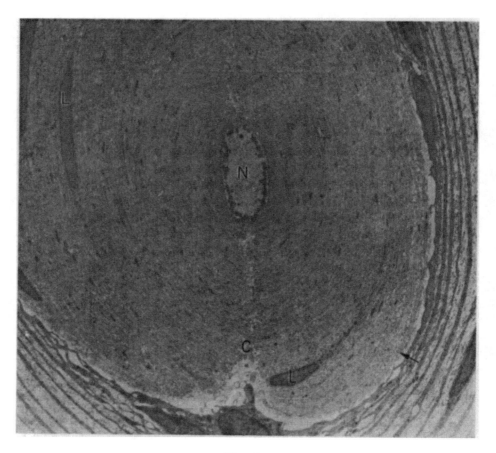

Fig. 6b

Most of the corpuscle is composed of the successive layers of simple squamous cellular lamellae separated from one another by material possessing the properties of a connective tissue ground substance admixed with collagen fibers (Fig. 6a, 6d). This material contains a strongly acid sulfated mucoprotein. Collagen fibrils and scattered fibroblasts are also found in the middle zone. By electron microscopy (MUNGER, *unpublished observations*), individual lamellae of the middle zone have been found to lack any connections between one another, in step sections through a cat Pacinian corpuscle.

The so-called "core" of the Pacinian corpuscle is a complex organization of lamellar cells and a neurite terminal (Fig. 6a, 6b). The lamellar cells form numerous concentric sheets of thin cytoplasm stacked one on top of the other. Collagen fibers are present between successive lamellae (Fig. 6c). The core is bilaterally

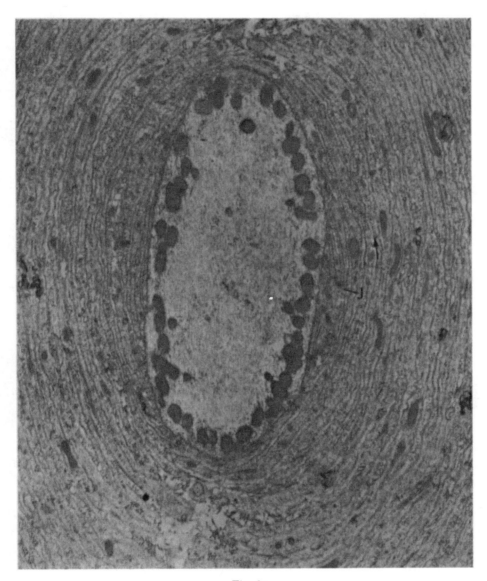

Fig. 6c

symmetrical (SCHUMACHER, 1911; CLARA, 1925), and this symmetry is due to a cleft in the stacked cytoplasmic lamellae (PEASE and QUILLIAM, 1957) (Fig. 6b, 6c, 6d). Several lamellar cells surround the neurite at any given point, unlike the situation in Herbst corpuscles.

The Herbst corpuscle is morphologically a first cousin of a Pacinian corpuscle. Malinovsky (1967, 1968) has recently provided a detailed study of avian sensory receptors, and he has described considerable variability in the form of Herbst corpuscles of various kinds. Malinovsky's work together with the descriptions of Merkel (1880) and Clara (1925) provide a complete and accurate description of the topography and histology of Herbst corpuscles. For purposes of simplicity, the present description is restricted to receptors found in the skin of the duck bill.

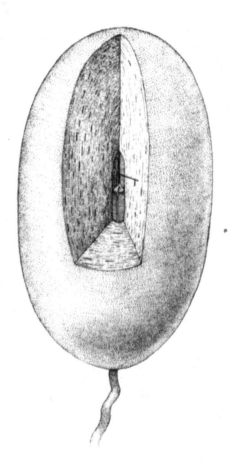

Fig. 6d

Although Pacinian and Herbst corpuscles are similar in appearance, both the size and ultrastructural organization of the two receptors are considerably different. Any mechanical model for transmission or transduction in Herbst or Pacinian corpuscles must consider both types of receptors. McIntyre and his group (Dorward et al., 1966) have found the frequency response and adaptive characteristics of Pacinian and Herbst corpuscles to be approximately equivalent to one another. The most significant difference between the two receptors is that the Herbst corpuscles will respond to a somewhat lower frequency (20 as com-

pared with 50 to 70 cycle/sec for Pacinian corpuscles). These differences in structure may be important in terms of evaluating the theories on the mechanism
of transduction (KATZ, 1950; GRAY and SATO, 1953; LOEWENSTEIN, 1961, 1965).

The Herbst corpuscle is much smaller than a typical Pacinian corpuscle; a
comparison of Figs. 6a and d with Figs. 7a and c will illustrate this difference.
The overall diameters are as follows: Pacinian corpuscles, approximately 500 μ
long; and Herbst corpuscles, approximately 200 μ. The Pacinian cores measure
approximately 30 μ in diameter, and Herbst, approximately 20 μ. The central
sensory neurites are similar in diameter, about 3 to 5 μ.

The Herbst corpuscle lacks cellular lamellae between the capsule and core
(Fig. 7a, 7c). The region between core and capsule is filled with connective tissue
components which by electron microscopy and staining characteristics appear
to be large fibers of elastica (MUNGER, *unpublished observation*) admixed with
scattered collagen fibrils (Fig. 7b). Scattered wisps of fibroblast cytoplasm are
present, but no cellular lamellae can be identified. The fibrous components form
an interlacing concentric meshwork. but they do not form continuous sheets or
lamellae (CLARA, 1925).

The core complexes are also somewhat different, but the common feature
of the thin concentric cytoplasmic lamellae, intimately associated with the sensory
neurite, is probably the important cytologic specialization. In the Herbst corpuscle, two lamellar cells straddle the neurite with numerous thin cytoplasmic
lamellae extending outward from the body of the lamellar cell (Fig. 7b). The
processes of one cell interdigitate with those of the opposite cell. Where the lamellae
abut the neurite, junctional complexes can be found, especially at the end of the
neurite and the first portion of the lamellar casing. The plane of bilateral symmetry
is in a plane transecting the nuclei of the lamellar cells. In Pacinian corpuscles,
the plane of bilateral symmetry is unrelated to the nuclear regions of the satellite
cells (PEASE and QUILLIAM, 1957). This difference may be related to the fact that
more than two cells are involved in producing the lamellar core in any given
region of a Pacinian corpuscle. The plane of bilateral symmetry can be correlated
with the ditectional sensitivity of compression producing depolarization or
hyperpolarization as described by NISHI and SATO (1968).

The Meissner corpuscle (MEISSNER, 1853) differs from Pacinian and Herbst
corpuscles in that the majority of this corpuscle is cellular (Fig. 8a; BOEKE,
1932; CAUNA, 1958; CAUNA and MANNAN, 1959; CAUNA and Ross, 1960). The
major cellular element is a lamellar cell, similar in many respects to the lamellar
cell of the core of Pacinian and Herbst corpuscles. The lamellar cell gives rise
to leaves of cytoplasm stacked one upon the other with scant, intervening,
connective tissue components (Fig. 8b). The sensory neurites, several of which
may enter a single corpuscle, follow a tortuous course through the stacked lamellae
(Fig. 8b, 8c; CAUNA and Ross. 1960), and the entire complex is located at the
apex of a dermal papilla. The neurite usually contains accumulations of mitochondria, as has been described in other sensory terminals. The neurite is always
intimately ensheathed by cytoplasmic processes of the lamellar cell.

The nature of the lamellar cells in Meissner corpuscles has been disputed in
the past (BOEKE, 1932; Discussion in Ciba Foundation Symposium, *Touch, Heat
and Pain*, pp., 132—137). By electron microscopy, they appear to be epithelial

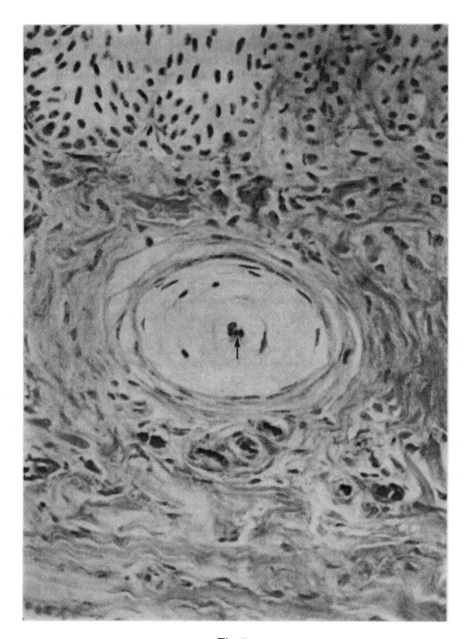

Fig. 7a

Fig. 7a—c. a) Herbst corpuscle from duck bill skin [silver method of SEVIER and MUNGER (1965)]. The corpuscle is much smaller than a Pacinian corpuscle (Fig. 6a), and the central blackened neurite (arrow) is encased on each side by the nucleus (darkened circle) and cytoplasm (not clearly depicted) of the central laminar cells. A few scattered fibroblasts are present in the corpuscle; the remainder of the contents appear amorphous. × 256. b) Herbst corpuscle from duck bill skin. The central neurite (N) is enveloped by successive sheets of thin cytoplasmic lamellae. No cleft is seen here as one would expect with Pacinian corpuscles. Outside the core, scattered fibroblasts (F) and strands of elastica (arrow) can be seen loosely enveloping the

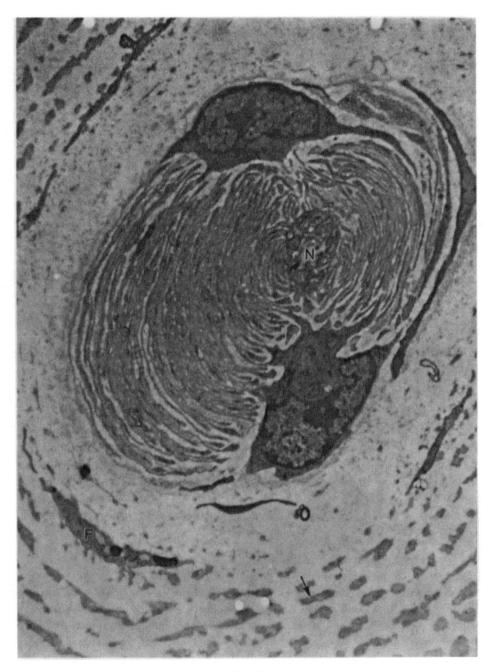

Fig. 7b

core. × 7,700. c) Idealized drawing of a Herbst corpuscle. This drawing illustrates the specialized features of a Herbst corpuscle as seen in the electron microscope. The central neurite is
seen in cross section, and the cells of the central core are disposed bilaterally. Cytoplasmic
extensions (arms) of these two cells produce the cellular lamellae of the core. The remainder
of the corpuscle consists of scattered cells, connective tissue fibers, and associated matrix

cells most likely derived from Schwann cells (KRAUSE, 1881; CAUNA and ROSS, 1960). The connective tissue space is well defined with scattered collagen fibrils as well as bundles of long-spacing collagen present. A basal lamina is prevent at the junction of the lamellar cell cytoplasm and connective tissue space. The entire corpuscle is invested with an indistinct capsule and connective tissue sheath.

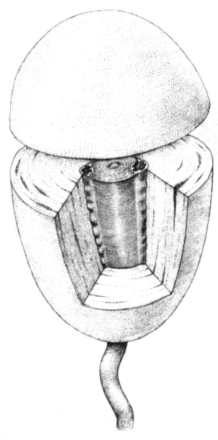

Fig. 7 c

When viewed in longitudinal section (MUNGER, *unpublished observations;* CAUNA and ROSS, 1960), the stacked cytoplasmic lamellae and neurites of a Meissner corpuscle appear amazingly similar to the imaginary appearance of a consolidated lanceolate (palisade) ending of a hair follicle. If the flattened neurites and associated flattened discs of Schwann cell cytoplasm, as present in the lanceolate endings on a hair follicle, were removed from the hair and stacked one upon the other, the result would appear remarkably similar to a Meissner corpuscle. Small regions of a Meissner corpuscle also appear superficially similar to genital end bulbs and the corpuscular endings found in glabrous skin of the snout of cats, dogs, opossums, and other animals (WINKELMANN, 1957); this is discussed below. These similarities emphasize the point of view expressed by WEDDELL *et al.* (1955) that gradations in the form of any given receptor can be found.

If a Meissner corpuscle is even vaguely regarded as morphologically similar to a lanceolate ending of a hair follicle, we should look for similar functional

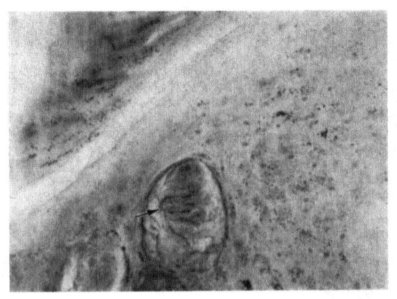

Fig. 8a

Fig. 8a—c. a) Meissner corpuscle from index finger skin of Macaque monkey [silver method of SEVIER and MUNGER (1965)]. The corpuscle is present high in a dermal papilla and consists of a compact mass of cells with the neurites (arrow) winding back and forth through the complex. × 512. b) Meissner corpuscle from Ateles finger pad skin. The entering neurite (lower left) enters the base of the corpuscle and becomes ensheathed with thin sheets of cytoplasm stacked one on top of the other. The neurites (*N*) are thus viewed in longitudinal section, and are seen to contain numerous mitochondria. Many pinocytotic vesicles are present in the lamellar cell cytoplasm (compare with Fig. 2b). The skin surface is to the top of the micrograph. × 20,000. c) Idealized drawing of a Meissner corpuscle. This illustrates the configuration of a Meissner corpuscle as seen in a silver-stained preparation. The window illustrates the flattened neurite and associated Schwann cells. The general arrangement of these two components is similar to that seen in Fig. 2d

capacity. Although to date the physiologic response of a single Meissner corpuscle has not been differentially analyzed, apart from other receptors in finger pad skin, numerous rapidly adapting receptors (possibly Meissner corpuscles) can be characterized in primate finger pad skin (IGGO, 1963b). The lanceolate ending of hair follicles are also rapidly adapting (FITZGERALD, 1940; IGGO, 1963b; WERNER and MOUNTCASTLE, 1965). We might speculate that a Meissner corpuscle is a morphologic specialization of glabrous skin functionally equivalent to the innervation of classical sensory hairs. This thesis has been expounded by DASTUR (1955), based on a somewhat different argument, as well as by IGGO (1963b), based on physiological experiments.

Genital end bulbs are somewhat simpler than Meissner corpuscles in terms of the basic pattern of organization (SETO, 1963; PATRIZI and MUNGER, 1966). A somewhat loosely coiled neurite (Fig. 9a) is intimately associated with a lamellar cell, which tightly invests the neurite terminal with a thin lamella of cytoplasm. Other cytoplasmic lamellae are stacked between successive coils of the neurite

Fig. 8b

(Fig. 9b). Basement membrane material and connective tissue surround the individual cytoplasmic lamellae. As is the case with other sensory receptors, accumulations of mitochondria are frequently found in expanded portions of the terminal. This receptor is similar in some respects to Meissner corpuscles and also to the core of Pacinian corpuscles in that the lamellar cell appears to be the distinct cytologic specialization and may implicate rapidly adapting characteristics.

Corpuscular endings resembling genital end bulbs have been described by many terms (KANTNER, 1952; SETO, 1963): *end bulbs of Krause* [KRAUSE, 1862 (despite their location); GAIRNS, 1953]; *Dogiel's end bulbs* (DOGIEL, 1893); *mucocutaneous end organ of Winkelmann* (WINKELMANN, 1957, 1960, 1967); *Ruffini end*

Fig. 8 c

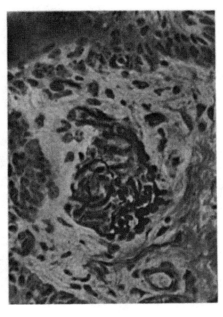

Fig. 9 a

Fig. 9a and b. a) Genital end bulb from rat penis [silver method of SEVIER and MUNGER (1965)]. A tangled skein of blackened neurites exhibits a compact appearance. Cellular details are difficult to decipher. × 512. b) Genital end bulb from rat penis. The neurites (*N*) are encased by and separated by successive sheets of thin cytoplasm of a lamellar cell. These sheets of cytoplasm evidence well-defined basal lamina (arrows). The capsular cell (*C*) lacks a distinct basement membrane. The neurites contain numerous mitochondria. × 24,000

bulbs, (DOGIEL, 1903); *innominate end organs* (QUILLIAM, 1966); and *Golgi-Mazzoni corpuscles* (DOGIEL, 1903). These names may designate receptors with minor variations on the theme as described structurally for genital end bulbs and Meissner corpuscles. In this regard, it is interesting note that DOGIEL (1893) described genital end bulbs as a form of Meissner corpuscle. We lack information on the electron microscopy of each of these units, mainly because of the difficulty in identifying and localizing them at the light microscopic level. WEDDELL and his group (WEDDELL *et al.*, 1955) and MALINOVSKY (1968) are strongly of the opinion that receptors cannot be considered unique entities; rather they merge into one another at a morphologic level: I share this opinion. The ultrastructure of a genital end bulb of the rat penis (PATRIZI and MUNGER, 1965) is quite similar to a coiled receptor found in the dermis of opossum and cat snout (glabrous) skin (Fig. 10a, 10b) (MUNGER, *unpublished observations*). The corpuscle found by ANDRES (1966)

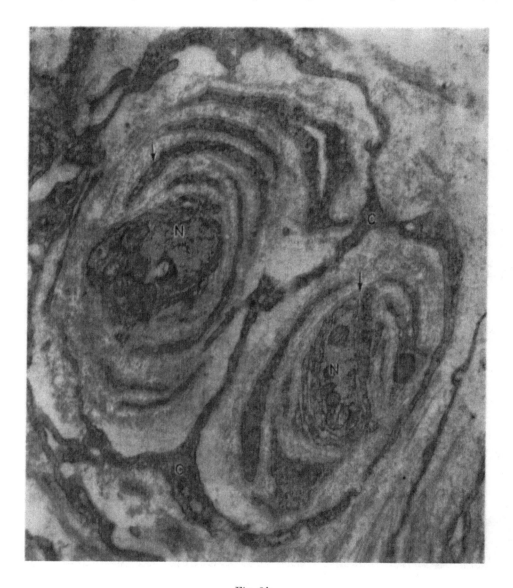

Fig. 9b

in the connective tissue surrounding a vibrissae is also similar to the above. All corpuscular endings described to date (except the Grandry epithelial cell-neurite complexes) are characterized by the association of lamellar cells (i. e., cells which produce stacks of thin cytoplasmic lamellae) enveloping the sensory neurite. This common denominator may prove to be *insignificant* on the basis of future studies, but it is suggestive of a distinct functional similarity. Since those corpuscular endings studied physiologically to date are rapidly adapting, the presence of lamellar cells may indicate rapidly adapting characteristics. The apparent histological and cytological differences of these numerous eponymic endings are then trivial.

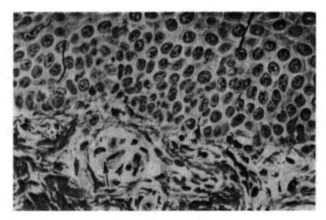

Fig. 10a

Fig. 10a and b. a) Corpuscular ending and intraepidermal neurites from opossum snout skin [silver method of SEVIER and MUNGER (1965)]. The arrow indicates a portion of a corpuscular ending. The blackened central neurite is surrounded by an ill-defined material presumed to be cytoplasm of the enveloping lamellar cell (see Fig. 10b). × 400. b) Corpuscular ending in the dermis beneath a rete peg from opossum snout skin. A central neurite is surrounded by successive cytoplasmic lamellae. The nucleus of a lamellar cell is present to the upper right. The lamellar cell and its cytoplasmic processes are associated with basal lamina material × 18,200

E. Muscle Spindles and Golgi Tendon Organs

Muscle spindles and Golgi tendon organs comprise a distinct type of sensory nerve ending which possesses a most intricate pattern of innervation. The present discussion is restricted to the sensory component. The description of MERRILLEES (1960) is most complete, and workers unfamiliar with this field should consult this source for a complete description. Based on the cytological observations of RUFFINI (1897, 1898–1899) and on the detailed studies of SHERRINGTON (1894–1895), HINSEY (1929), HINES and TOWER (1928), and KATZ (1961), two populations of sensory endings can be defined on a muscle spindle, one of which is also found in Golgi tendon organs (the "flower ending"). The two sensory endings on muscle spindles were termed primary and secondary by RUFFINI (1898–1899). The primary endings form a broad band that wraps itself around the muscle cell body in the equatorial region producing the so-called "annulo-spiral" ending. The neurites occupy an indentation in the surface of the intrafusal fiber (MERRILLEES, 1960; Fig. 11). The secondary endings overlap the primary to some degree and are described by RUFFINI as resembling a "flower-wreath" (RUFFINI, 1897; Fig. 11). The use of the term "flowerlike" has been retained in the literature (HINES and

Fig. 10b

TOWER, 1928) and does describe the appearance of metallic impregnations of both
Golgi tendon organs and the secondary ending in muscle spindles. Some secondary
endings have been described by BARKER (1967) to be annulo-spiral endings.

Electron microscopic studies to date (MERRILLEES, 1960; KARLSSON *et al.*, 1966) describe the primary endings as a relatively undifferentiated neurite closely applied to the muscle cell. The complex lacks a cellular investment. No significant cytologic specializations are seen in the neurite. The basal lamina of the

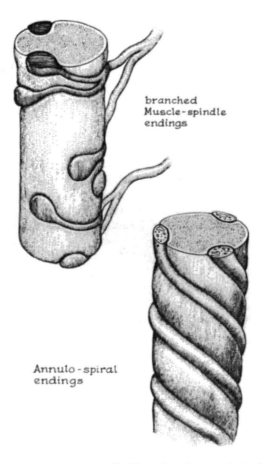

branched
Muscle-spindle
endings

Annulo-spiral
endings

Fig. 11. Idealized drawing of a muscle spindle. These drawings are derived from the description in the text. The two classes of sensory endings differ in cytologic detail, but they resemble one another in many ways; e. g., both consist of sensory neurites of similar appearance indented into the muscle fiber

muscle cell envelops the complex, and no basal lamina is present between neurite and muscle cell. The neurite contains numerous mitochondria, as do other sensory receptors, but other specific cytologic specializations have not been described. The "flower endings" of RUFFINI are described by MERRILLEES (1960) as somewhat broader expanses of neurite as compared to the annulo-spiral endings, and they frequently contain dense lipid material, as has been described in other sensory endings (MUNGER, 1965). Both annulo-spiral and flower endings lack Schwann cell investment (MERRILLEES, 1960; KARLSSON *et al.*, 1966), unlike other sensory endings. Golgi tendon organs have not been studied by electron microscopy to date. An interesting point for future study is the observation of Ruffini (1897), recently confirmed by Barker (1967), that small corpuscular endings (Paciniform endings) are sometimes found within the capsule of a Golgi tendon organ.

The sensory nature of these endings has been clearly established (Sherrington, 1894—1895; Ruffini, 1898—1899). The spindles and Golgi tendon organs have opposite effects in terms of sensing muscular contraction (reviewed by Granit, 1955). Stretch of the intrafusal fiber fires the spindle, and contraction fires the Golgi tendon organ. We unfortunately do not possess sufficient cytologic information even to attempt to correlate cell structure and function in these two highly specialized systems.

Receptors of similar form are present in joint capsules. This subject has recently been thoroughly reviewed by Poláček (1966). He describes three basic types of receptors associated with mammalian and avian joint capsules: (1) free nerve endings, (2) spray- or clew-like endings, and (3) Paciniform or Herbst endings (with a central core and distinct lamellae). Poláček also emphasized the extent of overlap and variability in form of any given type of receptor. Based on the studies of Wyke (Wyke, 1962; Freeman and Wyke, 1967; Kirchner and Wyke, 1965). The corpuscular receptors of joints have been characterized as rapidly adapting the spray endinas slowly adapting. Correlated ultrastructural studies have not been done.

F. Receptors Unknown

The specific receptors for temperature, pain, and some types of pressure cannot be identified at the present time. Hot and cold spots can definitely be identified, to the point that slight cooling or warming a small circumscribed area produces firing of a single afferent (Zotterman, 1959; Murray, 1962; Hensel, 1966). The response of single fibers to temperature change has also been studied in man (Hensel and Boman, 1960), where large myelinated fibers innervated skin areas responsive to cooling. The nature of the receptor involved is not known. So-called "warm" fibers, at least in some instances, are small C fibers. As such, they are more difficult to study and to characterize. We also known nothing of the receptors involved in this instance.

The classic "warm" or "infra-red" receptor is the facial pit of crotalid snakes (Bullock and Diecke, 1956; Bullock and Fox, 1957). This receptor is a thin highly innervated membrane stretched across the middle of a cup-shaped depression in the face of the snake. The ultrastructure of this membrane has been described by Bleichmar and De Robertis (1962). The neurites terminate as free nerve endings in the midst of an amorphous connective tissue matrix. No distinct structural specialization is present. The neurites are not intimately related to a specialized cell, and thus they appear to be truly a free nerve terminal. The neurites contain accumulations of mitochondria as described above for virtually every other receptive terminal. This feature is not distinctive.

However, this system does provide an example of a "free nerve ending" (Weddell and Miller, 1962), definitely characterized as such by electron microscopy. In the pit membrane, the naked neurite terminal is presumably responsible for the events of transduction (MURRAY, 1962, 1966). If such a terminal were present in mammalian skin, positive identification by electron microscopy would be virtually impossible, utilizing present technology. The description of BLEICH-MAR and DE ROBERTIS (1962) gives added emphasis to the concept of WEDDELL that the *undistinguished* array of neurites in skin may well be involved in modality-specific reception. How such a system functions has been considered hypothetically on numerous occasions.

Our discussion has completed a full circle. I originally proposed that our goal was the identification of modality-specific receptors in order to understand better the events of transduction and input of sensory information to the cenral nervous system. It was stated at the onset that the tangled skein of cutaneous nerves presented a bewildering, perhaps incomprehensible, array of neutral elements. In this neural network, certain distinctive receptors can be defined, and these receptors might provide clues as to common denominators of cellular function. I have closed finally with a brief discussion of modalities lacking defined receptors in mammalian skin. Future studies will be directed toward untangling further the array of neural elements involved in peripheral sensory fields.

References

ADRIAN, E. D., ZOTTERMAN, Y.: Impulses produced by sensory nerve endings, Part 3. Impulses set up by touch and pressure. J. Physiol. (Lond.) **61**, 465—483 (1926).

ANDERSEN, A. E., NAFSTAD, P. H. J.: An electron microscopic investigation of the sensory organs in the hard palate region of the hen *(Gallus domesticus)*. Z. Zellforsch. **91**, 391—401 (1968).

ANDRES, K. H.: Über die Feinstruktur der Rezeptoren an Sinushaaren. Z. Zellforsch. **75**, 339—365 (1966).

BARKER, D.: The innervation of mammalian skeletal muscle. In: Myotatic, Kinesthetic and Vestibular Mechanisms, Ciba Foundation Symposium. A. V. S. de Reuck and J. Knight, Eds. pp. 3—14. Boston: Little, Brown & Co. 1967.

BARKER, D. S., WELKER, W. I.: Receptive fields of first-order somatic sensory neurons innervating rhinarium in coati and racoon. Brain Res. **14**, 367—386 (1969).

BLEICHMAR, H., de ROBERTIS, E.: Submicroscopic morphology of the infra-red receptor of pit vipers. Z. Zellforsch. **56**, 748—761 (1962).

BOEKE, J.: Nerve endings, motor and sensory. In: Cytology and Cellular Pathology of the Nervous System. W. Penfield, Ed. pp. 243—315. New York: Hoeber 1932.

— Problems of Nervous Anatomy. Oxford: University Press 1940.

BOTEZAT, I. E.: Die Nerven der Epidermis. Anat. Anz. **33**, 45—75 (1908).

BULLOCK, T. H., DIECKE, F. P. J.: Properties of an infra-red receptor. J. Physiol. (Lond.) **134**, 45—87 (1956).

— Fox, W.: The anatomy of the infra-red sense organs in the facial pit of pit vipers. Quart. J. micr. Sci. **98**, 219—234 (1957).

CAUNA, N.: Structure of digital touch corpuscles. Acta anat. (Basel) **32**, 1—23 (1958).

— The fine motphology of the sensory receptor organs in the auricle of the rat. J. Comp. Neurol. **136**, 81—98 (1969).

— HINDERER, K. H., WENTGES, R. T.: Sensory receptor organs of the human nasal respiratory mucosa. Am. J. Anat. **124**, 187—210 (1969).

— MANNAN, G.: Development and postnatal changes of digital Pacinian corpuscles in the human hand. J. Anat. (Lond.) **93**, 271—286 (1959).

— Ross, L. L.: The fine structure of Meissner's touch corpuscles of human fingers. J. biophys. biochem. Cytol. 8, 467—482 (1960).

Ciba Foundation Symposium: Touch, Heat and Pain. A. V. S. de Reuck & J. Knight, Eds. London: J. & A. Churchill, Ltd. 1966.

CLARA, M.: Über den Bau des Schnabels der Waldschnepfe. Beiträge zur Kenntnis der Lamellenkörperchen. Z. mikro.-anat. Forsch. 3 (1925).

DASTUR, D. K.: Cutaneous nerves in leprosy. The relationship between histopathology and cutaneous sensibility. Brain 78, 615—633 (1955).

DOGIEL, A. S.: Die Nervenendigungen in der Haut der äußeren Genitalorgane des Menschen. Z. mikr.-anat. Forsch. 41, 585—612 (1893).

— Über die Nervendapparate in der Haut des Menschen. Z. Wiss. Zool. 75, 46—110 (1903).

DORWARD, P., McINTYRE, A. K., PROSKE, U.: Discussion in: Touch, Heat and Pain (Ciba Foundation Symposium). p. 112. London: J. & A. Churchill, Ltd. 1966.

DUTHIE, H. L., GAIRNS, F. W.: Sensory nerve endings and sensation in the anal region of man. Brit. J. Surg. 47, 585—595 (1960).

FITZGERALD, O.: Discharges from the sensory organs of the cat's vibrissae and the modification in their activity by ions. J. Physiol. (Lond.) 98, 163—178 (1940).

FREEMAN, M. A. R., WYKE, B.: The innervation of the knee joint. An anatomical and histological study in the cat. J. Anat. 101, 505—532 (1967).

GAIRNS, F. W.: Sensory endings other than taste buds in the human tongue. J. Physiol. (Lond.) 121, 33—34 (1953).

GRANDRY, M.: Recherches sur les corpuscles de Pacini. J. Anat. (Paris) 6, 390—395 (1869).

GRANIT, R.: Receptors and Sensory Perception. New Haven: Yale University Press 1955.

HENSEL, H.: Classes of receptor units predominantly related to thermal stimuli. In: Touch, Heat, and Pain (Ciba Foundation Symposium). pp. 275—290. London: J. & A. Churchill, Ltd. 1966.

— BOMAN, K. K. A.: Afferent impulses in cutaneous sensory nerves in human subjects. J. Neurophysiol. 23, 564—578 (1960).

HINES, M., TOWER, S. S.: Studies on the innervation of skeletal muscles. II. of muscle spindles in certain muscles of the kitten. Bull. Johns Hopkins Hosp. 42, 264—307 (1928).

HINSEY, J. C.: Some observations on the innervation of skeletal muscle of the cat. J. comp. Neurol. 44, 87—195 (1927—28).

IGGO, A.: New specific sensory structures in hairy skin. Acta neuroveg. (Wien) 24, 175—180 (1963a).

— An electrophysiological analysis of afferent fibers in primate skin. Acta neuroveg. (Wien) 24, 225—240 (1963b).

— MUIR, A. R.: A cutaneous sense organ in the hairy skin of cats. (abstract) J. Anat. (Lond.) 97, 151 (1963).

de IRALDI, A. P., RODRIGUEZ-PÉREZ, A. P.: Ultrastructura de las corpúsculos de Grandry. Trab. Inst. Cajal Invest. biol. 53, 185—188 (1961).

KADANOFF, D.: Beiträge zur Kenntnis der Nervenendigungen im Epithel der Säugetiere. Z. Anat. Entwickl.-Gesch. 73, 431—452 (1924).

— Über die intraepithelialen Nerven und ihre Endigungen beim Menschen und bei den Säugetieren. Z. Zellforsch. 7, 553—576 (1928).

KANTNER, M.: Studien über den sensiblen Apparat in der glans Penis. Anat. Anz. 99, 159—179 (1952).

KARLSSON, U., ANDERSSON-CEDERGREN, E., OTTOSON, D.: Cellular organization of the frog muscle spindle as revealed by serial sections for electron microscopy. J. Ultrastruct. Res. 14, 1—35 (1966).

KATZ, B.: Depolarization of sensory terminals and the initiation of nerve impulses in muscle spindles. J. Physiol. (Lond.) 111, 261—282 (1950).

— The terminations of the afferent nerve fibre in the muscle spindle of the frog. Phil. Trans. B. 243, 221—240 (1961).

KIRCHNER, J. A., WYKE, B. D.: Articular reflex mechanisms in the larynx. Ann. Oto. Rhino. Laryn. 74, 1—20 (1965).

KRAUSE, W.: Recherches sur L'anatomie et la physiologie de la conjonctive. J. Physiol. (Paris) 5, 296—307 (1862).

— Die Nervenendigung innerhalb der terminalen Körperchen. Arch. mikr. Anat. 19, 53—136 (1881).

LOEWENSTEIN, W. R.: Excitation and inactivation in a receptor membrane. Ann. N. Y. Acad. Sci. 94, 510—534 (1961).

LOEWENSTEIN, W. R.: Facets of a transducer process. Cold Spr. Harb. Symp. Quant. Biol. XXX, 29—43 (1965).
— MENDELSON, M.: Components of receptor adaptation in a Pacinian corpuscle. J. Physiol. (Lond.) 177, 377—397, (1965).
— RATHKAMP, R.: The sites for mechano-electric conversion in a Pacinian corpuscle. J. gen. Physiol. 41, 1245—1265 (1958).
— SKALAK, R.: Mechanical transmission in a Pacinian corpuscle. An analysis and a theory. J. Physiol. (Lond.) 182, 364—378 (1966).
MALINOVSKY, L.: Die Nervenendkörperchen in der Haut von Vögeln und ihre Variabilität. Z. mikr.-anat. Forsch. 77, 279—303 (1967).
— Types of sensory corpuscles common to mammals and birds. Folia Morph. (Warszawa) 16, 67—73 (1968).
MANN, S. J., STRAILE, W. E.: Tylotrich (hair) follicle: Association with a slowly adapting tactile receptor in the cat. Science 147, 1043—1045 (1963).
MEISSNER, G.: Beiträge zur Anatomie und Physiologie der Haut. Leipzig: Voss 1853.
MERKEL, F.: Die Tastzellen der Ente. Arch. mikr. Anat. 15, 415—427 (1878).
— Über die Endigungen der sensiblen Nerven in der Haut der Wirbelthiere. Rostock: H. Schmidt 1880.
MERRILLEES, N. C. R.: The fine structure of muscle spindles in the lumbrical muscles of the rat. J. biophys. biochem. Cytol. 4, 725—742 (1960).
MILLER, M. R., RALSTON, H. J. III, KASAHARA, M.: The pattern of cutaneous innervation of the human hand. Amer. J. Anat. 102, 183—217 (1958).
— — — The pattern of cutaneous innervation of the human hand, foot and breast. Advanc. Biol. Skin 1, 1—47 (1960).
MONTAGNA, W., ELLIS, R. A.: The vascularity and innervation of human hair follicles. In: The biology of hair growth. W. Montagna and R. A. Ellis, Eds. pp. 189—219. New York: Academic Press Inc. 1958.
MUNGER, B. L.: The intraepidermal innervation of the snout skin of the opossum. A light and electron microscopic study, with observations on the nature of Merkel's Tastzellen. J. Cell Biol. 26, 79—96 (1965).
MURRAY, R. W.: Temperature receptors. Advanc. comp. Physiol. Biochem. 1, 117—175 (1962).
— Nerve membrane properties and thermal stimuli. In: Touch, Heat and Pain. (Ciba Foundation Symposium) pp. 164—182. London: J. & A. Churchill, Ltd. 1966.
ORFANOS, C.: Elektronenmikroskopischer Nachweis epithelio-neuraler Verbindungen (Mechano-Receptoren) am Haarfollikelepithel des Menschen. Arch. klin. exp. Derm. 228, 421—429 (1967).
OZEKI, M., SATO, M.: Initiation of impulses at the non-myelinated nerve terminals in Pacinian corpuscles. J. Physiol. (Lond.) 170, 167—185 (1964).
PATRIZI, G., MUNGER, B. L.: The cytology of encapsulated nerve endings in the rat penis. J. Ultrastruct. Res. 13, 500—515 (1965).
— — The ultrastructure and innervation of rat vibrissae. J. comp. Neurol. 126, 423—436 (1966).
PEASE, D. C., QUILLIAM, T. A.: Electron microscopy of the Pacinian corpuscle. J. biophys. biochem. Cytol. 3, 331—342 (1957).
PINKUS, F.: Ueber einen bisher unbekannten Nebenapparat am Haarsystem des Menschen: Haarscheiben. Derm. Z. 9, 465—469 (1902).
— Über Hautsinnesorgane neben dem menschlichen Haar (Haarscheiben) und ihre vergleichend-anatomische Bedeutung. Arch. mikr. Anat. 65, 121—179 (1905).
QUILLIAM, T. A.: Unit design and array patterns in receptor organs. In: Touch, Heat and Pain (Ciba Foundation Symposium). pp. 86—112. London: J. & A. Churchill, Ltd. 1966.
RANVIER, L.: On the terminations of nerves in the epidermis. Quart. J. micr. Sci. 20, 456—458 (1880).
RETZIUS, G.: Über die Nervendigungen an den Haaren. Biol. Unters. 4, 45—48 (1892).
RICHARDSON, K. C.: Studies in the structure of autonomic nerves in the small intestine, correlating the silver-impregnated image in light microscopy with the permananate-fixed ultrastructure in electron microscopy. J. Anat. (Lond.) 94, 457—472 (1960).

Ruffini, A.: Observations on sensory nerve endings in voluntary muscles. Brain **20**, 368—374 (1897).
— On the minute anatomy of the neuromuscular spindles of the cat, and on their physiological significance. J. Physiol. (Lond.) **23**, 190—208 (1898—99).
Schumacher, V.: Beiträge zur Kenntnis des Baues und die Funktion der Lamellenkörperchen. Arch. mikr. Anat. **77**, 157—193 (1911).
Seto, H.: Studies on the sensory innervation. Tokyo: Isaku Shoin 1963.
Sevier, A. C., Munger, B. L.: A silver method applicable to paraffin sections of formol-fixed tissue. J. Neuropath. exp. Neurol. **24**, 130—135 (1965).
— — The use of Oxone to facilitate tissue stainability for light microscopy. Anat. Rec. **162**, 43—48 (1968).
Shanthaveerappa, T. R., Bourne, G. H.: A simple method for preparation and staining of the whole Pacinian corpuscle. Acta Anat. (Basel) **60**, 199—206 (1955).
— — New observations on the structure in Pacinian corpuscles and its relationship to the perineural epithelium of peripheral nerves. Amer. J. Anat. **112**, 97—109 (1963).
Sherrington, C. S.: On the anatomical constitution of nerves of skeletal muscles; with remarks on recurrent fibers in the ventral spinal nerve root. J. Physiol. (Lond.) **17**, 211—258 (1894—95).
Smith, K. R.: The structure and function of *Haarscheibe*. J. comp. Neurol. **131**, 459—474 (1967).
— The ultrastructure of the human *Haarscheibe* and Merkel cell. J. Invest. Derm. **54**, 150—159 (1970).
— Creech, B. J.: Effects of pharmacological agents on the physiological responses of hair discs. Expl. Neurol. **19**, 477—482 (1967).
Stöhr, P.: Das Peripherische Nervensystem. In: Handbuch mikroskopische Anatomie des Menschen. W. von Mollendorf IV, Ed. pp. 202—423. Berlin: J. Springer 1928.
Straile, W. E.: Sensory hair follicles in mammalian skin: The tylotrich follicle. Amer. J. Anat. **106**, 133—148 (1960).
Symonowicz, L.: Über die Nervenendigungen in den Haaren des Menschen. Arch. mikr. Anat. **74**, 622—634 (1909).
Tapper, D. A.: Stimulus-response relationships in the cutaneous slowly-adapting mechano-receptor in hairy skin of the cat. Expl. Neurol. **13**, 364—385 (1965).
Tretjakoff, D.: Zur Frage der Nerven der Haut. Z. wiss. Zool. **71**, 625—643 (1902).
— Die Nervendigungen an den Sinus Haaren des Kindes. Z. wiss. Zool. **97**, 314—416 (1911).
Vincent, S. B.: The tactile hair of the white rat. J. comp. Neurol. **23**, 1—34 (1913).
Weddell, G.: The anatomy of cutaneous sensibility. Br. Med. Bull. **3**, 167—172 (1945).
— Studies related to the mechanism of common sensibility. In: Advances in Biology of Skin W. Montagna, Ed. pp. 112—160. New York: Pergamon Press 1960.
— Miller, S.: Cutaneous sensibility. Ann. Rev. Physiol. **24**, 199—222 (1962).
— Palmer, E., Pallie, W.: Nerve endings in mammalian skin. Biol. Rev. **30**, 159—195 (1955).
Werner, G., Mountcastle, V. B.: Neural activity in mechanoreceptive cutaneous afferents: Stimulus-response relationships, Weber functions, and information transmission. J. Neurophysiol. **28**, 359—397 (1965).
Winkelmann, R. K.: The muco-cutaneous end-organ. Arch. Derm. **76**, 225—235 (1957).
— Nerve Endings in Normal and Pathologic Skin. Springfield, Ill.: C. C. Thomas 1960.
— Cutaneous nerves. In: Ultrastructure of Normal and Abnormal skin. A. S. Zelickson, Ed. pp. 202—227. Philadelphia: Lea and Febiger 1967.
Wyke, B.: The neurology of joints. Ann. Roy. College Surg. **41**, 25—50 (1962).
Yamamoto, T.: On the sensory innervation of the hair follicle in mice. In: Electron microscopy, Vol. II. p. 515. Ryozi Uyeda, Ed. Tokyo: Maruzen 1966.
Zotterman, Y.: Thermal sensations. In: Handbook of Physiology, Vol. 1, Sec. 1. pp. 431—458. Washington, D. C.: American Physiological Society (1959).
Zucker, E., Welker, W. I.: Coding of somatic sensory input by vibrissae neurons in the rat's trigeminal ganglion. Brain Res. **12**, 138—156 (1969).

Author Index

Page numbers in *italics* refer to the bibliography

Davis, H.. see Stevens, S. S.
231, *242*
— see Tasaki, I.   409, 427,
441
— see Teas, D. C.   420, *441*
Davis, L., jr., Lorente de no,
R.   137, *161*
Davison, A. N.   45, *93*
— see Aldridge, W. N.   44,
91
Dawson, W. W.   348, *364*
Deal, S. E., see Walsh, R. R.
86, *101*
Deal, W. J., Erlanger, B. F.,
Nachmansohn, D.   85, *93*
Deatherage, B. H., see Smith,
C. A.   421, *440*
DeFries, H., see Jielof, R.
399, 410, 417, *439*
Delbrück, M., Reichardt, W.
372, 373, 381, *392*
DelCastillo, J., Katz, B.   244,
266, 272, 287
DeLong, R. G., Coulombre, A.
J.   176, *185*
DeLorenzo, A. J. D.   158, *161*
— Dettbarn, W. D., Brzin, M.
29, *93*
Dennison, D. S.   374, *392*
Desmedt, J. E.   201, 204, 207,
223
Desroche, P.   370, 371, 374,
392
Dethier, V. G.   136, 146, *162*
Dethrage, B. H.   435, *437*
Dettbarn, W.-D.   30, 31, *93*
— Bartels, E., Hoskin, F. C.
G., Welsch, F.   46, *93, 94*
— Davis, F. A.   34, 35, *93*
— Rosenberg, P.   42, 87, *93*
— — Nachmansohn, D.   *37,
94*
— see Brzin, M.   29, 32, 38,
39, *92*
— see DeLorenzo, A. J. D.
29, *93*
— see Rosenberg, P.   *37, 100*
Detwiler, S. R.   174, *185*
Deutsch, S.   200, *223*
DeValois, R.   218, *223*
— Jacobs, G. H., Jones, A. E.
218, *223*
DeValois, R. L.   237, *241*
— Jacobs, G. H., Jones, A. E.
237, *241*

Diamant, H., see Borg, G.
199, 201, 203, 204, 205,
223, 233, 241
Diamond, A. L. 232, *241*
Diamond, J., Gray, J. A. B.,
Inman, D. R.   272, 277,
279, *287*, 472, 491, *496*
— Gray, J. A., Sato, M.   278,
287
— see Gray, J. A. B.   *288,
490, 497*
Diamond, R., see Perutz, M. F.
64, 73, *99*
Dickens, F., see Burgen, A. S.
V.   *130*
Diecke, E. P. J., see Bullock,
T. H.   552, *553*
Diecke, F. P. J., see Bullock,
F. H.   434, *436*
Diehn, B., Tollin, G.   369,
370, 371, 374, 382, 383,
385, 387, 388, *392, 393*
Diehn, D., s. Lindes, D.   370,
383, *394*
Dijkgraaf, S.   136, *162*, 401,
416, *437*
Dijkstra, C.   170, *186*
Dirac, P. A. M.   257, *266*
Dodge, F. A., see Purple, R. L.
139, 142, 151, 159, *164*,
198, *225*
Dodt, E., Echte, K.   352, *364*
Döving, K. B.   198, *223*
Dogiel, A. S.   526, 530, 547,
554
Dohlman, G.   423, *437*
— Ormerod, F. C., McLay, K.
423, *437*
Doolittle, R. F., see Singer, S.
J.   79, *100*
Doran, R., see Schuknecht, H.
F.   432, *440*
Dorward, P., McIntyre, A. K.,
Proske, U.   540, *554*
Dowling, J. A.   456, *496*
Dowling, J. E.   150, *162*, 213,
223
— Boycott, B. B.   139, 147,
160, *162*, 207, 213, *223*
— Brown, J. E., Major, D.
213, *223*
— Sidman, R. L.   355, *364*
— see Werblin, F. S.   207,
208, 212, *225*, 251, *268*
Dubin, D. T., Davis, B. D.
426, *437*
Dudel, J.   153, *162*

Duffy, P. E., see Brzin, M.   25'
92
Duthie, H. L., Gairns, F. W.
530, *554*
Duvall, A. J., Wersäll, J.   424,
426, *437*
— see Flock, A.   423, *437*
Duvall, J., Flock, A., Wersäll,
J.   419, 424, *437*

Easter jr., S. S.   239, *241*
Ebersold, W. T., see Levine,
R. P.   379, *394*
Ebert, J. D., Kaighn, M. E.
184, *186*
Ebon Arnelund, Lundegard,
K.   334
Ebrey, T. G.   348, *364*
— Cone, R. A.   350, 351, 352,
353, 357, *364*
— see Pak, W. L.   348, 350,
351, 352, 353, 354, *365*
Eccles, J. C.   88, *94*, 106, 125,
131, 147, *162*, 240, 244,
265, *266, 272, 278, 287*,
426, *437*, 492, *496*
— Eccles, R. M., Lundberg,
A.   176, 180, 181, *186*
— — Magni, F.   176, 180,
181, *186*
— — Shealy, C. M.   176,
180, *186*
— — — Willis, W. D.   176,
180, *186*
— Libet, B.   108, 109, *131*
— Lundberg, A.   180, *186*
— see Brock, L. G.   103, *130*
— see Coombs, J. S.   106,
131, 244, *266*
Eccles, J. G., Iggo, A., Lundberg, A.   113, *131*
Eccles, R. M., see Curtis, D. R.
106, 113, *131*
— see Eccles, J. C.   176, 180,
181, *186*
Echte, K., see Dodt, E.   352,
364
Edds, M. V.   167, *186*
— see Weiss, P.   170, *189*
Edwards, C.   283, *287*, 467,
496
— Hagiwara, S.   317, *337*
— Ottoson, D.   140, *162*,
278, *287*, 505, *521*
— Terzuolo, C. A., Washizu,
Y.   148, *162*, 518, 519,
521

Subject Index

Page numbers in parentheses refer to data from the figures

SPRINGER-VERLAG
BERLIN·HEIDELBERG·NEW YORK

Information Processing in the Nervous System

Proceedings of a Symposium held at the State University of New York at Buffalo, 21st—24th October 1968

Edited by K. N. Leibovic

With 96 figures
390 pages. 1969
Cloth DM 79.20
US $ 19.80

This volume presents both the formal papers and much of the stimulating discussion between experimentalists and theoreticians from different disciplines who participated in this symposium.

The thread which runs throughout the symposium is the concept of "neuronal machinery in relation to psychophysiology". It is explored from many different approaches: automata theory, linguistics, physiological psychology and classical neurophysiology. Too often anatomy, physiology and psychology are treated in isolation when in fact they refer to the same biological unity of form and function. The aim of this symposium was to attempt to bring these disciplines together and allow them to be enriched by theory and model building. Mathematicians, cyberneticists and communications theorists create formal models that more and more closely simulate experimental "reality" and help to clarify hypotheses developed by neurobiologists in their attempt to explain their experimental observations. These models must, in turn, be continuously redesigned in the light of experimental results so that they be of use in the design of future experiments.

The subject matter of this symposium covers the spectrum from the activity of the single cell to that of the organ and the responses of the whole man in psychophysical experiments. The concluding discussions relate the material to more general topics: information theory, neural coding, interdisciplinary communication, and education.

Reprint from

Handbook of Sensory Physiology

Editorial Board
H. Autrum · R. Jung · W. R. Loewenstein · D. M. MacKay · H. L. Teuber

Volume I

Principles of Receptor Physiology
Edited by W. R. Loewenstein

Springer-Verlag Berlin · Heidelberg · New York 1971

Mechano-Chemical Conversion

By

A. Katchalsky and A. Oplatka

With 11 Figures

Reprint from

Handbook of Sensory Physiology

Editorial Board

H. Autrum · R. Jung · W. R. Loewenstein · D. M. MacKay · H. L. Teuber

Volume I

Principles of Receptor Physiology
Edited by W. R. Loewenstein

Springer-Verlag Berlin · Heidelberg · New York 1971

Proteins in Bioelectricity.
Acetylcholine-Esterase and -Receptor

By

D. Nachmansohn

With 12 Figures

Reprint from

Handbook of Sensory Physiology

Editorial Board

H. Autrum · R. Jung · W. R. Loewenstein · D. M. MacKay · H. L. Teuber

Volume I

Principles of Receptor Physiology

Edited by W. R. Loewenstein

Springer-Verlag Berlin · Heidelberg · New York 1971

Transmission Action on
Synaptic Neuronal Receptor Membranes

By

L. Tauc

With 15 Figures

Reprint from

Handbook of Sensory Physiology

Editorial Board

H. Autrum · R. Jung · W. R. Loewenstein · D. M. MacKay · H. L. Teuber

Volume I

Principles of Receptor Physiology

Edited by W. R. Loewenstein

Springer-Verlag Berlin · Heidelberg · New York 1971

The General Electrophysiology
of Input Membrane in Electrogenic Excitable Cells

By

H. Grundfest

With 17 Figures

Reprint from

Handbook of Sensory Physiology

Editorial Board

H. Autrum · R. Jung · W. R. Loewenstein · D. M. MacKay · H. L. Teuber

Volume I

Principles of Receptor Physiology
Edited by W. R. Loewenstein

Springer-Verlag Berlin · Heidelberg · New York 1971

Formation of Neuronal Connections
in Sensory Systems

By

M. Jacobson

With 6 Figures

Reprint from

Handbook of Sensory Physiology

Editorial Board

H. Autrum · R. Jung · W. R. Loewenstein · D. M. MacKay · H. L. Teuber

Volume I

Principles of Receptor Physiology
Edited by W. R. Loewenstein

Springer-Verlag Berlin · Heidelberg · New York 1971

The Relation of Physiological and Psychological Aspects of Sensory Intensity

By

L. E. Lipetz

With 18 Figures

Reprint from

Handbook of Sensory Physiology

Editorial Board

H. Autrum · R. Jung · W. R. Loewenstein · D. M. MacKay · H. L. Teuber

Volume I

Principles of Receptor Physiology

Edited by W. R. Loewenstein

Springer-Verlag Berlin · Heidelberg · New York 1971

Sensory Power Functions and Neural Events

By

S. S. Stevens

With 4 Figures

Reprint from

Handbook of Sensory Physiology

Editorial Board

H. Autrum · R. Jung · W. R. Loewenstein · D. M. MacKay · H. L. Teuber

Volume I

Principles of Receptor Physiology
Edited by W. R. Loewenstein

Springer-Verlag Berlin · Heidelberg · New York 1971

Generation of Responses in Receptors

By

M. G. F. Fuortes

With 17 Figures

Reprint from

Handbook of Sensory Physiology

Editorial Board

H. Autrum · R. Jung · W. R. Loewenstein · D. M. MacKay · H. L. Teuber

Volume I

Principles of Receptor Physiology
Edited by W. R. Loewenstein

Springer-Verlag Berlin · Heidelberg · New York 1971

Mechano-electric Transduction in the Pacinian Corpuscle. Initiation of Sensory Impulses in Mechanoreceptors

By

W. R. Loewenstein

With 16 Figures

Reprint from

Handbook of Sensory Physiology

Editorial Board

H. Autrum · R. Jung · W. R. Loewenstein · D. M. MacKay · H. L. Teuber

Volume I

Principles of Receptor Physiology

Edited by W. R. Loewenstein

Springer-Verlag Berlin · Heidelberg · New York 1971

A Biophysical Analysis of
Mechano-electrical Transduction

By

T. Teorell

With 29 Figures

Reprint from

Handbook of Sensory Physiology

Editorial Board
H. Autrum · R. Jung · W. R. Loewenstein · D. M. MacKay · H. L. Teuber

Volume I

Principles of Receptor Physiology
Edited by W. R. Loewenstein

Springer-Verlag Berlin · Heidelberg · New York 1971

Responses of Nerve Fibers to Mechanical Forces

By

D. E. Goldman

Reprint from

Handbook of Sensory Physiology

Editorial Board

H. Autrum · R. Jung · W. R. Loewenstein · D. M. MacKay · H. L. Teuber

Volume I

Principles of Receptor Physiology
Edited by W. R. Loewenstein

Springer-Verlag Berlin · Heidelberg · New York 1971

The Early Receptor Potential

By

R. A. Cone and W. L. Pak

With 12 Figures

Reprint from

Handbook of Sensory Physiology

Editorial Board

H. Autrum · R. Jung · W. R. Loewenstein · D. M. MacKay · H. L. Teuber

Volume I

Principles of Receptor Physiology

Edited by W. R. Loewenstein

Springer-Verlag Berlin · Heidelberg · New York 1971

The Nature of the Photoreceptor in Phototaxis

By

M. E. Feinleib and G. M. Curry

With 6 Figures

Reprint from

Handbook of Sensory Physiology

Editorial Board
H. Autrum · R. Jung · W. R. Loewenstein · D. M. MacKay · H. L. Teuber

Volume I

Principles of Receptor Physiology
Edited by W. R. Loewenstein

Springer-Verlag Berlin · Heidelberg · New York 1971

Sensory Transduction in Hair Cells

By

Å. Flock

With 30 Figures

Reprint from

Handbook of Sensory Physiology

Editorial Board
H. Autrum · R. Jung · W. R. Loewenstein · D. M. MacKay · H. L. Teuber

Volume I

Principles of Receptor Physiology
Edited by W. R. Loewenstein

Springer-Verlag Berlin · Heidelberg · New York 1971

Transducer Properties and Integrative Mechanisms in the Frog's Muscle Spindle

By

D. Ottoson and G. M. Shepherd

With 39 Figures

Reprint from

Handbook of Sensory Physiology

Editorial Board

H. Autrum · R. Jung · W. R. Loewenstein · D. M. MacKay · H. L. Teuber

Volume I

Principles of Receptor Physiology
Edited by W. R. Loewenstein

Springer-Verlag Berlin · Heidelberg · New York 1971

Static and Dynamic Behavior
of the Stretch Receptor Organ of Crustacea

By

C. A. Terzuolo and C. K. Knox

With 19 Figures

Reprint from

Handbook of Sensory Physiology

Editorial Board
H. Autrum · R. Jung · W. R. Loewenstein · D. M. MacKay · H. L. Teuber

Volume I

Principles of Receptor Physiology
Edited by W. R. Loewenstein

Springer-Verlag Berlin · Heidelberg · New York 1971

Patterns of Organization of Peripheral Sensory Receptors

By

B. L. Munger

With 11 Figures

CPSIA information can be obtained at www.ICGtesting.com
Printed in the USA
LVOW11s0111031013

355202LV00004BA/100/P

9 783642 650659